KU-683-605

Protein Purification Protocols

METHODS IN MOLECULAR BIOLOGY™

John M. Walker, SERIES EDITOR

METHODS IN MOLECULAR BIOLOGY™

Protein Purification Protocols

Second Edition

Edited by

Paul Cutler

Genomic and Proteomic Sciences, Medicines Research Center,
GlaxoSmithKline Pharmaceuticals, Stevenage, UK

HUMANA PRESS ✳ TOTOWA, NEW JERSEY

© 2004 Humana Press Inc.
999 Riverview Drive, Suite 208
Totowa, New Jersey 07512

www.humanapress.com

All rights reserved. No part of this book may be reproduced, stored in a retrieval system, or transmitted in any form or by any means, electronic, mechanical, photocopying, microfilming, recording, or otherwise without written permission from the Publisher. Methods in Molecular Biology™ is a trademark of The Humana Press Inc.

All papers, comments, opinions, conclusions, or recommendations are those of the author(s), and do not necessarily reflect the views of the publisher.

This publication is printed on acid-free paper. ∞
ANSI Z39.48-1984 (American Standards Institute)
Permanence of Paper for Printed Library Materials.

Production Editor: Wendy S. Kopf.

Cover illustration: Paul Cutler.

Cover design by Patricia F. Cleary.

For additional copies, pricing for bulk purchases, and/or information about other Humana titles, contact Humana at the above address or at any of the following numbers: Tel.: 973-256-1699; Fax: 973-256-8341; E-mail: humana@humanapr.com; or visit our Website: humanapress.com

Photocopy Authorization Policy:
Authorization to photocopy items for internal or personal use, or the internal or personal use of specific clients, is granted by Humana Press, provided that the base fee of US $20.00 per copy is paid directly to the Copyright Clearance Center at 222 Rosewood Drive, Danvers, MA 01923. For those organizations that have been granted a photocopy license from the CCC, a separate system of payment has been arranged and is acceptable to Humana Press Inc. The fee code for users of the Transactional Reporting Service is: [0-58829-067-0/04 $20.00].

Printed in the United States of America. 10 9 8 7 6 5 4 3 2 1
e-ISBN 1-59259-655-X
Library of Congress Cataloging in Publication Data
Protein Purification Protocols.--2nd ed. / edited by Paul Cutler.
    p. ; cm -- (Methods in molecular biology ; v. 244)
Includes bibliographical references and index.
  ISBN 1-58829-067-0 (alk. paper) e-ISBN 1-59259-655-X
  ISSN 1064-3745
  1. Proteins--Purification--Laboratory manuals.
  [DNLM: 1. Proteins--isolation & purification--Laboratory Manuals. QU 25 P967 2004] I. Cutler, Paul, 1962- II. Methods in molecular biology (Totowa, N.J.) ; v. 244.
  QP551.P69756 2004
  547.7'5046--dc21
                        2003006803

Preface

The first edition of *Protein Purification Protocols* (1996), edited by Professor Shawn Doonan, rapidly became very successful. Professor Doonan achieved his aims of producing a list of protocols that were invaluable to newcomers in protein purification and of significant benefit to established practitioners. Each chapter was written by an experienced expert in the field. In the intervening time, a number of advances have warranted a second edition. However, in attempting to encompass the recent developments in several areas, the intention has been to expand on the original format, retaining the concepts that made the initial edition so successful. This is reflected in the structure of this second edition. I am indebted to Professor Doonan for his involvement in this new edition and the continuity that this brings.

Each chapter that appeared in the original volume has been reviewed and updated to reflect advances and bring the topic into the 21st century. In many cases, this reflects new applications or new matrices available from vendors. Many of these have increased the performance and/or scope of the given method. Several new chapters have been introduced, including chapters on all the currently used protein fractionation and chromatographic techniques. They introduce the theory and background for each method, providing lists of the equipment and reagents required for their successful execution, as well as a detailed description of how each is performed. The Notes section constitutes a reference guide on the issues and pitfalls that may be encountered and provides the means for circumventing or overcoming them effectively.

Around the time of the first edition, the concept of proteomics was being forged and has subsequently led to a rapid growth in a new and exciting area of protein isolation and analysis. Techniques such as two-dimensional gel electrophoresis have now entered the mainstream, not only in analysis, but also as a preparative technique for protein characterization. Even newer techniques combine with analytical chromatography as multidimensional separations of proteins and peptides. In combination with mass spectrometric techniques, these are now the most powerful methods for isolating proteins. *Protein Purification Protocols* reflects these developments, with chapters encompassing all the current thinking. In addition, since the advance of technology means that simple spectrometric detection is no longer the only option for separating proteins, the various methods for detecting proteins are covered.

Each chapter is designed to allow a particular step of a purification to be performed in isolation; however, it is understood that a number of steps may need to be run in sequence from initial sample fractionation (e.g., tissue homogenization) to chromatography and final polishing steps (e.g., buffer exchange). Our book's format allows for this, and the initial chapter addresses strategies that should place the various methodologies in context. At the end of the book it was also felt timely to include brief descriptions of how to scale-up purification methods and evaluate the purification of proteins for therapeutic use. These do not rigidly follow the regular pattern for the main body of protocols, but should give an insight into the strategies needed for different final applications.

Paul Cutler

Contents

Contents

Contributors

DAVID J. BELL • *Genomic and Proteomic Sciences, Medicines Research Center, GlaxoSmithKline Pharmaceuticals, Stevenage, UK*

ROBERT J. BEYNON • *Protein Function Group, Faculty of Veterinary Science, University of Liverpool, Liverpool, UK*

JULIAN BONNERJEA • *Lonza Biologics, Slough, UK*

CAROLINE G. BOWSHER • *School of Biological Sciences, University of Manchester, Manchester, UK*

PAUL D. BRIDGE • *School of Biological and Chemical Sciences, University of London, London, UK and Mycology Section, Royal Botanic Gardens, Kew, Richmond, UK*

SUDHIR BURMAN • *Biopharmaceutical Analytical Development, GlaxoSmithKline Pharmaceuticals, King of Prussia, PA*

PAUL CUTLER • *Genomic and Proteomic Sciences, Medicines Research Center, GlaxoSmithKline Pharmaceuticals, Stevenage, UK*

SHAWN DOONAN • *School of Biosciences, University of East London, UK*

MICHAEL J. DUNN • *Department of Neuroscience, Institute of Psychiatry, King's College London, London, UK*

ROGER J. FIDO • *Long Ashton Research Station, Department of Agricultural Sciences, University of Bristol, Bristol, UK*

OWEN GOLDRING • *North East Surrey College of Technology, Ewell, Surrey, IK*

T. WILLIAM HUTCHENS • *LumiCyte Inc., Fremont, CA*

JANE A. IRWIN • *Department of Veterinary Physiology and Biochemistry, University College Dublin, Dublin, Ireland*

NEIL A. JONES • *Genomic and Proteomic Sciences, Medicines Research Center, GlaxoSmithKline Pharmaceuticals, Stevenage, UK*

TETSUO KOKUBUN • *Biological Interactions Section, Royal Botanic Gardens, Kew, Richmond, UK*

TIMOTHY J. MANTLE • *Department of Biochemistry, Trinity College, Dublin, Ireland*

ANNE F. MCGETTRICK • *Department of Biochemistry Trinity College, Dublin, Ireland*

DEAN E. MCNULTY • *Department of Computational, Analytical, and Structural Sciences, GlaxoSmithKline Pharmaceuticals, King of Prussia, PA*

PAUL MATEJTSCHUK • *National Institute for Biological Standards and Control, South Mimms, Potters Bar, UK*

E. N. CLARE MILLS • *Institute of Food Research, Norwich Laboratory, Norwich, UK*

JACEK MOZDZANOWSKI • *Biopharmaceutical Analytical Development, GlaxoSmithKline Pharmaceuticals, King of Prussia, PA*

TIM NADLER • *Applied Biosystems, Framingham, MA*

WILLIAM A. NEVILLE • *Safety Assessment, GlaxoSmithKline Pharmaceuticals, Ware, UK*

PATRICIA NOONE • *Department of Biochemistry, Trinity College, Dublin, Ireland*

CIARÁN Ó'FÁGÁIN • *School of Biotechnology, Dublin City University, Dublin, Ireland*

PAUL A. O'FARRELL • *W. M. Keck Structural Biology Laboratory, Cold Spring Harbor Laboratory, Cold Spring Harbor, NY*

SIMON OLIVER • *Department of Biochemistry and Applied Microbiology, UMIST, Manchester, UK*

SIOBHAN O'SULLIVAN • *Department of Biochemistry and Analytical and Biological Chemistry Research Facility, University College Cork, Mardyke, Cork, Ireland*

KAY OHLENDIECK • *Department of Biology, National University of Ireland, Maynooth, Ireland*

WILLIAM S. PIERPOINT • *Rothamsted Research, Harpenden, UK*

KARL PRINCE • *Lonza Biologics, Portsmouth, NH*

NEIL M. RIGBY • *Institute of Food Research, Norwich Laboratory, Norwich, UK*

NORMA M. RYAN • *Department of Biochemistry, University College, Cork, Ireland*

PAUL SCHRATTER • *Millipore Corporation, Bedford, MA*

CHRIS SELKIRK • *Biotherapeutics Development Unit, Cancer Research UK, South Mimms, UK*

DAVID SHEEHAN • *Department of Biochemistry and Analytical and Biological Chemistry Research Facility, University College Cork, Mardyke, Cork, Ireland*

PETER R. SHEWRY • *Long Ashton Research Station, Department of Agricultural Sciences, University of Bristol, Bristol, UK*

J. MARK SKEHEL • *Genomic and Proteomic Sciences, Medicines Research Center, GlaxoSmithKline Pharmaceuticals, Stevenage, UK*

MONIQUE S. J. SIMMONDS • *Biological Interactions Section, Royal Botanic Gardens, Kew, Richmond, UK*

J. RANDALL SLEMMON • *Department of Genomics and Biotechnology, Pharmacia Corporation, Skokie, IL*

MARTIN SMITH • *Lonza Biologics, Slough, UK*

DAVID W. SPEICHER • *The Wistar Institute, Philadelphia, PA*

KEITH F. TIPTON • *Department of Biochemistry, Trinity College, Dublin, Ireland*

ALYSON K. TOBIN • *Plant Science Laboratory, School of Biology, Sir Harold Mitechell Building, University of St Andrews, St. Andrews, Scotland, UK*

ROD WATSON • *Applied Biosystems, Warrington, UK*

IRIS WEST • *North East Surrey College of technology, Ewell, Surrey, UK*

REINER WESTERMEIER • *Amersham Biosciences Europe GmbH, Freiburg, Germany*

D. MARGARET WORRALL • *Department of Biochemistry and Conway Institute for Biomolecular and Biomedical Research, University College Dublin, Dublin, Ireland*

TAI-TUNG YIP • *Ciphergen Biosystems Inc., Fremont, CA*

XUN ZUO • *The Wistar Institute, Philadelphia, PA*

1

General Strategies

Shawn Doonan and Paul Cutler

1. Defining the Problem

The chapters that follow in this volume give detailed instructions on how to use the various methods that are available for purification of proteins. The question arises, however, of which of these methods to use and in which order to use them to achieve purification in any particular case; that is, the purification problem must be clearly defined. What follows outlines the sorts of question that need to be asked as part of that definition and how the answers affect the approach that might be taken to developing a purification schedule. It should be noted here that the discussion concentrates mainly on laboratory-scale isolation of proteins. Special cases of purification of therapeutic proteins and isolation at industrial scale are covered in Chapters 43 and 44 *(1–5)*.

1.1. How Much Do I Need?

The answer to this question depends on the purpose for which the protein is required. For example, to carry out a full chemical and physical analysis of a protein may require several hundreds of milligrams of purified material, whereas a kinetic analysis of the reaction catalyzed by an enzyme could perhaps be done with a few milligrams and less than 1 mg would be required to raise a polyclonal antibody. At the extreme end of the scale, if the objective is to obtain limited sequence information from the N-terminus of a protein as a preliminary to the design of an oligonucleotide probe for clone screening, then using modern microsequencing techniques, a few micrograms will be sufficient. In the field of proteomics, previously analytical techniques have become preparative with mass spectrometry commonplace for sensitive protein characterization from spots on gels. Chapters 36 and 40–42 describe these methodologies. These different requirements for quantity may well dictate the source of the protein chosen (*see* **Subheading 1.4.**) and will certainly influence the approach to purification. Purification of large quantities of protein requires use of techniques, at least in the early stages, that have a high capacity but low resolving power, such as fractional precipitation with salt or organic solvents (*see* Chapter 13). Process only when the volume and protein content of the extract has been reduced to manageable levels, methods of medium resolution and capacity, such as ion-exchange chromatography (*see* Chapter 14) can be used leading on, if necessary, to high-resolution

From: *Methods in Molecular Biology, vol. 244: Protein Purification Protocols: Second Edition*
Edited by: P. Cutler © Humana Press Inc., Totowa, NJ

but generally lower-capacity techniques, such as affinity chromatography (*see* Chapter 16) and isoelectric focusing (*see* Chapter 24). On the other hand, for isolation of small to medium amounts of proteins, it will usually be possible to move directly to the more refined methods of purification without the need for initial use of bulk methods. Often the decision as to whether or not to expose a costly matrix to the system early in the strategy will rest on issues related to the stability and/or the value of the target protein. This is, of course, important because the fewer steps that have to be used, the higher the final yield of the protein will be and the less time it will take to purify it.

1.2. Do I Want to Retain Biological Activity?

If the answer to this is positive, then it restricts to some extent the range of techniques that can be employed and the conditions under which they can be performed. Most proteins retain activity when handled in neutral aqueous buffers at low temperature (although there are exceptions and these exceptions lend themselves to somewhat different approaches to purification). This consideration then rules out the use of those techniques in which the conditions are likely to deviate substantially from the above. For example, immunoaffinity chromatography is a very powerful method, but the conditions required to elute bound proteins are often rather severe (e.g., the use of buffers of low pH) because of the tightness of binding between antibodies and antigens (*see* Chapters 16 and 19 for a discussion of this problem). Similarly, reversed-phase chromatography (*see* Chapter 28) requires the use of organic solvents to elute proteins and rarely will be compatible with recovering an active species. Ion-exchange chromatography provides the most general method for the isolation of proteins with retention of activity unless the protein has special characteristics that offer alternative strategies (*see* **Subheading 2.4.**). With labile molecules, it is important to plan the purification schedule to contain as few steps as possible and with minimum requirement for changing buffers (*see* Chapter 11), as this will reduce losses of activity. Most proteins retain their activity better at lower temperatures, although it should be remembered that this is not absolute because some proteins are cryopreciptants and lose solubility at lower temperatures.

In some cases, retention of biological activity is not required. This would be the case, for example, if the protein is needed for sequence analysis or perhaps for raising an antiserum. There is then no restriction on the methods that can be used and, indeed, the very powerful separation method of polyacrylamide gel electrophoresis in the presence of sodium dodecyl sulfate (SDS-PAGE) followed by blotting or elution from the gel can be used to isolate small amounts of pure protein either from partially purified extracts or even from crude extracts (*see* Chapters 34 and 35). It is important in this context to differentiate between loss of biological activity arising from loss of three-dimensional structure, which will not be of concern in the applications outlined earlier, from loss of activity owing to modification of the chemical structure of the protein, which certainly would be a major concern. The most important route to chemical modification is proteolytic cleavage, and ways in which this can be detected and avoided are discussed in Chapter 9.

1.3. Do I Need a Completely Pure Protein?

The concept of purity as applied to proteins is not entirely straightforward. It ought to mean that the protein sample contains, in addition to water and things like buffer ions that have been purposefully added, only one population of molecules, all with identical covalent and three-dimensional structures. This is an unattainable goal and indeed an unnecessary one. Even therapeutic proteins will retain impurities all be it at the level of parts per million (*see* Chapter 43). What is required is a sample of protein that does not contain any species that will interfere with the experiments for which the protein is intended. This is not simply an academic point because it will usually become more and more difficult to remove residual contaminants from a protein sample as purification progresses. Extra purification steps will be required, which take time (effectively an increase in cost of the product) and will inevitably lead to decreasing yields. What is required is an operational definition of purity for the particular project in hand because this will not only define the approach to the purification problem but may also govern its feasibility. It may not be possible to obtain a highly purified sample of a labile protein, but it may be possible to obtain it in a sufficient state of purity for the purposes of a particular investigation.

The usual criterion of purity used for proteins is that a few micrograms of the sample produces a single band after electrophoresis on SDS-PAGE when stained with a reagent such as Coomasie blue or some similar nonspecific stain (*see* **ref. 6** for practical details of this procedure and other chapters in the same volume for many other basic protein protocols). This simple criterion begs several questions. The most important of these is that SDS-PAGE separates proteins effectively on the basis of size and it may be that whether the sample contains two or more components that are sufficiently similar not to be resolved; the answer here is to subject the sample to an additional procedure, such as nondenaturing PAGE *(7)* because it is unlikely that two proteins will migrate identically in both systems. It must always be kept in mind, however, that even if a single band is observed in two such systems, minor contaminants will inevitably become visible if the gel is more heavily loaded or if staining is carried out using a more sensitive procedure, such as silver staining *(8)*.

The major question is: Does it matter if the protein is 50%, 90%, or 99% pure? The answer is that it depends on the purpose of the purification. For example, a 50% pure protein may be entirely acceptable for use in raising a monoclonal antibody, but a 95% pure protein may be entirely unacceptable for raising a monospecific polyclonal antibody, particularly if the contaminants are highly immunogenic. Similarly, a relatively impure preparation of an enzyme may be acceptable for kinetic studies provided that it does not contain any competing activities; an affinity chromatography method might provide a rapid way of obtaining such a preparation. As a final example, a 95% pure protein sample is perfectly adequate for amino acid sequence analysis and, indeed, a lower state of purity is acceptable if proper quantitation is carried out to ensure that a particular sequence does not arise from a contaminant. The highest level of purity is needed for therapeutic proteins. In this instance, other criteria need to observed such as compliance with good laboratory practice (GLP) and good manufacturing practice (GMP), which is beyond the scope of most standard research laboratories.

The message here is that preparation of a sample of protein approaching homogeneity is difficult and may not always be necessary so long as one knows what else there is. By taking account of the purpose for which the protein is required, it may be possible to decide on an acceptable level of contaminants, and consideration of the nature of acceptable contaminants may suggest a purification strategy to be adopted.

1.4. What Source Should I Use?

The answer to this question may be partly or entirely dictated by the problem in hand. Clearly, if the objective is to study the enzyme ribulose bisphosphate carboxylase, then there is no choice but to isolate it from a plant, but the plant can be chosen for its ready availability, high content of the enzyme, ease of extraction of proteins (*see* Chapter 3), and low content of interfering polyphenolic compounds (*see* Chapter 8). Of course, if one is interested in, for example, comparative biochemistry or molecular evolution, then not only the desired protein but also its source may be completely constrained.

In general, however, plants will not be the source of choice for isolation of a protein of general occurrence and where species differences are not of interest. Microbial or fungal sources may be a better choice because they can usually be grown under defined conditions, thus assuring the consistency of the starting material and, in some cases, allowing for manipulation of levels of desired proteins by control of growth media and conditions (*see* Chapters 4 and 5). They have the disadvantage, however, of possesing tough cell walls that are difficult to break and, consequently, micro-organisms are not ideal for large-scale work unless the laboratory has specialized equipment needed for their disruption.

The most convenient source of proteins in most cases is animal tissue, such as heart and liver and, except for relatively small-scale work, the tissues will normally be obtained from a commercial abattoir. Laboratory animals provide an alternative for smaller-scale purifications. The content of a particular protein is likely to be tissue-specific, in which case the most abundant source will probably be the best choice. It is worth noting, however, that it is easier to isolate proteins from tissues, such as heart, than from liver and, hence, the heart may be the better bet even if the levels of the protein are lower than in liver.

A different sort of question arises if the protein of interest exists in soluble form in a subcellular organelle, such as the mitochondrion or chloroplast. Once the source organism has been chosen, there remains the decision as to whether to carry out a total disruption of the tissue under conditions where the organelles will lyse or whether to homogenize under conditions where the organelles remain intact and can be isolated by methods such as those described in Chapters 6 and 7. The latter approach will, of course, result in a very significant initial enrichment of the protein and subsequent purification will be easier because the range and amount of contaminating proteins will be much decreased. In the case of animal tissues, the decision will probably depend on the scale at which it is intended to work (assuming, of course, that access to the necessary preparative high-speed centrifuges is available). Subcellular fractionation of a few hundred grams of tissue is a realistic objective, but if it is intended to work with larger amounts, then the time required for organelle isolation probably will be prohibitive and is unlikely to compensate for the extra work that will be involved in purification from a total cellular extract. Subcellular fractionation of plants is a much more difficult operation in most cases (*see* Chapter 7). Hence, except in the most favorable cases and for small-scale work, purification from a total cellular extract will probably be the only realistic option.

In the case of membrane proteins, there again will be a considerable advantage in isolating as pure a sample of the membrane as possible before attempting purification. The ease with which this can be done depends on the organism and membrane system in question. Chapters 6 and 31 give some approaches to this problem for specific cases, but if it is intended to isolate a membrane protein from other sources, then a survey of the extensive literature on membrane purification is recommended (*see* **ref. 9**).

For proteins that are present in only very small quantities or found only in inconvenient sources, gene cloning and expression in a suitable host now provide an alternative route to purification (for a review of methods, *see* **ref. 10**). This is, of course, a major undertaking and is likely to be used only when conventional methods are not successful. Suffice it to say that once the protein is expressed and extracted from the host cell (*see* Chapter 4 for a method of extracting recombinant proteins from bacteria), the methods of purification are the same as those for proteins from conventional sources.

1.5. Has It Been Done Previously?

It is quite common to need to purify a protein whose purification has been reported previously, perhaps to use it as an analytical tool or perhaps to carry out some novel investigations on it. In this case, the first approach will be to repeat the previously described procedure. The chances are, however, that it will not work exactly as described because small variations in starting material, experimental conditions, and techniques (which are inevitable between different laboratories) can have a significant effect on the behavior of a protein during purification. This should not matter too much because adjustments to the procedures should be relatively easy to make once a little experience has been gained of the behavior of the protein. One pitfall to watch out for is the conviction that there ought to be a better way of doing it. It is possible to spend a great deal of time trying to improve on a published procedure, often to little avail.

Even if the particular protein of interest has not been isolated previously, it may be that a related molecule has been, for example, the same protein but from a different organism or a member of a closely related class of proteins. In the former case, particularly if the organisms are closely related, then the properties of the proteins should be quite similar and only minor variations in procedures (e.g., the pH used for an ion-exchange step) might be required. Even if the family relationships are more distant, significant clues might still be available, such as the fact that the target is a glycoprotein, which will provide valuable approaches to purification (*see* **Subheading 2.4.**). Much time and wasted effort can be saved by using information in the literature rather than trying to reinvent the wheel.

2. Exploiting Differences

Protein purification involves the separation of one species from perhaps 1000 or more species of essentially the same general characteristics (they are all proteins!) in a mixture of which it may constitute a small fraction of 1% of the total. It is, therefore, necessary to exploit to the full those properties in which proteins differ from one another in devising a purification schedule. The following lists the most important of those properties and outlines the techniques that make use of them with comments on their practical application. More details on each technique will be found in the chapters that follow.

2.1. Solubility

Proteins differ in the balance of charged, polar, and hydrophobic amino acids that they display on their surfaces and, hence, in their solubilities under a particular set of conditions. In particular, they tend to precipitate differentially from solution on the addition of species such as neutral salts or organic solvents and this provides a route to purification (*see* Chapter 13). It is, however, a rather gross procedure because precipitation will occur over a range of solute concentrations and those ranges necessarily overlap for different proteins. It is not to be expected, therefore, that a high degree of purification can be achieved by such methods (perhaps twofold to threefold in most circumstances), but the yield should be high and, most importantly, fractional precipitation can be carried out easily on a large scale provided only that a suitable centrifuge is available. It is, therefore, very common for this technique to be used at the stage immediately following extraction when working on a moderate to large scale. An important added advantage is that a substantial degree of concentration of the extract can be obtained at the same time, which, considering that water is the major single contaminant in a protein solution, is a considerable added benefit.

2.2. Charge

Proteins differ from one another in the proportions of the charged amino acids (aspartic and glutamic acids, lysine, arginine, and histidine) that they contain. Hence, they will differ in net charge at a particular pH or, another manifestation of them same difference, in the pH at which the net charge is zero (the isoelectric point). The first of these differences is exploited in ion-exchange chromatography, which is perhaps the single most powerful weapon in the protein purifier's armory (*see* Chapter 14). This makes use of the binding of proteins carrying a net charge of one sign onto a solid supporting material bearing charged groups of the opposite sign; the strength of binding will depend on the magnitude of the charge on the particular protein. Proteins may then be eluted from the matrix in exchange for ions of the opposite charge, with the concentration of the ionic species required being determined by the magnitude of the charge on the protein.

Ion-exchange chromatography is a technique of moderate to high resolution depending on the way in which it is implemented. For large-scale work (around 100 g of protein), use is generally made of fibrous cellulose-based resins that give good flow rates with large bed volumes but not particularly high resolution; this would normally be done at an early stage in a purification. Better resolution is available with the more advanced Sepharose-based materials but generally on a smaller scale. For small quantities (<10 mg), the technique of fast protein liquid chromatography (*see* Chapter 27) is available, which makes use of packing materials with very small diameters and correspondingly high resolving power; this, however, requires specialized equipment that may not be available in all laboratories. Because of the small scale, this method would usually be used at a late stage for final cleanup of the product. It should be kept in mind that two proteins that carry the same charge at a particular pH might well differ in charge at a different pH. Hence, it is quite common for a purification procedure to contain two or more ion-exchange steps either using the same resin at different pH values or perhaps using two resins of opposite charge characteristics (e.g., one carrying the negatively charged carboxymethyl [CM] group and the other the positively charged diethylaminoethyl [DEAE] group).

There are two main ways of exploiting differences in isoelectric points between proteins. Chromatofocusing is essentially an ion-exchange technique in which the proteins are bound to an anion exchanger and then eluted by a continuous decrease of the buffer pH so that proteins elute in order of their isoelectric points (*see* Chapter 25). It is a method of moderately high resolving power and capacity and is hence best used to further separate partially purified mixtures. The other technique is isoelectric focusing (*see* Chapter 24), in which proteins are caused to migrate in an electric field through a system containing a stable pH gradient. At the pH at which a particular protein has no net charge (the isoelectric point), it will cease to move; if it diffuses away from that point, then it will regain a charge and migrate back again. This method, although of low capacity, is capable of very high resolution and is frequently used to separate mixtures of proteins that are otherwise difficult to fractionate.

2.3. Size

This property is exploited directly in the techniques of size-exclusion chromatography (*see* Chapter 26) and ultrafiltration (*see* Chapter 12). In the former, the protein solution is passed through a column of porous beads, the pore sizes being such that large proteins do not have access to the internal space, small proteins have free access to it, and intermediate-sized proteins have partial access; a range of these materials with different pore sizes is available. Clearly, large proteins will pass through the column most rapidly and small proteins will pass through most slowly with a range of behavior in between. The method is of limited resolving power but is useful in some circumstances, particularly when the protein of interest is at one of the extremes of size. The capacity is low because of the need to keep the volume of solution applied to the column as small as possible.

In ultrafiltration, liquid is forced through a membrane with pores of a controlled size such that small solutes can pass through but larger ones cannot. It, therefore, can be used to obtain a separation between large and small protein molecules and also has the advantage that it is not limited by scale. Use of the method for protein fractionation is, however, restricted to a few special cases (*see* Chapter 12) and the principal value of the technique is for concentration of protein solutions.

A completely different approach to the use of size differences to effect protein separation is SDS-PAGE. In this method, the protein molecules are denatured and coated with the detergent so that they carry a large negative charge (the inherent charge is swamped by the charge of the detergent). The proteins then migrate in gel electrophoresis on the basis of size; small proteins migrate most rapidly and large ones slowly because of the sieving effect of the gel. The method has enormously high resolving power and its use in various forms for analytical purposes is one of the most important techniques in analytical protein chemistry *(6)*. The development of methods for recovery of the protein bands from the gel after electrophoresis (*see* Chapters 34 and 35) has enabled this resolving power to be exploited for purification purposes. Obviously, the scale of separation is small and the product is obtained in a denatured state, but a sufficient amount often can be obtained from very complex mixtures for the purposes of further investigation (*see* **Subheading 1.2.**). Combining isoelectric focusing and SDS-PAGE in two-dimensional gel electrophoresis also offers a very highly resolving preparatory technique (*see* Chapter 36) *(11,12)*.

2.4. Specific Binding

Most proteins exert their biological functions by binding to some other component in the living system. For example, enzymes bind to substrates and sometimes to activators or inhibitors, hormones bind to receptors, antibodies bind to antigens, and so on. These binding phenomena can be exploited to effect purification of proteins usually by attaching the ligand to a solid support and using this as a chromatographic medium. An extract or partially purified sample containing the target protein is then passed through this column to which the protein binds by virtue of its affinity for the ligand. Elution is achieved by varying the solvent conditions or introducing a solute that binds strongly either to the ligand or to the protein itself.

Various types of affinity chromatography, as the method is called, are described in detail in Chapters 16–20. Immunoaffinity chromatography, in particular, is capable of very high selectivity because of the extreme specificity of antibody–antigen interactions. As mentioned earlier and dealt with in more detail in Chapters 16 and 19, the most common problem with this technique is to effect elution of the target protein under conditions that retain biological activity *(13)*. Lectin-affinity chromatography (*see* Chapter 18) exploits the selective binding between members of this class of plant proteins and particular carbohydrates. It has therefore found widespread use both in the isolation of glycoproteins and in removal of glycoprotein contaminants from other proteins, and it is also capable of high specificity.

Affinity methods that rely on interactions of the target protein with low-molecular-weight compounds (e.g., enzymes with substrates or substrate analogs) are frequently less specific because the ligand may bind to several proteins in a mixture. For example, immobilized NAD^+ will bind to many dehydrogenases, and benzamidine will bind to most serine proteases; thus, a group of related enzymes rather than individual species may be isolated using these ligands. A novel application of affinity methods is provided by the use of bifunctional NAD^+ derivatives to selectively precipitate dehydrogenases from solution (*see* Chapter 23).

The use of organic dyes as affinity ligands (*see* Chapter 17) is interesting because these molecules seem to bind fairly specifically to nucleotide-binding enzymes, although from their structures, it is not at all clear why they should do so; it is likely that hydrophobic interactions between the dye and protein also contribute to binding. Use of the latter interaction has led to development of a specific form of chromatography that uses hydrophobic stationary phases (*see* Chapter 15); this method has elements of biospecificity in that some proteins have binding sites for natural hydrophobic ligands, but in the general case, it relies on the fact that all proteins have hydrophobic surface regions to a greater or lesser extent *(14)*.

Finally, many proteins are known that bind metal ions with varying degrees of specificity and this forms the basis of immobilized metal-ion affinity chromatography (*see* Chapter 20). Specific affinity of proteins for calcium ions may also be the basis, in part, for binding to hydroxyapatite but ion-exchange effects are probably also involved (*see* Chapter 21).

In summary, there are a variety of affinity methods available, ranging from medium to very high selectivity, and, in favorable cases, affinity chromatography can be used to obtain a single-step purification of a protein from an initial extract. Generally, however, the capacities of affinity media are not high and the materials can be very expensive, thus rendering their use on a large-scale unrealistic. For these reasons, affinity methods are usually used at a late stage in a purification schedule.

2.5. Special Properties

In a sense the specific binding properties discussed in the **Subheading 2.4.** are "special," but that is not what is meant here. Some proteins have, for example, the property of greater than normal heat stability and in those circumstances it may be possible to obtain substantial purification by heating a crude extract at a temperature at which the target protein is stable, but contaminants are denatured and precipitate from solution (*see* **ref. *15*** for an example of the use of this method). It is not likely, of course, that this approach will be useful in purification of proteins from thermophilic organisms because all or most of the proteins present would be expected to share the property of thermostability. Another possibility is that the protein of interest may be particularly stable at one or other of the extremes of pH; in this case, incubation of an extract at low or high pH might well lead to selective precipitation of contaminants. It is always worthwhile carrying out some preliminary experiments with an unknown protein to see if it possesses special properties of this kind that would assist in its purification.

Finally, mention should be made of the fact that it is now feasible, if the need is sufficiently great, to engineer special properties into proteins to assist in their purification. Typical examples include the addition of polyarginine or polylysine tails to improve behavior on ion-exchange chromatography, or of polyhistidine tails to introduce affinity on immobilized metal affinity chromatography *(16)*. It is, however, likely that these techniques would be used as a last resort if all other attempts to purify the protein failed unless recombinant DNA technology had been selected as the route to protein production and purification in the first place (*see* **Subheading 1.4.**).

3. Documenting the Purification

It is vitally important to keep an inventory at each stage of a purification of volumes of fractions, total protein content, and content of the protein of interest. The last of these is particularly important because otherwise it is very easy to end up with a vanishingly small yield of target protein and not to know at which step the protein was lost. If the protein has a measurable activity, then it is equally important to monitor this because it is also possible to end up with a protein sample that is inactive if one or more steps in the purification involves conditions under which the protein is unstable.

Measurement of the total protein content of fractions presents no problems. At early stages of a purification, it is usually sufficient to determine the absorbance of the solution at 280 nm (making sure that it is optically clear to avoid errors owing to light scattering) and to use the rough approximation that $A_{280nm}^{1\%} = 10$. At later stages, one of the more accurate methods, such as the Bradford procedure *(17)* or the bicinchoninic acid assay *(18)*, should be used unless the absorbance/dry weight correlation for the target protein happens to be known.

Measurement of the amount and/or activity of the protein of interest may or may not be straightforward. For example, many enzymes can be assayed using simple and rapid spectrophotometric methods. For other proteins, the assay may be more difficult and time-consuming, such as bioassay or immunoassay. (It should also be recognized that these are not necessarily the same thing; immunoassay frequently will not distinguish between inactive and active molecules, so care must be taken in the interpretation of results using this method.) In other situations, the protein of interest may have no measurable biological activity; in such cases, immunoassay can be used or, more commonly, quantitation of the appropriate band after separation of the protein on polyacrylamide gels *(19)*. Indeed, it may be that the target protein will only have been identified as a spot on two-dimensional polyacrylamide gels *(20)* and purification is being attempted as a preliminary to determining its biological activity.

Obviously, it is not possible to be prescriptive here about what methods of analysis and quantitation to use in any specific case. What must be said, however, is that it is very unwise to embark on an attempted purification without first devising a method for quantitation of the protein of interest. Not to do so is courting failure.

4. An Example

To give the newcomer to protein purification a "feel" for what the process might look like in practice, **Table 1** shows the fully documented results of the isolation of a particular enzyme starting from 5 kg of pig liver. All techniques used are described in detail in subsequent chapters and are only summarized here.

The strategy was to start by totally homogenizing the tissue in 10 L of buffer and, after removal of cell debris by centrifugation, to carry out an initial crude purification by fractional precipitation with ammonium sulfate. This had the added advantages of removing residual insoluble material from the extract (this precipitated in the first ammonium sulfate fraction) and achieving a very large reduction in volume of the active fraction. Ammonium sulfate was removed from the active fraction by dialysis.

Because of the large amount of protein remaining in the active fraction, the next step was a relatively crude ion-exchange separation using a large column (7 × 50 cm) of CM–cellulose CM23 (this has a high capacity and good flow rates but is of only moderate resolving power). Conditions were chosen so that the enzyme was absorbed onto the column and then, after washing off unbound contaminants, it was eluted with a single stepwise increase in ionic strength to 0.1 M using sodium chloride.

Previous trial experiments had shown that the enzyme bound to an affinity matrix in a buffer at the same pH and salt content as that with which it was eluted from CM–cellulose, and so affinity chromatography was used for the next step without changing the buffer and without prior concentration. The enzyme was eluted by applying a linear salt gradient up to a concentration of 1 M.

At this stage, electrophoresis of the active fraction under nondenaturing conditions showed the presence of two major contaminants, both of them more basic than the protein of interest. Hence, the sample was applied to a column of DEAE–Sepharose under conditions where the target protein was absorbed, but the majority of the contaminating protein was not; the sample was equilibrated in starting buffer by dialysis before application to the column. The target protein was eluted from the column using a linear salt gradient and was found to be homogeneous by the usual techniques (*see* **Subheading 1.3.**).

The results in **Table 1** show that the purification procedure was quite successful in that a high yield (50% overall) of enzyme activity was obtained; this was achieved by using a small number of steps each of which gave a good step yield. There will inevitably be losses on any purification step and the important point is that these and the number of steps should be kept as low as possible (a 5-step schedule in which the yield

Table 1
Example Protein Purification Schedule

Fraction	Volume (mL)	Protein concentration (mg/mL)	Total protein (mg)	Activity[a] (U/mL)	Total activity (U)	Specific activity (U/mg)	Purification factor[b]	Overall yield[c] (%)
Homogenate	8,500	40	340,000	1.8	15,300	0.045	1	100
45–70% $(NH_4)_2SO_4$	530	194	103,000	23.3	12,350	0.12	2.7	81
CM–cellulose	420	19.5	8,190	25	10,500	1.28	28.4	69
Affinity chromatography	48	2.2	105.6	198	9,500	88.4	1,964	62
DEAE–Sepharose	12	2.3	27.6	633	7,600	275	6,110	50

[a]The unit of enzyme activity is defined as that amount which produces 1 μmol of product per min under standard assay conditions.
[b]Defined as follows: purification factor = specific activity of fraction/specific activity of homogenate.
[c]Defined as follows: overall yield = total activity of fraction/total activity of homogenate.

from each step is 50% will give an overall yield of 3%; a 10-step schedule with 80% step yield will give a final yield of 11%). It can also be seen from the final purification factor that the amount of this particular enzyme in the liver was low (about 0.016% of soluble protein) and, hence, a relatively large amount of tissue had to be used to obtain the required amount of product. This was an important factor in deciding the first two steps in the schedule (*see* **Subheading 1.1.**).

The purification in its final form can be completed in 5–6 working days. It must be kept in mind, however, that each step has been optimized and that development of the procedure took several months of work. This is common when working out a new purification schedule and it is always necessary to be conscious of the time commitment when deciding to embark on purifying a protein.

References

1. Asenjo, J. A. and Patrick, I. (1990) Large-scale protein purification, in *Protein Purification Applications: A Practical Approach* (Harris, E. L. V. and Angal, S., eds.), IRL, Oxford, pp. 1–28.
2. Bristow, A. F. (1990) Purification of proteins for therapeutic use, in *Protein Purification Applications: A Practical Approach* (Harris, E. L. V. and Angal, S., eds.), IRL, Oxford, pp. 29–44.
3. Levison, P. R., Hopkins, A. K. and Hathi, P. (1999) Influence of column design on process-scale ion-exchange chromatography. *J. Chromatogr. A* **865**, 3–12.
4. Lightfoot, E. N. (1999) The invention and development of process chromatography: interaction of mass transfer and fluid mechanics. *Am. Lab.* **31**, 13–23.
5. Prouty, W. F. (1993) Process chromatography in production of recombinant products. *ACS Sympo. Ser.* **529**, 43–58.
6. Smith, B. J. (1994) SDS polyacrylamide gel electrophoresis of proteins, in *Methods in Molecular Biology, Vol. 32: Basic Protein and Peptide Protocols* (Walker, J. M., ed.), Humana, Totowa, NJ, pp. 23–34.
7. Walker, J. M. (1994) Nondenaturing polyacrylamide gel electrophoresis of proteins, in *Methods in Molecular Biology, Vol. 32: Basic Protein and Peptide Protocols* (Walker, J. M., ed.), Humana, Totowa, NJ, pp. 17–22.
8. Dunn, M. J. and Crisp, S. J. (1994) Detection of proteins in polyacrylamide gels using an ultrasensitive silver staining technique, in *Methods in Molecular Biology, Vol. 32: Basic Protein and Peptide Protocols* (Walker, J. M., ed.), Humana, Totowa, NJ, pp. 113–118.
9. Graham, J. M. and Higgins, J. A. (eds.) (1993) Methods in Molecular Biology, Vol. 19: Biomembrane Protocols: I. Isolation and Analysis, Humana, Totowa, NJ.
10. Murray, E. J. (ed.) (1991) Methods in Molecular Biology, Vol. 7: Gene Transfer and Expression Protocols, Humana, Totowa, NJ.
11. Figeys, D. (2001) Two dimensional gel electrophoresis and mass spectrometry for proteomic studies: state of the art, in *Biotechnology*, 2nd ed., Wiley–VCH, Weinheim, pp. 241–268.
12. Unlu, M. and Minden, J. (2002) Proteomics: difference gel electrophoresis, in *Modern Protein Chemistry* (Howard, J. C. and Brown, W. E. eds.), CRC, Boca Raton, FL, pp. 227–244
13. Porath, J. (2001) Strategy for differential protein affinity chromatography. *Int. J. Biochromatogr.* **6**, 51–78.
14. Querioz, J. A., Tomaz, C. T., and Cabral, J. M. S. (2001) Hydrophobic interaction chromatography of proteins. *J. Biotechnol.* **87**, 143–159.
15. Banks, B. E. C., Doonan, S., Lawrence, A. J., and Vernon, C. A. (1968) The molecular weight

and other properties of aspartate aminotransferase from pig heart muscle. *Eur. J. Biochem.* **5,** 528–539.

16. Brewer, S. J. and Sassenfeld, H. M. (1990) Engineering proteins for purification, in *Protein Purification Applications: A Practical Approach* (Harris, E. L. V. and Angal, S., eds.), IRL, Oxford, pp. 91–111.

17. Kruger, N. J. (1994) The Bradford method for protein quantitation, in *Methods in Molecular Biology, Vol. 32: Basic Protein and Peptide Protocols* (Walker, J. M., ed.), Humana, Totowa, NJ, pp. 9–15.

18. Walker, J. M. (1994) The bicinchoninic acid (BCA) assay for protein quantitation, in *Methods in Molecular Biology, Vol. 32: Basic Protein and Peptide Protocols* (Walker, J. M., ed.), Humana, Totowa, NJ, pp. 5–8.

19. Smith, B. J. (1994) Quantification of proteins on polyacrylamide gels (nonradioactive), in *Methods in Molecular Biology, Vol. 32: Basic Protein and Peptide Protocols* (Walker, J. M., ed.), Humana, Totowa, NJ, pp. 107–111.

20. Pollard, J. W. (1994) Two-dimentional polyacrylamide gel electrophoresis of proteins, in *Methods in Molecular Biology, Vol. 32: Basic Protein and Peptide Protocols* (Walker, J. M., ed.), Humana, Totowa, NJ, pp. 73–85.

2

Preparation of Extracts From Animal Tissues

J. Mark Skehel

1. Introduction

The initial procedure in the isolation of an protein, a protein complex, or a subcellular organelle is the preparation of an extract that contains the required component in a soluble form. Indeed, when undertaking a proteomic study, the production of a suitable cellular extract is essential. Further isolation of subcellular fractions depends on the ability to rupture the animal tissues in such a manner that the organelle or macromolecule of interest can be purified in a high yield, free from contaminants and in an active form. The homogenization technique employed should, therefore, stress the cells sufficiently enough to cause the surface plasma membrane to rupture, thus releasing the cytosol; however, it should not cause extensive damage to the subcellular structures, organelles, and membrane vesicles. The extraction of proteins from animal tissues is relatively straightforward, as animal cells are enclosed only by a surface plasma membrane (also referred to as the limiting membrane or cell envelope) that is only weakly held by the cytoskeleton. They are relatively fragile compared to the rigid cell walls of many bacteria and all plants and are thus susceptible to shear forces. Animal tissues can be crudely divided into soft muscle (e.g., liver and kidney) or hard muscle (e.g., skeletal and cardiac). Reasonably gentle mechanical forces such as those produced by liquid shear may disrupt the soft tissues, whereas the hard tissues require strong mechanical shear forces provided by blenders and mincers. The homogenate produced by these disruptive methods is then centrifuged in order to remove the remaining cell debris.

The subcellular distribution of the protein or enzyme complex should be considered. If located in a specific cellular organelle such as the nuclei, mitochondria, lysosomes, or endoplasmic reticulum, then an initial subcellular fractionation to isolate the specific organelle can lead to a significant degree of purification in the first stages of the experiment *(1)*. Subsequent purification steps may also be simplified, as contaminating proteins may be removed in the centrifugation steps. In addition, the deleterious affects of proteases released as a result of the disruption of lysosomes may also be avoided. Proteins may be released from organelles by treatment with detergents or by disruption resulting from osmotic shock or ultrasonication. Although there is clearly an ad-

From: *Methods in Molecular Biology, vol. 244: Protein Purification Protocols: Second Edition*
Edited by: P. Cutler © Humana Press Inc., Totowa, NJ

vantage in producing a purer extract, yields of organelles are often low, so consideration has to be made to the acceptability of a lower final yield of the desired protein.

Following production of the extract, some proteins will inevitably remain insoluble. For animal tissues, these generally fall into two categories: membrane-bound proteins and extracellular matrix proteins. Extracellular matrix proteins such as collagen and elastin are rendered insoluble because of extensive covalent crosslinking between lysine residues after oxidative deamination of one of the amino groups. These proteins can only be solubilized following chemical hydrolysis or proteolytic cleavage.

Membrane-bound proteins can be subdivided into integral membrane proteins, where the protein or proteins are integrated into the hydrophobic phospholipid bilayer, or extrinsic membrane proteins, which are associated with the lipid membrane resulting from interactions with other proteins or regions of the phospholipid bilayer. Extrinsic membrane proteins can be extracted and purified by releasing them from their membrane anchors with a suitable protease. Integral membrane proteins, on the other hand, may be extracted by disruption of the lipid bilayer with a detergent or, in some cases, an organic solvent. In order to maintain the activity and solubility of an integral membrane protein during an entire purification strategy, the hydrophobic region of the protein must interact with the detergent micelle. Isolation of integral membrane proteins is thought to occur in four stages, where the detergent first binds to the membrane, membrane lysis then occurs, followed by membrane solubilization by the detergent, forming a detergent–lipid–protein complex. These complexes are then further solubilized to form detergent–protein complexes and detergent–lipid complexes. The purification of membrane proteins is, therefore, not generally as straightforward as that for soluble proteins *(2,3)*.

The principal aim of any extraction method must be that it be reproducible and disrupt the tissue to the highest degree, using the minimum of force. In general, a cellular disruption of up to 90% should be routinely achievable. The procedure described here is a general method and can be applied, with suitable modifications, to the preparation of tissue extracts from both laboratory animals and from slaughterhouse material *(4,5)*. In all cases tissues, should be kept on ice before processing. However, it is not generally recommended that tissues be stored frozen prior to the preparation of extracts.

2. Materials

The preparation of extracts from animal tissues requires normal laboratory glassware, equipment, and reagents. All glassware should be thoroughly cleaned. If in doubt, clean by immersion in a sulfuric–nitric acid bath. Apparatus should then be thoroughly rinsed with deionized and distilled water. Reagents should be Analar grade or equivalent. In addition, the following apparatuses are required:

1. Mixers and blenders: In general, laboratory apparatus of this type resemble their household counterparts. The Waring blender is most often used. It is readily available from general laboratory equipment suppliers and can be purchased in a variety of sizes, capable of handling volumes from 10 mL to a few liters. Vessels made from stainless steel are preferable, as they retain low temperatures when prechilled, thus counteracting the effects of any heat produced during cell disruption.
2. Refrigerated centrifuge: Various types of centrifuge are available, manufacturers of which are Beckman, Sorval-DuPont, and MSE. The particular centrifuge rotor used depends on

the scale of the preparation in hand. Generally, for the preparations of extracts, a six-position fixed-angle rotor capable of holding 250-mL tubes will be most useful. Where larger-scale preparations are undertaken, a six-position swing-out rotor capable of accommodating 1-L containers will be required.

3. Centrifuge tubes: Polypropylene tubes with screw caps are preferable, as they are more chemically resistant and withstand higher *g* forces than other materials such as polycarbonate. In all cases, the appropriate tubes for the centrifuge rotor should be used.

3. Methods

All equipment and reagents should be prechilled to 0–4°C. Centrifuges should be turned on ahead of time and allowed to cool down.

1. First, trim fat, connective tissue, and blood vessels from the fresh chilled tissue and dice into pieces of a few grams (*see* **Note 1**).
2. Place the tissue in the precooled blender vessel (*see* **Note 2**) and add cold extraction buffer using 2–2.5 vol of buffer by weight of tissue (*see* **Note 3**). Use a blender vessel that has a capacity approximately that of the volume of buffer plus tissue so that the air space is minimized; this will reduce aerosol formation.
3. Homogenize at full speed for 1–3 min depending on the toughness of the tissue. For long periods of homogenization, it is best to blend in 40-s to 1-min bursts with a few minutes in between to avoid excessive heating. This will also help reduce foaming.
4. Remove cell debris and other particulate matter from the homogenate by centrifugation at 4°C. For large-scale work, use a 6 × 1000-mL swing-out rotor operated at about 600–3000g for 30 min. For smal-scale work (up to 3 L of homogenate), a 6 × 250-mL angle rotor operated at 5000g would be more appropriate (*see* **Note 4**).
5. Decant the supernatant carefully, avoiding disturbing the sedimented material, through a double layer of cheesecloth or muslin. This will remove any fatty material that has floated to the top. Alternatively, the supernatant may be filtered by passing it through a plug of glass wool placed in a filter funnel. The remaining pellet and intermediate fluffy layer may be re-extracted with more buffer to increase the yield (*see* **Note 5**) or discarded.

The crude extract obtained by the above procedure will vary in clarity depending on the tissue from which it was derived. Before further fractionation is undertaken, additional clarification steps may be required (*see* **Note 6**).

4. Notes

1. The fatty tissue surrounding the organ/tissue must be scrupulously removed prior to homogenization, as it can often interfere with subsequent protein isolation from the homogenate.
2. Where only small amounts of a soft tissue (1–5 g) such as liver, kidney, or brain are being homogenized, then it may be easier to use a hand-held Potter–Elvehjem homogenizer *(6)*. This will release the major organelles; nuclei, lysosomes, peroxisomes, and mitochondria *(7)*. The endoplasmic reticulum, smooth and rough, will vesiculate, as will the Golgi if homogenization conditions arc too severe. On a larger scale, these soft tissues are easily disrupted/homogenized in a blender. However, tissues such as skeletal muscle, heart, and lung are too fibrous in nature to place directly in the blender and must first be passed through a meat mincer, equipped with rotating blades, to grind down the tissue before homogenization *(8,9)*. As the minced tissue emerges from the apparatus, it is placed directly into an approximately equal volume by weight of a suitable buffer. This mixture is then squeezed

Table 1
Protease Inhibitors

Inhibitor	Target proteases	Effective concentrations	Stock solutions
EDTA	Metalloproteases	0.5–2.0 mM	500 mM in water, pH 8.0
Leupeptin	Serine and thiolproteases	0.5–2 µg/mL	10 mg/mL in water
Pepstatin	Acid proteases	1 µg/mL	1 mg/mL in methanol
Aprotinin	Serine proteases	0.1–2.0 µg/mL	10 mg/mL in phosphate-buffered saline
PMSF	Serine proteases	20–100 µg/mL	10 mg/mL in isopropanol

through one thickness of cheesecloth, to remove the blood, before placing the minced tissue in the blender vessel.

3. Typically, a standard isotonic buffer used for homogenization of animal tissues is of moderate ionic strength and neutral pH. For instance, 0.25 M sucrose and 1 mM EDTA and buffered with a suitable organic buffer: Tris, MOPS, HEPES, and Tricine at pH 7.0–7.6 are commonly employed. The precise composition of the homogenization medium will depend on the aim of the experiment. If the desired outcome is the subsequent purification of nuclei, then EDTA should not be included in the buffer, but KCl and a divalent cation such as MgCl$_2$ should be present *(10)*. MgCl$_2$ is preferred here when dealing with animal tissues, as Ca^{2+} can activate certain proteases. The buffer used for the isolation of mitochondria varies depending on the tissue that is being fractionated. Buffers used in the preparation of mitochondria generally contain a nonelectrolyte such as sucrose *(4,11)*. However, if mitochondria are being prepared from skeletal muscle, then the inclusion of sucrose leads to an inferior preparation, showing poor phosphorylating efficiency and a low yield of mitochondria. The poor quality is the result of the high content of Ca^{2+} in muscle tissue, which absorbs to the mitochondria during homogenization; mitochondria are uncoupled by Ca^{2+}. The issue of yield arises from the fact that when skeletal muscle is homogenized in a sucrose medium, it forms a gelatinous consistency, which inhibits the disruption of the myofibrils. Here, the inclusion of salts such as KCl (100–150 mM) are preferred to the nonelectrolyte *(8,12)*.

In order to protect organelles from the damaging effect of proteases, which may be released from lysosomes during homogenization, the inclusion of protease inhibitors to the homogenization buffer should also be considered. Again, their inclusion will depend on the nature of the extraction and the tissue being used. Certain proteins are more susceptible to degradation by proteases than others, and certain tissues such as liver contain higher protease levels than others. A suitable cocktail for animal tissues contains 1 mM phenylmethylsulfonyl fluoride (PMSF) and 2 µg/mL each of leupeptin, antipain, and aprotinin (*see* **Table 1**). These are normally added from concentrated stock solutions. Further additions to the homogenization media can be made in order to aid purification. A sulfhydryl reagent, 2-mercaptoethanol or dithiothreitol (0.1–0.5 mM), will protect enzymes and integral membrane proteins with reactive sulfhydryl groups, which are susceptible to oxidation. The addition of a cofactor to the media, to prevent dissociation of the cofactor from an enzyme or protein complex, can also assist in maintaining protein stability during purification.

4. Centrifugation is the application of radial acceleration by rotational motion. Particles that have a greater density than the medium in which they are suspended will move toward the outside of the centrifuge rotor, wheras particles lighter than the surrounding medium will move inward. The centrifugal force experienced by a particle will vary depending on its

distance from the center of rotation. Hence, values for centrifugation are always given in terms of g (usually the average centrifugal force) rather than as revolutions per minute (rpm), as this value will change according to the rotor used. Manufacturers provide tables that allow the relative centrifugal fields at a given run speed to be identified. The relative centrifugal field (RCF) is the ratio of the centrifugal acceleration at a certain radius and speed (rpm) to the standard acceleration of gravity (g) and can be described by the following equation:

$$RCF = 1.118r \, (rpm/1000)^2 \tag{1}$$

where r is the radius in millimeters.

Centrifuges should always be used with care in order to prevent expensive damage to the centrifuge drive spindle and, in some instances, to the rotor itself. It is important that centrifuges and rotors are cleaned frequently. Essentially, this means rinsing with water and wiping dry after every use. Tubes must be balanced and placed opposite one another across the central axis of the rotor. Where small volumes are being centrifuged, the tubes can usually be balanced by eye to within 1 g. When the volumes are >200 mL, the most appropriate method of balancing is by weighing. Consideration should be given to the densities of the liquids being centrifuged, especially when balancing against water. A given volume of water will not weigh the same as an equal volume of homogenate. The volume of water used to balance the tubes can be increased, but it is better practice to divide the homogenate between two tubes. The tubes may well be of equal weight, but their centers of gravity will be different. As particles sediment, there will also be an increase in inertia and this should always be equal across the rotor. Care should also be taken not to over fill the screw-cap polypropylene tubes. Although they may appear sealed, under centrifugation the top of the tube can distort, leading to unwanted and potentially detrimental leakage of sample into the rotor. Fill tubes such that when they are placed in the angled rotor, the liquid level is just below the neck of the tube.

5. Following centrifugation of the homogenate, a large pellet occupying in the region of 25% of the tubes volume will remain. The pellet contains cells, tissue fragments, some organelles, and a significant amount of extraction buffer and, therefore, soluble proteins. If required, this pellet can be resuspended/washed in additional buffer. Disperse the pellet by using a glass stirring rod against the wall of the tube or, if desired, a hand-operated homogenizer. The resuspended material is centrifuged earlier and the supernatants combined. This washing will contribute to an increased yield but inevitably will also lead to a dilution of the extract. Therefore, the value of a repeat extraction needs to be assessed. For instance, when preparing liver or kidney mitochondria, washing the pellet in this way not only increases the yield, it also improves the integrity of the preparation, by allowing the recovery of the larger mitochondria.

6. The procedure outlined in this chapter is of general applicability and will, in some cases, produce extracts of sufficient clarity to proceed immediately to the next set of fractionation experiments. This is particularly true for cardiac muscle. However, for other tissues, the extract produced may require further steps to remove extraneous particulate matter before additional fractionations can be attempted. Colloidal particles made up of cell debris and fragments of cellular organelles are maintained as a suspension that will not readily sediment by increasing the run length and RCF applied. In these cases, it is often appropriate to bring about coagulation in order to clarify the extract. Coagulation may be induced in a number of ways, all of which alter the chemical environment of the suspended particles. The extract can be cooled or the pH may be adjusted to between pH 3.0 and 6.0. Indeed, rapidly altering the pH can be quite effective. Surfactants that alter the hydration of the particles

may also be used. In some situations, the presence of excessive amounts of nucleic acid can cause turbidity and increased viscosity of the extract. In these situations, it may be appropriate to precipitate with a polycationic macromolecule such as protamine sulfate in order to cause aggregation of the nucleic acid (addition to a final concentration of 0.1% w/v). The agglutinated particles will now sediment more easily when the mixture is recentrifuged.

Conditions for the clarification of an extract by coagulation should be arrived at through a series of small-scale tests, such that coagulation is optimized, whereas any detrimental effects such as denaturation are minimized. The coagulant should be added to the extract that is being stirred at high speed, thus maximizing particle interactions. Reducing the speed at which the mixture is stirred will then aid coagulation.

References

1. Claude, A. (1946) Fractionation of mammalian liver cells by differential centrifugation: II. Experimental procedures and results. *J. Exp. Med.* **84**, 61–89.
2. Rabilloud, T. (1995) A practical guide to membrane protein purification. *Electrophoresis* **16(3)**, 462–471.
3. Arigita, C., Jiskoot, W., Graaf, M. R., and Kersten, G. F. A. (2001) Outer membrane protein purification. *Methods Mol. Med.* **66**, 61–79.
4. Smith, A. L. (1967) Preparation, properties and conditions for assay of mitochondria: slaughterhouse material, small scale. *Methods Enzymol.* **10**, 81–86.
5. Tyler, D. D. and Gonze, J. (1967) The preparation of heart mitochondria from laboratory animals. *Methods Enzymol.* **10**, 75–77.
6. Dignam, J. D. (1990) Preparation of extracts from higher eukaryotes. *Methods Enzymol.* **182**, 194–203.
7. Völkl, A. and Fahimi, H. D. (1985) Isolation and characterization of peroxisomes from the liver of normal untreated rats. *Eur. J. Biochem.* **149**, 257–265.
8. Ernster, L. and Nordenbrand, K. (1967) Skeletal muscle mitochondria, *Methods Enzymol.* **10**, 86–94.
9. Scarpa, A., Vallieres, J., Sloane, B., and Somlyo, A. P. (1979) Smooth muscle mitochondria. *Methods Enzymol.* **55**, 60–65.
10. Blobel, G. and Potter, V. R. (1966) Nuclei from rat liver: isolation method that combines purity with high yield. *Science* **154**, 1662–1665.
11. Nedergaard, J. and Cannon, B. (1979) Overview—preparation and properties of mitochrondria from different sources *Methods Enzymol.* **55**, 3–28.
12. Chappell, J. B. and Perry, S. V. (1954) Biochemical and osmotic properties of skeletal muscle mitochondria. *Nature* **173**, 1094–1095.

3

Protein Extraction From Plant Tissues

Roger J. Fido, E. N. Clare Mills, Neil M. Rigby, and Peter R. Shewry

1. Introduction

Plant tissues contain a wide range of proteins, which vary greatly in their properties, and require specific conditions for their extraction and purification. It is therefore not possible to recommend a single protocol for extraction of all plant proteins.

The scale of the extraction must be considered at an early stage, and suitably sized extraction equipment must be used. For large amounts, a polytron or similar equipment will be needed, but for a small weight of tissue, then a small-scale homogenizer or simple pestle and mortar is quite suitable.

Plant tissues do pose specific problems, which must be taken into account when developing protocols for extraction. The first is the presence of a rigid cellulosic cell wall, which must be sheared to release the cell contents. Breaking up fresh tissue can be achieved with acid-washed sand (Merck/BDH) added with the extraction buffer and grinding in a pestle and mortar or adding liquid nitrogen to rapidly freeze the material before blending. The second is the presence of specific contaminating compounds that may result in protein degradation or modification, and, where the protein of interest is an enzyme, the subsequent loss of catalytic activity. Such compounds include phenolics and a range of proteinases. It is sometimes possible to avoid these problems or partially control them by using a specific tissue (e.g., young tissue rather than old leaves) or using a particular plant species. However, in other cases (e.g., enzymes involved in secondary product synthesis), this is not possible and the biochemist must find ways to remove or inactivate the active contaminants. The removal of phenolics is dealt with in Chapter 8. Because many plant proteinases are of the serine type, it is often convenient to include the serine protease inhibitor phenylmethylsulfonylfluoride (PMSF) in extraction buffers on a routine basis (*see* Chapter 9 for a general discussion of protease inhibition).

Animals have many highly specialized tissues (e.g., liver, muscle, brain) that are rich sources of specific enzymes, thus facilitating their purification. This is not usually the case with plant enzymes, which may be present at low levels in highly complex protein mixtures. An exception to this is storage organs, such as seeds, tubers, and tap roots. These organs contain high levels of specific proteins whose role is to act as a store of nitrogen, sulfur, and carbon. These storage proteins are among the most widely studied

From: *Methods in Molecular Biology, vol. 244: Protein Purification Protocols: Second Edition*
Edited by: P. Cutler © Humana Press Inc., Totowa, NJ

proteins of plant origin, because of their abundance, ease of purification, and their economic and nutritional importance as food, feed for livestock, and raw material in the food and other industries. Indeed, seed proteins were among the earliest of all proteins to be studied in detail, with wheat gluten being isolated in 1745 *(1)*, the Brazil nut globulin edestin crystallized in 1859 *(2)*, and a range of globulin storage proteins being subjected to ultracentrifugation analysis by Danielsson in 1949 *(3)*.

Comparative studies of the extraction and solubility of plant proteins also formed the basis for the first systematic attempt to classify proteins. Osborne, working at the Connecticut Agricultural Experiment Station between about 1880 and 1930, compared and characterized proteins from a range of plant sources, including the major storage proteins of cereal and legume seeds *(4)*. He defined four groups that were extracted sequentially in water (albumins), dilute salt solutions (globulins), alcohol–water mixtures (prolamins), and dilute acid or alkali (glutelins). These "Osborne groups" still form the basis for studies of seed storage proteins, and the terms albumin and globulin have become accepted into the general vocabulary of protein chemists.

Four detailed protein extraction protocols are given. The first two are for the extraction of enzymically active proteins ribulose 1,5-bisphosphate carboxylase/oxygenase (Rubisco) (E.C. 4.1.1.39) and nitrate reductase (E.C. 1.6.6.1.) from vegetative tissues. Rubisco is a hexadecameric protein (eight subunits of approx Mr 50,000–60,000 and eight subunits of Mr 12,000–20,000) with an Mr of 500,000, which catalyzes the fixation of carbon in the chloroplast stroma. It often represents more than 50% of the total chloroplast protein and is recognized as the most abundant protein in the world. In contrast, the complex enzyme nitrate reductase that has a Mr of approx 200,000, is present in plant tissues at less than 5 mg/kg fresh weight *(5)*. This low abundance, combined with susceptibility to proteolysis and loss of functional prosthetic groups during extraction and purification, often leads to a very low recovery of the enzyme. The third protocol is a specialized procedure for the extraction of seed proteins from cereals, based on the classical Osborne fractionation. In addition, two rapid methods are described for the extraction of leaf and seed proteins for sodium dodecyl sulfate-polyacrylomide gel electrophoresis (SDS-PAGE) analysis. These are suitable for monitoring the expression of transgenes in engineered plants.

Finally, a protocol is given for the extraction of a moderately abundant protein from apple tissues for immunoassay. This is the allergen known as Mal d 1, which is a homolog of the major birch pollen allergen, Bet v 1. The function of the Bet v 1 family in plant tissues is not known, but they may be synthesised as part of the response of the plant to stress and pathogen attack, and as such, they have been termed PR (pathogenesis-related) proteins. Mal d 1 is unstable in apple extracts and may become modified by interactions with plant polyphenols and pectins, which affect its immunoreactivity.

2. Materials

1. Buffer A (Rubisco): 20 mM Tris-HCl, pH 8.0, 10 mM NaHCO$_3$, 10 mM MgCl$_2$, 1 mM EDTA, 5 mM dithiothreitol (DTT), 0.002% (w/v) Hibitane, and 1% (w/v) polyvinylpolypyrolidone.
2. Buffer B (nitrate reductase) (NR): 0.5 M Tris-HCl, pH 8.6, 1 mM EDTA, 5 μM Na$_2$MoO$_4$, 25 μM FAD, 5 mM PMSF, 5 μg/mL pepstatin, 10 μM antipain, and 3% (w/v) bovine serum albumin (BSA).

3. Buffer C: 0.0625 *M* Tris-HCl, pH 6.8, 2% (w/v) SDS, 5% (v/v) 2-mercaptoethanol or 1.5 (w/v) DTT, 10% (w/v) glycerol, 0.002% (w/v) bromophenol blue.
4. Buffer D: 0.1 *M* Tris-HCl, pH 8.0, 0.01 *M* MgCl$_2$, 18% (w/v) sucrose, 40 m*M* 2-mercaptoethanol.
5. Buffer E: 0.02 *M* sodium phosphate buffer, pH 7.0, 0.002 *M* EDTA, 0.01 *M* sodium deithyldithiocarbamate, 2% (w/v) polyvinylpolypyrolidone.
6. Buffer F: Phosphate-buffered saline (PBS), 0.14 *M* NaCl, 0.0027 *M* KCl, 0.0015 *M* KH$_2$PO$_4$, 0.008 *M* Na$_2$HPO$_4$, pH 7.4.

3. Methods

3.1. Extraction of Enzymically Active Preparations From Leaf Tissues

All procedures are carried out at 0–4°C with precooled reagents and apparatus. Tissue can be used fresh, or after rapid freezing using liquid nitrogen, and stored at −20 to −80°C or under liquid nitrogen. Tissue homogenization can be accomplished in a pestle and mortar or a ground-glass homogenizer (for small volumes) or a Waring blender or Polytron for larger initial weights.

The method for the extraction of Rubisco from wheat leaves is taken from the work of Keys and Parry *(6)*. It is reported that the extraction procedure and extraction buffers used are important in affecting the initial rate and total activities of the enzyme (*see* **Note 1**). It is also important for initial activity measurements to maintain the extract at a temperature of 2°C.

1. Cut 3-wk-old wheat leaves into 1-cm lengths and homogenize in an ice-cold buffer using a ratio of 6:1.
2. Filter the homogenate through four layers of muslin and then add sufficient solid (NH$_4$)$_2$SO$_4$ to give 35% saturation.
3. After 20 min, centrifuge the suspension at 20,000*g* for 15 min. Discard the pellet.
4. Add additional solid (NH$_4$)$_2$SO$_4$ to give 55% saturation. After centrifugation, dissolve the pellet in 20 m*M* Tris-HCl containing 1 m*M* DTT, 1 m*M* MgCl$_2$, and 0.002% Hibitane (*see* **Note 2**) at pH 8.0. After clarification, the Rubisco can then be fractionated by sucrose density centrifugation.

The method for NR extraction, using a complex extraction buffer (see buffer B), is taken from the work of Somers et al. *(7)*, who attempted to identify whether barley NR was regulated by enzyme synthesis and degradation or by an activation–inactivation mechanism.

1. Both root and shoot tissues were excised at different ages (days), weighed, frozen in liquid nitrogen and stored at −80°C.
2. Pulverize the frozen tissue in a pestle and mortar under liquid nitrogen. Extract with 1 mL/g fresh weight of buffer B (*see* **Note 3**).
3. Filter the homogenate through two layers of cheesecloth and centrifuge 30,000*g* to clarify. The supernatant can be used directly for enzyme activity measurements (*see* **Note 4**).

3.2. Extraction of Cereal Seed Proteins, Using a Modified Osborne Procedure

The procedure is based on the work of Shewry et al. *(8)*. Air-dry grain (approx 14% water) is milled to pass a 0.5-mm mesh sieve. The meal is then extracted by stirring (*see* **Note 5**) with the following series of solvents: 10 mL of solvent is used per gram of meal and each extraction is for 1 h. Extractions are carried out at 20°C and repeated as stated.

1. Water-saturated 1-butanol (twice) to remove lipids.
2. 0.5 *M* NaCl to extract salt-soluble proteins (albumins and globulins) and nonprotein components (twice) (*see* **Note 6**).
3. Distilled water to remove residual NaCl.
4. 50% (v/v) 1-Propanol containing 2% (v/v) 2-mercaptoethanol (or 1% [w/v] DTT) and 1% (v/v) acetic acid (three times) to extract prolamins (*see* **Note 7**).
5. 0.05 *M* Borate buffer, pH 10.0, containing 1% (v/v) 2-mercaptoethanol and 1% (w/v) SDS to extract residual proteins (glutelins) (*see* **Note 8**).

The supernatants are separated by centrifugation (20 min at 10,000*g*) and treated as follows:

6. Supernatants 2 and 3 from **steps 2** and **3**, respectively, are combined and dialyzed against several changes of distilled water at 4°C over 48 h. Centrifugation removes the globulins, allowing the soluble albumins to be recovered by lyophilization.
7. Supernatants from **step 4** are combined, and the prolamins recovered after precipitation, either by dialysis against distilled water or addition of 2 vol of 1.5 *M* NaCl followed by standing overnight at 4°C.
8. Supernatants from **step 5** are combined and glutelins recovered by dialysis against distilled water at 4°C followed by lyophilization (*see* **Note 9**). SDS can be removed from the protein using standard procedures.

3.3. Extraction of Proteins for SDS-PAGE Analysis

The methods described in **Subheadings 3.1.** and **3.2.** are suitable for the bulk extraction of proteins for purification of individual components. However, in some situations (e.g., analysis of transgenic plants or studies of seed protein genetics), it is advantageous to extract total proteins for direct analysis by SDS-PAGE. The following methods are specially designed for this purpose.

3.3.1. Extraction of Leaf Tissues

The method, based on the work of Nelson et al. *(9)*, gives good results with chlorophyllous tissues.

1. Freeze tissue in liquid N_2.
2. Grind for about 30 s in a mortar with 3 mL of buffer D per gram of tissue (*see* **Note 10**).
3. Filter through muslin and centrifuge for 15 min in a microfuge.
4. Dilute to about 2 mg protein/mL, ensuring that the final solution contains about 2% (w/v) SDS, 0.002% (w/v) bromophenol blue, and at least 6% (w/v) sucrose (*see* **Note 11**).
5. Separate aliquots by SDS-PAGE.

3.3.2. Extraction of Seed Proteins

1. Grind in a mortar with 25 µL of buffer C/mg meal.
2. Transfer to an Eppendorf tube and allow to stand for 2 h.
3. Suspend in a boiling water bath for 2 min.
4. Allow to cool, and then spin in a microfuge.
5. Separate 10- to 20-µL aliquots by SDS-PAGE.

3.4. Extraction of the Soluble Protein, Mal d 1 From Apples for ELISA (see Notes 12–18)

In general, an immunoassay requires a protein to be quantitatively extracted from a tissue in its native form, preferably using a buffer compatible with the immunoassay. Wherever possible, extraction procedures should be kept simple in order to maximize the benefit of using high-throughput methodology such as enzyme-linked immunosorbent assay (ELISA). This extraction procedure is based on methodology developed by Bjorksten et al. *(10)*.

1. Peel and core apples, chopping flesh into 0.5-cm-thick slices and either freeze in liquid nitrogen and store at $-40°C$ until required or homogenize immediately for 2 min in 10 vol of buffer E using a Waring blender, followed by gentle shaking for 1 h at 1°C.
2. Centrifuge extracts for 30 min at 30,000g to clarify the extract.
3. Dilute Mal d 1 extract either 1:2 or 1:10 (v/v) in PBS containing 0.05%(v/v) Tween-20 (PBST) and add to the ELISA-coated plate.

4. Notes

1. A wide range of buffers can be used, depending on the pH range required for optimal enzyme activity and the preference (or prejudice) of the operator. However, Tris is very widely used.
2. A range of specific additions can be made in order to help preserve the activity of the enzyme under consideration. For example, with NR, it is advantageous to add a flavin compound (i.e., FAD) in order to maintain the endogenous levels needed for catalytic activity. The inclusion of both CO_2 and Mg^{2+} ions in the extraction buffer of Rubisco has been reported to be necessary *(11)*. Polyethylene glycol has also been included in a complex extraction buffer and was described as the most successful of a number of buffers tested in the extraction of Rubisco from *Kalanchoe* *(12)*.

   There is no single simple method to guarantee activity, the operator should consult published protocols for the extraction of related enzymes and be prepared to carry out exploratory extractions using different buffer compositions:
   a. 1 mM Dithiothreitol or 10 mM 2-mercaptoethanol to preserve sulfhydryl groups.
   b. 1 mM EDTA to chelate metals, especially with phosphate buffers that commonly contain inhibitory concentrations of ferrous ions.
   c. 50 mM Sodium fluoride to inhibit phosphatases that inactivate phosphoenzymes.
   d. 25 g/kg Fresh weight of polyvinylpolypyrrolidone (PVPP). This is an insoluble form of polyvinylpyrrolidone that binds phenolic compounds. It forms a slurry and can be removed by centrifugation.
   e. 0.1 mM PMSF to inhibit serine proteinases. This is readily dissolved in a small volume of 1-propanol prior to mixing with the buffer. **It is highly toxic**.
   f. Glycerol (up to 30%) or other organic alcohols (ethylene glycol, mannitol) may help to stabilize some highly labile enzymes.
   g. The addition of exogenous proteins (e.g., casein and BSA) has been used to stabilize enzymes by preventing hydrolysis because of protease activity.
   h. Antibacterial agents, such as Hibitane, can also be added.
3. Similar methods can be used to extract enzymes from seed tissues, either by direct homogenization or after milling. Lipid-rich tissues can either be defatted with cold (4°C) acetone or an acetone powder *(13)*. *Note:* Extreme care should be taken because of the low flash point of acetone: Operations should be carried out in a fume cupboard and electrical sparks avoided.

4. The supernatant may be concentrated by precipitation with $(NH_4)_2SO_4$ (**12**) and assayed directly after desalting on a column of Sephadex G25.
5. Extraction may be carried out by stirring magnetically or with a paddle; the mechanical grinding that occurs may assist extraction.
6. It may be advantageous to extract the salt-soluble proteins at 4°C and include 1.0 m*M* PMSF (*see* **Note 2**) to minimize proteolysis.
7. It is sometimes of interest to extract the prolamins in two fractions. Extraction twice with 50% (v/v) 1-propanol gives monomeric prolamins and alcohol-soluble disulfide-stabilized polymers, whereas subsequent extraction twice with 50% (v/v) 1-propanol with 2% (v/v) 2-mercaptoethanol and 1% (v/v) acetic acid gives reduced subunits derived from alcohol-insoluble disulfide-bonded polymers.
8. It is usual to determine the amounts of extracted proteins by Kjeldahl N analysis of aliquots removed from the supernatants. The values can then be multiplied by a factor of 5.7 for prolamins or 6.25 for other fractions to give the amount of protein.
9. SDS-PAGE is used to monitor the compositions of the fractions.
10. Addition of 2% (w/v) SDS to buffer D allows the extraction of membrane and other insoluble proteins.
11. If required, soluble and insoluble proteins can be extracted in two sequential fractions. Soluble proteins are initially extracted in buffer D (3 mL/g) and insoluble proteins by re-extracting the pellet with 0.05 vol (relative to the original homogenate) of 2% (w/v) SDS, 6% (w/v) sucrose, and 40 m*M* 2-mercaptoethanol.
12. Some antibody preparations are able to recognize both native and denatured proteins. In such instances, harsher denaturing extraction buffers employing 1–2% (w/v) SDS may be used, which allow better extraction or recovery of the protein to be analyzed from a plant. If a sufficiently concentrated extract is used and the ELISA has a reasonable degree of sensitivity, the extract can be diluted sufficiently (1:50 or 1:100, v/v) into the ELISA assay buffer so that the concentration of SDS is low enough not to affect the immunoassay. However, such high dilutions can pose problems when ELISAs are being used quantitatively, as any ELISA errors are multiplied by the dilution factor when calculating back to the original tissue concentration of a protein.
13. Although the ELISA assay buffer described here is PBST, other immunoassay assay buffers can be substituted, such as Tris-buffered saline.
14. For certain proteins, it may be necessary to employ extraction buffers at extreme pHs (<4.5 and >8.0). For example, seed storage globulins may be more soluble at pH 8.8 or only efficiently extracted with high concentrations of salt (10% w/v NaCl for sesame seed globulins). Such extreme pH values and high ionic strengths can disrupt antibody-binding reactions, necessitating dilution into assay buffers prior to analysis. In some instances, it may be necessary to increase the strength of the immunoassay buffer to ensure that the pH of the diluted extract is near neutral.
15. Lipids from oil-rich seeds and other tissues may cause interference in an immunoassay and such tissues may require prior extraction with 10 vol of a solvent such as hexane.
16. Polyphenols may modify protein immunoreactivity; hence, when analyzing tissues particularly rich in these compounds, it is advisable to include additives, such as PVPP.
17. Soluble plant polysaccharides, such as pectins, can cause problems by binding proteins or affecting immunoassay performance by altering sample viscosity. These problems can be overcome by the addition of 10 m*M* Ca^{2+} to precipitate the pectins, allowing their removal by centrifugation.
18. Coextraction of proteinases can present problems, both by degrading the protein to be analyzed and also digesting the adsorbed protein that constitutes the solid phase of the immunoassay. In general, they do not present a problem in the analysis of freshly prepared

extracts, but can have a dramatic effect on extract stability at $-20°C$ even for only a few days. In such instances, it is advisable to add a cocktail of proteinase inhibitors (*see* Chapter 9).

References

1. Beccari, J. B. (1745) De Frumento, in De Bononiensi Scientiarum et Artium atque Academia Commentarii, Tomi Secundi. Bononia, Bologna.
2. Matschke, O. (1859) Ueber den Bau und die Bestandtheile der Kleberbläschen in Bertholletia, deren Entwickelung in Ricinus, nebst einigen Bemerkunger über Amylonbläschen. *Botan. Z.* **17**, 409–447.
3. Danielsson, C. E. (1949) Seed globulins of the Gramineae and Leguminosae. *Biochem. J.* **44**, 387–400.
4. Osborne, T. B. (1924) *The Vegetable Proteins*, Longmans Green, London.
5. Wray, J. L. and Fido, R. J. (1990) Nitrate and nitrite reductase, in *Methods in Plant Biochemistry* (Lea, P. J., ed.), Academic, New York, Vol. 3, pp. 241–256.
6. Keys, A. J. and Parry, M. A. J. (1990) Ribulose bisphosphate carboxylase/oxygenase and carbonic anhydrase, in *Methods in Plant Biochemistry* (Lea, P. J., ed.), Academic, New York, Vol. 3, pp. 1–14.
7. Somers, D. A., Kuo, T.-M., Kleinhofs, A., Warner, R. L., and Oaks, A. (1983) Synthesis and degradation of barley nitrate reductase. *Plant Physiol.* **72**, 949–952.
8. Shewry, P. R., Franklin, J., Parmar, S., Smith, S. J., and Miflin, B. J. (1983) The effects of sulphur starvation on the amino acid and protein compositions of barley grain. *J. Cereal Sci.* **1**, 21–31.
9. Nelson, T., Harpster, M. H., Mayfield, S. P., and Taylor, W. C. (1984) Light regulated gene-expression during maize leaf development. *J. Cell. Biol.* **98**, 558–564.
10. Bjorksten, F., Halmepuro, L., Hannuksela, M., and Lahti, A. (1980) Extraction and properties of apple allergens. *Allergy* **35,** 671–677.
11. Servaites, J. C., Parry, M. A. J, Gutteridge, S., and Keys, A. J. (1986) Species variation in the predawn inhibition of ribulose-1,5-bisphosphate carboxylase/oxygenase. *Plant Physiol.* **82**, 1161–1163.
12. Maxwell, K., Borland, A. M., Haslam, R. P., Helliker, B. R., Roberts, A. and Griffiths, H. (1999) Modulation of Rubisco activity during the diurnal phases of the Crassulacean acid metabolism plant Kalanchoe daigremontiana. *Plant Physiol.* **121**, 849–856.
13. Nason, A. (1955) Extraction of soluble enzymes from higher plants. *Methods Enzymol.* **1**, 62–63.
14. Green, A. A. and Hughes, W. L. (1955) Protein fractionation on the basis of solubility in aqueous solutions of salts and organic solvents. *Methods Enzymol.* **1**, 67–90.

4

Extraction of Recombinant Protein From Bacteria

Anne. F. McGettrick and D. Margaret Worrall

1. Introduction

The use of bacteria for overexpression of recombinant proteins is still a popular choice because of lower cost and higher yields when compared to other expression systems *(1,2)*, but problems can arise in the recovery of soluble functionally active protein. In some cases, secretion of recombinant proteins by bacteria into the media has eliminated the need to lyse the cells, but most situations still require lysis of the bacterial cell wall in order to extract the recombinant protein product. A number of methods based on enzymatic methods and mechanical means are available for breaking open the bacterial cell wall, and the choice will depend on the scale of the process *(3,4)*. Enzymatic methods include lysozyme hydrolysis, which cleaves the glucosidic linkages in the bacterial cell-wall polysaccharide. The inner cytoplasmic membrane can then be disrupted easily by detergents, osmotic pressure, or mechanical methods.

Overexpression of the recombinant proteins from strong promoters on multiple-copy plasmids can result in expression levels of up to 40% of the total cell protein. However, in many cases, this results in the formation of insoluble protein aggregates known as inclusion bodies *(5)*. Inclusion bodies are cytoplasmic granules seen as phase bright under the light microscope and can contain most or all of the protein of interest. Scanning electron micrographs of *Escherichia coli* containing inclusion bodies and isolated inclusions are shown in **Fig. 1**.

Inclusion bodies were first reported by Williams et al. *(6)* on overexpression of proinsulin in *E. coli*. Formation of inclusion bodies is not only found on overexpression of foreign eukaryotic proteins, but is also on overexpression of bacterial proteins that are normally soluble *(7)*. The nature of the expressed protein, the rate of expression, and the level of expression all influence the formation of inclusion bodies. This is presumed to be the result of insufficient time for the nascent polypeptide to fold into the native conformation. Proteins that contain strongly hydrophobic or highly charged regions are more likely to form inclusions *(8)*.

A number of parameters have been found to effect the partitioning of the overexpressed protein between the cytosol and inclusion body fractions. Soluble protein can be increased in some cases by lowering the growth temperature, decreasing concentra-

From: *Methods in Molecular Biology, vol. 244: Protein Purification Protocols: Second Edition*
Edited by: P. Cutler © Humana Press Inc., Totowa, NJ

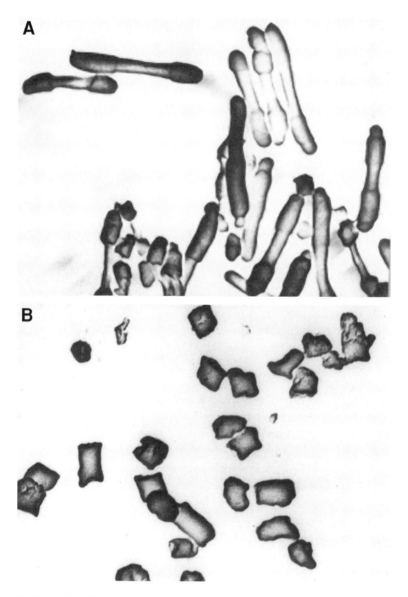

Fig. 1. (A) Scanning electron micrograph of *E. coli* cells containing inclusion bodies. The preparation process has shrunk the surrounding cell but not the inclusions, allowing their outline to be clearly seen. **(B)** Isolated washed inclusion bodies, which still retain a rigid cylindrical shape.

tion of the inducing agent, or increasing aeration. Fusion proteins with a highly soluble protein can also increase the solubility of the protein of interest and a comparative study suggests that maltose-binding protein is considerably better than glutathione-*S*-transferase or thioredoxin for this purpose *(9)*.

Coexpression with chaperone proteins such as the GroEL/ES and the DnaK-DanJ-GrpE systems or folding catalysts such as protein disulfide isomersase may also help to facilitate correct protein folding *(10)*. A number of expression protocols now include a heat-shock step to induce expression of endogenous *E. coli* chaperone proteins.

Formation of inclusion bodies can be advantageous in that they generally allow greater levels of expression and they can be easily separated from a large proportion of bacterial cytoplasmic proteins by centrifugation, giving an effective purification step. If a particular protein is harmful to bacteria in its native form, then insoluble expression may be the preferred method to obtain significant yields.

The major disadvantage of inclusion bodies is that extraction of the protein of interest generally requires the use of denaturing agents. This can cause problems where native folded protein is required, because refolding methods are rarely 100% effective and may be difficult to scale up.

Some inclusion bodies can be solubilized by extremes of pH and temperature, but most require strong denaturing agents. Certain proteins, such as DNase1, can be refolded after solubilization with sodium dodecyl sulfate (SDS), but detergents are difficult to remove from most proteins and can interfere with subsequent refolding. The most commonly used solubilizing agents are water-soluble chaotropic agents, such as urea and guanidinium hydrochloride, which are more compatible with protein refolding. Most inclusions will be soluble in 8 *M* urea, and a reducing agent, such as dithiothreitol (DTT), is generally required in order to prevent the formation of disulfide bonds between aggregates or denatured polypeptide chains.

There are many protocols for refolding proteins (for reviews, *see* **refs. *11*** and ***12***). Some advocate slow removal of the denaturant; others maintain that rapid dilution is important to prevent aggregation of partially folded intermediates.

2. Materials

All reagents are analytical grade.

1. Prepare a stock solution of 100 m*M* phenylmethylsulfonylfluoride (PMSF) in isopropanol and store at −20°C. Add PMSF to buffers just before use. (*Note:* PMSF is a hazardous chemical and should be treated with caution.)
2. Lysis buffer: 50 m*M* Tris-HCl, pH 8.0, 1 m*M* EDTA, 50 m*M* NaCl, 1 m*M* PMSF (*see* **Note 1**).
3. Hen egg lysozyme (Sigma): 10 mg/mL stock solution.
4. DNase 1 (Boehringer Mannheim): 1 mg/mL stock solution (*see* **Note 2**).
5. Solubilization buffer: 50 m*M* Tris-HCl, pH 8.0, 8 *M* urea, 1 *M* DTT.
6. Refolding buffer: 50 m*M* Tris-HCl, pH 8.0, 1 m*M* EDTA, 100 m*M* NaCl, 0.5 m*M* oxidized glutathione, 5 m*M* reduced glutathione, 200 m*M* L-arginine.

Urea solutions should be used within 1 wk of preparation and stored at 4°C, in order to reduce the formation of cyanate ions, which can react with protein amino groups, forming carbomylated derivatives.

3. Methods

3.1. Lysis of Escherichia coli *and Harvesting of Inclusion Bodies*

1. Harvest the bacterial cells by centrifugation at 1000*g* for 15 min at 4°C and pour off the supernatant. Weigh the wet pellet. This is easiest to do if you have preweighed the centrifuge tubes, which can then be deducted from the total weight.

2. Add approx 3 mL of lysis buffer for each wet gram of bacterial cell pellet and resuspend. Add lysozyme to a concentration of 300 µg/mL and stir the suspension for 30 min at 4°C (*see* **Notes 3** and **4**).

3. Add Triton X-100 to a concentration of 1% (v/v) and apply ultrasound sonication for three bursts of 30 s followed by cooling.

4. Place at room temperature and add DNase1 to a concentration of 10 mg/mL and MgCl$_2$ to 10 mM. Stir suspension for a further 15 min to remove the viscous nucleic acid (*see* **Note 2**).

5. Centrifuge the suspension at 10,000g for 15 min at 4°C. Resuspend the pellet in lysis buffer to the same volume as the supernatant and analyze aliquots of both for the protein of interest on SDS-polyacrylamide gel electrophoresis (SDS-PAGE). If the bulk of a normally soluble protein is found in the insoluble pellet fraction, then inclusion bodies are likely to have formed.

3.2. Washing of Inclusion Bodies

Washing of inclusion bodies prior to solubilization can remove further contaminant proteins, and using solutions other than water or buffer can increase the purification obtained *(13)*. It is advisable to carry out a small-scale trial to optimize the buffer and to ensure that the protein of interest is not solubilized.

1. Centrifuge 200-µL aliquots of the resuspended cell pellet in microfuge tubes at 12,000g for 10 min at 4°C. Resuspend the pellets in a range of test solutions. Lysis buffer containing 1, 2, 3, and 4 M urea and 0.5% Triton X-100 are suggested. Mix and incubate for 10 min at room temperature. Centrifuge as earlier in a microfuge and resuspend in 200 µL H$_2$O.

2. Take equal volumes of the supernatant and the resuspended pellet and add to the SDS boiling buffer. Analyze samples for the protein of interest on SDS-PAGE. The best washing buffer will contain the most contaminant proteins and little or none of the protein of interest.

3. Scale this procedure up and wash the inclusion bodies twice with the optimum buffer. An example of the purification achieved on washing of inclusion bodies of plasminogen activator inhibitor-2 (PAI-2) is shown on **Fig. 2**.

3.3. Solubilization of Recombinant Protein From Inclusion Bodies

It is also important to optimize the solubilization solution, because a number of factors will affect solubility depending on the nature of the protein of interest. These include the nature and strength of the solubilization agent, the temperature and time taken to obtain efficient solubilization, protein purity and concentration, and presence or absence of reducing agents.

1. Resuspend the washed inclusion bodies in the solubilization buffer. Stir this suspension for 1 h at room temperature to ensure complete solubilization.

2. Centrifuge the solution at 100,000g for 10 min at 4°C to remove any remaining insoluble material. Check this pellet for the protein of interest on SDS-PAGE. If a substantial proportion of the protein has remained insoluble, then increase the incubation time with the solubilization buffer or try a different agent to solubilize the inclusions.

3.4. Refolding of Protein

The extracted recombinant protein can be refolded from the urea at this stage or may be purified under denaturing conditions prior to refolding (*see* **Note 5**). Refolding success is extremely protein-specific, but the following protocol may be a useful starting point. Additives such as L-arginine can increase yields (*see* **Note 6**).

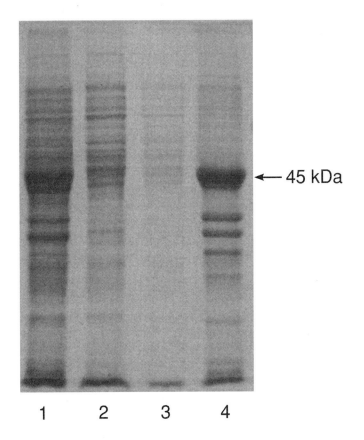

Fig. 2. Purification and solubilization of inclusion bodies containing recombinant PAI-2. *Lane 1*: insoluble *E. coli* fraction containing harvested inclusions; *lane 2*: first 2 *M* urea wash; *lane 3*: second 2 *M* urea wash ; *lane 4*: 8 *M* urea solubilized material.

1. Add the denatured protein sample dropwise to a stirred solution of refolding buffer at 4°C. Dilute the denaturant by 25- to 50-fold, with a final protein concentration not exceeding 0.05 mg/mL.
2. Continue stirring for 2 h at 4°C. The remaining denaturant can then be removed by dialysis against any suitable buffer. Filter the refolded material through a 0.2-μm filter to remove any aggregates.
3. Concentrate by centricon ultrafiltration or similar method and determine the recovery of refolded protein by functional assays or biophysical methods (e.g., circular dichroism or fluoresence)

4. Notes

1. The lysis buffer for enzymatic digestion can be critical. Hen egg lysozyme has a pH optimum of between 7.0 and 8.6 and works best in ionic strength of 0.05 *M*.
2. Bacterial extracts roughly consist of protein (40–70%), nucleic acid (10–30%), polysaccharide (2–10%), and lipid (10–15%). The nucleic acid fraction can often cause high viscosity. In addition to DNase treatment detailed in **Subheading 3.1., step 4**, this nucleic acid can also be removed from soluble protein solutions by precipitation with positively charged compounds, such as polyethyleneamine *(14)*. Precipitation methods should obviously not be

used with inclusion body preparations, because the precipitate will cocentrifuge with the inclusions.

3. Inclusion of extra protease inhibitors, such as aprotinin or leupeptin, in the lysis buffer may also be advantageous if proteolysis of the target protein is occurring. This is generally not necessary when the protein is packaged in inclusion bodies, but inhibitors may be required in the solubilization and refolding buffers, because proteins in semifolded states are more susceptible to proteolysis.

4. Sonication is suitable for smaller-scale purifications, but the generation of heat during sonication can be difficult to control and may cause denaturation of proteins. For larger-scale processing, the French press and the Mantin Gaulin press are most commonly used. These devices lyze the cells by applying pressure to the cell suspension, followed by a release of pressure, which causes a liquid shear and, thus, cell disruption. Multiple passes of the cells through the presses are generally necessary to obtain adequate lysis *(3)*.

5. It is often desirable to carry out some purification of the protein under denaturing conditions before refolding. Urea solutions are compatible with ion-exchange chromatography, metal-ion affinity chromatography, gel filtration, and reversed-phase high-performance liquid chromatography (HPLC). Owing to its charge, guanidium hydrochloride should not be used in ion-exchange purification steps.

6. Additives used to promote protein folding include amino acids (L-arginine), ionic and nonionic detergents (e.g., CHAPS, SDS, sarkosyl), salts, and sugars (sucrose, glycerol). Nondetergent sulfobetaines (NDSB) are reported to be particularly efficient in preventing aggregation *(15)*.

7. Proteins used for animal immunization purposes often do not require refolding unless antibodies to tertiary epitopes are required. It is also possible to use inclusion bodies directly injected, because particulate antigens are highly immunogenic *(16)*. Sonication of the inclusions into smaller particles is recommended prior to injection.

References

1. Balbas, P. (2001) Understanding the art of producing protein and nonprotein molecules in *Escherichia coli. Mol. Biotechnol.* **19**, 251–267.
2. Baneyx, F. (1999) Recombinant protein expression in *Escherichia coli. Curr. Opin. Biotechnol.* **10**, 411–421.
3. Cull, M. and McHenry, C. S. (1990) Preparation of extracts from prokaryotes. *Methods Enzymol.* **182**, 147–153.
4. Hopkins, T. R. (1991) Physical and chemical cell disruption for the recovery of intracellular proteins. *Bioprocess. Technol.* **12**, 57–83.
5. Kane, J. F. and Hartley, D. L. (1991) Properties of recombinant protein-containing inclusion bodies in *Escherichia coli. Bioprocess. Technol.* **12**, 121–145.
6. Williams, D. C., Van Frank, R. M., Muth, W. L., and Burnett, J. P. (1982) Cytoplasmic inclusion bodies in *Escherichia coli* producing biosynthetic human insulin proteins. *Science* **215(4533)**, 687–689.
7. Worrall, D. M. and Goss, N. H. (1989) The formation of biologically active beta-galactosidase inclusion bodies in *Escherichia coli. Aust. J. Biotechnol.* **3**, 28–32.
8. Mukhopadhyay, A. (1997) Inclusion bodies and purification of proteins in biologically active forms. *Adv. Biochem. Eng. Biotechnol.* **56**, 61–109.
9. Kapust, R. B. and Waugh, D. S. (1999) *Escherichia coli* maltose-binding protein is uncommonly effective at promoting the solubility of polypeptides to which it is fused. *Protein Sci.* **8**, 1668–1674.
10. Thomas, J. G., Ayling, A., and Baneyx, F. (1997) Molecular chaperones, folding catalysts, and the recovery of active recombinant proteins from *E. coli*. To fold or to refold. *Appl. Biochem. Biotechnol.* **66(3)**, 197–238.

11. Misawa, S. and Kumagai, I. (1999) Refolding of therapeutic proteins produced in *Escherichia coli* as inclusion bodies. *Biopolymers* **51,** 297–307.
12. Lilie, H., Schwarz, E., and Rudolph, R. (1998) Advances in refolding of proteins produced in *E. coli. Curr. Opin. Biotechnol.* **9,** 497–501.
13. Schoner, R. G., Ellis, L. F., and Schoner, B. E. (1992) Isolation and purification of protein granules from *Escherichia coli* cells overproducing bovine growth hormone. *Biotechnology* **24,** 349–352.
14. Burgess, R. R. and Jendrisak, J. J. (1975) A procedure for the rapid, large-scale purification of *Escherichia coli* DNA-dependent RNA polymerase involving Polymin P precipitation and DNA-cellulose chromatography. *Biochemistry* **14,** 4634–4638.
15. Goldberg, M. E., Expert-Bezancon, N., Vuillard, L., and Rabilloud, T. (1996) Non-detergent sulphobetaines: a new class of molecules that facilitate in vitro protein renaturation. *Fold Des.* **1,** 21–27.
16. Harlow, E. and Lane, D. (1988) *Antibodies: A Laboratory Manual.* Cold Spring Harbor Laboratory, Cold Spring Harbor , NY, pp. 88–91.

5

Protein Extraction From Fungi

Paul D. Bridge, Tetsuo Kokubun, and Monique S. J. Simmonds

1. Introduction

The fungi encompass a wide variety of organisms ranging from simple single-celled yeasts, such as *Saccharomyces cerevisiae*, to highly differentiated macrofungi that can be up to a meter or more in diameter (e.g., *Rigidoporus ulmarius* and *Langermannia gigantea* [*1*]). Fungi contain many different proteinaceous materials and these may comprise up to 31% of the dry weight of a mushroom (*2*). Protein extraction can be undertaken from almost any type of fungal material, including fresh fruiting bodies (*3*). This chapter will consider some methodology for protein extraction from yeasts and filamentous fungi growing in liquid laboratory culture.

In order to study proteins from yeasts and filamentous fungi, it is important to consider a number of basic features of the organisms. First, filamentous fungi undergo a growth cycle that includes differentiation and compartmentalization. In addition, both the filamentous fungi and yeasts will age during growth, and older cultures will undergo autolysis. As a result, particular proteins may only be associated with one part of the growth cycle, such as sporulation or autolysis, and this must be taken into account in determining growth conditions and sampling times.

Second, many of the enzymes produced during the growth period are sequential and may either be subject to significant repression or require induction by a substrate or substrate component. Examples of this include the requirement for chitin or chitinlike components to induce chitinases (*4*) and the repression of some fungal proteases by glucose (*5*).

Third, fungi possess rigid cell walls and complex cell wall/membrane systems (*6*). As a result, the fungi produce many extracellular enzymes for the degradation of large molecules and have extensive transport protein systems for the movement of materials across the walls and membranes. It is therefore important to ascertain the potential location of proteins prior to their extraction, because cell-wall-associated and extracellular proteins will be lost during intracellular extractions. In the natural environment, fungi commonly utilize large organic molecules such as lignin, cellulose, and pectin, and these are broken down to simple components by extracellular enzymes. Such enzymes are generally inducible, and once produced, they diffuse into the growth medium

From: *Methods in Molecular Biology, vol. 244: Protein Purification Protocols: Second Edition*
Edited by: P. Cutler © Humana Press Inc., Totowa, NJ

or environment. As a result, they are generally not subject to significant repression, and in the presence of an appropriate inducer, they can be produced in sufficient concentrations to be purified and characterized directly from the spent growth medium *(7,8)*.

A simple growth and extraction procedure on an analytical scale is described here. This is a standard regime that will allow the extraction of intracellular proteins from a wide range of filamentous fungi and has been used successfully with many fungal genera, including *Fusarium, Ganoderma, Aspergillus, Colletotrichum, Beauveria, Phoma, Verticillium*, and *Metarhizium (9,10)*. The method has not been optimized towards any particular fungal group and has proven suitable for filamentous ascomycetes and basidiomycetes as well as yeasts *(10–12)*. The major variation that will be needed for different fungal groups is the growth medium and the length of the growth period (*see* **Notes 1** and **2**). Although a crude method, extracts produced in this way retain sufficient integrity and activity for enzyme assays and isoenzyme electrophoresis. The physical disruption of the cells and/or endogenous protease activity may result in poor recovery of large ($>$150 kDa) proteins. Recovery of large proteins may be improved by the inclusion of protease inhibitors (*see* **Note 3**) or by the use of specific buffers (*see* **Note 4**). It should also be remembered that cell-wall and membrane associated proteins may be retained in the cell debris fraction (*see* **Note 5**). An additional feature of this method is that the spent culture fluid may be retained for the detection of extracellular enzymes. Initially, this will only contain a small number of glucose-independent enzymes, but as the culture grows and the free-glucose concentration decreases, further enzymes can be detected or extracted *(10,13)*.

2. Materials

Fungal growth media and buffers should be sterilized prior to use. Growth media and Tris-glycine buffer can routinely be sterilized at 10 psi for 10 min in a benchtop autoclave. In complex media and buffers, individual components may break down or react during autoclaving and so may need to be individually filter-sterilized. This is particularly true of media containing high glucose concentrations (as the glucose may "caramelize") and buffers containing significant quantities of acetate or urea (both of which may break down on heating).

1. Malt extract agar (MEA): 20 g malt extract (Oxoid, Basingstoke, UK), 1 g peptone (Oxoid; Bacteriological), 20 g glucose, 15 g agar, 1 L distilled water *(14)*.
2. Glucose yeast medium (GYM): 1g $NH_4H_2PO_4$, 0.2 g KCl, 0.2 g $MgSO_4 \cdot 7H_2O$, 10 g glucose, 1 mL of 0.5% aqueous $CuSO_4 \cdot 5H_2O$, 1 mL 1% aqueous $ZnSO_4 \cdot 7H_2O$, distilled water to 1 L *(10)*.
3. Tris-glycine buffer: 3 g Trizma (Sigma, Poole, UK), 14.4 g glycine, 1 L deionized water, pH 8.3.
4. 4-MU substrate buffer: 0.05 M Na acetate, pH 5.4 (adjusted with glacial acetic acid).
5. Sterile deionized water.

3. Methods

The methods presented here will enable the extraction of intracellular proteins and extracellular enzymes from filamentous fungi and yeasts (*see* **Fig. 1**). The Notes section details further considerations that may be needed for specific organisms or extractions.

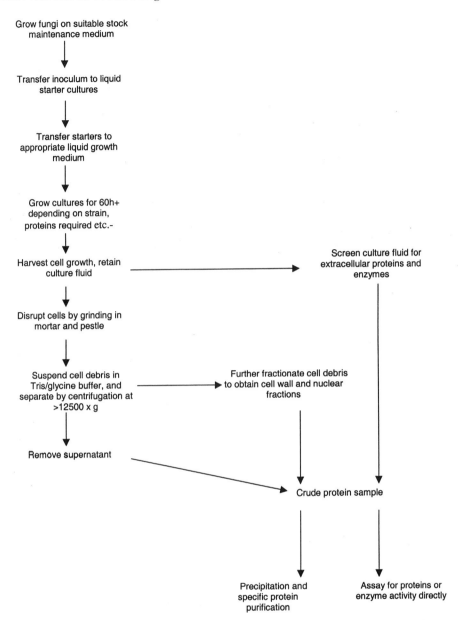

Fig. 1. Schematic diagram of basic steps in the extraction of proteins from fungi.

3.1. Extraction of Intracellular Proteins From Metarhizium anisopliae

The following protocol describes the extraction of cytoplasmic proteins from the fil amentous fungus *Metarhizium anisopliae*. Further details regarding growth and extraction conditions for other filamentous fungi are given in **Notes 2** and **6–10**.

1. Grow *Metarhizium* culture on malt extract agar for 7 d at 25–28°C.
2. Remove a plug (approx 0.5 cm in diameter) of culture from the agar plate with a flamed cork borer or scalpel. Cut the plug into at least 10 smaller pieces with a flamed scalpel

and inoculate into 10-mL sterile GYM in a 28-mL Universal bottle or small flask (*see* **Note 6**).

3. The 10-mL culture is a starter culture for the subsequent main growth period and is required to ensure an actively growing inoculum. Incubate the starter culture at 25–28°C in an incubator, with shaking, for 60–72 h.

4. Aseptically transfer the starter culture into a 250-mL conical flask containing 60 mL of fresh sterile GYM. Replace the flask in the incubator and continue incubation with shaking for a further 60–72 h (*see* **Note 7**).

5. Harvest the mycelium from the flask by vacuum-assisted filtration. A Buchner funnel containing Whatman No. 3 filter paper is suitable for this purpose (*see* **Note 8**). Take care to ensure that any aerosols that may be formed from the filtration or the vacuum pump are minimized and contained.

6. Wash the mycelium once in the Buchner funnel with sterile deionized water and transfer the harvested mycelium from the filter paper to a plastic Petri dish with a flamed or alcohol-sterilized spatula.

7. Freeze the mycelium at −20°C for storage prior to extraction.

8. Prepare the mycelium for extraction by freeze-drying for 24 h. This is one of several possible methods of obtaining cell breakage (*see* **Notes 9** and **10**).

9. Disrupt the freeze-dried mycelium by briefly grinding it in a mortar and pestle (*see* **Note 9**). The ground mycelium should be collected in sterile 1.5-mL microcentrifuge tubes in approx 500-mg amounts. This is roughly equivalent to the conical portion of the tube. These tubes of ground mycelium can be stored at −20°C for at least 3 mo (*see* **Note 11**).

10. Rehydrate 500 mg of ground mycelium in 1 mL of Tris-glycine buffer (*see* **Note 12**). This will need mixing with a micropipet tip or Pasteur pipet to form an even slurry.

11. Clarify the slurry by centrifugation at 12,500g for 40 min at 4°C. After centrifugation, collect the supernatant into another sterile microcentrifuge tube (*see* **Note 13**).

12. The collected supernatant will contain the total cytoplasmic proteins. The samples as prepared here typically contain 15–100 mg/mL protein (*see* **Note 14**). The extracts can be used directly in gel electrophoresis or column chromatography. Alternatively, standard precipitation techniques can be used to purify specific proteins further (*see* **Notes 8** and **13**).

3.2. Screening for Extracellular Enzymes

As detailed earlier, when filamentous fungi are grown in liquid media, extracellular enzymes are excreted into the culture media. These enzymes can be present in considerable amounts and may retain their activity for several days or more depending on the culture conditions used. In some cases, spent culture media has been used directly in electrophoresis to characterize extracellular enzymes, and two examples of this are amylase and pectinase enzymes, which can be detected in this way after growth on starch or pectin-containing media, respectively *(7,9)*. Such electrophoretic approaches can be very useful in specific circumstances, but for more general applications, a wide range of fungal extracellular enzymes can be detected by rapid screening with simple dye or other indicator-substituted substrates (*see* **Note 15**).

The following method describes such a simple screening method, based on the use of the fluorophore 4-methylumbelliferone (7-hydroxy-4-methylcoumarin). This is a compound that is highly fluorescent under ultraviolet (UV) light and can be combined with a wide range of sugars and other compounds to produce simple enzyme substrates. These substituted compounds can then be used to screen either crude or clarified spent culture broths for enzyme activity. Chromophores such as *o*- and *p*-nitrophenyl-substituted compounds may also be used, however, with spent fungal culture media, their usefulness may be limited by the production by many fungi of extracellular pigments. In general, methods based on the fluorescent compounds are highly sensitive and rarely affected by the metabolites produced and released into the growth medium by the fungi. The method given here was originally described by Barth and Bridge *(13)* and is based on earlier methods described for testing bacterial material *(15)*. The method uses standard 96-well microtiter plates for reactions, and small quantities of culture fluid and substrate are mixed in these. After an incubation period, enzyme activity is detected by fluorescence under UV light. The method was developed for rapid screening of either multiple samples or multiple enzymes, and pretreated plates can be maintained for some months. A simple method for this is to lyophilize the enzyme substrates in the wells of the microtiter plates in a shelf freeze-drier; these plates can then be stored in vacuum-sealed bags at −20°C. The method has been used successfully without modifications with *Beauveria*, *Fusarium*, *Penicillium*, *Colletotrichum*, *Ganoderma*, and *Rhizoctonia*.

1. Cultures should be grown in an appropriate liquid medium (*see* **Note 1**) for up to 14 d, depending on the fungal species.
2. Cultures are harvested by filtration or centrifugation and the spent culture broth (SCB) is collected (*see* **Note 16**).
3. Stock preparations of 4-methylumbelliferyl-substituted substrates are made up in dimethylformamide to give a final concentration of 50 mM in 1.6 mL (*see* **Note 17**). These stock solutions can generally be stored for several weeks at 4°C.
4. Stock substrates are diluted for use at 0.15 mL in 9.85 mL of 0.05 M sodium acetate, pH 5.4.
5. Fifty microliters of substrate are placed in the appropriate number of wells in a microtiter plate. Fifty microliters of SCB are then added to this and mixed briefly with the micropipet tip.
6. Controls of substrate without SCB, SCB alone, and acetate buffer alone should be included on all plates in case of extraneous fluorescence from other compounds in SCB or breakdown of substrate.
7. Plates are incubated at 37°C for 4 h (*see* **Note 17**). After incubationm 50 µL of saturated sodium bicarbonate solution is added to each well.
8. Enzyme activity is visualised by fluorescence under long-wave UV light (*see* **Note 18**).

4. Notes

1. The growth medium selected will vary depending on the requirements of the fungus under study, and general fungal liquid growth media such as malt extract and potato dextrose broths are available commercially. The GYM medium given in **Subheading 2.** is a general-purpose medium suitable for the growth of many filamentous ascomycetes and basidiomycetes. For organisms that may be more fastidious, a richer medium, such as peptone yeast extract glucose *(16)*, can be very useful (*see* **ref. 9**). Although Oomycetes are not currently classified as fungi, these techniques have been used for these organisms, and species of *Pythium* and *Phytophora* may require either a source of sterol or a more complex organic medium, such as V8 medium *(14)*. Organic components in growth media can vary between suppliers and this variation may affect the biochemical properties of the fungus *(17)*. It is therefore important to standardize organic components, such as yeast extract, by only using a single source of supply.
2. Growth conditions are important in obtaining a reliable and constant source of protein, and it is usually desirable to extract from cultures consisting of material of a constant age. This

is obtained by first inoculating fungal cultures into small starter cultures of the main growth medium. These starter cultures are incubated, usually as shaken liquids, and provide the inoculum for the main growth phase. Although the exact conditions will vary depending on the organism studied, a typical starter culture would consist of 10 mL of GYM medium in a 25- to 30-mL vessel, incubated on an orbital shaker at 25°C for 60 h. The entire starter culture is then used as inoculum for 60 mL of GYM medium in 250-mL conical flasks, which is then incubated at 25°C on the shaker for 3–5 d. Growth periods and temperatures will depend on the rate of growth and differentiation of the fungus studied. Most yeast cultures will produce sufficient suitable biomass after incubation at 30°C under a growth regime of 24 h in a starter culture and 24 h in a shaken flask. However, many filamentous fungi require 60 h in a starter culture followed by 3–5 d in shaken flasks at 25°C. In order to maintain reproducibility, incubation temperatures should be kept constant.

3. Fungi can exhibit strong proteinase activity and this can increase with the age of the culture *(18)*. Many different protease inhibitors have been incorporated into fungal protein extraction procedures, including pepstatin, EDTA, thiourea, and polyvinylpyrrolidone *(19)*. Probably the most commonly used protease inhibitor has been phenylmethylsulfonylfluoride (PMSF) *(20)*.

4. Various buffer systems have been suggested for the extraction of fungal proteins, and some of these were recently reviewed by Hennebert and Vancanneyt *(19)*. More recently, Osherov and May *(21)* compared different buffers for the extraction of high-molecular-weight proteins from *Aspergillus nidulans* and reported good recovery of large molecules with a 9 *M* urea buffer.

5. Some fungal proteins are intrinsically associated with the cell walls, membranes, and organelle membranes. These include the major cytoskeletal proteins and the enzymes involved in transport processes. In a relatively crude extraction method as detailed here, many of these proteins will remain in the cell debris and so will not be recovered in either the intracellular or extracellular preparations. If protein extracts are required for subsequent enzymatic screening, it may also be necessary to consider whether cofactors are present in the same fraction as the enzymes of interest.

6. The inoculum used for the initial starter cultures is important and should contain a large number of potential "growing" points; that is, the inoculum should ideally be particulate so that mycelial or yeast growth can occur at many different points. This presents no problems if the culture is naturally particulate, such as yeast cells, or if conidial suspensions can be used (e.g., **ref. 8**). However, if the fungus grows only as a mycelial mat, an inoculum made from broken mycelia will give a more homogenous and greater biomass than one derived from plugs or "lumps" of culture.

7. In many cases, the proteins of interest will be associated with active growth of the fungus. It is therefore often necessary to establish some form of growth curve for the organism prior to harvesting the growth. The growth rate of yeasts and fungi growing in a yeastlike phase can be estimated by sampling and measuring the turbidity of the growth medium over a period of time *(22)*. However, with filamentous organisms, this becomes impractical. One alternative is to assay the culture medium for the extracellular enzymes β-glucosidase, β-galactosidase, and diacetyl-chitobiosidase. In many fungi, these enzymes are each associated with particular stages of the growth cycle when grown in glucose-containing media. β-Glucosidase is generally produced in the early phases of growth, β-galactosidase is not produced until the glucose concentration has been reduced, and diacetyl-chitobiosidase is generally produced during autolysis. Proteins associated with major cellular differentiation such as the formation of sporocarps may not be expressed during liquid culture.

8. When the required phase of growth has been reached, the fungal biomass must be harvested from the growth medium. Again, methods differ depending on the organism, but, in gen-

eral, yeasts can be harvested by centrifugation at 3000–7000g, whereas filamentous fungi are better harvested by vacuum filtration onto Whatman No. 3 filter papers in a Buchner funnel. In both cases, the resulting biomass should be washed in sterile distilled water to remove any residues from the media. If extracellular proteins, such as extracellular enzymes, are required, these can be extracted directly from the culture fluid after the mycelium has been harvested. Typically, the culture fluid is clarified by centrifugation and proteins extracted by one of the general protein extraction techniques, such as precipitation with acetone or methanol/ammonium or by immunoprecipitation *(23,24)*. Alternatively, solid-phase extraction (SPE) methods may be considered. In this technique, a sample solution is passed through the bed of adsorbent, usually packed in a syringe barrel, which selectively retains components in the sample mixtures. By choosing a correct adsorbent, the interfering substances (cleanup) or desired analyte (concentration) can be retained, and this is then recovered with a suitable elution buffer. Prepacked columns based on size-exclusion, ion-exchange and reversed-phase mechanisms are commercially available.

9. Efficient cell breakage is necessary to ensure a good recovery of proteins from filamentous fungi. A wide variety of cell breakage methods have been reported in the literature, varying from enzymatic digestion of the cell wall to physical disruption procedures, such as grinding mycelium in a mortar and pestle *(9)* or homogenizing thick cell pastes *(25)*. Enzymatic digestion has been used to generate protoplasts, which can be harvested from the remaining cell debris. The protoplast preparations can then be lysed to release intracellular proteins *(26,27)*. Many fungi produce active intracellular and extracellular proteinases, and these are usually active under the conditions used for enzymatic digestion of cell walls (*see* **Note 3**). As a result, physical disruption of an "inert" sample is generally preferred. Fungal cell walls can be disrupted by briefly grinding freeze-dried mycelium or a fragment of a sporocarp in a mortar and pestle. Additional abrasives are not usually required, although carborundum may be added if required. This procedure should not be performed on an open bench because of the possible hazardous nature of the dust produced. A single flask of 60 mL actively growing culture will give about 500–2000 mg of freeze-dried material. The method of grinding a freeze-dried culture as described here is quick and reliable, but it requires access to freeze-drying equipment. A commonly used alternative is to grind the harvested mycelium in liquid nitrogen *(23)*. However, appropriate safety measures should be considered to avoid any potential contact or splashing from the liquid nitrogen.

10. Yeast cells can be disrupted by passage through a pressure cell, such as a French press *(28)*, although three or more passages may be necessary to achieve 70% breakage. Yeast cells can also be broken by shaking frozen cultures with glass beads. The glass beads used are generally larger than the Ballotini beads used for bacteria and 2- to 2.5-mm beads can produce adequate disruption after 10–15 min of shaking. This method can generate sufficient local heating to denature some proteins, so effective cooling of the system is required.

11. Ground fungal material sealed in microcentrifuge tubes will generally not rehydrate significantly when maintained at −20°C. However, repeated exposure to air through opening the tube and repeated freezing and thawing may cause damage, particularly to the higher-molecular-weight proteins. If samples are likely to be required on a number of occasions over time, it is advisable to aliquot the dried material into a number of tubes prior to storage.

12. In some cases, buffers containing detergents, urea, or high EDTA concentrations (*see* **Note 4**) have been shown to give higher yields of total protein *(29,30)*, but the simple Tris-glycine buffer used here is adequate for general purposes.

13. The total aqueous extract will contain the cytoplasmic proteins. Normally, the pellet of unbroken cells and debris is discarded, although this can be saved if the cell-wall fraction is required. This extract will, however, contain many other components, including polysaccharides. The degree of further purification necessary will depend on the protein and level

of activity required, as well as the level of polysaccharide contamination and the analytical methods that are used following the extraction. As mentioned, the supernatant contains the total cytoplasmic proteins and may be used directly. Proteins in these crude extracts can be readily separated by polyacrylamide gel electrophoresis *(10,11)*, and this technique can be used to provide cell-free total protein patterns or, with specific stains, to demonstrate particular enzymes. Alternatively, it may be necessary to clarify the supernatant further by a second centrifugation or filtration through a 0.45-μm filter. Specific proteins can then be purified through standard precipitation techniques, as mentioned in **Note 8**. Further clarification and separation, generally by differential centrifugation in sucrose density gradients, will be required to extract the cell-wall fraction *(30)*.

14. The protein content of the supernatant should be estimated by one of the standard protein determination methods, such as Lowry et al.'s determination *(31)*.

15. More recent developments for protein analyses are the use of various detection methods coupled with particular forms of chromatography. Both high-performance liquid chromatography (HPLC) and capillary-(zone-) electrophoresis (CE, or CZE) offer very high resolving power in a relatively short period of operating time, with a high degree of automation. Additionally, with the CE instrument, an isoelectric point (p*I*) determination *(32)* and isotachophoresis *(33)* may be performed for proteins occurring at a low concentration. Although not yet in widespread use, combining these separation methods with mass spectrometry (MS) has been achieved, and instruments designed specifically for such a purpose have appeared on the market (Amersham, Finnigan, etc.). Highly accurate molecular-weight measurement and structure information may be obtained through this approach. Readers are encouraged to explore these new techniques that have become available in recent years.

16. Enzymic activity can be assessed during a growth period by aseptically removing small aliquots from the growing culture. These should be clarified by filtration or centrifugation prior to assay.

17. The substrate concentration, incubation time, and temperature listed were arrived at empirically and will give satisfactory results for common enzymes from many fungi. Substrate concentrations can often be reduced, and in our experience, one-half and one-fourth strength concentrations can give acceptable results. However, it may occasionally be necessary to optimize substrate concentration and reaction conditions for the particular enzyme system being considered and the interference by the background.

18. The enzymatically released 4-methylumbelliferone is excited by long-wavelength UV light and emits in the visible spectrum; as an example, the excitation and emission wavelengths for 4-methylumbelliferyl phosphate are 360 and 440 nm, respectively.

References

1. Hawksworth, D. L., Kirk, P. M., Sutton, B. C., and Pegler, D. N. (1995) *Ainsworth & Bisby's Dictionary of the Fungi*, 8th ed., CAB International, Wallingford, UK.

2. Rammeloo, J. and Walleyn, R. (1993*) The Edible Fungi of Africa, South of the Sahara*, National Botanic Garden of Belgium, Meise, Belgium.

3. Rosendahl, S. and Banke, S. (1998) Use of isozymes in fungal taxonomy and population studies, in *Chemical Fungal Taxonomy* (Frisvad, J. C., Bridge, P. D., and Arora, D. K., eds.), Marcel Dekker, New York, pp. 107–120.

4. Clarkson, J. H. (1992) Molecular biology of filamentous fungi used for biological control, in *Applied Molecular Genetics of Filamentous Fungi* (Kinghorn, J. R. and Turner, G., eds.), Blackie Academic and Professional, Glasgow, pp. 175–190.

5. St. Leger, R. J., Cooper, R. M., and Charnley, A. K. (1986) Cuticle-degrading enzymes of

entomopathogenic fungi: regulation of production of chitinolytic enzymes. *J. Gen. Microbiol.* **132,** 1509–1517.

6. Peberdy, J. F. (1990) Fungal cell walls—a review, in *Biochemistry of Cell Walls and Membranes in Fungi* (Kuhn, P. J., Trinci, A. P. J., Jung, M. J., Goosey, M. W., and Copping, L. G., eds.), Springer-Verlag, Berlin, pp. 5–30.

7. Cruickshank, R. H. and Wade, G. C. (1980) Detection of pectin enzymes in pectin acrylamide gels. *Anal. Biochem.* **107,** 17–181.

8. Elad, Y., Chet, I., and Henis, Y. (1982) Degradation of plant pathogenic fungi by *Trichoderma harzianum. Can. J. Microbiol.* **28,** 719–725.

9. Paterson, R. R. M. and Bridge, P. D. (1994) *Biochemical Techniques for Filamentous Fungi,* CAB International, Wallingford, UK.

10. Mugnai, L., Bridge, P. D., and Evans, H. C. (1989) A chemotaxonomic evaluation of the genus *Beauveria. Mycol. Res.* **92,** 199–209.

11. Jun, Y., Bridge, P. D., and Evans, H. C. (1991) An integrated approach to the taxonomy of the genus *Verticillium. J. Gen. Microbiol.* **137,** 1437–1444.

12. Monte, E., Bridge, P. D., and Sutton, B. C. (1990) Physiological and biochemical studies in Coelomycetes. *Phoma. Studies Mycol.* **32,** 21–28.

13. Barth, M. G. and Bridge, P. D. (1989) 4-Methylumbelliferyl substituted compounds as fluorogenic substrates for fungal extracellular enzymes. *Lett. Appl. Microbiol.* **9,** 177–179.

14. Smith, D. and Onions, A. H. S. (1994) *The Preservation and Maintenance of Living Fungi,* 2nd ed., CAB International, Wallingford, UK.

15. O'Brien, M. and Colwell, R. R. (1987) A rapid test for chitinase activity that uses 4-methylumbelliferyl-*N*-acetyl-β-D-glucosaminide. *Appl. Environ. Microbiol.* **53,** 1718–1720.

16. Conti, S. F. and Naylor, H. B. (1959) Electron microscopy of ultrathin sections of *Schizosaccharomyces octosporus.* I. Cell division. *J. Bacteriol.* **78,** 868–877.

17. Filtenborg, O., Frisvad, J. C., and Thrane, U. (1990) The significance of yeast extract composition on secondary metabolite production in *Penicillium,* in *Modern Concepts in* Penicillium *and* Aspergillus *Classification* (Samson, R. A. and Pitt, J. I., eds.), Plenum, New York, pp. 433–441.

18. Petäistö, R. L., Rissanen, T. E., Harvima, R. J., and Kajander, E. O. (1994) Analysis of the protein of *Gremmeniella abietina* with special reference to protease activity. *Mycologia* **86,** 242–249.

19. Hennebert, G. L. and Vancanneyt, M. (1998) Proteins in fungal taxonomy, in *Chemical Fungal Taxonomy* (Frisvad, J. C., Bridge, P. D., and Arora, D. K., eds.), Marcel Dekker, New York, pp. 77–106.

20. Kim, W. K. and Howes, N. K. (1987) Localization of glycopeptides and race-variable polypeptides in uredosporling walls of *Puccinia graminis tritici;* affinity to concalvin A, soybean agglutinin, and *Lotus* lectin. *Can. J. Bot.* **65,** 1785–1791.

21. Osherov, N. and May, G. S. (1998) Optimization of protein extraction from *Aspergillus nidulans* for gel electrophoresis. *Fung. Gen. Newslett.* **45,** 38–40.

22. Barnett, J. A., Payne, R. W., and Yarrow, D. (1990) *Yeasts: Characteristics and Identification,* 2nd ed., Cambridge University Press, Cambridge.

23. St. Leger, R. J., Staples, R. C., and Roberts, D. W. (1991) Changes in translatable mRNA species associated with nutrient deprivation and protease synthesis in *Metarhizium anisopliae. J. Gen. Microbiol.* **137,** 807–815.

24. Kim, K. K., Fravel, D. R., and Papavizas, G. C. (1990) Production, purification and properties of glucose oxidase from the biocontrol fungus *Talaromyces flavus. Can. J. Microbiol.* **36,** 199–205.

25. Hien, N. H. and Fleet, G. H. (1983) Separation and characterization of six (1→3)-β-glucanases from *Saccharomyces cerevisiae. J. Bacteriol.* **156,** 1204–1213.

26. Sambrook, J., Fritsch, E. F., and Maniatis, T. (1989) *Molecular Cloning: A Laboratory Manual*, Cold Spring Harbor Laboratory, Cold Spring Harbor, NY, Vol. 3.

27. Messner, R. and Kubicek, C. P. (1990) Synthesis of cell wall glucan, chitin and protein by regenerating protoplasts and mycelia of *Trichoderma reesei*. *Can. J. Microbiol.* **36,** 211–217.

28. Schnaitman, C. A. (1981) Cell fractionation, in *Manual of Methods for General Bacteriology* (Gerhardt, P., Murray, R. G. E., Costilow, R. N., et al., eds.), American Society for Microbiology, Washington, DC, pp. 52–61.

29. Kim, W. K., Rohringer, R., and Chong, J. (1982) Sugar and amino acid composition of macromolecular constituents released from walls of uredosporlings of *Puccinia graminis triticii. Can. J. Plant Pathol.* **4,** 317–327.

30. Fèvre, M. (1979) Glucanase, glucan synthases and wall growth in *Saprolegnia monoica,* in *Fungal Walls and Hyphal Growth* (Burnett, J. H. and Trinci, A. P. J., eds.), Cambridge University Press, British Mycological Society Symposium 2, Cambridge, pp. 225–263.

31. Lowry, O. H., Rosebrough, N. J., Farr, A. L., and Randall, R. J. (1951) Protein measurement with the Folin phenol reagent. *J. Biol. Chem.* **193**, 265–275.

32. Pritchett, T. J. (1996) Capillary isoelectric focusing of proteins. *Electrophoresis* **17**, 1195–1201.

33. Foret, F., Szoko, E., and Karger, B. L. (1993) Trace analysis of proteins by capillary zone electrophoresis with on-column transient isotachophoretic preconcentration. *Electrophoresis* **14**, 417–428.

6

Subecllular Fractionation of Animal Tissues

Norma M. Ryan

1. Introduction

A number of techniques exist which exploit various physical parameters or biological properties of cells as means of investigating the complexity of cellular organelles and membranes. Subcellular fractionation using the centrifuge is the basis of traditional methods for separating cellular components. Even when other methods are used to effect the separation, it is often the case that better results will be obtained if the material is first purified or partially purified by centrifugal methods (*see* **refs.** *1* and *2* for a recently published discussion of the theory and applications of the use of the centrifuge in preparative procedures). In this chapter some of the basic centrifugal techniques used to fractionate a typical animal cell are described.

Subcellular fractionation involves three successive steps (for a detailed discussion of the principles involved *see* **refs.** *3* and *4*). The first step converts a tissue or cell suspension into a homogenate. The second step reintroduces a new kind of order into the system by grouping together, in separate fractions, those components of the homogenate of which certain physical properties, such as density or sedimentation coefficient, fall between certain limits set by the investigator. The third step consists of the analysis of the isolated fractions. Interpretation of the results involves a retracing of these three steps. It is essential to perform all steps of this process, regardless of whether the objective of the work is to study the spatial organization of components within a cell (i.e., subcellular fractionation itself) or to use the procedures as a means of obtaining a partially purified component which then can be more readily purified using other procedures.

Tissue disruption or homogenization may be accomplished by a variety of means such as grinding, ultrasonic vibrations, and by making use of the osmotic properties of cells. The grinding of tissues to form homogenates is usually accomplished with the Potter–Elvejhem homogenizer, which is highly effective with soft tissues, such as liver and kidney. Tough tissues, such as muscle, generally require the use of a mincer and blender of which a common example is the Waring Blender. In recent years the Chaikoff Press, which entails the forcing of the material through fine holes under high pressure, has found much use. Osmotic methods of tissue disruption are most useful when employed in connection with red blood cells and reticulocytes. The use of ultrasonic vibra-

From: *Methods in Molecular Biology vol. 244: Protein Purification Protocols: Second Edition*
Edited by: P. Cutler © Humana Press Inc., Totowa, NJ

tions is useful for the disruption of bacterial cells and some animal cells. In such systems cooling tends to cause a problem because very high temperatures are generated at the point of disintegration. Homogenization should be carried out so as to leave the subcellular organelles virtually intact, although somewhat distorted, and yet shear the plasma membrane, endoplasmic reticulum, and other endomembranous systems into fragments which form spherical vesicles.

The suspension so prepared contains whole cells, partially broken cells, nuclei, mitochondria, and so on. Separation of the various components, in theory, can be achieved by the application of any method that exploits the differences between the physical and/or chemical properties of the constituents. Methods such as electrophoresis, counter-current distribution, and centrifugation are but a few of the methods available. The method most routinely used is centrifugation in one form or another, both because of the wide availability of the equipment necessary to carry out the procedures and also the facility of fractionating relatively large amounts of material with easy recovery of sample when fractionation is complete.

The behavior of particles in a centrifugal field and the various factors that affect their rate of sedimentation have been studied in great detail *(5,6)*. The rate of movement of an ideal (spherical) particle in a centrifugal field is described by the following equation:

$$V = dr/dt = \{[a^2(Dp - Dm)]/18\eta\} \times \omega^2 r \qquad (1)$$

where V is the velocity of particle in cm/s $= dr/dt$; a is the particle diameter; Dp is the particle density; Dm is the density of the medium; ω is the angular velocity in radian/s, which is equal to the number of revolutions per second $\times 2$; r is the distance between the particle and the center of rotation in centimeters; and η is the viscosity of the medium.

The centrifugal force (CF) on a particle in a spinning rotor is given by:

$$CF = m\omega^2 r \qquad (2)$$

where m is the mass of the particle, r is the distance between the particle and the center of rotation in centimeters, and ω is the angular velocity in radian/s.

When centrifugation conditions are reported it is a common practice to list the relative centrifugal force (RCF). RCF is the force on a particle at some r value (generally for a point midway down the centrifuge tube, r_{av}) divided by the force on that particle in the earth's gravitational field:

$$RCF = \omega^2 r/g \qquad (3)$$

where g is the gravitational constant, 980 cm/s.

Thus:

$$RCF = \omega^2 r/980 \qquad (4)$$

Since temperature affects both the density and viscosity of the medium, it thus affects the rate of sedimentation of the particles.

In this chapter the subcellular fractionation of rat liver tissue is described as a fairly typical example of a fractionation procedure; **ref. 7** should be consulted for information on variations in procedures appropriate for other tissues. The separation of the various constituents of an homogenate formed from rat liver tissue, using differential centrifugation, is described. The method described here does not yield an absolute purification of the individual components within a rat liver cell, but rather a partial purification of the cellular components results which will enable alternative/subsequent purification procedures to be carried out with a far higher chance of success.

2. Materials

All reagents are of Analar grade.

1. Homogenization buffer medium: 0.25 *M* sucrose, 5 m*M* imidazole-HCl, pH 7.4 (*see* **Notes 1** and **2**).
2. 0.9% Saline for perfusion of the liver.
3. Resuspension buffer medium: 0.25 *M* sucrose, 5 m*M* imidazole-HCl, pH 7.4 (*see* **Note 3**).
4. Refrigerated laboratory centrifuge, such as Sorvall RC5B fitted with an 8 × 50 mL rotor (*see* **Note 2**).
5. Potter–Elvejhem homogenizer.
6. Dounce homogenizer.
7. Beckman L65 ultracentrifuge fitted with an 8 × 35 mL rotor or equivalent (*see* **Note 2**).

3. Methods

3.1. Preparation of the Homogenate

1. Remove the liver from a rat large enough to yield approx 10 g liver (wet weight), having first perfused the liver with 0.9% saline. The perfusion should be carried out immediately after the sacrifice of the rat by appropriate means (*see* **Note 4**).
2. Wash the liver free of blood, hairs, and so on, by suspending it in ice-cold homogenizing medium.
3. Blot the tissue lightly with filter paper to dry it.
4. Mince the liver finely in a preweighed, chilled beaker and weigh again (*see* **Note 5**).
5. Homogenize the liver in homogenizing medium to form a 25% (w/v) homogenate, using six passes of a Potter–Elvejhem homogenizer (*see* **Note 6**).
6. Dilute homogenate to 12.5% (w/v).

3.2. Subcellular Fractionation (see Note 7)

1. Retain a small portion of the 12.5% (w/v) homogenate for analysis and centrifuge the remainder at 4°C at a speed of 600*g* for 10 min in a refrigerated laboratory centrifuge (*see* **Note 8**).
2. Resuspend the pellet in the same volume of homogenization medium as previously and centrifuge again at 600*g* for an additional 10 min (*see* **Note 9**).
3. Combine the post-600 g-minute supernatants and centrifuge at 15,000*g* for 10 min in order to prepare a fraction rich in mitochondria and lysosomes.
4. Resuspend the resulting pellet in the same volume of homogenization medium as previously and centrifuge again at 15,000*g* for another 10 min (*see* **Note 10**).
5. At the end of the washing steps pool all the supernatants and use in subsequent procedures (*see* **Note 11**). If the material retained in the pellet is required for further studies, resuspend it in homogenizing medium or other appropriate medium (*see* **Note 3**).
6. Pool the post-15 × 10^4 g-minute supernatants and centrifuge at 100,000*g* for 60 min in an ultracentrifuge at 4°C. The resulting complex micrososmal pellet contains vesicles derived from the plasma membrane, endoplasmic reticulum—some containing ribosomes and some ribosome-free—peroxisomes, polysomes, endosomes, Golgi stacks, and other such membranous systems from within the cell, while the supernatant will contain the remainder of the cellular components, that is, the soluble components and smaller elements, such as free ribosomes (*see* **Note 12**).
7. Resuspend each of the pellets in a volume of 0.25 *M* sucrose, 5 m*M* imidazole-HCl, pH 7.4 using a loose Dounce homogenizer. It may be more appropriate to use a different resuspension medium depending on the ultimate objectives (*see* **Note 3**).
8. Store all fractions in a deep freeze (*see* **Note 13**) until required for analysis (*see* **Note 14**) or further purification (*see* **Note 15**).

4. Notes

1. Homogenization is the first essential step in any fractionation procedure. It involves the disruption of an ordered system and results in a loss of some morphological information, but homogenization is necessary in order to apply the techniques of biochemistry to a study of subcellular components. The choice of an adequate homogenization medium is a critical one. The homogenization medium most suited to the biological material involved can only be elucidated by a process of trial and error. Sucrose is very widely used, but such details as concentration of sucrose, pH of the buffer, traces of specific cations, and so on, vary considerably. The best homogenate is the one which lends itself most successfully to fractionation and will result in a satisfactory resolution of the disrupted components of the homogenate. It is also the one which enables the next experiment to be carried out.

2. For best results the rat liver and every preparation made from the tissue must be kept cold (0–4°C) from the moment the organ is removed from the rat. Thus, all solutions must be prechilled to 0°C before addition to tissue, as should all glassware which comes in contact with the preparations. Centrifugation must be carried out in refrigerated centrifuges at 0–4°C, ensuring that the rotors are prechilled.

3. Depending on the final objectives of the purification procedures and the restrictions on the techniques following the differential centrifugation, it may be more appropriate to use a different resuspension medium than the one described here. Sucrose interferes with many assays and is not always easy to eliminate from preparations.

4. The method chosen for sacrifice of the rat will depend on the aims of the experiments. Most often this will be cervical dislocation. Anesthetics are occasionally used but these could have undesirable side effects on the tissue of interest, and therefore should only be employed if it is certain that the side effects will not affect the subcellular component/protein of interest.

5. Heavy metal ions are powerful enzyme inhibitors, so avoid sticking scissors, forceps, spatulas, and so on, into the tissue preparations. Manipulations, such as stirring of solutions or resuspension of pellets, should be carried out using glass rods.

6. This step is critical and the clearance of the homogenizer should be considered with care. Once again it is a matter of trial and error until precisely the right conditions are determined to suit the purposes of the experiment.

7. Subcellular fractionation is accomplished by the stepwise process of differential centrifugation, which separates particles from a supernatant in the form of a pellet. Differential pelleting is the simplest method for obtaining a crude separation which exploits the mass of the major organelles and membrane systems. All steps are carried out in the temperature range 0–4°C. A measured portion of the original homogenate is retained for further analysis. It is extremely important to keep an accurate record of all volumes and weights used in preparing the fractions. Otherwise it will be impossible to interpret the results in a meaningful way.

8. This first centrifugation step is designed to remove all nuclei, whole cells, partially intact cells, and plasma membrane sheets, and is very effective in this.

9. The objective of this washing step is to reduce contamination of the fraction by membrane components.

10. This washing step can be repeated two or three times, depending on how critical the degree of contamination by other membranous components is to the objectives of the experiment.

11. This fraction is not a pure preparation of mitochondria and lysosomes, but rather it is enriched in mitochondria and lysosomes with a reduction in the level of contaminating membranous components.

12. The ultimate location of the Golgi Apparatus and vesicles derived from the Golgi stacks and associated "trafficking" vesicles will be determined by the extent to which the homogenization procedures disrupt the networks of the Golgi. Sometimes these will be found in the post-6×10^3 g-minute supernatant, but more often they are detected primarily in the post-15×10^4 supernatant. The ribosomes may also be pelleted by subjecting the supernatant to sedimentation at $100,000g$ at $4°C$ in an ultracentrifuge for a minimum of 3 h.

13. In normal procedures, aging studies to determine the lability of the enzyme markers to be studied should be carried out. Freezing may damage some particles resulting in a loss of activity or of a constituent. The interference of components of the media used in the fractionation process with the assay of the markers should also be thoroughly assessed before proceeding with the analysis and interpretation.

14. After fractionation of the tissue each fraction should be assayed, in addition to the original homogenate, for selected markers for the purpose of following each constituent throughout the fractionation procedure. de Duve *(2,8)* laid down two criteria for the selection of an enzyme as a marker which can also be applied to a chemical constituent:
    a. The enzyme must have a specific location, that is, be present only in one type of particle; and
    b. That all subunits of a given population have the same enzyme content as related to their mass or total protein.

The second criterion is not absolutely essential for a marker. Known markers for specific organelles are used to appreciate the efficiency of the fractionation procedure and thereby the quality of the homogenate. If two markers show the same distribution pattern it is an indication, but not proof, that they may be associated with the same particle. By a comparison of the specific activity of the markers in the different fractions, an estimation of the contamination or degree of purity of the particles may be obtained.

The markers selected are usually biochemical (chemical constituents, enzyme activities, immunological), but morphological and sedimentation coefficient analysis can also prove very informative. A list of some of the classical enzymic markers normally employed in fractionation studies is provided in **ref. 7**. The amount of biochemical constituents, such as DNA, RNA, cholesterol, total protein, and lipids should be assayed and calculated in each of the fractions. The absolute activities and specific activities of each of the relevant enzymes should be assayed. The concentration of the components or enzyme activities should be expressed as a percentage of the total constituent or enzyme activity found in the homogenate, that is, the percent recovery of each marker is calculated for every fractionation experiment carried out. Quantitative recovery of each marker is important and before any interpretation of the data is made, quantitative recovery of any enzyme activity/constituent must be established. The results of the analysis are presented as distribution patterns or as frequency histograms if possible.

It is essential to prepare a balance sheet of the constituents and enzyme activities in the differing fractions compared with those in the original homogenate. This is achieved by:
a. Retaining a portion of the original homogenate;
b. Recording accurately the volume of the homogenate used for centrifugation and also the volumes of buffered medium in which the particles are resuspended; and
c. Carrying out all analyses on the homogenate as well as on the prepared fraction. Concentration of a constituent or enzyme activity should be expressed as a percentage of that which was present in the original homogenate.

A particular enzyme activity may be recovered in the fractions at values considerably less than or more than 100%. In the former case this may be caused by the handling of the materials or the removal of an influencing factor (i.e., an activator or a cofactor) which may have been separated into another fraction. Recoveries in excess of 100% would tend to indicate the removal of some inhibiting substance.

15. The subcellular fractionation that has been described here applies principally to soft tissues, especially liver tissue. However, the properties described do apply to most animal cells, although different methods of homogenization, in particular, may be necessary to effect the desired subcellular fractionation. What has been described is the crude preparation of crude subcellular fractions which are partially purified and which may form the basis for purification of a specific membrane fraction or a particular protein by either techniques, such as density gradient centrifugation or other protein purification procedures as described in Chapters 14–30.

References

1. Rickwood, D. (ed.) (1992) *Preparative Centrifugation: A Practical Approach,* IRL, Oxford, UK.
2. de Duve, C. (1967) General principles, in *Enzyme Cytology* (Roodyn, D. B., ed.), Academic, London, pp. 1–26.
3. Birnie, G. D. and Rickwood, D. (eds.) (1978) *Centrifugal Separations in Molecular and Cell Biology,* Butterworths, London.
4. de Duve, C. and Beaufay, H. (1981) A short history of tissue fractionation. *J. Cell Biol.* **91,** 293s–299s.
5. Schachman, H. K. (1959) *Ultracentrifugation in Biochemistry,* Academic, London.
6. Svedberg, T. and Pederson, K. O. (1940) *The Ultracentrifuge,* Clarendon, Oxford (Johnson Reprint Corporation, New York).
7. Evans, W. H. (1992) Isolation and characterisation of membranes and cell organelles, in *Preparative Centrifugation: A Practical Approach* (Rickwood, D., ed.), IRL, Oxford, UK, pp. 233–270.
8. de Duve, C. (1971) Tissue fractionation-past and present. *J. Cell Biol.* **50,** 20d–55d.

7

Subcellular Fractionation of Plant Tissues

Isolation of Plastids and Mitochondria

Alyson K. Tobin and Caroline G. Bowsher

1. Introduction

Plant tissue is not the easiest material from which to isolate good quality, functional, and intact organelles. Leaves are often covered in waxy cuticles, frequently contain silica (as in grasses) and toxic components such as phenolics, proteolytic enzymes, and high concentrations of acids and salts in the vacuole. In addition, all higher plants have one major barrier in common—the presence of a rigid, cellulose cell wall, which has to be broken in order to release the organelles. Mechanical isolation methods (i.e., where leaf material is macerated in a mechanical homogenizer) are likely to succeed only for a limited number of species (e.g., pea and spinach) of which the leaves do not contain large amounts of tough, thickened tissue. Otherwise, the prolonged homogenization required to release significant numbers of organelles from silicaceous leaves, such as those of wheat or barley, results in most of them being broken and inactive. For this reason, the only viable method for obtaining good quality organelles from this type of tissue is to isolate them from protoplasts. Although chloroplasts can be isolated successfully from protoplasts, this technique is generally inappropriate for mitochondrial isolation because of the very low yield. This means that there are many plant species from whose leaves it has proved impossible to isolate good quality mitochondria using existing techniques.

A major factor in the success, or otherwise, of all of these methods is the quality of the plant material. Even the best technician cannot isolate good plastids or mitochondria from poor quality plants and it is essential both to optimize and standardize the growing conditions if reproducible results are to be obtained.

It is also important to note that isolated organelles are easily damaged and must be handled gently. Detergents and volatile solvents are extremely harmful, as they disrupt the lipid-rich membranes. Detergents must never be used to wash any apparatus used for organelle isolation. Steps should also be taken to avoid exposure to volatile solvents (particularly phenol) as even the vapor can be disruptive to organelles.

Given the variation in composition of different plant material, there is no universally applicable technique. In this chapter, we describe methods that have been used with suc-

From: *Methods in Molecular Biology, vol. 244: Protein Purification Protocols: Second Edition*
Edited by: P. Cutler © Humana Press Inc., Totowa, NJ

cess in our laboratories to isolate plastids (chloroplasts, amyloplasts, and root plastids) and mitochondria from various species. Although they have been used successfully on the species listed here, it may be necessary to adapt these techniques when using other plant material. As with all new techniques, we recommend using the simpler, mechanical method for organelle isolation at first and to use the protoplast approach only if mechanical isolation proves unsuitable.

2. Materials

Media 1–20 will keep for a maximum of 2 wk at 4°C providing that bovine serum albumin (BSA), sodium pyrophosphate, ATP, NADP, glucose 6-phosphate dehydrogenase (G6PDH), 6-phosphogluconate dehydrogenase (6PGDH), fructose 6-phosphate, $MnSO_4$, Nycodenz, and Percoll are omitted and added to the stocks, where required, immediately before use. It is advisable to dialyze Percoll overnight against distilled water before use. Wheat and barley digestion media are made immediately before use by adding cellulysin, pectolyase, and cellulase, as appropriate, to the surface of the medium. This is left to stand, without stirring, until the dry powders have been fully absorbed into the medium, preventing the formation of lumps of dry powder that are difficult to disperse. BSA is added to solutions in the same way.

Make the potassium ferricyanide [$K_3Fe(CN)_6$] solution in water and store in the dark at −20°C; protect from bright light. $KHCO_3$ and $NaHCO_3$ should be freshly made in the assay buffers.

The mitochondrial isolation medium and wash medium will both keep at 4°C for up to 2 wk if stored in the absence of polyvinyl pyrrolidone (PVP), BSA, and cysteine, which should be added immediately before use, where required. Solution I (minus BSA) will keep for 2 wk, provided BSA is added fresh on the day, and solutions II and III should be freshly made.

2.1. Wheat Protoplast, Chloroplast, and Amyloplast Preparation

1. Medium 1: 0.5 *M* Sucrose, 5 m*M* (2[*N*-morpholino] ethanesulfonic acid) (MES) (pH 6.0), 1 m*M* $CaCl_2$.
2. Medium 2: 0.4 *M* Sucrose, 0.1 *M* sorbitol, 5 m*M* MES, pH 6.0, 1 m*M* $CaCl_2$.
3. Medium 3: 0.5 *M* Sorbitol, 5 m*M* MES, pH 6.0, 1 m*M* $CaCl_2$.
4. Medium 4: 0.4 *M* Sorbitol, 10 m*M* EDTA, 25 m*M* Tricine, pH 8.4.
5. Medium 5: 0.8 *M* Sorbitol, 50 m*M* HEPES, pH 7.5, 1 m*M* KCl, 2 m*M* $MgCl_2$, 1 m*M* EDTA, 4 m*M* dithiolthreitol (DTT).
6. Medium 6: 3% (w/v) Nycodenz in medium 5.
7. Wheat digestion medium: 0.6 g Cellulysin (Sigma), 6 mg Pectolyase (Sigma) in 30 mL medium 3.

2.2. Barley Protoplast and Chloroplast Preparation

1. Medium 7: 0.4 *M* Sorbitol, 10 m*M* MES, pH 5.5, 1 m*M* $CaCl_2$, 1 m*M* $MgSO_4$.
2. Barley digestion medium: 1.5% (w/v) Cellulase (Onozuka R10) in medium 7.
3. Medium 8: 35% (v/v) Percoll in medium 10.
4. Medium 9: 25% (v/v) Percoll in medium 10.
5. Medium 10: 0.4 *M* Sorbitol, 25 m*M* Tricine, pH 7.2, 1 m*M* $CaCl_2$, 1 m*M* $MgSO_4$.
6. Medium 11: 0.33 *M* Sorbitol, 50 m*M* Tricine, pH 7.8, 2 m*M* EDTA, 1 m*M* $MgSO_4$, 1 m*M* $MnSO_4$.

2.3. Pea Chloroplast and Root Plastid Preparation

1. Medium 12: 0.33 M Sorbitol, 50 mM Tricine, pH 7.9, 1 mM EDTA, 2 mM MgCl$_2$, 0.1% (w/v) BSA.
2. Medium 13: 40% (v/v) Percoll, 0.33 M Sorbitol, 50 mM Tricine, pH 7.9, 0.1% (w/v) BSA.
3. Medium 14: 10% (v/v) Percoll, 0.33 M Sorbitol, 50 mM Tricine, pH 7.9.

2.4. Chloroplast and Plastid Assays

1. Medium 15: 0.33 M Sorbitol, 20 mM HEPES, pH 7.6, 10 mM EDTA, 0.2 mM KH$_2$PO$_4$, 30 mM MgCl$_2$.
2. Medium 16: Double-strength medium 15.
3. Medium 17: 0.33 M Sorbitol, 50 mM Tricine, pH 8.2, 2 mM EDTA, 1 mM MgSO$_4$, 1 mM MnSO$_4$.
4. Medium 18: 0.33 M Sorbitol, 50 mM Tricine, pH 8.2, 10 mM KCl, 5 mM sodium pyrophosphate, 2 mM EDTA, 2 mM ATP.
5. 0.1 M K$_3$Fe(CN)$_6$.
6. 1.0 M NH$_4$Cl.
7. 1.0 M KHCO$_3$.
8. Medium 19: 0.33 M Sorbitol, 0.25 M glycylglycine, pH 7.5, 10 mM MgCl$_2$, 0.39 mM NADP, 1 unit G6PDH/6PGDH (Sigma).
9. Medium 20: 0.33 M Sorbitol, 0.25 M glycylglycine, pH 7.5, 50 mM fructose-6-phosphate.
10. Molybdate solution: 500 mg Molybdate, 5 mL of 10 N sulfuric acid, 2.25 g iron sulfate, and 50 mL water.

2.5. Mitochondrial Isolation

1. Mitochondrial isolation medium: 0.3 M Sucrose, 50 mM MOPS, pH 7.5, 2 mM EDTA, 1 mM MgCl$_2$, 1% (w/v) PVP 40 (soluble), 0.4% (w/v) BSA, 4 mM cysteine.
2. Mitochondrial wash medium: 0.3 M Sucrose, 20 mM MOPS, pH 7.5, 0.1% (w/v) BSA.
3. Solution I: 0.6 M Sucrose, 20 mM KH$_2$PO$_4$, 0.2% (w/v) BSA, pH 7.5.
4. Solution II: 20 mL Solution I, 11.2 mL Percoll, 8.8 mL of 40% (w/v) PVP 40.
5. Solution III: 20 mL Solution I, 11.2 mL Percoll, 8.8 mL H$_2$O.

3. Methods

General note: All procedures are carried out at 4°C unless otherwise stated. All solutions and apparatus should be prechilled when working at this temperature. Use detergent-free glassware and avoid exposure to volatile solvents.

3.1. Isolation of Chloroplasts From Protoplasts

3.1.1. Wheat Protoplast Preparation

1. Take primary leaves from 7- to 10-d-old plants (*see* **Note 1**). Finely chop the leaves into thin sections, approximately 1–2 mm thick. Use sharp, single-sided razor blades. Spread the leaf sections onto the surface of 30 mL of wheat digestion medium (*see* **Note 2**) in a shallow dish that provides a large surface area of medium (a crystallizing dish is suitable). Add sufficient leaf material to completely cover the surface of the medium. Place the dish under a white light (*see* **Note 3**) and incubate for 3 h at 28°C; do not shake or stir the medium during this period. Illumination helps to maintain a pool of Calvin cycle intermediates within the chloroplast and reduces the lag time for maximal photosynthetic activity to occur in the isolated chloroplasts.
2. Following digestion, carefully remove the digestion medium from the dish, leaving the leaf sections in place. Add 20 mL of medium 3 to the leaf sections and gently swirl to release

the protoplasts into the medium (*see* **Note 4**). Filter through coarse nylon mesh (e.g., a plastic tea strainer) and retain the filtrate. Return the leaf sections to the dish, add a further 20 mL of medium 3, swirl again, filter, and combine the two filtrates.

3. Centrifuge the filtrates in a swing-out rotor at 150*g* for 5 min (*see* **Note 5**). Discard the supernatant and gently resuspend the pellets in a total volume of 20 mL of medium 1. Divide the suspension equally among four tubes and overlay 2 mL of medium 2, followed by 1 mL of medium 3 onto each suspension to form a discontinuous gradient. Centrifuge at 250*g* for 5 min in a swing-out rotor.

4. Intact protoplasts collect at the interface between medium 2 and medium 3. Using a wide-bore Pasteur pipet, carefully remove the layer of protoplasts, transfer to a clean centrifuge tube, and add 2 vol of medium 3. Centrifuge at 150*g* for 5 min in a swing-out rotor. Resuspend the pellet in medium 4 to give a chlorophyll concentration of approx 1 mg/mL.

3.1.2. Wheat Chloroplast Preparation (see **Note 6**).

1. Resuspend the protoplasts (prepared in **Subheading 3.1.1.**) in 5 mL of medium 4. Pour the suspension into a modified 25-mL disposable plastic syringe (*see* **Note 4**).
2. Break the protoplasts by passing the suspension through 20-μm-pore-size nylon mesh and immediately centrifuge at 250*g* for 1–2 min in a swing-out rotor (*see* **Note 5**).
3. Gently resuspend the pellet in medium 4 to give a chlorophyll concentration of approx 1 mg/mL.

3.1.3. Barley Protoplast Preparation

1. Take primary leaves from 7- to 10-d-old barley plants (*see* **Note 1**). Carefully remove the lower epidermis and lay the leaves, lower side down, onto the surface of 50 mL of barley digestion medium (*see* **Note 2**) in a clear plastic sandwich box. Cover the box with cling film and incubate at 30°C for 2 h without disturbance. Illuminate with white light as described for wheat protoplast isolation (*see* **Note 3**).
2. Following digestion, harvest the protoplasts by gently swirling the leaves within the digestion medium. Decant the suspension through nylon mesh (e.g., a plastic tea strainer) into 50-mL centrifuge tubes.
3. Centrifuge at 250*g* for 5 min in a swing-out rotor (*see* **Note 5**), discard the supernatant, and resuspend the pellet in 10 mL of medium 8.
4. Divide the suspension equally between two 20-mL centrifuge tubes. Overlayer each with 2 mL of medium 9 followed by 1 mL of medium 10 to form a discontinuous gradient.
5. Leave the gradients to stand on ice for 1 h and then centrifuge at 300*g* for 5 min in a swing-out rotor.
6. Intact protoplasts collect at the interface between medium 9 and medium 10. Carefully remove the protoplasts using a wide-bore Pasteur pipet, transfer to a clean 20-mL centrifuge tube, and add 2 vol of medium 11.
7. Centrifuge at 150*g* for 5 min in a swing-out rotor, discard the supernatant, and gently resuspend the pellet in medium 11 to give a chlorophyll concentration of approx 1 mg/mL.

3.1.4. Barley Chloroplast Preparation (see **Note 6**)

1. After removing the protoplasts from the gradient (as in **step 6** of **Subheading 3.1.5.**), dilute (approximately fivefold) in medium 11 and transfer to a modified 25-mL disposable plastic syringe (*see* **Note 4**).
2. Break the protoplasts by passing the suspension through 20-μm-pore-size nylon mesh and immediately centrifuge at 150*g* for 2 min in a swing-out rotor (*see* **Note 5**).
3. Gently resuspend the pellet in medium 11 to give a chlorophyll concentration of approx 1 mg/mL (*see* **Note 7**).

3.2. Mechanical Isolation of Plastids (see *Note 8*)

3.2.1. Wheat Amyloplast Preparation (see *Note 11*)

1. Harvest 20–50 wheat ears from fully grown wheat plants, 10–15 d postanthesis (grains 18–30 mg). Stems may be kept in water while grains are harvested. Using a razor blade, the embryo tip should be removed from the grain. The endosperm can then be extracted by squeezing the grain and placing it in medium 5 in a glass Petri dish on ice. The tissue is left to plasmolyse in this medium for between 20 min and 3 h.
2. The medium in which the endosperm plasmolyse is removed and replaced by 5 mL of medium 5. Endosperm are then chopped by hand using two single-sided razor blades held together. Razor blades should be replaced regularly, as they blunt.
3. Filter the resulting brei through 10 layers of muslin (prewetted in medium 5). Up to 30 mL of this crude extract is layered onto 20 mL of medium 6 in a 50-mL centrifuge tube and centrifuge at 100g for 25 min.
4. The pellet should be carefully resuspended in 1 mL of medium 5.

3.2.2. Pea Root Plastid Preparation (see *Note 6*)

1. Grow pea plants on vermiculite for 5–6 d in the dark at 20°C. Remove plants, rinse roots in water, and, using scissors, cut root material onto aluminum foil placed on ice. Homogenize 60–200 g of roots in 1 vol of medium 12 using the steel blade of a food processor at maximum speed (1500 rpm) for two 10-s pulses. Smaller amounts of tissue may be used (1–20 g) and chopped finely using a single-sided razor blade rather than a food processor.
2. Filter brei through six layers of muslin (prewetted in medium 12) and centrifuge at 500g for 1 min.
3. Centrifuge supernatant at 4000g for 3 min in an 8 × 50-mL fixed-angle rotor. Gently re-suspend pellet in medium 12 (without BSA).
4. Underlay 10-mL aliquots of the supernatant from **step 3** with medium 14. Centrifuge the suspension in a 4 × 50 mL swing-out windshield rotor at 4000g for 5 min. Gently resuspend resulting pellet in a small volume (1–3 mL) of medium 12 (without BSA).

3.2.3. Chloroplasts From Pea Leaves (see *Note 6*)

1. Take 10 g (fresh weight) of pea leaves, add to 40 mL of ice-cold medium 12, and cut into small pieces using single-sided razor blades or sharp scissors.
2. Homogenize (5 s full speed) using an Ultraturrax or Polytron homogenizer. Filter the homogenate through four layers of muslin (prewetted with medium 12).
3. Divide the filtrate equally between two 50-mL centrifuge tubes. Underlay each portion of filtrate with 10 mL of medium 13.
4. Centrifuge at 2500g for 1 min in a swing-out rotor.
5. The chloroplasts form a soft, green pellet at the bottom of the centrifuge tube. Carefully remove the supernatant layers by aspiration without disturbing the pellet. Gently resuspend the pellet (e.g., using a fine paint brush) in 1 mL of medium 12.

3.3. Assay of Plastid Intactness

3.3.1. Chloroplast Intactness (see *Note 9*)

Chloroplast intactness is assayed according to the method of Lilley et al. (*1*). The rate of ferricyanide-dependent oxygen evolution is measured using an oxygen electrode (Hansatech, Norfolk, UK).

1. The assay medium, final volume 1.0 mL, is added to an oxygen electrode and consists of medium 15 and 2 mM $K_3(FeCN)_6$ (final concentration, added from a 0.1 M stock solution).

LIVERPOOL
JOHN MOORES UNIVERSITY
AVRIL ROBARTS LRC
TEL 0151 231 4022

Intact chloroplasts are added (final concentration of approx 50 µg chlorophyll/mL assay) and the assay is illuminated (photosynthetically active radiation [PAR] of 1000 µmol· $m^{-2} \cdot s^{-1}$). Following the addition of 5 mM NH_4Cl (final concentration), the rate of O_2 evolution is measured and this gives the "intact rate" of ferricyanide-dependent O_2 evolution.

2. To determine the "broken rate," chloroplasts (the same amount as used in the intact assay) are added to 0.5 mL of H_2O in the oxygen electrode. After 2 min, 0.5 mL of medium 16 is added, followed by 2 mM $K_3(FeCN)_6$ (as in **step 2**). The sample is illuminated and the rate of O_2 evolution measured after adding NH_4Cl (as in **step 1**).

3. Chloroplast intactness is calculated as

$$\frac{(\text{Broken rate} - \text{Intact rate})}{\text{Broken rate}} \times 100 = \% \text{ intactness}$$

3.3.2. Root Plastid Intactness

Root plastid intactness is assayed according to the method of MacDonald and ap Rees *(2)* using a hexose phosphate isomerase assay.

1. The assay medium (medium 19), final volume 0.90 mL, is placed in a cuvet. Intact root plastids are added (final concentration of approx 10 µg protein/mL assay) and the assay is monitored at 340 nm for 5 min.

2. One-tenth milliliter of medium 20 is added and the production of NADPH monitored for a further 5 min to give the "intact rate" of activity.

3. To determine the "broken rate," 0.1 mL of 1% (v/v) Triton X-100 is added to the cuvet, and NADPH production monitored for a further 5 min.

4. Root plastid intactness is calculated as described for chloroplasts.

3.3.3. Amyloplast Intactness

Amyloplast intactness may be assayed using an alkaline pyrophosphatase assay adapted from Gross and ap Rees *(3)*.

1. The assay medium, final volume 0.2 mL consisting of 800 mM sucrose, 100 mM Tris-HCl, pH 8.0, 2 mM $MgCl_2$, and 20 µL extract (representing "intact rate"), is placed in a tube. The assay is started by the addition of $Na_4P_2O_7$ to give a final concentration of 1.5 mM.

2. Four-tenths milliliter of 25 mM trichloroacetic acid is added at 0 or 20 min to stop the reaction.

3. Particulate matter is removed by centrifugation for 30 s at 11,000g.

4. Pi production in the supernatant is determined by the addition of 0.3 mL molybdate solution.

5. Absorbance at 710 nm is recorded after a 15-min period. The concentration of Pi is calculated using a standard curve (0–200 nmol Pi/mL).

6. To determine the "broken rate," amyloplasts should first be deliberately ruptured by freeze–thawing three times in liquid nitrogen prior to **steps 1–5** being performed.

7. Percentage intactness is calculated as described for chloroplasts.

3.4. Assay of Photosynthetic Activity (CO_2-Dependent Oxygen Evolution) of Chloroplasts (see Notes 8 and 9)

3.4.1. Wheat

1. The assay medium, final volume 1.0 mL, is added to an oxygen electrode (Hansatech) and consists of medium 4, 10 mM $KHCO_3$ (final concentration), 0.15 mM KH_2PO_4 (final concentration), and chloroplasts (approx 50 µg chlorophyll/mL assay).

2. The rate of oxygen evolution is measured (*see* **Note 9**) under illumination (minimum PAR of 1000 µmol·$m^{-2} \cdot s^{-1}$) at 20°C.

3.4.2. Barley

1. The assay medium, final volume 1.0 mL, is added to an oxygen electrode (Hansatech) and consists of medium 17, 20 mM KHCO$_3$ (final concentration, added from a 1.0-M stock), 0.2 mM KH$_2$PO$_4$ (final concentration), and chloroplasts (approx 50 µg chlorophyll/mL assay).
2. Oxygen evolution is measured as for wheat chloroplasts.

3.4.3. Pea

1. The assay medium, final volume 1.0 mL, is added to an oxygen electrode (Hansatech) and consists of medium 18, 10 mM NaHCO$_3$ (added from a 1.0-M stock), and chloroplasts (approx 50 µg chlorophyll/mL assay).
2. Oxygen evolution is measured as for wheat chloroplasts.

3.5. Assay of Plastid Purity

Marker enzymes are assayed in aliquots of the original, crude homogenate and in the final chloroplast preparation to determine contamination by cytosol, peroxisomes, and mitochondria, the main contaminants of concern.

3.5.1. Cytosolic Contamination

Phosphoenol pyruvate (PEP) carboxylase, exclusive to the cytosol, is assayed as follows: 50 mM HEPES (pH 8.0), 5 mM MgCl$_2$, 0.08 mM NADH, 1 unit malate dehydrogenase, and 10–20 µL protoplasts, plastids, or crude extract are added to a cuvet (final volume 1.0 mL). The assay is carried out at 25°C and the reaction is started with the addition of 2 mM PEP and measured as the decrease in absorbance at 340 nm resulting from the oxidation of NADH. The method is modified from that of Foster et al. *(4)*.

3.5.2. Mitochondrial Contamination

Citrate synthase, exclusive to the mitochondria, is assayed as follows: 50 mM Tricine, pH 8.0, 1.0 mM MgCl$_2$, 1.0 mM oxaloacetate (OAA), 0.2 mM acetyl coenzyme A, 1.5 mM of 5,5′-dithio-bis-(2-nitrobenzoic acid) (DTNB), and 50 µL extract (or more, depending on the activity) are added to a cuvet (final volume 1.0 mL). The assay is carried out at 25°C and the reaction is started with the addition of OAA and measured as the increase in absorbance at 412 nm as a result of the formation of the 2-nitro-5-thiobenzoate anion. The assay is based on the method of Cooper and Beevers *(5)*.

3.5.3. Peroxisomal Contamination

Glycolate oxidase, exclusive to the peroxisome in leaves *(6)*, is assayed as follows: 50 mM MOPS, pH 7.8, 3 mM EDTA, 0.008% (v/v) Triton X-100, 0.66 mM reduced glutathione, 0.2 mM flavin mononucleotide (FMN), 0.033 % (v/v) phenylhydrazine, and 50 µL extract are added to a cuvet in a final volume of 3.0 mL. The assay is carried out at 25°C and the reaction is started with the addition of 5 mM sodium glycolate and measured as the increase in absorbance at 324 nm resulting from the formation of glyoxylate phenylhydrazine. The assay is based on the method of Behrends et al. *(7)*.

3.6. Isolation of Mitochondria From Pea Leaves (see Note 12)

1. Use sharp razor blades to remove fully expanded leaves by cutting through the petiole. All subsequent steps are carried out at 4°C.
2. Add 600 mL of mitochondrial isolation medium per 100 g of leaves.

3. Cut the leaves into approx 1-cm^2 pieces, using razor blades or sharp scissors.
4. Homogenize (in 150-mL aliquots) using an Ultraturrax homogenizer (Orme Scientific, UK) using 2X 5-s bursts at full speed (mark 10).
5. Filter through four layers of muslin, prewetted with chilled isolation medium.
6. Centrifuge at 2960g for 5 min (*see* **Note 13**) in a fixed-angle rotor in 250-mL centrifuge tubes; discard the pellets.
7. Centrifuge the supernatant at 17,700g for 15 min in a fixed angle rotor in 250-mL centrifuge tubes. Gently, using a fine paint brush, resuspend the pellet in a small volume of wash medium and dilute to a final volume of 30 mL in the same medium.
8. Centrifuge at 1940g for 10 min in a fixed-angle rotor in 50-mL centrifuge tubes. Gently resuspend the pellet (as earlier) in a small volume (approx 1 mL) of wash medium.
9. Load this onto a continuous Percoll/PVP gradient made up as follows: Add 17 mL of solution III to the mixing chamber (i.e., nearest the outlet) of a gradient former and 17 mL of solution II to the left-hand chamber. Form the gradient into a 50-mL centrifuge tube. Centrifuge at 39,200g for 40 min in a fixed-angle rotor. Intact mitochondria form a straw-colored band near the bottom of the gradient.
10. Remove the top half of the gradient, which contains broken chloroplasts (green color) and discard. Using a wide-bore Pasteur pipet, carefully remove the mitochondrial fraction without disturbing the rest of the gradient.
11. Add approx 30 mL of wash medium to the mitochondria and centrifuge at 12,100g for 10 min in a fixed-angle rotor in 50-mL centrifuge tubes.
12. Carefully remove the supernatant by aspiration, taking care not to disturb the soft mitochondrial pellet. Repeat **step 11**.
13. Carefully remove the supernatant as earlier and gently resuspend the purified mitochondria in a minimal volume of wash medium.

3.6.1. Assay of Mitochondrial Activity

3.6.1.1. Measurement of Respiratory Control Ratios and ADP/O Ratios

The assay is carried out in an oxygen electrode (Rank Bros, Cambridge, UK; Hansatech, Norfolk, UK) at 25°C. The reaction mixture consists of 0.3 M sucrose, 10 mM MOPS (pH 7.2), 10 mM KH$_2$PO$_4$, 2 mM MgCl$_2$, 0.1% (w/v) BSA (defatted), mitochondria (0.2–1.0 mg protein/mL assay), and 10 mM substrate (e.g., malate, glycine) in a final volume of 1.0 mL. Oxygen uptake is measured continuously. The State 3 rate of respiration is the rate of oxygen uptake in the presence of ADP. This is determined by adding 100 μM ADP to the above assay and measuring oxygen uptake. The State 4 rate of oxygen uptake is the rate following State 3 when all of the ADP has been converted to ATP (i.e., State 4 is the "ADP-limited" rate). The respiratory control ratio is the ratio of State 3 to State 4 rates (*see* **Fig. 1**). The ADP/O ratio (also called P/O ratio) is calculated from the amount of oxygen consumed during the phosphorylation of a known amount of added ADP. The ADP/O differs for different substrates. For NAD-linked substrates, the ADP/O ratio is 3.0; for NADH or succinate, it is 2.0.

3.6.1.2. Assay of Mitochondrial Purity and Intactness (*see* **Notes 14** and **15**)

To determine mitochondrial purity, the same marker enzymes are assayed as for chloroplasts. Although there are valid assays for mitochondrial intactness, the preferred method of determining mitochondrial quality is to measure the respiratory control ratio (RCR) and ADP/O ratios (*see* **Note 16**). For purified leaf mitochondria, a RCR in

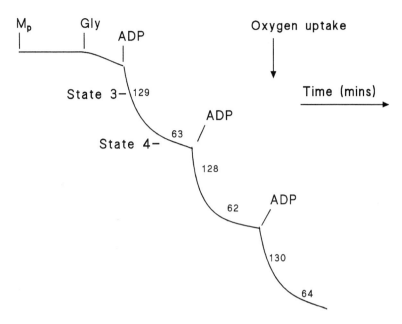

Fig. 1. Oxygen consumption by pea leaf mitochondria. The assay was carried out in an oxygen electrode as described in the text. The following additions were made: Purified mitochondria (M_p, 0.4 mg protein), 10 mM glycine (gly), and 100 μmol ADP (where indicated). State 3 refers to the rate of oxygen uptake in the presence of saturating concentrations of ADP. State 4 occurs when ADP has been depleted. The numbers on the traces refer to the rate of oxygen uptake in nmol/min/mg mitochondrial protein. The respiratory control ratio is the State 3 rate divided by the State 4 rate. The ADP/O ratio is the amount of oxygen consumed (in natoms) during the phosphorylation of a given amount of ADP (in nmol).

excess of 2.0 is acceptable, whereas ADP/O ratios should approach theoretical maximum values (2.0 for NADH or succinate, 3.0 for glycine, malate, etc.).

4. Notes

1. The age of the plant is an important factor in determining the yield of protoplasts, the yield of intact protoplasts decreases rapidly beyond the age ranges presented here. It may be necessary to vary the concentration of digestive enzymes and/or the digestion time if older or younger material is to be used.

2. Different species may well require different digestive enzymes and different periods of digestion and it is important to optimize these conditions. A simple method is to carry out small-scale digestions with 5 mL of digestion medium. Approximate protoplast yields may be assessed by measuring chlorophyll content of the medium after the incubation period and by visual inspection of the extract under a microscope. The chlorophyll content will increase with increasing periods of digestion, but eventually the number of intact protoplasts (readily seen as completely circular in outline under the light microscope) will decrease with prolonged digestion. As the cellulase and cellulysin preparations are not generally very pure, it is advisable to use the minimum amount during digestion.

3. The white light used to illuminate the tissue during digestion for protoplast preparation can be provided by a 60-W light bulb in an angle poise bench lamp shining a few centimeters above the tissue. Place a shallow clear tray over the dish and fill this with water to produce

a heat trap. Illumination is not essential, but it does seem to reduce the time taken for the isolated chloroplasts to attain maximal rates of photosynthesis (*see also* **Note 11**).

4. To break the protoplasts, use a 25-mL disposable syringe adapted as follows: Cut off the end section to which the syringe needle is usually attached, so that the barrel becomes an open cylinder. Cover the end with a piece of nylon mesh, pore size 20 μm. Remove the syringe plunger and pour the protoplast suspension into the open end of the syringe barrel. Carefully replace the plunger; displace it slowly so that the protoplasts are forced through the mesh into a collecting beaker on ice. This is usually sufficient to completely break the protoplasts. Check by inspecting an aliquot under a light microscope. Broken protoplasts will no longer be circular in outline. If intact protoplasts remain, repeat the passage through the mesh. This procedure has to be carried out with care, as rough handling at this stage will result in chloroplast breakage.

5. All centrifugations for protoplast and chloroplast isolation are carried out with the brakes off, whereas the brakes are kept on for root plastid and amyloplast preparation.

6. Isolated protoplasts are stable for many hours, whereas chloroplasts will only remain active for a short period of time. Barley and wheat chloroplasts should be used as soon as possible after isolation and will rarely remain viable for more than 2 h. Pea chloroplasts will generally last longer, but it is always good practice to carry out experiments on isolated organelles as soon as possible after isolation. If a large number of experiments are planned, it is possible to make one large protoplast preparation in the morning and to break aliquots at intervals during the day to provide freshly prepared chloroplasts. The remaining protoplasts may be kept on ice until required. It is good practice to determine the rate of CO_2-dependent O_2 evolution at the beginning and end of the series of experiments to determine whether deterioration has occurred. Root plastids and amyloplasts will generally keep for up to 4 h.

7. The most serious problem during chloroplast isolation is "clumping," where the organelles aggregate into sticky lumps. Whenever this occurs, the quality of the chloroplasts is poor. One contributory factor is the age of the plant material; the older the leaves, the more likely clumping will occur. We have tried varying the EDTA and Mg ion concentration in the resuspension medium, as suggested by other authors, yet this does not always solve the problem. There also appears to be a difference between cultivars, the barley variety Klaxon is particularly problematic, whereas Maris Mink is more reliable. The peeling of the epidermis prior to digestion was found to significantly reduce the problem of chloroplast clumping in barley and it also increased the yield of intact protoplasts.

8. A high stromal starch concentration may result in poor quality chloroplasts when using mechanical isolation, as starch grains are thought to physically disrupt the chloroplast envelope during homogenization. To minimize this effect, leaves should be harvested close to the end of the dark period, when starch content will be at a minimum.

9. Chloroplasts isolated from protoplasts should be at least 90% intact and capable of CO_2-dependent O_2 evolution at rates comparable to those of the isolated protoplasts. For wheat and barley, depending on the growing conditions, rates of at least 40 μmol O_2 evolved h^{-1} mg chlorophyll^{-1} of chlorophyll should be achieved. Rates for pea will be higher than this, at least 50 μmol O_2 evolved per hour per milligram of chlorophyll. The rate of CO_2-dependent O_2 evolution is a better determination of chloroplast quality than is the intactness assay, as it is possible for chloroplasts to break, release their stromal contents, and then reseal. These chloroplasts would appear to be intact, according to the ferricyanide assay, and yet would be incapable of photosynthesis.

10. The conditions required for chloroplasts to carry out CO_2-dependent O_2 evolution are often quite critical, hence the difference in assay media for the different species. The pH, divalent cation concentration, and phosphate concentration should all be optimized for the species being studied. ATP, ADP, and pyrophosphate may also be required to varying extents.

11. The phenomenon of "induction" is frequently observed in isolated chloroplasts, where a significant lag occurs between the start of illumination and the onset of CO_2-dependent O_2 evolution. This lag may last for several minutes and is the result of the depletion of Calvin cycle intermediates from the chloroplast. Preillumination of the leaves with bright light for 30 min prior to mechanical extraction will reduce the induction time if this is thought to be a problem.

12. The method described for the isolation of mitochondria from pea leaves will also work for spinach.

13. All centrifugations for mitochondrial isolation are carried out with the brake on.

14. A major source of contamination of the mitochondria is from broken chloroplasts. Ideally, the mitochondrial band that forms on the density gradient should be straw colored and not green. A high chlorophyll content is the result of thylakoid contamination, often caused by too harsh homogenization or to overloading of the gradient with too much extract.

15. The methods described in this chapter produce good quality plastids and mitochondria suitable for proteomic analysis. It is essential to check the purity, intactness, and physiological competence of the organelle preparation prior to carrying out further analysis so that the quality of the sample is verified before obtaining a proteomic profile. Two recent publications *(8,9)* provide good examples of methods and proteomic analysis of organelles isolated from Arabidopsis.

16. Isolated leaf mitochondria will only remain active for approx 2 h. It is good practice to determine the RCR at the start and end of the series of experiments to ascertain whether there has been significant deterioration.

References

1. Lilley, R. McC., Fitzgerald, M. P., Rienits, K. G., and Walker, D. A. (1975) Criteria of intactness and the photosynthetic activity of spinach chloroplast preparations. *New Phytol.* **75**, 1–10.

2. MacDonald, F. D. and ap Rees, T. (1983) Enzyme properties of amyloplasts from suspension cultures of soybean. *Biochim. Biophys. Acta* **755**, 81–89.

3. Gross, P. and ap Rees, T. (1986) Alkaline inorganic pyrophosphatase and starch synthesis in amyloplasts. *Planta* **167**, 140–145.

4. Foster, J. G., Edwards, G. E., and Winter, K. (1982) Changes in levels of phosphoenol pyruvate carboxylase with induction of Crassulacean acid metabolism in *Mesembryanthemum crystallinum* L. *Plant Cell Physiol.* **23**, 585–594.

5. Cooper, T. G. and Beevers, H. (1969) Mitochondria and glyoxysomes from castor bean endosperm. Enzyme constituents and catalytic capacity. *J. Biol. Chem.* **244**, 3507–3513.

6. Tolbert, N.E. (1981) Microbodies—peroxisomes and glyoxysomes. *Annu. Rev. Plant Physiol.* **26**, 45–73.

7. Behrends, W., Rausch, U., Loffler, H. G., and Kindl, H. (1982) Purification of glycolate oxidase from greening cucumber cotyledons. *Planta* **156**, 566–571.

8. Millar, A. H., Sweetlove, L. J., Giege, P., and Leaver, C. J. (2001) Analysis of the Arabidopsis mitochondrial proteome. *Plant Physiol.* **127**, 1711–1727.

9. Prime, T. A., Sherrier, D. J., Mahon, P., Packman, L. C., and Dupree, P. (2000) A proteomic analysis of organelles from *Arabidopsis thaliana*. *Electrophoresis* **21**, 3488–3499.

8

The Extraction of Enzymes From Plant Tissues Rich in Phenolic Compounds

William S. Pierpoint

1. Introduction

1.1. The Problems

Many enzyme extracts must, of necessity, be made from green leaves, fruits, and other vegetable tissues that contain large amounts of phenols and polyphenols. These may hinder or, unless precautions are taken, completely prevent a successful extraction. In spite of this, the processes by which the polyphenols interfere are rarely studied and still very incompletely understood. Some principles are clear, mostly from work done in past decades, and are applied on an ad hoc basis to current problems. They can usually be adapted to devise a successful if rather complicated procedure, but there is seldom the time or interest for researchers to establish what the problems really were and if they could have been better overcome. This is unfortunate, but wholly understandable. The reactions involved are often very complex, demand specialist investigations, and may be only relevant to extraction from a particular plant species.

Part of the difficulty is caused by the vast range of phenolic compounds that may be involved. Harborne (1) estimated that several thousand structures are known, and he summarized the vagaries of their distribution. Simple phenols, such as catechol, are comparatively rare or are present only in traces. Phenolic acids, such as p-OH benzoic and syringic acid, are almost ubiquitous, as are the phenyl-propanoids (C_6-C_3-), including caffeic acid and its quinic acid ester, chlorogenic acid. Thousands of flavonoid compounds are known, differing in hydroxylation pattern, oxidation state of the heterocyclic ring, and degree of glycosylation: Some of them are restricted to particular species and particular tissues, whereas others, such as quercetin and cyanidin, are widespread. Perhaps the most notorious phenolic compounds in the present context are the polymeric, astringent tannins, whose protein-binding properties have been appreciated in food technology and the leather industry for centuries. Characteristic of some "advanced" orders of cotyledonous plants are the "hydrolyzable" tannins, based on gallic acid residues linked, often as esterified chains, to glucose or some other polyhydric alcohol. More widespread are the "condensed" tannins, oligomers of the flavonoid cat-

From: *Methods in Molecular Biology, vol. 244: Protein Purification Protocols: Second Edition*
Edited by: P. Cutler © Humana Press Inc., Totowa, NJ

echin linked by C_4–C_8 interflavan bonds, which occur in most orders of the vascular plants. These polymers have many of the features of monomeric phenols that predispose them to combine with proteins, but they have more of them and often in a disposition that facilitates and strengthens this reaction.

The initial binding of phenolic compounds, both monomers and polymers, to enzymes and other proteins is via noncovalent forces, which may, initially, be reversible. It is believed that hydrophobic, ionic, and H bonds may be involved, depending on the specific phenol and protein involved. Fully methoxylated phenolics with a high content of aromatic rings are, of course, more likely to be bound hydrophobically. Free phenolic groups, especially vicinal dihydroxy groups, may form hydrogen bonds with, for instance, the CO and NH of peptide bonds. Such complexes have only been investigated thoroughly in a few simple model cases (2,3) and shown to involve a variety of these linkages and also coordination attachments involving cations. Nevertheless, Haslam and his colleagues have extrapolated from such information to produce a model for the binding of polymeric tannins to proteins. It involves two stages. In the first, the tannin is attached to hydrophobic sites on the protein surface via its aromatic residues. This bonding is then reinforced by the formation of H bonds between the phenolic groups and nearby polar functions on the protein. The final product is thus a dissociable complex in which the surface of the protein is rendered more hydrophobic and more susceptible to aggregation and precipitation. If the polyphenol is large enough to interact with more than one protein molecule, the likelihood of aggregation and precipitation will, of course, be much increased. More recent developments of this work (4) have emphasized the structural features of proteins, including those of the proline-rich salivary proteins of herbivores, that predispose them to such coupling with tannins (*see* **Fig. 1**): For most enzymes, this complexing facilitates inactivation.

Phenolic compounds form irreversible covalent linkages with proteins primarily as a consequence of the oxidation of their vicinal dihydroxygroups to quinones or semiquinones. These oxidations may occur nonenzymically in alkaline conditions, especially in the presence of metal ions. They are, however, catalyzed by a variety of enzymes, including *o*-diphenol oxidases (polyphenoloxidases [PPO]), monophenoloxidases (laccases), and by peroxidases in the presence of H_2O_2. The resulting quinone molecules are highly reactive, not only polymerizing with each other but oxidizing other phenolic compounds and, most relevant in the present context, combining with reactive groups in proteins causing aggregation, crosslinking, and precipitation. These reactions are described as "enzymic browning" because the products, although complex and poorly defined, are usually brown in color. Leaf extracts that brown rapidly are generally regarded (5) as poor sources of active enzymes. Insights on aspects of the browning reactions that affect proteins have been gained from studies of single proteins in simple model-oxidizing systems (6). They emphasize the vulnerability of nucleophilic groups, such as the NH_2- and SH- groups in amino acid side chains, to substitution reactions with quinone rings to give protein *N*- or *S*-substituted phenols. These may be reoxidized by excess quinone to the *o*-quinone state, when they have the potential to react with other nucleophiles producing intraprotein or interprotein crosslinks or, in more alkaline conditions, react with quinones giving more complex, greenish, protein *N*-substituted hydroxyquinone polymers. Other reactions may occur depending on the conditions and especially on the phenols being oxidized and the nature of the oxidizing system. It would

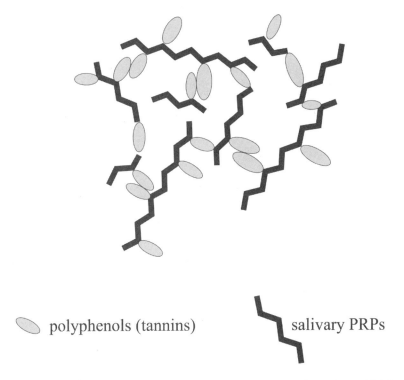

polyphenols (tannins)   salivary PRPs

Fig. 1. Illustration of the complexes formed between proteins and tannins, which lead to precipitation and enzyme inactivation. The molecular shapes represent the helical proline-rich proteins (PRPs) of saliva and disk-shaped hydrolyzable tannins and the complexes illustrate the multivalent interaction between tannin and protein as well as tannin and tannin. (Modified, with permission, from an as yet unpublished diagram by Professor E. Haslam in Bioactive Compounds in Plant Foods—proceeding of the Final COST 916 Conference, Tenerife 2001.)

take a major analytical effort to characterize satisfactorily the products formed when leaf proteins are exposed to the "natural" enzymic browning reactions of leaf extracts.

1.2. Some Solutions

In most cases, where it is required to extract enzymes from phenol-rich tissue, the specific phenolic compounds and oxidizing systems that are present are unknown or can only be guessed. An ideal approach would be to establish their nature so that the simplest method of preventing interference could be established. Thus, Gray *(7)* described how the phenolic compounds of bean leaves could be simply extracted and a suitable adsorbent for them chosen. A more usual approach is to follow a general procedure that has worked for other tissues and deals with as many eventualities as possible. Generally, these procedures involve disrupting fresh or deeply frozen (−70°C) tissue as quickly as possible in the presence of polymers that adsorb phenolic compounds, both monomers and polymers, and in conditions that minimize the oxidative reactions that produce quinonoid compounds.

Many polymers, natural and synthetic, have been used to adsorb phenols. They include albumins, hide powders, powdered nylon (ultramid), polyvinylpyrrolidones, both soluble (PVP) and insoluble (polyvinylpolypyrrolidone [PVPP]), relatively uncharged polystyrene and polyacrylic resins, such as Amberlites XAD-2, -4, and -7, and ion-ex-change resins based on polystyrenes, both anion exchangers (Bio-Rad AG1-X8, AG2-X8, and Dowex-1) and cation exchangers (Dowex-50). These polymers are listed in the reviews and articles by Loomis *(5)*, Loomis et al. *(8)*, Rhodes *(9)*, and Smith and Montgomery *(10)*. They interact with phenolic compounds in different ways so that PVP, for example, is thought to form stable H bonds to phenol groups via its -CO-N< linkages, whereas the porous polystyrene resins present large, adsorptive, hydrophobic surfaces. As a consequence, their affinities for different phenols differ. The adsorbents may either be soluble, as are the tannin-binding proteins and PVP, or, more usually, insoluble like PVPP and the resins, so that they can be readily removed from the tissue extracts.

An obvious way of preventing the oxidative reactions is to work in anaerobic conditions, but this is both difficult and cumbersome. It is much easier to make extracts in the presence of low-molecular-weight compounds that form unoxidizable complexes with phenols or that inhibit oxidases, trap quinones, or reduce quinones back to phenols *(5,6,11)*. Borate and germanate have been used to complex phenols, and copper-chelating agents, such as diethyldithiocarbamate (Dieca), are used to inhibit copper-dependent PPOs. Quinone-trapping agents that have been used include benzene sulfinic acid, and a range of substances, including ascorbate, metabisulfite, and 2-mercaptoethanol have been used as quinone reductants. However, detailed studies (e.g., **ref.** *12*) have made it clear that the action of many of these compounds cannot be simply explained and that they may act in more than one manner. Thus, thioglycollate both inhibits oxidases and reduces quinones, and Dieca inhibits oxidases and reacts with quinones; even PVP, primarily thought of as a phenol adsorbent, is also an oxidase inhibitor.

The complexity of leaf extracts and of the reactions that may occur in them thus makes it difficult, if not impossible, to write a procedure suitable for extracting enzymes from all phenol-rich plant tissues. The method described here is the first few stages in the procedure used for extracting the photosynthetic enzyme ribulose bisphosphate carboxylase (Rubisco; E. C. 4.1.1.39) from green tissue *(13)*. It has been used routinely by Keys and his colleagues for many years and, with a few adaptions, applied successfully to a wide range of plants, including some ferns and mosses. More recently, with the modifications mentioned later, it has been used to extract active Rubisco from the "difficult" leaves of Mediterranean tree species *(14)*. The subsequent purification steps specific to Rubisco have been omitted; the extracts, however, should be a suitable starting material for the purification of many other soluble enzymes by appropriate, specific procedures.

2. Materials

Use reagents of analytical (AR) quality wherever possible, and otherwise of the highest standard available.

1. Ammonium sulfate: Use the grade (BDH-Merck, Lutterworth, Leicestershire, UK) especially low in heavy metals that is suitable for enzyme work.
2. PVPP (Polyclar AT; Sigma-Aldrich, Poole, Dorset, UK.): Free from metal ions and other contaminants by boiling in 10% HC1 for 10 min and then washing extensively with glass-distilled water *(5)*. Air-dry for storage and rehydrate for at least 3 h before using: Hydration increases the weight of the polymer about fivefold *(8)*.

3. Extraction buffer: 100 mM HEPES/NaOH, 10 mM NaHCO$_3$, 10 mM MgCl$_2$, 0.1 mM EDTA, 10 mM 2-mercaptoethanol (2-Me), 10 mM sodium diethyldithiocarbamate (Dieca), 0.1% Tween-80. Prepare freshly overnight, adjust its pH to 7.0 *(see* **Note 1**) before the addition of NaHCO$_3$ and Dieca, and readjust it afterward if necessary. Add 2-Me just before use. NaHCO$_3$ is added specifically to protect Rubisco activity and is unnecessary in extracting most other enzymes.
4. Resuspension buffer: Tris-HCl, pH 7.0, containing 1.0 mM dithiothreitol (DTT).

3. Methods

3.1. Extraction of Proteins *(see Note 2)*

1. Cut leaf material (about 100 g), without mincing, into small pieces, and immerse in 500 mL of cold extraction medium contained in the bowl of a chilled Ato-Mix (MSE) or similar blender *(see* **Note 3**).
2. Add insoluble PVPP (4–8 g), which has been soaked overnight in 250 mL of extraction medium, and homogenize the mixture at high speed for four periods of 15 s with a 30-s interval between the periods.
3. Filter the extract through four layers of muslin into a measuring cylinder and allow to stand to settle any froth *(see* **Note 4**).
4. Treat the extract with a further addition of 4 g of PVPP, which has been soaked overnight in the extraction buffer, stir gently for 10 min, and clarify by centrifugation at 10,000g for 30 min to remove PVPP and cell residues. The resulting supernatant fluid provides a suitable starting material for purification procedures appropriate for particular enzymes *(see* **Notes 5** and **6**).

For enzymes like Rubisco, which can be further purified and concentrated by ammonium sulfate precipitation, this second treatment with PVPP can conveniently be included in the precipitation schedule described in **Subheading 3.2.**, and the PVPP removed along with a fraction of inactive protein. The ammonium sulfate concentrations are those found to be suitable for Rubisco; other enzymes may require different concentrations.

3.2. Ammonium Sulfate Fractionation of the Extract

1. After the second addition of PVPP, transfer the extract into a beaker and add ammonium sulfate (20 g/100 mL; 35% saturation) in portions with continuous stirring until it has dissolved.
2. Transfer the extract to centrifuge bottles, allow to stand for 30 min, and centrifuge at 25,000g for 15 min.
3. Discard the sediment containing PVPP and inactive protein. Filter the supernatant through eight layers of muslin, treat with a further 12 g/100 mL of ammonium sulfate, allow it to stand for 30 min, and centrifuge as in **step 2**.
4. Resuspend this protein fraction, precipitated between 35% and 55% saturation, in 25 mL of 20 mM Tris-HCl, pH 7.0, containing 10 mM DTT, and treat for a third time with 3 g of PVPP that has been soaked in this resuspension buffer overnight.
5. After allowing it to stand for 10 min with occasional stirring, centrifuge the suspension at 100,000g for 30 min and filter the supernatant liquid, containing active Rubisco, through a 50-µm mesh nylon gauze. This extract can be used for further stages of Rubisco's purification *(see* **Note 7**).

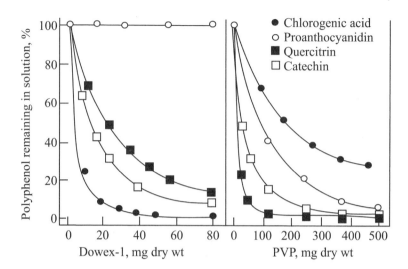

Fig. 2. Adsorption of different types of polyphenols by Dowex-1 (**left**) and PVPP (labeled PVP, **right**). Two milligrams of polyphenol in 3 mL phosphate buffer (50 m*M*; pH 7.5) were blended for 30 s with different amounts of hydrated adsorbents. The mixture was centrifuged and unadsorbed polyphenol in the supernatant estimated from $A_{280\ nm}$. (Reprinted from **ref. 7** by kind permission of the author and the editor on behalf of Pergamon Press.)

4. Notes

1. The pH of the extraction medium will affect some interactions of phenols with proteins and adsorbents. Hydrophobic interactions will presumably be unaffected. Ionization of phenolic groups in alkaline conditions will prevent the formation of H bonds with proteins, but it will also prevent the formation of H bonds with PVPP. Alkaline conditions (pH >7.5) will increase the auto-oxidation of dihydroxyphenols to quinonoids and also the auto-oxidation of protective thiol reagents. In the instances cited by Loomis *(5)*, enzyme extraction was generally optimal at pH values between 7.0 and 6.5. The optimal pH to extract an uncharacterized enzyme from an unfamiliar, phenol-containing plant, is a matter for experimentation.

2. All operations were performed as far as possible at 5°C either in a cold room or on a cold bench. Apparatus and solutions were cooled overnight. Operations were performed as quickly as possible to minimize exposure of enzymes to phenolic compounds.

3. The use of relatively large amounts of extraction liquid, usually five to seven times the weight of the leaf material, ensures that cells are disrupted while they are submerged, that liberated phenols and oxidative enzymes are immediately diluted, and that there is an adequate supply of phenol adsorbent and oxidase inhibitors. The volume of the medium should be such that the speed of the blender blades, which is adequate to disrupt the leaves, does not produce undue vortexing and the sparging of air into the homogenate.

4. The original extraction medium contained 10 times as much Tween-80, but this often gave rise to excessive frothing.

5. Choice of polyphenol-adsorbents:
    a. Insoluble adsorbents: Polyvinylpolypyrrolidone was chosen, as it is an excellent adsorbent of preformed tannins as well as many monomeric phenols. Its affinity for different classes of compounds is illustrated in **Fig. 2**, taken from the experiments of Gray *(7)*. It is less effective in absorbing hydroxy-cinnamic acids, such as chlorogenic acid, that are

common, widespread substrates for the browning reaction. In this respect, ion exchange resins, such as the anion-exchanger Dowex-1 (*see* **Fig. 2**) or the uncharged polystyrene Amberlites XAD-2 and XAD-4 (Rohm and Haas Co., Philadelphia, PA) are more effective. Dowex-1,X10 (Bio-Rad, Hercules, CA) was the preferred adsorbent used for the extraction of mevalonate kinase from the leaves of French beans *(Phaseolus vulgaris)* (**15**); it was pre-equilibrated with the buffered (sodium phosphate; 50 m*M*, pH 7.5) extraction medium and used at the rate of 20 g/100 mL of extraction medium. The Amberlite resins, especially XAD-4, appear to have been used more widely, presumably because of their great porosity and hydrophobic surface area and relative freedom from charged groups. They require extensive washing with water and methanol following manufacturer's instructions *(5)*, but this can be done in bulk so that they may be stored, ready for use, in a moist condition. They have been used in specific extraction procedures in quantities ranging from 0.1 g *(16)* to 1–2 g *(8)* hydrated weight per gram of fresh tissue. It must be remembered that these resins may absorb protein as well as polyphenols, although tests using bovine serum albumin (BSA) and plant peroxidases suggest that this is negligible with the Amberlite resins at broadly neutral pH values *(8)*. The ability of these resins to adsorb leaf components other than polyphenols, which interfere with enzyme extraction, such as the monoterpenes from peppermint leaves and the isothiocyanates of horseradish *(8)*, are additional advantages of their use.

A possible disadvantage of the presence of either PVPP or polystyrene polymers during leaf disruption, especially when blenders are used, is a small degree of breakdown. Thus, PVPP is reported to produce soluble PVPP–protein complexes in the extract, and the resins release ultraviolet (UV)-absorbing compounds, such as benzoic acid, following blending. The adsorbents could, of course, be added after leaf disintegration or used as columns through which the extracts could be filtered. Any advantages of these procedures would be offset by the disadvantage of not having adsorbents present during leaf disruption, when cell proteins are first exposed to liberated polyphenols.

Because the adsorptive abilities of PVPP and polystyrene resins are different and to some extent complementary (*see* **Fig. 2**), they have often been used together when extracting refractory plant tissues. Thus, PVPP and XAD-4 (5 and 1 g hydrated weight, respectively, per gram of tissue) have been used successfully in extracting active enzymes from such unpromising material as apple fruits, peppermint leaves, and even the hulls of walnuts *(8)*. More recently, a mixture of PVPP and XAD-4 was shown to be most effective in extracting polyphenoloxidases from strawberry fruits (**17**).

Polyvinylpolypyrrolidone (5%) has been found to prevent the inhibition of β-glucuronidase (GUS) by the phenols present in carnation leaves (**18**). This is a salutary reminder, if such were needed, that phenolics may inhibit the extraction and activity of transgenically introduced enzymes as well as native ones. Surprisingly, the same phenols had no effect on the extraction and measurement of chloramphenicol acetyltransferase (CAT), the enzymic product of a gene which, like the GUS gene, is frequently used as a "reporter" gene in genetic modification.

b. Soluble polyphenol absorbents: A recent procedure for extracting Rubisco from the leaves of Mediterranean trees uses casein (2% [w/v]) as well as PVPP (1% [w/v]) in the extraction medium *(14)*. A more usual protein additive is bovine serum albumin (BSA) used at about 1%. These proteins are known to react with tannins as well as with enzyme-generated quinones *(19)*. BSA is especially helpful in the extraction of organelles, such as mitochondria *(see* **ref. 5***)*, when the protein, along with adsorbed and reacted polyphenols, is effectively removed during centrifugal sedimentation of the organelles. The separation of added casein from Rubisco is, apparently, satisfactorily accomplished in the subsequent chromatographic stages of Rubisco purification. Rubisco, however, is the major protein component of leaves. Researchers attempting to purify enzymes present in very small amounts are often reluctant to add large amounts of additional protein to their extracts.

Soluble forms of PVP have been used and, indeed, were the first forms of this type of polymer to be used as polyphenol adsorbents. They adsorb phenolic compounds, but are unlikely to react with generated quinones. They are commercially available in three forms, with molecular weights about 10,000, 40,000, and 360,000, respectively, so that a size suitable for separation by gel filtration from a specific enzyme can be conveniently chosen. Earlier publications *(5,8)* emphasized the necessity of using only pharmaceutical grades of these chemicals: The grades supplied by a major supplier in the UK (Sigma) are presumed to be satisfactory, as they have been tested for suitability in such demanding processes as plant tissue culture and nucleic acid manipulation.

Soluble forms of poly(ethylene oxide), more usually called poly(ethylene glycol) or PEG, of general formula $H(OCH_2CH_2)_n OH$, have also been used to complex phenols in leaf extracts. They were initially introduced, under the trade name Carbowax, as an "osmoticum" in the isolation of chloroplasts and other organelles from leaves. Their phenol-complexing potential *(20)* has often been dismissed (*see* **ref. *11***), but they have proved especially useful in extracting enzymes from fruits, such as bananas *(21)* which contain condensed tannins. Downton and Hawker *(22)* have used PEG 4000 in extracting enzymes from the tanniniferous leaves and berries of vines. Recent and as yet unpublished results (Keys and Parry, personal communication) confirm PEG's usefulness with these leaves; it is effective in extracting active Rubisco from vine leaves when other methods, including those refered to here, failed. Preliminary studies on the interaction of PEGs with phenols *(23)* emphasize conformational requirements; PEGs that form helical structures in solution maximize the potential to form hydrogen bonds with phenols. Their specific binding with the tannins of grapes and vines seems worthy of further study.

6. Choice of oxidase inhibitor/antioxidant: The procedure described uses both 2-Me (10 m*M*) and Dieca (10 m*M*) to prevent the browning reactions. They both inhibit PPO, and this may be the principal cause of their effectiveness, although they also combine with quinones, either reducing them, complexing them, or both. A more recent modification of the procedure *(14)* omits the Dieca and increases the concentration of 2-Me to 100 m*M*. This modification also includes casein (2% [w/v]) in the extraction medium and this, as mentioned, also reacts with quinones.

   A brief survey of extraction procedures described in the then current volumes of *Phytochemistry (6)*, suggested that 2-Me was a common choice of antioxidant, as were dithiothreitol (DTT) and dithioerythritol (DTE), the two sterioisomers of Cleland's reagent. Ascorbate was used less commonly. DTT, like 2-Me, appears to act as an enzyme inhibitor and also as a quinone scavenger, but the action of neither DTT nor DTE has been studied in detail. Ascorbate probably acts principally by reducing quinones back to reoxidizable phenols, and so is readily removed by low concentrations of enzyme-produced quinones *(12)*. Therefore, it should be used in relatively high (>50 m*M*) concentrations. The thiol reagents, although they may be very effective in many, if not most cases, may be much less so in others. 2-Me activates unwanted proteases in extracts of tobacco leaves *(6)*, and the technical literature on DTT and DTE lists a number of enzymes that are inhibited by these reagents.

   Many other compounds have proven useful as inhibitors/antioxidants but have been less generally used. Mercaptobenzathiazole, a copper chelator that powerfully inhibits PPO, has been advocated by Palmer (cited in **ref. 5**), and at low concentrations (0.1 m*M*), it aids in extracting enzymes from tobacco leaves and mitochondria from potato tubers. Inorganic sulfites, especially metabisulfite, are effective in the extraction of enzymes from many leaves *(9,24)*, although their potential to cleave bonds in proteins demands a careful choice of concentration (approx 4 m*M).* Benzene sulfinic acid is a milder reagent that is thought principally to be a quinone scavenger, and it is effective in extracting acylases from tobacco leaves *(25).* KCN powerfully inhibits some oxidases, but is, understandably, now seldom used for this purpose. At 15 m*M,* it is reported to be a less effective inhibitor of the PPO from apple peel than Dieca, potassium metabisulfite, 2-Me, or ascorbic acid. It was used (1 m*M),* in conjunction with sodium metabisulfite (10 m*M),* in successfully extracting mevalonic acid kinase from the phenol-rich leaves of French beans *(15).*

7. All of the extracts, including the final one, are likely to contain active proteinases, which may modify or inactivate sensitive enzymes. Rubisco in extracts from young green tissues is apparently insensitive to them. Another photosynthetic enzyme, the phosphatase that hydrolyzes 2'-carboxy-D-arabinitol 1-phosphate, an inhibitor of Rubisco that is induced in leaves during darkness, is very sensitive. During its purification from the leaves of French beans, using a method similar to that described in **Subheading 3.** *(26),* it is necessary to include the proteinase inhibitors phenylmethylsulfonylfluoride (PMSF) (1 m*M)* and either benzamidine hydrochloride (1 m*M)* or *p*-amino-benzamidine dihydrochloride (1 m*M)* in the extraction buffer, in the resuspension buffer, and also in the liquid (1 L) against which the resuspended ammonium sulfate precipitate is dialyzed overnight before chromatography on an affinity column. These inhibitors are added to buffer solutions and dialysis water along with 2-Me (50 m*M)* just before use. PMSF is usually dissolved in the minimal amount of ethanol before addition to the liquids.

5. Envoi

The enzyme extraction procedures that were published in *Phytochemistry* during 1988 were briefly surveyed *(6);* this has now been repeated for procedures published in 2001. 2-Me and DTT are still the antioxidants of choice, with 2-Me, the less expensive one, being more popular. PVP or PVPP are the polymers most commonly used to absorb phenols. It is confirmed that when tissue that is very rich in phenols was investigated, complicated extraction media may be used. Thus, in extracting enzymes involved in the lignification of woody stems of *Robinia pseudoacacia*, Magel et al. *(27)* used the four adsorbents PVPP, crosslinked Dowex-1, BSA, and PEG 2O,OOO as well as the reductant 2-Me. It is not clear whether this comprehensive cocktail was designed by rule-of-thumb reasoning or whether all the ingredients are individually necessary.

Acknowledgment

The author acknowledges the help and advice of A. J. Keys, M. A. J. Parry, and Professor E. Haslam.

References

1. Harborne, J. B. (1980) Plant phenolics, in *Encyclopedia of Plant Physiology* (Bell, E. A. and Charlwood, B. V., eds.), Springer-Verlag, Berlin, Vol. 8, pp. 329–402.
2. Spencer, C. M., Cai, Y., Martin, R., et al. (1988) Polyphenol complexation-some thoughts and observations. *Phytochemistry* **27**, 2397–2409.
3. Haslam, E. (1989) *Plant Polyphenols: Vegetable Tannins Revisited,* Cambridge University Press, Cambridge.
4. Luck, G., Liao, H., Murray, N. J., et al. (1994) Polyphenols, astringency and proline-rich proteins. *Phytochemistry* **37,** 357–371.
5. Loomis, W. D. (1974) Overcoming problems of phenolics and quinones in the isolation of plant enzymes and organelles. *Methods Enzymol.* **31,** 528–544.

6. Jervis, L. and Pierpoint, W. S. (1989) Purification technologies for plant proteins. *J. Biotechnol.* **11,** 161–198.

7. Gray, J. C. (1978) Absorption of polyphenols by polyvinylpyrrolidone and polystyrene resins. *Phytochemistry* **17,** 495–497.

8. Loomis, W. D., Lile, J. D., Sandstrom, R. P., and Burbott, A. J. (1979) Adsorbent polystyrene as an aid in plant enzyme isolation. *Phytochemistry* **18,** 1049–1054.

9. Rhodes, M. J. C. (1977) The extraction and purification of enzymes from plant tissue, in *Regulation of Enzyme Synthesis and Activity in Higher Plants* (Smith, H., ed.). Academic, London, pp 245–269.

10. Smith, D. M. and Montgomery, M. W. (1985) Improved methods for the extraction of polyphenol oxidase from d'Anjou pears. *Phytochemistry* **24,** 901–904

11. Anderson, J. W. (1968) Extraction of enzymes and subcellular organelles from plant tissues. *Phytochemistry* **7,** 1973–1988

12. Pierpoint, W. S. (1966) The enzymic oxidation of chlorogenic acid and some reactions of the quinone produced. *Biochem. J.* **98,** 567–580.

13. Bird, I. F, Cornelius, M. J., and Keys, A. J. (1982) Affinity of RuBP carboxylases for carbon dioxide and inhibition of the enzymes by oxygen. *J. Exp. Bot.* **33,** 1004–1013.

14. Delgado, E., Medrano, H., Keys, A. J., and Parry, M. A. J. (1995) Species variation in Rubisco specificity factor. *J. Exp. Bot.* **46,** 1775–1777.

15. Gray, J. C. and Kekwick, R. G. 0. (1973) Mevalonate kinase in green leaves and etiolated cotyledons of the French bean *Phaseolus vulgaris. Biochem. J.* **133,** 335–347.

16. Hallahan, D. L., Dawson, G. W., West, J. M., and Wallsgrove, R. M. (1992) Cytochrome P-450 catalysed monoterpene hydroxylation in *Nepeta mussinii. Plant Physiol. Biochem.* **30,** 435-443.

17. Wesche-Ebeling, P. and Montgomery, M. W. (1990) Strawberry polyphenol-oxidase: extraction and partial characterization. *J. Food Sci.* **55,** 1320–1324.

18. Vainstein, A., Fisher, M., and Ziv, M. (1993) Applicability of reporter genes to carnation transformation. *Hort. Sci.* **28,** 1122–1124.

19. Pierpoint, W. S. (1969) *o*-Quinones formed in plant extracts: their reaction with bovine serum albumin. *Biochem. J.* **112,** 619–629.

20. Jones, D. E. (1965) Banana tannin and its reaction with polyethylene glycols. *Nature* **206,** 299–300.

21. Young, R. E. (1965) Extraction of enzymes from tannin-bearing tissue. *Arch. Biochem. Biophys.* **111,** 174–180.

22. Downton, W.J.S. and Hawker, J.S. (1973) Enzymes of starch metabolism in leaves and berries of *vitis vinifera. Phytochemistry* **12,** 1557–1563.

23. McManus, J. P., Davis, K. G., Beart, J. E., Gaffney, S. H., Lilley, T. H., and Haslam, E. (1985) Polyphenol interactions. Part 1. Some observations on the reversible complexation of polyphenols with proteins and polysaccharides. *J. Chem. Soc. Perkin Trans.* II, 1429–1438.

24. Anderson, J. W. and Rowan, K. S. (1967) Extraction of soluble leaf enzymes with thiols and other reducing agents. *Phytochemistry* **6,** 1047–1056.

25. Pierpoint, W. S. (1973) An N-acylamino acid acylase from *Nicotiana tabacum* leaves. *Phytochemistry* **12,** 2359–2364.

26. Kingston-Smith, A. H., Major, I., Parry, M. A. J., and Keys, A. J. (1992) Purification and properties of a phosphatase in French bean *(Phaseolus vulgaris)* leaves that hydrolyzes 2'-carboxy-D-arabinitol 1-phosphate. *Biochem. J.* **287,** 821–825.

27. Magel, E. A., Hillinger, C., Wagner, T., and Holl, W. (2001) Oxidative pentose phosphate pathway and pyridine nucleotides in relation to heartwood formation in *Robinia pseudoacacia* L. *Phytochemistry* 57, 1061–1068.

9

Avoidance of Proteolysis in Extracts

Robert J. Beynon and Simon Oliver

1. Introduction

Proteolytic enzymes (or proteases, peptidases, or proteinases) hydrolyze the peptide bond in proteins and peptides. The nomenclature is imprecise, but there is a broad acceptance that endopeptidases break bonds that are "internal" in the primary sequences, whereas exopeptidases trim one, two, or perhaps three amino acids from the amino or carboxy terminus of the substrate. Every cell and subcellular compartment has its own complement of proteolytic enzymes, and in normal circumstances, the activities of the proteolytic enzymes are well regulated. When a tissue is disrupted, however, this control is lost, and the proteinases may then attack proteins at a rate that leads to a loss of those proteins within the time scale of the study *(1)*. Adventitious proteolysis is a technical problem that may require modification to methodology to minimize the assault on the protein of interest *(2–5)*.

The isolation of almost every protein will incur losses, and in some circumstances, those losses become unacceptable. Exactly what constitutes "unacceptable" will be dictated by individual needs and will depend on downstream requirements and timescale *(6,7)*. Few purification protocols generate pure protein, but rather, reduce contaminant proteins to the level where they cannot be detected by routine methods—consider the difference between a Coomassie-blue-stained gel and a silver-stained gel—the additional sensitivity of the latter will usually identify many contaminants in a preparation that looked homogenous by the former. Even apparently pure proteins may contain low levels of contaminating peptidases, which may explain the tendency to assume that any loss of protein or of biological activity associated with that protein is the result of proteolytic destruction. Loss of material or activity can, of course, be the result of other, nonproteolytic processes; a strategy based on suppression of proteolysis will only be effective if proteolysis is really the culprit. In this chapter, therefore, we advocate the use of methods to prove that proteolysis is a problem and to measure proteolytic activity directly in the samples, as a prerequisite to prevention.

As mentioned previously, loss of material, as protein or as a biological activity, has many potential causes. Eliminate the possibility of, for example, thermal denaturation, oxidative damage, adsorption onto surfaces, or persistent binding to the column matrix.

From: *Methods in Molecular Biology, vol. 244: Protein Purification Protocols: Second Edition*
Edited by: P. Cutler © Humana Press Inc., Totowa, NJ

More subtle reasons for losses of activity might be complexation with an inhibitor in a time-dependent process or loss of an essential cofactor or accessory protein as a consequence of a purification step. In crude extracts, and particularly when the protein has yet to be purified, discriminating between proteolysis and other events is difficult, but it is as well to be aware of the range of traumas that a protein can undergo after it has been removed from the normal cellular milieu, substantially diluted, and removed from the protective effects of other proteins and low-molecular-weight ligands, such as natural reductants. Even if recoveries are initially good, proteolysis can become a problem during storage. Slow proteolytic degradation might destroy the material—or worse, allow it to retain some biological functions, but change others. Awareness of this possibility is the first step in its control.

Proteolysis of a protein may also occur in vivo before the cells are disrupted. This is sometimes evident in heterologous expression systems, where a protein is exposed to different proteolytic systems in the host cell. Strategies for prevention of this type of intracellular degradation are very different from those for the prevention of postextraction degradation and are not discussed here.

1.1. Recognition of Proteolytic Attack

1.1.1. Why Are Proteins Susceptible to Digestion?

The classical studies of Linderström-Lang and colleagues, over 40 yr ago, opened the way to an understanding of the attack by proteinases on native structures *(8,9)*. Two types of proteolysis were defined: (1) a "zipper" mechanism wherein the proteinase "nicks" one substrate molecule and then attacks a further molecule, often at the same site and (2) "all or none" proteolysis, in which the initial proteolytic attack renders the products much more likely to be further digested, such that they are quickly and completely digested to limit peptides. In the first mode, the products are resistant to further digestion, and they accumulate as partially degraded proteins that may retain some biological function—the Klenow fragment of DNA polymerase, for example. In the second mode, the products of the first proteolysis are destabilized and become more susceptible to further digestion, such that transient intermediate products are unlikely to be observed.

Compared to unfolded polypeptides, most proteins are, in fact, relatively resistant to attack by endopeptidases and exopeptidases *(10)*. A denatured protein is usually much easier to digest than the native counterpart; denaturation and enhanced digestibility are, in part, the function of the low pH of the stomach and the lysosome. Folding of a protein chain into a compact tertiary or quaternary structure protects many peptide bonds because they are internalized in the protein. Moreover, there is accumulating evidence that proteolytic susceptibility can, in part, be attributed to the ability of the whole or a segment of a protein chain to be flexible *(11,12)*. The more flexible a loop, the more likely it is to be accommodated in the active site of the proteinases in a productive interaction. Thus, proteolytic attack can be dependent on the ability to cut a specific site on the protein that is flexible and susceptible. After the first nick, the behavior of the protein can adopt either of the two extreme models described by Linderström-Lang. Treatments that diminish this flexibility will protect the protein and thus achieve the same goal as inhibition or removal of the proteinases.

1.1.2. Experimental Evidence for Proteolytic Degradation

Evidence for proteolytic degradation of a protein in a crude extract or during the course of a protein purification comes from several sources. A few guidelines may be provided herein, but we stress that each case must be considered unique and that there is no general solution that will apply in all cases.

1. Loss of biological activity. If the only parameter that can be measured is biological activity, proving that proteolysis is responsible may be difficult. As discussed in **Subheading 1.**, there are many reasons why the activity of a protein may be diminished. It may be necessary to conduct some analyses of the time-course of inactivation. Is the inactivation first order? Is the rate of inactivation reduced by the addition of stabilizers, such as glycerol or nonionic detergents? Can inactivation be enhanced or prevented by the addition of proteinase activators or inhibitors (see **Subheading 1.3.**)?

2. Loss of protein. Loss of the protein of interest is harder to assess, particularly if the protein is in low abundance in a crude mixture, such as a cell extract. If it is naturally abundant or is expressed heterologously at high levels, it may be possible to identify the protein on one-dimensional gel electrophoresis. Under circumstances in which, for example, 50% of the biological activity is lost, it ought to be possible to see a similar loss of material on the gel. Of course, this strategy cannot discriminate proteolytic destruction of the protein (e.g., from absorption to glass or plastic surfaces). Experiments should be conducted with siliconized glassware, in the presence of a protective protein of different size, such as albumin, or in the presence of a nonionic detergent, such as Tween-80 at 0.1% (w/v).

   Monitoring of proteolysis is simpler if an antibody to the protein can be used for Western blotting. The protein can be identified at low concentrations, and degradation will be manifest as loss of the band of the correct molecular weight, possibly with the appearance of degradation products. Reducing and nonreducing gels may give different results, depending on the relative disposition of nick sites and disulfide-linked cysteine residues, and both types of gel should be used if the protein is suspected or known to contain disulfide bonds.

3. Microheterogeneity. Proteolysis can manifest itself as a degree of heterogeneity in a finished product. Multiply-sized bands on sodium dodecyl sulfate–polyacrylamide gel electrophoresis (SDS-PAGE) or on isoelectric focusing may indicate limited proteolytic attack. A blotting antibody can help to confirm that the heterogeneity is the result of the protein of interest.

   If sufficient material is available for electrospray or MALDI-TOF mass spectrometry, then the precise mass of the isolated protein and the extent of heterogeneity can be checked directly. It will be important to know the extent of posttranslational modification that the protein might have undergone.

1.2. Strategies for Prevention of Proteolysis

Although individual cases will require unique solutions, there are some general principles that can be addressed. First, cells differ in the intracellular concentrations of proteinases, and artifactual proteolysis can be diminished by the use of a cell/tissue in which proteinase levels are low. For example, the liver and kidney contain much higher amounts of proteolytic activity than skeletal or cardiac muscle. In many single-cell systems, mutant strains, defective in the expression of one or several proteinases, have been developed *(13)*.

Many proteinases coexist with naturally occurring inhibitors. Complexation between proteinase and inhibitor may reduce the proteolytic activity in crude broken cell preparations and this can be advantageous. However, in a subsequent purification step, the proteinase and inhibitor may be separated, and the proteolytic degradation may manifest itself late in the purification protocol.

Proteins offer protection to each other by competing for the active site of the proteinase. The early stages of a protein purification, in which the goal is to eliminate much of the bulk unwanted protein, may therefore have the effect of rendering a protein more readily digested by a contaminating proteinase. Proteolytic losses can therefore, paradoxically, become more of a problem as the protein is purified.

Unlike many procedures for purification of nucleic acids, which use potent denaturants to destroy nucleases, the need to preserve the native structure of proteins means that such denaturants cannot be used. Also, many proteinases are more stable to a range of denaturants than their substrates. A denaturing step, which has the aim of inactivating the proteinase to diminish proteolysis, may, in fact, have the opposite effect and render the protein more susceptible to digestion. This behavior is often seen in sample preparation for SDS-PAGE: When the sample is heated in SDS-containing buffer, the protein is unfolded and undergoes rapid destruction, with apparent loss of material from the expected position on the gel. We routinely precipitate samples with trichloroacetic acid (TCA) before SDS-PAGE. This concentrates the samples, eliminates a great deal of buffer variability, and denatures proteinases and substrates in a rapid step that does not allow time for extensive artifactual proteolysis. Typically, a final concentration of TCA of 5% (w/v) at 4°C for 10 min will be adequate. The precipitated proteins are recovered by brief centrifugation (10,000g for 1 min, in a benchtop microcentrifuge), and the TCA is poured away. Residual TCA, which would otherwise change the pH of the sample buffer, is removed by repeated gentle washing with acetone or diethyl ether, and after three washes, the protein pellet is redissolved in SDS-PAGE sample buffer.

Routine additions to buffers can modify the activity of proteinases in the preparation. For example, reducing agents, such as 2-mercaptoethanol, activate cysteine proteinases, but inhibit metalloproteinases. Chelators also inhibit metalloproteinases. Some proteinases are activated by the presence of calcium ions. Thus, the choice of buffer may influence proteinase activity.

Thus, judicious choice of appropriate methodology may help to diminish the risk of proteolysis. However, there are many circumstances in which it is not possible to separate proteins and potentially damaging proteinases. Under these circumstances, the only real option is to inhibit the proteinases *in situ*.

1.3. Proteinases and Their Inhibition

A strategy of prevention of proteinase activity is usually based on a combination of two approaches: inhibition of the proteinases *in situ* and separation of the proteinases from the protein of interest. Proteinases bring about the hydrolysis of peptide bonds by several catalytic mechanisms, and inhibition of their action will be different for each mechanistic class *(14)*. Mechanism-based inhibitors will comprise a reactive group that is susceptible to attack by the active-site residues, and high-affinity interaction of the reactive group with the enzyme can be achieved by additional functional groups that bind, for example, to the specificity pocket or the extended subsites of the proteinases. Most of these inhibitors bring about covalent modification of the active-site residues and are thus irreversible. Once the proteinase has been inhibited, it should not be necessary to add further inhibitor.

Another class of inhibitor capitalizes on the affinity of the proteinases for a substrate analog, without covalent interaction between them. These noncovalent inhibitors can be proteins or low-molecular-weight peptide analogs. Such inhibitors may also be usefully immobilized on columns for removal of the proteinases from the protein solution, although this is rarely used as an option. The complex formed between the proteinase and the inhibitor is reversible and depends on the concentration of inhibitor in solution. If that inhibitor concentration is reduced by, for example, inactivation or dialysis, then the proteinase will recover activity. Strategies that add inhibitor to a crude preparation, which then undergoes precipitation (by ammonium sulfate or polyethylene glycol), column chromatography, or dialysis, must ensure that fresh inhibitors are added when appropriate.

In a few instances, the noncovalent binding is so tight that to all practical intents and purposes, the inhibitor can be considered as practically irreversible, and it will only be necessary to sustain low concentrations of inhibitor in buffers.

Serine proteinases use a nucleophilic serine residue to attack the carbonyl carbon of the scissile bond. The serine residue is a much stronger nucleophile than other serine residues in proteins, and this extreme property is facilitated by input of electrons from an aspartate residue via a histidine residue—the charge relay system. The serine and the histidine residues are both targets for inhibition. The inhibitors used most commonly are sulfonyl fluorides of which the most common is phenylmethylsulfonylfluoride (PMSF). All sulfonylfluorides are unstable in solution, and a common mistake is to prepare stock solutions of PMSF in buffer or to assume that the inhibitor continues to work for extended times. At neutral pH values and 25°C, PMSF has a half-life of <1 h. Provided all of the serine proteinases are inhibited within this time-scale, PMSF is effective, but exposure of cryptic proteinase activity later in a purification would need PMSF to be added afresh. Stock solutions are normally stored at −20°C in dry organic solvents, in which state PMSF is stable for weeks. Sulfonylfluorides are also potent inhibitors of acetylcholinesterase and are thus toxic. A second class of proteinase inhibitors are derived from coumarins, of which 3,4 dichloroisocoumarin (3,4-DCI) is best known. Unlike PMSF, 3,4-DCI does not react quickly with acetylcholinesterase and is relatively nontoxic. Although 3,4-DCI is inactivated quite quickly in aqueous buffers (half-life about 20 min at neutral pH values), stock solutions in organic solvents are relatively stable.

Cysteine proteinases use a cysteine residue as the nucleophile to attack the scissile bond, and most inhibitors target this cysteine residue. The reactivity of the cysteine residue means that it reacts well with general-purpose -SH reactive reagents, such as iodoacetamide or iodoacetic acid, but these also modify free -SH groups in other proteins. If a cysteine proteinase is suspected, it may be preferable to use the epoxide inhibitor E-64 (L-*trans*-epoxysuccinyl-leucylamide-[4-guanidino]-butane, *N*-[*N*-L–3-*trans*carboxyrane-2-carbonyl]-L-leucyl-agmatine). This is a potent irreversible inhibitor of cysteine proteases that does not react with other -SH groups in proteins and that will not react with, and be consumed by, other low-molecular-weight thiols. Stock solutions are stable for days in aqueous buffers.

Some peptide aldehydes, such as leupeptin, chymostatin, and elastatinal, have been used to inhibit serine or cysteine proteases. These are reversible inhibitors and will need to be maintained at a reasonable working concentration in the buffers. Also, these peptides are prone to inactivation, and it is preferable to develop strategies based on irreversible inhibition.

Aspartic proteinases use a pair of aspartic residues to polarize the scissile bond. In general, aspartic proteinases are less of a problem than other proteinases, and they often have acidic pH optima. The nearest to a general-purpose inhibitor of aspartic proteinases is pepstatin, a reversible but tight-binding inhibitor.

Metalloproteinases use a bound zinc ion as an electrophile to polarize a water molecule, which then attacks the scissile bond. The most general inhibitors of metalloproteinases are chelators, such as EDTA or 1,10 phenanthroline. If chelators cannot be included in the sample buffers, more complex strategies, based on some awareness of the type of metalloproteinase, may dictate an inhibition strategy. For example, some metalloproteinases are inhibited by phosphoramidon.

Method 1 provides a recipe for an inhibitor cocktail that ought to prevent proteolysis under many circumstances. However, it should be stressed that such cocktails are not guaranteed to work under all conditions. Detailed literature, the mechanism of action, and manipulation of a range of commercially available proteinase inhibitors can be found elsewhere (*see* **refs. *4*,*5,* and *14*).

1.4. Assay of Endopeptidase Activity

There is some virtue in adopting a simple, yet sensitive assay for proteolytic activity, for tracking proteolytic activity during sample workup, or for monitoring the efficiency of inhibition strategies. In a previous publication *(4),* the use of a radioiodinated peptide was advocated, but it is appreciated that few laboratories would be willing to take on this method unless they were already equipped for this type of radiochemical work. Accordingly, we present here a different assay, using a protein labeled with fluorescein isothiocyanate (FITC) based on a previously published protocol *(15,16)* (*see* **Subheadings 2.2.** and **2.3.**). The fluorescent-labeled protein is digested with a proteinase containing sample, and the undigested and large FITC peptides are separated from small FITC products by precipitation with TCA. The FITC peptides in the supernatant fraction are monitored in a fluorimeter.

A second method (Method 4), which does not work for all endopeptidases, is zymography. In this method, a proteinase sample is electrophoresed in an SDS-PAGE into which has been copolymerized a protein, such as gelatin or casein. After the gel has been run, it is incubated in a nonionic detergent to remove the SDS, and then in a "refolding" buffer that may allow the proteinases to recover their structure and activity. If the proteinase is active, the gelatin is digested and converted to low-molecular-weight peptides that are washed out of the gel. Subsequent staining with Coomassie blue indicates the zone of lysis as a clear region on a uniform blue background.

Zymography is quite sensitive, and a visible zone of lysis can be seen when nanogram amounts of enzyme are loaded. If excess proteinase is loaded, the zone of lysis can become large and diffuse. Also, if the proteinase is small, excess loading can show up as an unstructured zone of staining because the proteinase can diffuse out of the gel and digest the gelatin over all or part of the surface. In general, zymography does not work well if the gel or sample is reduced, possibly because of the opportunities for incorrect disulfide pairing and the role of preformed disulfide bonds in directing folding. Not all proteinases will refold correctly and, thus, will not be detectable by this method.

2. Materials

2.1. Method 1: Stock Inhibitor Solutions

1. PMSF stock solution: 0.2 M in dry methanol or propanol. Dissolve 38 mg ($M = 174.2$) of PMSF in 1.0 mL of solvent. *PMSF is toxic!* Weigh this compound in a fume hood, and wear disposable gloves and a mask. Store at $-20°C$.
2. 3,4-DCI stock solution: 10 mM in dimethylsulfoxide (DMSO). Dissolve 2.2 mg ($M = 215$) of 3,4-DCI in 1.0 mL of DMSO. Store at $-20°C$.
3. Iodoacetic acid stock solution: 200 mM in water. Dissolve 42 mg ($M = 207.9$) of sodium iodoacetate in 1.0 mL of water. Use immediately.
4. E64-c stock solution: 5 mM in water. Dissolve 1.8 mg of E64-c ($M = 357.4$) in 1.0 mL water. Store at $-20°C$.
5. 1,10 Phenanthroline stock solution: 100 mM in methanol. Dissolve 19.8 mg 1,10 phenanthroline ($M = 198.2$) in 1.0 mL of methanol. Store tightly capped at room temperature or 4°C.
6. EDTA stock solution: 0.5 M in water. Dissolve 18.6 g EDTA (disodium salt, dihydrate, $M = 372.2$) into 70 mL water, titrate to pH 7.0 or 8.5, and make up to 100 mL. Stable at room temperature or 4°C.
7. Pepstatin stock solution: 10 mM in DMSO. Dissolve 6.9 mg pepstatin ($M = 685.9$) in 1.0 mL of DMSO. Store at $-20°C$.

2.2. Method 2: Preparation of FITC–Casein Substrate

1. Casein (purified powder; Sigma [St. Louis, MO]; cat. no. C-5890).
2. Fluorescein isothiocyanate (Sigma; cat. no. F-7250).
3. 0.05 M Sodium carbonate, 0.15 M NaCl, pH 9.5.
4. 0.05 M Tris-HCl, pH 8.6.
5. Sephadex G-25 column, equilibrated with assay buffer.

2.3. Method 3: Assay of Enzyme Activity With FITC–Casein

1. Stock assay buffer (5X final concentration) (e.g., 0.1 M HEPES, pH 7.5).
2. FITC–casein: 1 mg/mL, in water or buffer, or casein fluorescein isothyocyanate (Sigma; cat. no. C-0528).
3. TCA: 5% (w/v) in water.
4. Neutralizing buffer: 0.5 M Tris-HCl, pH 8.6.

2.4. Method 4: Zymography

1. Stock gelatin solution: 1.2% (w/v) in water. Store at 4°C and microwave gently on the defrost setting to melt before use. Use electrophoresis-grade gelatin, Type A (Sigma; cat. no. G-8150).
2. Triton-X100: 2.5% (w/v) in water.
3. 10X Refolding buffer: 0.5 M Tris-HCl, 2 M NaCl, 5.5% CaCl$_2$, 0.67% (w/v) Brij35, pH 7.6. Adjust pH before adding Brij35.

3. Methods

3.1. Method 1: Working Inhibitor Cocktails

1. From the stock solutions described in **Subheading 2.1.**, make a working inhibitor cocktail in water (not buffer, because some buffer compounds accelerate decomposition of the inhibitors). For 1.0 mL of working solution and for each class of proteinases, proceed as follows:

   a. Serine: 200 µL PMSF (20 mM final) or 200 µL 3,4-DCI (2 mM final).
   b. Cysteine: 200 µL iodoacetate (40 mM final) or 200 µL E64c (1 mM final).
   c. Metallo: 100 µL 1,10 phenanthroline (10 mM final) or 100 µL EDTA (50 mM final).
   d. (Optional) Aspartic: 100 µL pepstatin (1 mM final).
       Make up the final volume to 1.0 mL with water.
2. Use the working cocktail within 1 h of preparation. Dilute this by 20-fold into the sample (*see* **Notes 1** and **2**).

3.2. Method 2: Preparation of FITC–Casein Substrate

1. Dissolve 2 g of casein in 100 mL of sodium carbonate buffer, pH 9.5. Add 100 mg of FITC. Mix gently for 1 h at room temperature.
2. Dialyze the FITC–casein against several changes of 2 L of 0.05 M Tris-HCl, pH 8.5, followed by the buffer that is preferred for routine assay. Alternatively, dialyze against Tris-HCl, and then distilled water to exchange the substrate into any buffer in the future. The Tris buffer is used to consume excess reagent.
3. Determine the protein concentration by standard procedures. If needed, calculate the number of FITC residues/casein molecule from the $A_{490\ nm}$ of the conjugate at pH 8.6. The extinction coefficient is 61,000 M^{-1}cm^{-1}. The conjugate can now be frozen at −20°C.
4. To change the buffer in which the substrate is dissolved and to remove residual unbound FITC, apply the conjugate to a Sephadex G-25 column, equilibrated and eluted with a suitable assay buffer. Typically, the column volume should be about 10 times greater than the volume of FITC–casein for good buffer exchange. Monitor the elution profile at 280 nm and collect the protein peak.
5. This substrate, approx 5 mg/mL, can be diluted to 1 mg/mL and stored in aliquots at −20°C.

3.3. Method 3: Assay of Enzyme Activity With FITC–Casein

1. Combine 10–50 µL of enzyme sample with 10 µL FITC–casein in a microfuge tube. Add further assay buffer to make a reaction volume of 100 µL. Include a blank consisting of the buffer only.
2. Incubate at the desired temperature for 1–24 h.
3. Stop the reaction by adding 200 µL of 5% (w/v) TCA with mixing. Incubate the tubes at 4°C for 1 h to allow proteins to flocculate. Sediment the precipitated proteins by centrifugation at approx 10,000g for 10 min in a benchtop centrifuge at room temperature.
4. Add 100 µL of the supernatant to 2.9 mL of 0.5M Tris-HCl, pH 8.6. This strong buffer neutralizes the TCA, because the fluorescence of FITC is pH dependent.
5. Mix thoroughly and measure the fluorescence with the excitation wavelength set at 490 nm and the emission wavelength set at 525 nm. Slit widths of 10 nm or less should be used.
6. The fluorescence of a sample, incubated in the absence of a proteinase and substituting water for the TCA, but otherwise processed identically, gives the total fluorescence of the substrate and allows quantitation of the casein digestion as weight solubilized/time (*see* **Notes 3–5**).

3.4. Method 4: Zymography

1. If necessary, desalt samples into a low-salt buffer (e.g., 0.02 M HEPES, pH 7.5) using spun columns. This will improve the sharpness of the bands.
2. Add 5 µL of nonreducing SDS-PAGE sample buffer to 20 µL of the enzyme sample and incubate at 37°C for 1 h. Do not heat the sample in a boiling water bath.
3. While the sample is incubating, cast the gel in a standard kit, including gelatin (*see* **Note 6**) at a final concentration of 0.6% (w/v).

4. Typically, load between 1 and 20 μL of each sample. In early experiments, two different loadings (e.g., 1 and 15 μL) could usefully be included.

5. Run the electrophoresis as usual. There may need to be some adjustment of running time. At the end of the run, remove the gel without touching the surface.

6. Soak the gel in 2.5% Triton-X 1 00 for 1 h at room temperature to wash out the SDS.

7. Rinse the gels in deionized water until foaming ceases (at least three times).

8. Incubate the gels in 1X refolding buffer, overnight at 37°C (*see* **Notes** 7 and **8**).

9. Rinse the gels three times in deionized water.

10. Stain the gels with Coomassie brilliant blue for 1 h.

11. Destain the gels (several hours of destaining may be needed before the bands can be seen) (*see* **Note 9**).

4. Notes

1. The inhibitor cocktail can introduce salt (NaEDTA) and organic solvents to the sample. The 20-fold dilution is designed to minimize the effects of these constituents, but there may be circumstances in which this could still be problematic.

2. Although the serine and cysteine proteinase inhibitors are irreversible, the aspartic and metalloproteinase inhibitors are reversible. Even if the inhibitors are added at an early stage, they may be lost during purification. Alternatively, cryptic proteinases may be exposed at a later step by, for example, zymogen activation or dissociation of an endogenous inhibitor. Be prepared to add further inhibitor cocktail throughout the purification scheme.

3. The FITC–casein assay is very sensitive (nanogram to sub-nanogram) amounts of proteinase. The substrate should be stored at −20°C, because even a low level of bacterial contamination will give higher blank values. Always include a blank sample in FITC assays. If a standard is needed, a protease solution of 20 ng/mL trypsin will give a strong fluorescence in 2–3 h.

4. FITC–casein, like casein, is not very soluble at pH values below 4.0. This assay is not suitable for proteinases that have pH optima at low pH values.

5. This assay uses small amounts of substrate, which are usually far below saturating concentrations for the enzyme. As such, linearity over time will be lost if more than 10–20% of the substrate is digested. If digestion is limited, then the fluorescence signal is proportional to the amount of enzyme added.

6. For zymography, other proteins, such as fibronectin or casein, can be copolymerized into the gel. The presence of the protein in the gel may alter the mobility of the proteinase, and a molecular-weight estimate obtained by zymography should be carefully checked. It is possible to include a lane of molecular-weight markers at sufficiently high a concentration that they can be seen as even darker bands on the uniform blue background.

7. The "refolding" step seems to be important, and a temperature of 37°C gives much improved recovery of activity over an incubation at room temperature. For metalloproteinases, there is apparently no necessity to add zinc ions to the refolding buffer, and this should be discouraged because many metalloproteinases are inhibited by excess zinc.

8. Zymography is not routinely used for analysis of cysteine proteases because of the need to add reducing agents to activate the enzyme. It might be worthwhile adding a reducing agent (10 m*M* dithiothreitol) for the last few hours of incubation in refolding buffer.

9. Because zymography has the potential to separate noncovalently bound inhibitors from proteinases, it may indicate proteolytic activity when none is apparent in soluble extracts.

References

1. Lutkemeyer, D., Ameskamp, N., Tebbe, H., Wittler, J., and Lehmann J. (1999) Estimation of cell damage in bench- and pilot-scale affinity expanded-bed chromatography for the purification of monoclonal antibodies. *Biotechnol. Bioeng.* **65**, 114–119.

2. Pringle, J. R. (1975) Methods for avoiding proteolytic artifacts in studies of enzymes and other proteins from yeast, in *Methods in Cell Biology* (Prescott, D. M., ed.), Academic, New York, Vol. 12 pp. 149–184.

3. Pringle, J. R. (1978) Proteolytic artifacts in biochemistry, in *Limited Proteolysis in Microorganisms* (Cohen, G. N. and Holzer, H., eds.), US Department of Health, Washington, DC, pp. 191–196.

4. Beynon, R. J. (1988) Prevention of unwanted proteolysis, in *Methods in Molecular Biology, Vol. 3: New Protein Techniques* (Walker, J., ed.), Humana, Clifton, NJ, pp. 1–23.

5. North, M. J. (1989) Prevention of unwanted proteolysis, in *Proteolytic Enzymes—A Practical Approach* (Beynon, R. J. and Bond, J. S., eds.), IRL, Oxford, pp. 105–124.

6. Diano, M. and Le Bivic, A. (1996) Production of highly specific polyclonal antibodies using a combination of 2D electrophoresis and nitrocellulose-bound antigen. *Protein Protocols Handbook* (Walker J. M., ed.), Humana, Totowa, NJ, pp. 703–710.

7. Saijo-Hamano, Y., Namba, K., and Oosawa, K. (2000) A new purification method for overproduced proteins sensitive to endogenous proteases. *J. Struct. Biol.* **132,** 142–146.

8. Neurath, H. (1978) Limited proteolysis—an overview, in *Limited Proteolysis in Microorganisms* (Cohen, G. N. and Holzer, H. eds.), US Department of Health, Washington, DC, pp. 191–196.

9. Price, N. C. and Johnson, C. M. (1989) Proteinases as probes of conformation of soluble proteins, in *Proteolytic Enzymes—A Practical Approach* (Beynon, R. J. and Bond, J. S., eds.), IRL, Oxford, pp. 163–191.

10. Simon, A., Dosztanyl, Z., Rajnavolgyl, E., and Simon I. (2000) Function-related regulation of the stability of MHC proteins. *Biophys. J.* **79**, 2305–2313.

11. Hubbard, S. J., Campbell, S. F., and Thornton J. M. (1991) Molecular recognition: conformational analysis of limited proteolytic sites and serine proteinase inhibitors. *J. Mol. Biol.* **220,** 507–530.

12. Hubbard, S. J., Eisenmenger, F., and Thornton, J. M. (1994) Modeling studies of the change in conformation required for cleavage of limited proteolytic sites. *Protein Sci.* **3,** 757–768.

13. Burgers, P. (1996) Preparation of extracts from yeast and the avoidance of proteolysis. *Methods Mol. Cell. Biol.* **5**, 330–335.

14. Salvesen, G. and Nagase, H. (1989) Inhibition of proteolytic enzymes, in *Proteolytic Enzymes—A Practical Approach* (Beynon, R. J. and Bond, J. S., eds.), IRL, Oxford, pp. 25–55.

15. Sarath, G., De La Motte, R. S., and Wagner, F. W. (1989) Protease assay methods, in *Proteolytic Enzymes—A Practical Approach* (Beynon, R. J. and Bond, J. S., eds.), IRL, Oxford, pp. 83–104.

16. Twining, S. S. (1984) Fluorescein isothiocyanate-labeled casein assay for proteolytic enzymes. *Anal. Biochem.* **143,** 30–34.

10

Concentration of Extracts

Shawn Doonan

1. Introduction

In a typical purification starting from 1 kg of tissue, the volume of the initial homogenate might be 2 L and the final product of the procedure might be 2 mL of pure protein solution; that is, a volume decrease of 1000-fold is required. This emphasizes the fact that water is a major contaminant during protein purification, and at several steps during the procedure, the need will arise for concentration of the active extract or fraction. Concentration may occur as a concomitant of a step in purification (e.g., fractional precipitation [*see* Chapter 13] or ion-exchange chromatography [*see* Chapter 14]); in the latter case, protein from a dilute solution may be absorbed onto the resin and subsequently eluted in a smaller volume by application of a salt gradient. In general, however, one or more steps specifically aimed at concentration of the protein solution will need to be done.

For concentration of initial cell extracts, particularly when working on a large scale (1–5 L of homogenate), the method of choice is precipitation by ammonium sulfate. Virtually all proteins are precipitated from solution at sufficiently high salt concentrations. This arises from the fact that protein surfaces tend to have hydrophobic patches, which, in solution, are surrounded by ordered water molecules. When salt is added to the protein solution, water is recruited to solvate the ions of the dissociated salt, thus progressively exposing the hydrophobic regions of the protein surface. At some point, these patches will start to interact, leading to aggregation and, ultimately, to precipitation of the proteins from solution (for a more detailed discussion, *see* **ref. *1***). The requirements for the salt are that it be highly soluble in water, that its component ions be innocuous to proteins, and that it has a low heat of solution; ammonium sulfate satisfies these criteria most completely. The technique is, then, to precipitate the proteins from solution by the addition of ammonium sulfate, recover the proteins by centrifugation, resuspend in the minimum amount of water or buffer, and remove residual ammonium sulfate by dialysis.

A second technique for protein concentration that is also applicable to relatively large volumes, but that is usually used at later stages in a purification protocol (e.g., after column chromatography) is forced dialysis or ultrafiltration. In this, use is made of semipermeable membranes, which allow passage of water and other small molecules but not

From: *Methods in Molecular Biology, vol. 244:Protein Purification Portocols: Second Edition*
Edited by: P. Cutler © Humana Press Inc., Totowa, NJ

of proteins. If a protein solution is placed in a bag of such material and pressure is applied, then small molecules will be forced out and the protein molecules retained. The rate of passage of small molecules will, of course, be severely reduced by precipitation of protein onto the walls of the dialysis bag; this is quite likely to occur with crude protein solutions, and it is for this reason that the method is not generally used at early stages in a purification procedure.

The particular version of membrane filtration described in the present chapter requires a minimal amount of specialized equipment and can be used to concentrate relatively large volumes of solution in an overnight experiment. A variant of the method that requires the use of commercial pressure cells but has the advantage that it can be used for fractionation as well as for concentration is described in Chapter 12. For small protein samples, drying by lyophilization is an alternative means of removal of water but has the disadvantage that buffer ions (unless volatile) will remain in the sample; this technique is described in Chapter 32.

2. Materials

2.1. Precipitation With Ammonium Sulfate

1. Ammonium sulfate (preferably Analar grade).
2. Electric paddle stirrer.
3. Refrigerated centrifuge and rotor (e.g., a 6×250-mL angle rotor).
4. Screw-top plastic tubes for the rotor to be used.
5. Visking dialysis tubing (14- or 19-mm inflated diameter; 26- or 31-mm flat width). This is produced by Serva and available from most suppliers of laboratory materials.

2.2. Forced Dialysis

1. Heavy-walled Buchner flask (in the range 500-mL to 5-L capacity depending on the volume of solution to be concentrated) (*see* **Note 1**).
2. Rubber bung to fit the flask and bored to accommodate the stem of a glass funnel.
3. Glass funnel (*see* **Fig. 1** and **Note 2**).
4. Visking dialysis tubing (6-mm inflated diameter; 10-mm flat width).
5. Efficient water vacuum pump.

3. Methods

3.1. Precipitation With Ammonium Sulfate

1. Measure the volume of the protein solution to be concentrated and pour it into a glass beaker of capacity about twice the measured volume of solution.
2. Weigh out 0.5 g of solid ammonium sulfate for every 1 mL of protein solution. If the ammonium sulfate contains lumps, then these should be broken up using a pestle and mortar (*see* **Note 3**).
3. Place the beaker containing the protein on ice and stir either manually or with a paddle stirrer. A slow rate of stirring should be used to avoid foaming.
4. Add the ammonium sulfate to the protein solution in small batches over a period of several minutes, ensuring that one lot of ammonium sulfate has dissolved before adding the next.
5. After addition is complete, leave to stand for 10 min to ensure complete precipitation and then recover precipitated protein by centrifugation at about 5000*g* for 30 min using screw-top plastic tubes (*see* **Note 4**).

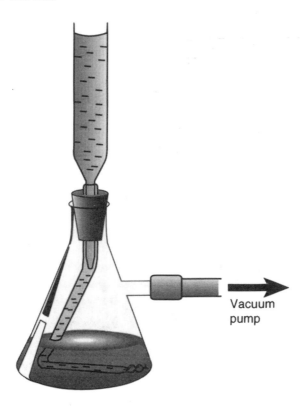

Fig. 1. Experimental arrangement for concentration by forced dialysis. The dialysis tubing should extend up the stem of the funnel so that it is above the top of the bung. In this way, it will be gripped tightly and will not slip off when the vacuum is applied.

transfer
6. Decant off the supernatant solution from each tube and suspend the protein pellets in the minimum volume of water or of an appropriate buffer (*see* **Note 5**).
7. Transfer the protein suspension to a dialysis sack and dialyze against at least two changes of buffer using 100 times the volume of the sample and allowing 3–4 h for equilibration (*see* **Note 6**).
8. After dialysis, remove any precipitated material by centrifugation.

3.2. Forced Dialysis

1. Cut a length of the 6-mm-diameter dialysis tubing (*see* **Note 7**) about twice as long as the height of the flask (*see* **Note 1**).
2. Soak the dialysis tubing in the buffer to be used for a few minutes (*see* **Note 7**). Tie two firm knots in one end. Open the other end by rubbing gently between the fingers and then push the open end onto the stem of the funnel (*see* **Note 2**) so that the tubing extends 3–4 cm up the stem. The stem of the funnel should be moistened with buffer to facilitate this and a towel used to grip the dialysis tubing. Care must be taken not to tear the dialysis tubing.
3. Pour buffer (*see* **Note 9**) into the Buchner flask to a depth of about 5 cm. Thread the knotted end of the dialysis tubing through the hole in the rubber bung, pull through, and then push the stem of the funnel through the hole until the end has projected through. Place the bung in the flask (*see* **Fig. 1**).

4. Evacuate the flask using a water pump (*see* **Note 10**), seal the outlet, and test for leaks in the system by rotating and inverting the flask so that all parts of the dialysis tubing are sequentially immersed in the buffer. Pinhole leaks will be apparent from a stream of air bubbles emerging from the tubing. If any are present, reject the piece of tubing and start again.

5. Release the pressure and pour protein solution into the funnel. Air will usually be trapped in the tubing. To remove this, take the bung out of the flask and gently squeeze the tubing to expel air bubbles.

6. Replace the tubing and bung in the flask, evacuate, and then seal off the outlet. Place the flask in a cold room overnight or until the volume of solution has decreased to the desired extent (*see* **Note 11**).

7. Release the vacuum, cut off the dialysis tubing from the funnel, and transfer the concentrated solution to an appropriate container (*see* **Note 12**).

4. Notes

1. The optimum size of flask and length of tubing to be used depend on the scale on the experiment to be conducted. For volumes of the order of 100–200 mL, a 500-mL flask is ideal and the total length of tubing in this case would be about 25 cm. A final volume of concentrate of 2–3 mL is easy to achieve. For very large volumes, a 5-L flask should be used with correspondingly long lengths of dialysis tubing (otherwise, concentration will take a very long time). With large flasks, it is perfectly feasible to mount up to three funnels in the bung and thus increase the capacity and speed of the system.

2. Funnels of the type shown in **Fig. 1** can be made from heavy glass tubing of about 25-mm diameter drawn down to a stem of about 7-mm diameter so that the dialysis tubing is a tight push fit; to do this requires an experienced glass-blower. An alternative is to purchase cylindrical dropping funnels of the appropriate size from a laboratory supplier. These will have a tap between the stem and the cylindrical part, but this is not a problem. Whatever the type of funnel used, it is crucial that the end of the stem be fire-glazed; otherwise, the sharp glass edge will inevitably cut the dialysis tubing.

3. This amount of ammonium sulfate gives a concentration of about 75% saturation at 0°C, which should be high enough to precipitate most proteins. An initial experiment should be carried out to confirm that the protein of interest is indeed precipitated at this concentration. Higher values can be used (up to a maximum of 0.7 g/mL, which corresponds to 100% saturation), but this increases the density of the solution and makes it more difficult to sediment the protein.

    Note that strong solutions of ammonium sulfate are acidic, so it is important that the protein solution be buffered at a neutral pH and a buffer concentration of 0.05–0.1 *M*. Note also that ammonium sulfate (even Analar grade) contains traces of metal ions; hence, if the protein of interest is metal sensitive, EDTA (10 m*M*) should be added to the protein solution.

4. Care must be taken with this step to protect both the centrifuge and the operator! The tubes must be balanced to within 0.1–0.2 g across the rotor axis (*see* **Note 3** of Chapter 2), and it is particularly important not to counterbalance a tube full of 75% ammonium sulfate with a tube full of water because of the large difference in density. Either the ammonium sulfate/protein suspension should be divided between two tubes or the suspension should be balanced using a solution of ammonium sulfate of the same concentration. It is also most important to avoid spillage of ammonium sulfate solutions into the centrifuge head. Such solutions are extremely corrosive to the materials of which rotors are constructed and can lead to irreversible damage, rendering the rotors unsafe to use. After centrifugation of ammonium sulfate (or other salt) solutions, rotors should always be removed and washed in warm water.

Note that proteins precipitated in 70% ammonium sulfate are usually stable and, hence, it is often convenient to store the suspension overnight at 4°C in this form before proceeding with the purification schedule. Indeed, pure proteins are often stored at −20°C in 70% ammonium sulfate for long periods.

5. This step is crucial to achieving a satisfactiory degree of concentration. Add a very small volume of water or buffer (about 10 mL if using a 250-mL tube) and resuspend the protein pellet using a glass rod. If several centrifuge tubes have been used, then transfer the protein suspension from tube to tube, resuspending the pellet each time; only add more water or buffer if the suspension becomes too thick to transfer readily. After transfer of the final suspension to the dialysis bag, a further small aliquot of water or buffer can be used to rinse out the tubes and the rinsings combined with the suspension. A common error is to attempt to redissolve the protein pellets after centrifugation; because the pellets contain considerable quantities of ammonium sulfate, this takes a large volume of buffer and it is easy to end up with a volume comparable to that at the outset!

6. Visking tubing comes in several sizes. For a given volume of solution, the larger the diameter of the dialysis tubing used, the shorter the piece that will be required; attainment of equilibrium, however, will be slower with short, fat bags than with long, thin ones. The sizes recommended are a compromise between these two factors.

For most applications, it is only necessary to soak the dry tubing in water or in the buffer and it is ready to be used. The tubing does, however, contain significant quantities of sulfur compounds and of heavy metal ions; the latter may be a problem if the protein of interest is metal sensitive. They may be removed by boiling the dialysis tubing in 2% (w/v) sodium bicarbonate 0.05% (w/v) EDTA for about 15 min, washing with distilled water, and then boiling in distilled water twice for 15-min periods. Prepared tubing can be stored indefinitely at 4°C in water or buffer containing 0.1% (w/v) sodium azide.

For most applications, Visking tubing is perfectly adequate, but problems will arise if the protein of interest is small. The pores in this type of tubing are of such a size that the nominal molecular-weight cutoff (NMWC) is about 15,000, although larger proteins may still pass through the pores if they have an elongated shape. As a rule of thumb, Visking tubing can be used with confidence if $M_r > 20,000$. For dialysis of smaller proteins, Spectropor tubings with a range of NMWC values starting at 1000 are available (Serva), and the appropriate tubing should be selected. These tubings are much more expensive than Visking tubing, and the rate of dialysis decreases with pore size; hence, they should only be used if strictly necessary.

To remove ammonium sulfate from the protein suspension, a piece of dialysis tubing with a volume about twice that of the suspension is taken and securely closed with a double knot at one end. The suspension is then poured into the bag using a funnel, air is removed from the top part of the bag by running it between the fingers, and the top secured with a double knot. It is very important to have this space in the bag to allow for expansion, because water will flow in while the internal salt concentration is high. If insufficient space is left, the bag can become very tight because of this inflow. Bags rarely burst because the membranes are quite strong, but they are tricky to open in this state; the best way is to insert one end into a measuring cylinder and then prick the bag with a scalpel.

Equilibrium will be reached in about 3–4 h, but only if the system is stirred; otherwise, about 6 h should be allowed. Care should be taken when stirring to ensure that the magnetic pellet or stirring paddle does not tear the dialysis bag.

The volume of dialysis solution to be used depends, of course, on the sample volume and the final concentration of ammonium sulfate desired. If a ratio of 1:100 (sample:dialysis fluid) is used, then, at equilibrium, the ammonium sulfate concentration will have been reduced 100-fold. (This is not strictly true because it ignores the Donnan effect, but will do as an approximation.) A second dialysis will then result in a total decrease of 10,000-fold and so on.

The buffer to be used for dialysis is usually dictated by the requirements of the next step in the purification schedule. For example, if this is to be ion-exchange chromatography, then column equilibration buffer is the logical choice.

7. The question of NMWC is also important here and if the protein of interest has an M_r <20,000, then one of the Spectropor tubings must be used. This will decrease the rate of concentration, but reduce losses of material.

    Passage of low-molecular-weight material through the dialysis tubing can sometimes be used to advantage if it is required to carry out a crude separation of large and small components of a mixture. For example, an effective separation of the peptide components from the protein components of bee venom was achieved using forced dialysis through Visking tubing *(2)*.

8. If desired, the tubing can be pretreated as described in Note 6 to remove metal ions.

9. The buffer chosen will usually be that for the next step in the purification schedule. It should be noted, however, that the composition of the buffer will change during ultrafiltration because of the efflux of buffer ions originally in the sample so that it cannot be asssumed that the concentrated sample is properly equilibrated for further purification.

10. Do not use an oil pump; otherwise, it is likely that the pressure will be reduced too much and the dialysis bag may break (or the flask may implode, but this is not likely if it is made of heavy glass). As a useful guide, evacuate the flask until bubbles start to form in the buffer in the flask; this will occur at about 15 mm Hg depending on the ambient temperature.

    Care should always be taken when using glass vessles under low pressure. Ideally, a cage should be used, but as a minimum, safety glasses should be worn. Do not drop evacuated flasks!

11. Using a single flask of 2-L capacity with a single funnel and about 40 cm of tubing, it should be possible to reduce 500 mL of protein solution to about 5 mL overnight. If the capacity of the funnels available is too small to take the amount of protein solution to be concentrated, then it is simple to set up a syphon between the solution in the funnel and the extra solution in an external reservoir. Using a single 5-L flask with three funnels, each one being replenished from a reservoir by means of a syphon, it is possible to concentrate 2 L of dilute protein solution down to about 20 mL overnight.

12. If left too long, the retentate may reduce in volume to such an extent that it is necessary to wash the protein out of the tubing with a small amount of buffer. On occasion, pure proteins may even crystalize in the tubing (*see* **ref.** *3* for an example); in this case, care should be exercised in handling the tubing so that the crystals do not puncture it.

References

1. Englard, S. and Seifter, S. (1990) Precipitation techniques. *Methods Enzymol.* **182,** 285–300.
2. Shipolini, R. A., Callewaert, G. L., Cottrell, R. C., Doonan, S., Vernon, C. A., and Banks, B. E. C. (1971) Phospholipase A from bee venom. *Eur. J. Biochem.* **20,** 459–468.
3. Barra, D., Bossa, F., Doonan, S., Fahmy, H. M. A., Martini, F., and Hughes, G. J. (1976) Large-scale purification and some properties of the mitochondrial aspartate aminotransferase from pig heart. *Eur. J. Biochem.* **64,** 519–526.

11

Making and Changing Buffers

Shawn Doonan

1. Introduction

Control of pH is a central consideration in handling proteins, and this requires the use of buffer solutions. Furthermore, a typical protein purification schedule will contain several steps generally needed to be carried out in buffers of different pH values and ionic strengths (*see* **Note 1**). Hence, it is necessary to have available methods for changing buffers. The present chapter reviews the properties of buffers and preparation of buffer solutions and then deals with ways in which a protein solution can be changed from one buffer to another.

Buffer solutions consist of a weak acid and a salt of that acid, or of a weak base and a salt of that base. For example, in a solution of a weak acid HA and its sodium salt, the following occur:

$$HA \rightleftharpoons H^+ + A^- \quad \text{(incomplete)} \tag{1}$$

$$NaA \rightarrow Na^+ + A^- \quad \text{(complete)} \tag{2}$$

The species A^- is referred to as the conjugate base of the acid HA.

The addition of H^+ to the buffer moves equilibrium (1) to the left using A^- supplied by equilibrium (2), whereas added OH^- (or other base) combines with H^+ provided by equilibrium (1) moving to the right; in either case, change of H^+ concentration and, hence, of pH is resisted. The Henderson–Hasselbalch equation describes the relationship among the pH, the pK_a of the buffer, and the relative concentrations of the free acid and of the salt as follows (*see* **Note 2**):

$$pH = pK_a - \log[(\text{acid})/(\text{salt})] \tag{3}$$

Hence, if, for example, a solution of a weak acid of concentration (a) M is partially neutralized by addition of a strong base to a concentration of (b) M, then the result is a buffer solution where the pH is given by

$$pH = pK_a - \log\{[(a) - (b)]/(b)\} \tag{4}$$

It is immediately apparent from Eq. 4 that when (b) = 0.5(a), then pH = pK_a, that is, the pH is numerically equal to the pK_a value. The equation also suggests that the pH of

From: *Methods in Molecular Biology, vol. 244: Protein Purification Protocols: Second Edition*
Edited by: P. Cutler © Humana Press Inc., Totowa, NJ

Table 1
pH and Buffer Values of a 0.1 *M* Solution of a Weak Acid (pK_a = 7)
as a Function of Concentration of an Added Base (b)

(b) *M*	pH	Buffer value
0.005	5.72	0.0109
0.010	6.05	0.0207
0.020	6.40	0.0368
0.030	6.63	0.0484
0.040	6.82	0.0553
0.050	7.00	0.0576
0.060	7.18	0.0553
0.070	7.37	0.0484
0.080	7.60	0.0368
0.090	7.95	0.0207
0.095	8.28	0.0109

a buffer is unaffected by dilution; this is true to a first approximation for some buffers, but with others, the pH changes quite markedly with dilution (*see* **Note 3**).

What is not so obvious is that the ability of the buffer to resist changes of pH is maximum at the pK_a and falls off on either side such that the buffering power is small to negligible outside the range p$K_a \pm 1$. This can be seen by considering the parameter β (the buffer capacity or buffer value), which is defined as

$$\beta = \frac{d(b)}{d\mathrm{pH}} \tag{5}$$

and is given by

$$\beta = 2.303\left\{\frac{(a)K_a(H^+)}{[K_a + (H^+)]^2}\right\} + 2.303[(H^+) + (OH^-)] \tag{6}$$

where (a) is the total concentration of the free acid plus the salt. In the pH range of approx 3–11, the value of β is determined entirely by the first term in the equation, and under those circumstances, a more convenient form is

$$\beta = 2.303\{(b)[(a) - (b)]/(a)\} \tag{7}$$

This is obtained by taking the first derivative of the Henderson–Hasselbalch equation and inverting it. The buffer value is a measure of the amount of base needed to produce a unit change in the pH of the buffer, and the larger its value, the greater the the resistance to pH change. **Table 1** shows the buffer values for a solution of a weak acid to which various quantities of base have been added and emphasizes the restricted range of pH values over which the solution has good buffering properties. Obviously, the stronger the buffer, the greater the buffer value; however, for practical reasons, buffers of strength >0.1 *M* are not usually used.

There are, then, a variety of considerations in selecting a buffer for a particular application. These include the following:

1. The desired pH: The pK_a of the buffer must be as close to this as possible and certainly not outside the range pH $\pm$ 1.
2. The pH of the buffer should change as little as possible with temperature, with dilution, and with added neutral salt (*see* **Note 4**).
3. The buffer should be chemically unreactive.
4. The buffer should not absorb light at 280 nm, particularly if it is to be used for chromatographic procedures where column monitoring will usually be carried out by absorbance measurements.
5. For cation ion-exchange chromatography, an anionic buffer should be used and vice versa.
6. Buffers that might interact with components of the protein mixture (e.g., borate with glycoproteins) should be avoided.

   Table 2 gives a list of commonly used buffer compounds with their pK_a values; values for other substances are given in **ref. 1**. Included in the table are a variety of "Good" buffers (MES, ADA, PIPES, ACES, BES, MOPS, HEPES, TAPS, CHES, CAPS). These zwitterionic buffers are chemically unreactive, do not absorb at 280 nm, and their pH values vary only slightly with temperature and dilution *(2)*. They are, therefore, ideal buffers in many respects, but are very expensive (*see* **Note 5**) and their use is often not feasible when large volumes of buffer are required for procedures such as ion-exchange chromatography and dialysis.

   As mentioned, there will frequently be a need during protein purification for changing the buffer in which a protein mixture is dissolved; this may be to change the pH, the buffer ionic strength, or the concentration of a neutral salt. There are two major ways in which this can be done. The first of these is dialysis, in which the protein solution is enclosed in a bag of a semipermeable membrane (i.e., one that allows the passage of small molecules but not of large ones) and is then equilibrated in two or more changes of a large excess of the target buffer solution. Ultimately, equilibrium will be reached where the internal and external buffers may approximate to the same pH and concentration (*see* **Note 6**). A related technique uses membrane ultrafiltration (UF); this is described in Chapter 12.

   The second method is gel filtration. This uses a column of porous beads designed such that water and low-molecular-weight solutes have access to the interior of the beads, whereas larger molecules do not. In gel filtration, the solution of protein is applied to such a column equilibrated in the target buffer and the proteins are then eluted with target buffer. The protein molecules will move ahead of the buffer in which they were originally dissolved and will emerge in the target buffer. This technique can also be used as a method for fractionation of protein mixtures on the basis of size, as described in Chapter 26.

2. Materials

2.1. Preparation of Buffers

1. pH meter capable of accuracy at two decimal places and with temperature compensation (*see* **Note 7**).
2. Magnetic stirrer and pellet.
3. Standard buffer solutions with pH values bracketing that of the buffer to be made. These can either be prepared using the information in **Table 3** or purchased from most suppliers of chemicals.
4. Analar HCl or NaOH pellets, depending on the buffer to be made.

Table 2
Buffer Compounds and Their pK_a Values at 25°C

Compound	Trivial name	pK_a
Phosphoric acid (pK_1)	—	2.15
Glycine (pK_1)	—	2.35
$Na_2 B_4 O_7$ (pK_1)	—	3.02
Citric acid (pK_1)	—	3.13
Formic acid	—	3.75
Succinic acid (pK_1)	—	4.21
Fumaric acid (pK_2)	—	4.38
Citric acid (pK_2)	—	4.76
Acetic acid	—	4.76
Succinic acid (pK_2)	—	5.64
2-(*N*-Morpholino)ethanesulfonic acid	MES	6.10
Carbonic acid (pK_1)	—	6.35
[Bis(2-hydroxyethyl)imino]-tris(hydroxymethyl)methane	Bis-Tris	6.46
N-2-Acetamidoiminodiacetic acid	ADA	6.59
Piperazine-*N*,*N'*-bis(2-ethanesulfonic acid)	PIPES	6.76
N-2-Acetamido-2-hydroxyethanesulfonic acid	ACES	6.78
Imidazole	—	6.95
N,*N*-bis-(2-hydroxyethyl)-2-aminoethanesulfonic acid	BES	7.09
3-(*N*-Morpholino)propanesulfonic acid	MOPS	7.20
Phosphoric acid (pK_2)	—	7.20
N-2-Hydroxyethylpiperazine-*N'*-2-ethanesulfonic acid	HEPES	7.48
N-Ethylmorpholine	—	7.67
Triethanolamine	—	7.76
Tris(hydroxymethyl)aminomethane	Tris	8.06
N-[Tris(hydroxymethyl)methyl]glycine	Tricine	8.05
N,*N*-bis(2-hydroxyethyl)glycine	Bicine	8.26
3-{Tris(hydroxymethyl)methyl]-amino}propanesulfonic acid	TAPS	8.40
2-Amino-2-methylpropan-1,3-diol	—	8.79
2-Aminoethylsulfonic acid	Taurine	9.06
Boric acid	—	9.23
Ammonia	—	9.25
Ethanolamine	—	9.50
Cyclohexylaminoethanesulfonic acid	CHES	9.55
3-Aminopropanesulfonic acid	—	9.89
β-Alanine	—	10.24
Carbonic acid (pK_2)	—	10.33
3-(Cyclohexylamino)propanesulfonic acid	CAPS	10.40
γ-Aminobutyric acid	—	10.56
Piperidine	—	11.12
Phosphoric acid (pK_3)	—	12.33

2.2. Dialysis

1. Visking dialysis tubing (either 14- or 19-mm inflated diameter; 26- or 31-mm flat width). This is produced by Serva and is available from most suppliers of laboratory materials.
2. Target buffer (at least 200 times the volume of protein solution is required).

Table 3
Standard Buffer Solutions

Buffer	Composition (g/L)	Concentration (M)	pH		
			5°C	15°C	25°C
Phthalate	10.12 g of KHC$_8$H$_4$O$_4$	0.05	4.00	4.00	4.01
Phosphate	3.39 g of KH$_2$PO$_4$ + 3.53 g of Na$_2$HPO$_4$	0.025	6.95	6.90	6.87
Borate	3.80 g of Nap$_2$B$_4$O$_7$·10 H$_2$0	0.01	9.40	9.28	9.18

2.3. Gel Filtration

1. Sephadex G-25 (medium grade) (*see* **Note 8**). About 1 g will be required for every 5-mL column volume. This material is produced by Amersham Biosciences but is available from general suppliers.
2. Appropriate size chromatography column. The packed volume will need to be about five times greater than the volume of protein solution to be treated (*see* **Note 9**).
3. Peristaltic pump, fraction collector, and absorbance detector (optional—*see* Chapter 38).
4. Target buffer (about 10 times the volume of the column bed).

3. Methods

3.1. Preparation of Buffers

It is usually convenient to make stock solutions of buffers that can then be diluted for use (*see* **Note 3**). Recipes are available for many of the more common buffers (*see* **ref. 1**), or compositions can be calculated using the Henderson–Hasselbalch equation. More usually, however, buffers are made by weighing out the required amount of the buffering substance, dissolving in water, and then adjusting the pH to the desired value by adding HCl or NaOH as appropriate; other acids or bases may, of course, be used. For example, to make 1 L of 2 *M* sodium acetate buffer, pH 5.0, proceed as follows.

1. Standardize the pH meter using standard buffers of pH 4.01 and 6.87, assuming that the buffer is to be made at room temperature.
2. Weigh out 2 mol (120 g) of glacial acetic acid and transfer to a 1-L beaker.
3. Add about 800 mL of distilled water, place on a magnetic stirrer, and insert the pH meter electrodes.
4. Slowly add solid NaOH pellets (preferably from a freshly opened bottle to ensure that they are not contaminated with sodium carbonate), making sure that one lot dissolves before adding the next. When the pH has reached about 4.8, allow the buffer to cool to room temperature (heat will have been generated by reaction of acetic acid with the NaOH), and then, finally, adjust the pH to 5.0 using a solution (about 4 *M*) of NaOH.
5. Transfer the solution quantitatively to a graduated flask or measuring cylinder (the latter is sufficiently accurate) and make up the volume to 1 L.
6. Store in a glass container at 4°C. Preservatives are not necessary because the high concentration inhibits bacterial growth.

3.2. Changing Buffers by Dialysis

It should be kept in mind that proteins are polyelectrolytes whose state of ionization varies with pH because of their content of ionizable side chains. Hence, particularly if the protein concentration is high and the solution is at a pH far removed from that of the target buffer, it is not possible to equalize the pH and ionic strength values of the internal and external buffers by dialysis because of the Donnan effect (*see* **Note 10**). In the following, it is assumed that dialysis will be carried out in a cold room or refrigerator at about 5°C using a diluted stock buffer.

1. Calibrate the pH meter with standard buffers (*see* **Table 3**) cooled to 5°C.
2. Dilute stock buffer solution to the desired concentration using cold distilled water and check that the pH has not been altered by dilution or the decrease in temperature. If it has, adjust it with the acidic or basic component of the buffer as appropriate. The volume of buffer should be about 100 times that of the sample to be dialyzed.
3. If the target buffer is to contain a neutral salt, then add this before checking and adjusting the pH.
4. Take a length of Visking dialysis tubing able to hold about 1.5 times the volume of the protein solution and soak it in the buffer for a few minutes. For some applications, it may be desirable to pretreat the dialysis tubing and/or use tubing with a lower nominal molecular-weight cutoff (NMWC) (*see* **Note 6** of Chapter 10).
5. Tie two tight knots in one end of the tubing and then pour in the protein solution with the aid of a funnel. Squeeze out air above the protein solution to allow room for expansion and seal the bag with two more knots.
6. Place the dialysis bag in the buffer solution and leave for at least 4 h in the cold to reach equilibrium. If a magnetic stirrer is used, be sure that the pellet does not tear the dialysis bag.
7. Replace the buffer with a fresh lot and repeat the dialysis for another 4 h.
8. Check the pH of the dialyzed solution and, if necessary, adjust it to the target value by very careful addition of dilute acid or base (*see* **Note 10**). Use constant stirring to ensure that local high concentrations of acid or base do not occur.
9. Remove any insoluble material from the dialyzed sample by centrifugation at about 5000*g* for 15 min.

3.3. Changing Buffers by Gel Filtration

This is a more rapid procedure (if a column is already available) and, hence, to be preferred particularly if time is important (e.g., if the protein of interest is unstable). It should not be used, however, with crude protein mixtures because there is a strong possibility that protein will precipitate during the procedure and ruin the column.

1. Take 1 g of Sephadex G-25 (medium grade) for every 5 mL of desired column volume and stir carefully into 10 vol of the target buffer. Leave at 5°C overnight for the gel to swell to its maximum extent. Resuspend by stirring, allow to settle, and remove any fine suspended material by aspiration.
2. Resuspend the gel in about 2 vol of buffer and pour into the chromatography column (*see* **Note 11**) with the outflow blocked. Allow a few centimeters of settled bed to form, and then start flow through the column at the rate to be used for gel filtration (*see* **Note 12**). As clear liquid forms above the gel suspension, remove it by aspiration and replace it with fresh suspension. After the gel bed has reached the desired height, attach a buffer reservoir and pass two column volumes of buffer through the column to ensure equilibration of the bed.
3. Apply the sample to the column and then continue elution with about one column volume of the target buffer. Collect appropriate sized fractions (about 1/20th of the column volume).

4. If using an automatic absorbance monitor, the fractions of interest will be obvious. Otherwise, measure the absorbance of individual fractions at 280 nm and combine fractions containing protein (*see* **Note 13**).
5. Check the pH of the combined fractions and, if necessary, adjust as described in **Subheading 3.2., step 8** (*see* **Note 14**).
6. Re-equilibrate the column by passage of two column volumes of buffer.

4. Notes

1. Sometimes, by judicious choice of the sequence of steps, it is possible to avoid changing the buffer between one step and the next. This is desirable because it saves time and can improve the overall yield of the purification. An example of where this approach has been used is given in **ref. *3***.
2. This equation is strictly only valid in the pH range 3–11. Outside this range, the self-ionization of water becomes significant.
3. The reason for this is that the concentration terms in the Henderson–Hasselbalch equation should properly be thermodynamic activities; these are concentration-dependent to varying extents.
4. K_a values are temperature dependent and, hence, so are the pH values of buffer solutions. The magnitude of the effect depends on the particular buffer substance chosen. Similarly, added salt can affect the pH of a buffer because of differential effects on the thermodynamic activities of the component buffer ions. If possible, buffers should be selected where these effects are minimal (*see* **ref. *1***).
5. Prices of "Good" buffers are in the range $300–750/kg. Common organic acids and bases used as buffers cost about one-tenth of this, and inorganic buffers are even less expensive.
6. This is not strictly true because of the so-called Donnan equilibrium. The protein, which is a nonpermeant polyelectrolyte, will affect the distribution of the permeant ionic species. Consider a protein solution volume V_1 at a pH such that it is negatively charged and suppose that the concentration of charges on the protein is $C_1 M$; these negative charges will be neutralized by positive ions (e.g., Na^+) also at a concentration of $C_1 M$. If this solution is enclosed in a dialysis bag and placed in a volume V_2 of NaCl solution concentration $C_2 M$, then NaCl will cross the membrane until equilibrium is attaine (i.e., until the activities of the NaCl in the two compartments are equal). It can be shown that, to a first approximation, the equilibrium condition is

$$(Na^+)_{in} \times (Cl^-)_{in} = (Na^+)_{out} \times (Cl^-)_{out} \tag{8}$$

where in and out refer to inside and outside, respectively, of the dialysis bag. If the change of concentration of NaCl inside the dialysis bag is $+x M$, then it follows that the change outside will be $-x (V_1/V_2) M$ and the equilibrium condition is

$$x(C_1 + x) = \{C_2 - [x(V_1/V_2)]\}^2 \tag{9}$$

If, for example, $C_1 = 20$ mM, $V_1 = 10$ mL, $C_2 = 100$ mM, and $V_2 = 1000$ mL, then $x = 89.6$ mM; that is, the NaCl concentration inside the dialysis bag at equilibrium will be 89.6 mM and that outside of the bag will be 99.1 mM. Repeating dialysis with a fresh NaCl solution will result in only a marginal increase of internal NaCl concentration to 90.5 mM. This example is, perhaps, a little extreme because it would require a 50-mg/mL solution of protein, average $M_r = 50,000$, each molecule of which carried a net 20 negative charges. It does, however, serve to illustrate the point that the distribution of salts, including buffer ions, in dialysis will be affected by the presence of protein and that even very extensive dialysis will not achieve equality of internal and external concentrations. One consequence of this is that dialyzed solutions of high and low concentrations of the same protein mixture will not behave identically on, for example, ion-exchange chromatography because their ionic strengths and pH values will differ.

7. Note that the temperature compensator corrects for the conversion of measured electromotive force (emf) to pH and not for the variation of buffer pH with temperature. If Tris buffers are to be used and very accurate pH values are required, then special glass electrodes are needed that do not respond to Tris itself; these are available from several suppliers. Care must be taken of glass electrodes if good pH measurements are to be made. They must never be allowed to dry out (store in saturated KCl solution), and protein must not be allowed to accumulate on them (wash with detergent solution if this occurs).

8. This exclusion limit is appropriate for proteins of $M_r > 20,000$. For smaller proteins, Sephadex G-10 should be used. This latter material produces a smaller bed volume per gram dry weight and hence is more expensive to use.

9. For desalting or changing buffers with small volumes of protein solution (up to about 2 mL), Amersham Biosciences markets prepacked Sephadex G-25 columns in either disposable or reusable form. These are very convenient, but moderately expensive. Information can be obtained from the company's technical publications.

10. The situation is complex and it is very difficult to analyze precisely because of the change in net charge of a protein with changing pH. However, it should be clear from the discussion in **Note 6** that, because proteins will carry a net charge at all pH values except the isoelectric point, the presence of protein inside the dialysis bag will affect the distribution of ions, including those of the buffer. Therefore, for example, dialysis of a negatively charged protein mixture with a buffer formed from a weak acid will lead to a situation where the concentration of conjugate base is lower inside than outside and, hence, the pH inside will be lower than in the bulk buffer. The higher the protein concentration and the greater the difference between the starting pH of the protein solution compared with the target pH, the greater the effect will be. Hence, it is not possible to obtain complete equality of pH and buffer concentration by dialysis even for relatively dilute protein solutions. Given that for most applications the pH rather than the ionic strength is the key factor, it is best to dialyze to equilibrium and then to adjust the pH to the target value if necessary by the addition of acid or base.

11. The ratio of the length to the diameter of the column should be in the range 10:1–20:1. Longer but narrower columns may give slow flow rates, whereas short, fat columns may give poor separations and greater dilutions of the emergent protein solution because of imperfections in the packing. Pharmacia Biotech markets columns of these dimensions, as do other companies, but homemade columns provide a less cheaper satisfactory alternative (*see* Chapter 38).

12. Flow rates should be approx 0.1–0.2 mL/min/cm² of cross-sectional area. Faster flow rates will lead to incomplete exchange between the bulk solvent and the beads, whereas slower values will result in more band spreading and dilution.

13. If a column is to be used several times for desalting or for exchange of buffers, then it may be convenient to calibrate it by passing through a mixture of a standard protein (e.g., bovine serum albumin) and a salt (e.g., NaCl) and determining the volume ranges over which the protein and the salt emerge. For detecting the protein, absorbance values at 280 nm are used, whereas detection of the presence of NaCl could be done by the addition of silver nitrate to aliquots of the fractions and observing the precipitation of silver chloride; if the buffer contains chloride or phosphate ions (silver phosphate is insoluble also), then some alternative low-molecular-weight substance such as acetone (which can be determined spectroscopically) would have to be used. Thereafter, provided the same equilibration buffer is used, the protein fraction can be collected simply on the basis of elution volume.

14. The point here is that protein solutions are themselves buffers by virtue of the ionizable side chains that they contain. In gel filtration, the protein may be transferred from a buffer at one pH to an only slightly larger (about 1.5 times) volume of a buffer at a different pH. Unless the second buffer is very strong, it is unlikely to have sufficient buffering capacity to effect the required change in pH, particularly if the starting and target pH values are widely different. For example, a change from pH 8.0 to 6.0 will require protonation of all or most of the histidine residues in a protein, thus withdrawing protons from the target buffer with a consequent rise in pH. Hence, again, the pH should be checked after gel filtration and adjusted if necessary.

References

1. Dawson, R. M. C., Elliot, D. C., Elliot, W. H., and Jones, K. M. (1986) *Data for Biochemical Research,* 3rd ed., Oxford University Press, Oxford, pp. 417–448.
2. Good, N. E., Winget, G. D., Winter, W., Connolly, T. N., Izawa, S., and Singh R. M. M. (1966) Hydrogen ion buffers for biological research. *Biochemistry* **5,** 467–477.
3. Cronin, V. B., Maras, B., Barra, D., and Doonan, S. (1991) The amino acid sequence of the aspartate aminotransferase from baker's yeast *(Saccharomyces cerevisiae). Biochem. J.* **277,** 335–340.

12

Purification and Concentration by Ultrafiltration

Paul Schratter

1. Introduction

Membrane ultrafiltration (UF) is a pressure-modified, convective process that uses semipermeable membranes to separate species in aqueous solutions by molecular size, shape, and/or charge. It separates solvents from solutes (i.e., the dissolved species) of various sizes. The result of removing solvent from a solution is solute concentration or enrichment. Repeated or continuous dilution and reconcentration are used to remove salts or exchange solvent (in such applications as buffer exchange). Definitions of some terms used in UF are given in **Table 1**.

Ultrafiltration is a low-pressure procedure, generally more gentle for the solutes than nonmembrane processes. It is more efficient than such processes and can simultaneously concentrate and desalt solutions. It does not require a phase change, which often denatures labile, species and can be performed at cold-room temperatures.

Ultrafiltration should be viewed as an excellent tool for efficient separation of biological substances into groups, according to molecular weight and size. For a finer separation, it must be followed by a more selective process, such as chromatography or electrophoresis.

1.1. Membrane Processes

Ultrafiltration is one of a spectrum of membrane separation techniques that include reverse osmosis, dialysis, and microfiltration. Reverse osmosis separates solvents from low-molecular-weight solutes (typically <100 Daltons) at relatively high pressures. Reverse osmosis membranes are normally rated by their retention of sodium chloride, whereas UF membranes are characterized according to the molecular weight of retained solutes.

Dialysis is a diffusive process employing a second liquid (dialysate) on the opposite side of the membrane from the sample. The permeation rate of molecules from sample to dialysate is in direct ratio to their concentration and inversely proportional to their molecular weights. Therefore, the rate of transport of a salt through the dialysis membrane diminishes as salt concentration in the sample declines during the process; thus, desalting by dialysis tends to be quite slow. In UF, all completely membrane-permeating species pass equally with the solvent, independent of their concentration.

From: *Methods in Molecular Biology, vol. 244: Protein Purification Protocols: Second Edition*
Edited by: P. Cutler © Humana Press Inc., Totowa, NJ

Table 1
Some Definitions of Terms Used in UF

Concentration
   Enrichment of a solution by solvent removal.
   The relative amount of a molecular species in a solution, expressed in percent.
Concentration polarization
   Accumulation of rejected solute on the membrane surface; depends on interactions of
      pressure, viscosity, crossflow (tangetial) velocity, fluid flow conditions, and temperature.
Cutoff (MWCO)
   The molecular weight at which at least 90% of a globular protein is retained by the
      membrane.
Fouling
   Irreversible decline in membrane flux because of deposition and accumulation of submicron
      particles and solutes on the membrane surface; also, crystallization and precipitation of
      small solutes on the surface and in the pores of the membrane, not to be confused with
      concentration polarization.
Permeate
   The solution passing through the membrane, containing solvent and solutes not retained by
      the membrane.
Rejection
   The fraction of solute held back by the membrane.
Retentate
   The solution containing the retained (rejected) species.
Yield
   Amount of species recovered at the end of the process as a percentage of the amount
   present in the sample

Source: Courtesy of Amicon, Inc.

Microporous membranes (microfilters) are generally rigid, continuous meshes of polymeric material with pore diameters that are two or three orders of magnitude larger than those of ultrafilters. Species are either retained on the membrane surface or trapped in its substructure (*see* **Fig. 1**). Microfilters retain bacteria, colloids, and particles upward of 0.025 μm in diameter, depending on the rated pore size. Pore sizes typically used for microfiltration are 0.22 and 0.45 μm.

1.2. UF Materials and Devices

Ultrafiltration is fundamentally a very simple process. It marries the selective permeability of a membrane structure to a device or system that applies the required pressure, minimizes buildup of retained material on the filter, and provides for access and egress of the fluid. The surface of the UF membrane contains pores with diameters small enough to distinguish between the sizes and shapes of dissolved molecules. Those above a predetermined size range are rejected, whereas those below that range pass through the membrane with the solvent flow.

1.2.1. Membrane Ultrafilters

Ultrafiltration membranes are made of various polymers. They generally have two distinct layers. On the side in contact with the sample or fluid stream is a very thin (0.1–1.5 μm) dense "skin" with extremely fine pores whose diameters are in the range of 10–400 Å (1×10^{-6} to 4×10^{-5} mm). Below this, is a much thicker (50–250 μm)

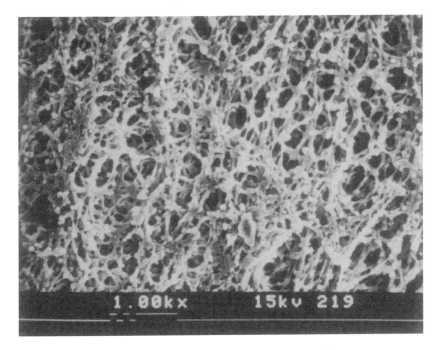

Fig. 1. Cross-section of microporous membrane. Particles are trapped on its surface or within pores. Electron micrograph.

open-celled substructure of progressively larger voids, largely open to the filtrate side of the ultrafilter (*see* **Fig. 2**). Any species capable of passing through the pores of the skin (whose sizes are precisely controlled in manufacture) can, therefore, freely pass the membrane. That arrangement provides a unique combination of selectivity and exceptional throughput at modest pressure. It resists clogging because retained substances are rejected at the smooth membrane surface.

Most membranes are cast on tough, porous substrates for improved handling qualities and repeated use. They offer dependably controlled retention, water permeability, and solute transport. The best membranes are inert, noncytotoxic, and do not denature biological materials.

Some membranes with high-flow characteristics are made of an inert, nonionic polymer. They do not adsorb ionic or inorganic solutes, but they may adsorb steroids and hydrophobic macromolecules. Advanced hydrophilic membranes have exceptionally low nonspecific protein-binding properties. They should be used where maximum solute recovery is of special importance.

1.2.2. Devices for UF

1.2.2.1. Stirred Cells

Pressurized cells are a convenient means of UF for volumes in the range of 3 mL to 2 L. They are capable of final concentrate volumes of 50 μL to 60 mL, with concentration factors typically in the range of 60- to 80-fold.

A stirred cell (*see* **Fig. 3**) is generally a vessel containing the solution to be filtered with a means of installing the membrane at the bottom, supported by a polymer grid.

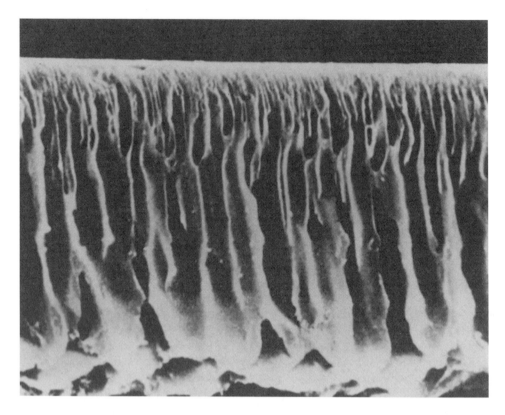

Fig. 2. Cross-section of anisotropic UF membrane. UF takes place in the top layer. Open-celled structure is highly permeable. Electron micrograph.

An access port permits pressurization, normally with nitrogen, in the range of 2.7–3.4 atm (40–50 psi). The pressure on the surface of the liquid forces the sample through the ultrafilter, where separation between retained and passing solutes takes place. Formation of a layer of retained material on the membrane surface is minimized by means of a magnetic stirrer that is propelled by mounting the cell on a magnetic stirring table. The gentle stirring action assures minimal exposure of labile solutes to degradation by shear effects. The cell can also be connected to a pressurized reservoir that continuously refills the cell as filtrate flows from it, for desalting or buffer exchange.

Stirred cells accommodate membranes of various diameters, normally from 25–90 mm. Choice of membrane diameter involves two conflicting aspects. The larger the membrane (and therefore the cell), the faster the run will be accomplished. However, if the molecules to be concentrated are dilute and their maximum recovery is very important, a large membrane and cell-wall surface area may expose them to nonspecific adsorption and, possibly, significant loss. Therefore, if speed is important, a large cell should be used; if high recovery from a dilute solution is vital, the smallest possible cell (and membrane area) should be employed. Because large membranes and cells are more costly, that may also be a factor in size selection. Normally, the choice is a trade-off between the two extremes.

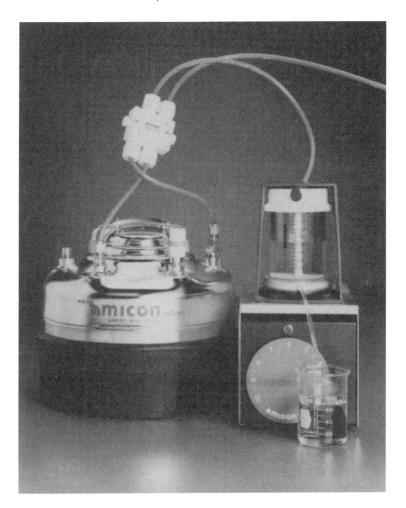

Fig 3. Stirred UF cell on magnetic stirrer. The reservoir and selector valve are used for desalting or buffer exchange.

Stirred cells are excellent for solutions with up to 10% solute concentration. They are used widely for concentration or desalting of dilute proteins, enzymes, polypeptides, viruses, yeasts, bacteria, and so forth.

1.2.2.2. CENTRIFUGAL DEVICES

A selection of devices is offered by UF equipment manufacturers that employs centrifugal force to exert the needed pressure on the sample to obtain UF. The smallest of these devices—for initial volumes in the microliter range and up to 2 mL—consist of small, capped sample reservoir tubes with the UF membrane sealed across the bottom, supported by a polymer grid. The tube fits into a vial to capture the filtrate (*see* **Fig. 4**). These units only require ordinary laboratory centrifuges, with centrifugal force of up to 14,000g. Use of a fixed-angle centrifuge rotor, rather than a swinging-bucket type, can reduce the time required for a separation run. Because centrifuge rotors can hold multiple units, multiple samples can be processed quickly at the same time. Some devices

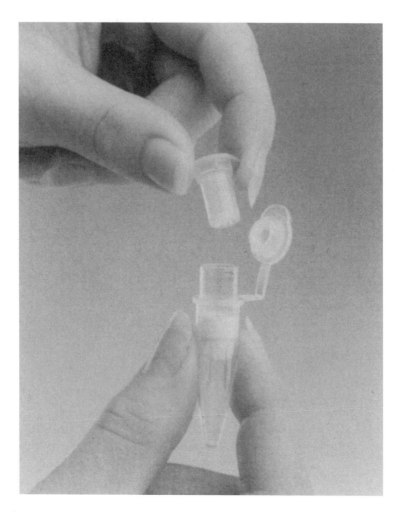

Fig. 4. Centrifugal microconcentrator with microporous insert (Micropure, top) for simultaneous separation of protein from electrophoresis concentration or desalting.

include a brief extra centrifugation step with the UF element reversed so that the side containing the retained material faces the filtrate vial. This drives every bit of the retentate into the vial for maximal recovery.

Centrifugal ultrafilters presently range in volume capacity from 0.5 to 15 mL. The smallest allow concentration from 0.5 mL initial volume to as little as 5 BL (100-fold concentration). Normal spin time for concentration at room temperature is 6–60 min, depending on the selected molecular-weight cutoff (MWCO) as well as solute viscosity. Use of low-adsorption membranes and plastics makes these devices very efficient, typically delivering solute recoveries of over 90%. They concentrate the product without change in ionic environment or denaturation.

Some larger centrifugal devices include more elaborate mechanisms for separation. One, for example, is so constructed that centrifugal force drives dense material in the sample away from the ultrafilter surface. It is, therefore, effective in concentrating par-

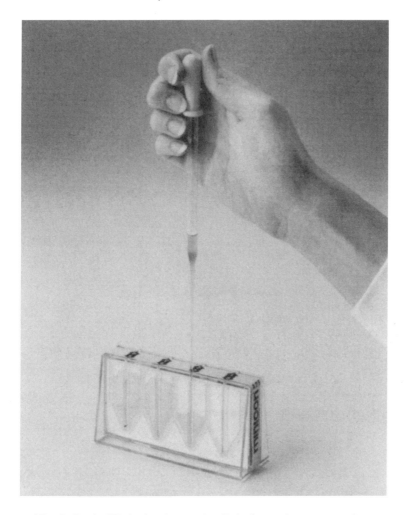

Fig. 5. Static UF device for use in clinical sample concentration.

ticle-laden samples with high recovery. This is useful, for example, for the concentration of cell supernatants, lysates, and extracts.

1.2.2.3. STATIC DEVICES

For concentration of samples without any centrifuge, pressurization, or other accessory equipment, some devices use capillary action as the driving force to transport the liquid through the membrane. By letting the membrane form the wall of one or several chambers and backing it with an absorbent material, the solvent (water) in the samples in the chambers is pulled through the membrane by capillary action. This causes the individual samples to be reduced in volume and the retained molecules to be concentrated (*see* **Fig. 5**). This type of device is widely used in clinical laboratories for the concentration of urine and cerebrospinal fluid in order to make it easier to detect very dilute disease-indicating species in those samples. Samples are loaded into individual chambers. The unit is left unattended for 1 or 2 h when the concentrated samples can be withdrawn with a pipet. A treatment near the bottom of the membrane prevents accidental

Table 2
Typical Solute Rejection With UF Membranes

Solute	M_r	YCO5	YM1	YM3	YM10	YM30
		\multicolumn{5}{c}{Membrane type}				
NaCl	58	<20	0	0	—	—
Dextrose	180	>70	0	0	—	—
Sucrose	342	>85	45	20	—	—
Raffinose	504	95	65	25	—	—
Bacitracin	1400	98	92	>80	20	—
Inulin	5000	>98	95	—	25	—
Cytochrome-*c*	12,400	>98	>98	>98	>95	<15
Myoglobin	17,000	>98	>98	>98	>98	—
α-Chymotripsinogen	24,500	>98	>98	>98	>98	>80
Albumin	67,000	>98	>98	>98	>98	>98
IgG	160,000	>98	>98	>98	>98	>98
Apoferritin	480,000	>98	>98	>98	>98	>98
IgM	960,000	>98	>98	>98	>98	>98

Source: Courtesy of Amicon, Inc.

drying of the sample. Graduation lines indicate the degree of concentration, which may be up to 200-fold. These devices can also be used to desalt the sample by repeated dilution and concentration.

1.3. Operating Parameters

1.3.1. Molecular-Weight Cutoff

The MWCO of a membrane is defined as the molecular weight of hypothetical globular solutes (proteins) that will be 90% rejected by the membrane (*see* **Table 2**). For example, a 10,000-MWCO ultrafilter will nominally reject 90% of molecules with a molecular weight of 10,000 Dalton. Because rejection is actually a function of physical size, shape, and electrical characteristics of the molecule, the MWCO is only a convenient indicator, based on model solutes. Linear molecules, such as polysaccharides, will tend to slip through a membrane that would reject globular molecules of the same molecular-weight.

Although two membranes can be claimed to have the same cutoff, they can exhibit quite different rejection behaviors because of distribution of pore diameters. Solute retention is not absolute. A "sharp" cutoff membrane will have minimal retention for species below its nominal MWCO rating. A "diffuse" cutoff membrane can significantly retain species of a size below the nominal MWCO or allow passage of some species above its cutoff. For concentration of retained species, either sharp or diffuse cutoff membranes will generally work equally well, but where the permeate is of interest, the final product may be markedly different. To determine cutoff sharpness, solute rejection tables for different membranes should be compared (*see* **Note 1**).

1.3.2. Solute Retention

Retention or rejection of a solute by an UF membrane defines the degree to which given molecules will be held back from the passing solution by the membrane and, hence, remain in the "retentate." For each membrane of a given MWCO, there is a specific degree of rejection of biomolecules. For example, cytochrome-*c* ($M_r = 12,400$) may be rejected >98% by a 3000-MWCO membrane, >95% by a 10,000-MWCO membrane, and <15% by a 30,000-MWCO membrane. Concentration proceds in direct proportion to volume reduction; that is, solute concentration doubles at 50% volume reduction. For guidance, manufacturers provide rejection tables or curves in their literature.

For freely membrane-permeating species, such as sodium chloride, the concentration of solute in the retentate and the permeate will be equal. If, for example, a solution containing 1% protein and 2% salt is processed with a membrane that is totally retentive to the protein, doubling the protein concentration by reducing the starting solution by 50% will result in a retentate containing 2% protein and 2% salt.

Many biological macromolecules tend to aggregate, or change conformation, under varying conditions of pH and ionic strength, so that their effective size may be much larger than that of the "native" molecule. This will cause increased retention at the membrane. The degree of hydration, counterions, and steric effects can also cause molecules with similar molecular weights to exhibit very different retention behaviors.

Solute–solvent and solute–solute interactions in the sample can also change the effective molecular size. For example, some proteins will polymerize under certain concentration and buffer conditions, whereas others (such as heme proteins) may dissociate into corresponding subunits. Ionic interactions or p–p stacking can cause small molecules to behave similarly to molecules of greater M_r. When this occurs, as in the case of phosphate ions with a 500-MWCO membrane, the small molecules may not effectively permeate the membrane.

1.3.3. Concentration Polarization

As solute concentration increases during the process of enrichment, solute at the membrane surface forms a gel that is permeable to solvent under pressure. This effect is called concentration polarization. At moderate to high concentrations of retained solutes, the flow resistance of the gel layer will reach a level where it significantly exceeds that of the membrane, in effect forming a secondary membrane. As solute continues to accumulate at the membrane–liquid interface, resistance grows. When net transport of solute by convection equals the back diffusion of solute toward the bulk solution, because of the concentration gradient, further increase in transmembrane pressure will not increase flow through the membrane and may cause it to decrease.

All efficient UF devices or systems must provide the means to minimize the effect of concentration polarization. The most important of these are magnetic stirring, pumped tangential flow across the membrane surface in narrow channels, and positioning the membrane surface at an acute angle with respect to the force vector acting on the fluid. The latter is achieved in centrifugal UF devices by using an angle-head rotor. Centrifugal force causes the gel film to slide across the angled surface, keeping the rest of the membrane surface relatively clear for permeation by solvent and membrane-passing molecules at relatively high rates. Cone-shaped membranes also employ the force vector at an angle during centrifugal separation.

Flow rate decrease owing to concentration polarization should not be confused with the effect of membrane fouling. Fouling is the deposition and accumulation of submicron particles and solute on the membrane surface, or crystallization and precipitation of smaller solutes on or within the pores of the membrane. There may, in addition, be chemical interaction with the membrane.

1.3.4. Maximizing Solute Recovery

Although UF membranes are normally inert, adsorptive losses may occur. Additional losses, caused by formation of concentrated solute gel or cake on the membrane, can be counteracted by polarization control, as indicated in **Subheading 1.3.3.**, by operating at modest pressures, and by a final agitation cycle at zero transmembrane pressure.

Effects of adsorption are more noticeable with dilute solutions, where adsorption may severely diminish the amount of the desired product. Because adsorption is largely a function of membrane and device surface area, the relation of sample concentration and volume to surface area should be considered before choosing a system. Small, dilute samples should be concentrated by using membranes or devices with minimum surface area while maintaining reasonable flow characteristics.

The buffer can affect membrane adsorption. Phosphate buffers can cause increased losses. Tris or succinate buffers allow better recovery. This may relate to lyotropic effects on hydrophobic bonding to the membrane or device.

Although rejection is used to characterize membrane performance, it does not always directly correlate with solute recovery from a sample or volume. Actual solute recovery—the amount of original material recovered after UF—is generally based on mass balance calculations (*see* **Note 2**).

1.3.5. Temperature and pH

Raising the operating temperature normally increases UF rates. Higher temperature increases solute diffusivity (typically 3–3.5%/°C for proteins) and decreases solution viscosity. One normally operates at the highest temperature tolerated by the solutes and the equipment. However, UF equipment is often used in cold rooms.

Changing solution pH often changes molecular structure. This is especially true for proteins. At the isoelectric point, the protein begins to precipitate in some cases, causing a decrease of filtrate flow.

1.4. Applications

1.4.1. Concentration

In macromolecular concentration, the membrane enriches the content of a desired biological species or provides filtrate cleared of retained substances. Microsolute (e.g., salt) is removed convectively with the solvent. Pressure, created by external means, forces liquid through the ultrafilter. Solutes larger than the nominal MWCO of the membrane are retained. The required pressure can be generated by use of compressed gas, pumping, centrifugation, or capillary action. With dilute solutions (1 mg/mL or less), flow rates are directly proportional to applied pressure. At higher concentrations, increased viscosity and polarization concentration act to reduce the flow, requiring steps to reduce concentration polarization (*see* **Note 3**).

1.4.2. Salt Removal or Buffer Exchange (Diafiltration)

Removal of small molecules from a solution by alternating UF and redilution or by continuous UF and dilution to maintain constant volume is called diafiltration.

Ultrafilters are ideal for removal or exchange of salts, sugars, nonaqueous solvents,

Table 3
Desalting by Repetitive Concentration and Redialysis

Spin number[a]	Protein recovery	NaCl concentration
Start point	—	500 mM
1	95.1%	140 mM
2	94.2%	25 mM
3	94.0%	5 mM

[a]Start: 15 mL of c-globulin. Spin 1: reduced to 3 mL and rediluted to 15 mL. Spins 2 and 3: repetition of reduction to 3 mL and dilution to 15 mL. After third spin, salt concentration reduced 100-fold.
Source: Courtesy of Amicon, Inc.

separation of free from protein-bound species, removing materials of low molecular weight, or rapid change of ionic and pH environment. In contrast to dialysis, the rate of microsolute removal or "washout" by diafiltration is a function of the UF rate and independent of microspecies (e.g., salt) concentration. This greatly reduces desalting times by using convective salt removal or exchange at flow rates equal to those of solvent passage. Diafiltration is also used for efficient microsolute exchange or "washin."

Membrane-permeating solutes (i.e., those significantly smaller than the cutoff, especially salts) pass through the membrane pores at the same rate as water. Transport through the membrane is by convection, not by diffusion, so that the rate of permeation is independent of molecular size.

Diafiltration washes microspecies from the solution, purifying the retained species (*see* **Notes 4** and **5**). In the discontinuous method, the sample is diluted before concentration or it is diluted after concentration and reconcentrated; this can be repeated one or more times, each time obtaining further desalting or solvent exchange (*see* **Table 3**). Small volumes may be easily desalted in one step by sample dilution before concentration. The continuous method of diafiltration is to connect the UF device (such as a stirred cell) to a pressurized reservoir-containing solvent, normally buffer or water. As filtration proceeds, solvent automatically flows from the reservoir into the device, at the same rate as the rate of filtration.

Ultrafiltration does not change salt concentration or buffer composition. A solution volume with 100 mM salt still contains 100 mM salt after concentration. Discontinuous diafiltration (rediluting the retentate with water and concentrating again) effectively decreases the salt concentration of the sample by the concentration factor of the UF. To achieve more complete salt removal, multiple concentration and redilution are required. For example, if a 1-mL sample containing 100 mM salt is diluted to 2 mL before concentration in a centrifugal device, the salt concentration in the 2-mL sample will be 50 mM. When reduced to 25 BL (80 times), the concentrate will still contain 50 mM salt. If more complete salt removal is desired, the sample can be rediluted with water or buffer to 2 mL before reconcentration. At this point, the salt concentration will have been reduced to 0.625 mM (50 mM/80), which will remain after the second concentration to 25 BL. Each further such dilution and reconcentration step would reduce salt concentration by 1/80, in the present example. For most samples, three concentration/reconstitution/reconcentration cycles will remove about 99% of the initial salt content. With very small sample volumes, dilution of the sample before the initial concentration step can often decrease salt concentration to an acceptable level.

LIVERPOOL JOHN MOORES UNIVERSITY
LEARNING SERVICES

1.4.3. Separation of Free From Protein-Bound Microsolute

In the past, free-drug levels in serum or plasma samples were not widely measured, partly for want of a convenient means of separating free from protein-bound drug. The chosen technique was generally equilibrium dialysis, a time-consuming procedure, subject to effects of dilution and buffers. However, this method does not directly indicate the free-drug concentration in the sample. Such other techniques as ultracentrifugation and gel filtration are no less time-consuming and there is inadequate standardization of results.

Today's better alternative for free/bound separation is UF with a centrifugal UF device specially designed for filtrate recovery. Spun in a standard laboratory centrifuge (preferably angle-head), free drugs readily pass the membrane for collection and analysis. The sample is not diluted in the process. Multiple samples are conveniently handled, typically in 10-min runs/set.

1.4.4. Recovery From Electrophoresis Gels

The availability of microporous inserts for centrifugal UF devices offers an easy method of recovering proteins (or other macromolecules) from electrophoretic gels. The gel piece containing the protein of interest is crushed or macerated to increase its surface area, then placed into the insert, and mixed with an elution buffer. During subsequent incubation, protein diffuses out of the gel. Centrifugation of the combined devices causes the buffer containing the protein to flow through the microporous membrane of the insert, which retains the gel particles. At the membrane surface of the UF device, the protein is retained for recovery while the buffer and any salts pass into the filtrate vial. An added new device, placed into the insert described (Gel Nebulizer), makes the process even easier by converting pieces of gel into a spray of fine particles during centrifugation (*see* **Fig. 6**). This makes extraction of the proteins more efficient.

1.4.5. Purification and Fractionation

Macromolecular mixtures may be separated into size-graded classes by UF, provided the species to be separated have at least a 10-fold difference in molecular weight (*see* **Note 6**). This can be accomplished either by direct UF, where the permeating solute is obtained in its initial concentration in the ultrafiltrate, or by diafiltration, where the retained solute is obtained in its initial concentration in the retentate and the diluted permeating solute in the ultrafiltrate. Normally, polarization effects require predilution or multiple dilution and reconcentration of the sample.

1.4.6. Detergent Removal

Ultrafiltration membranes efficiently remove detergents from protein solutions. The chemical nature of most detergents causes micelle formation above a critical concentration limit (critical micelle concentration), causing aggregation of the detergent and leading to gross changes in molecular structure. This affects the amount of the detergent that can be removed from solution with membranes of specific cutoff (*see* **Note 7**).

2. Materials

All UF equipment and membranes may be obtained from Amicon (Beverly, MA). In addition, the following may be required depending on the application:

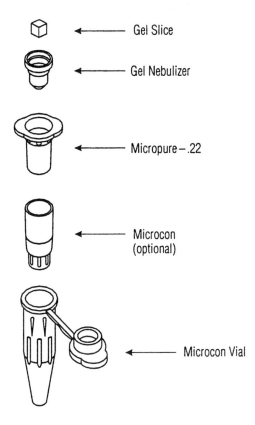

Gel Slice

Gel Nebulizer

Micropure – .22

Microcon (optional)

Microcon Vial

Fig. 6. Stacked elements of device for extraction of DNA, RNA, or proteins from gel.

1. Oxygen-free nitrogen with pressure regulator (for stirred cells) (*see* **Note 8**).
2. Microcentrifuge (variable speed) and microcentrifuge tubes.
3. Homogenizer (e.g., Eppendorf fitting pestle and pestle mixer motor).

The following solutions are required for staining and recovering proteins from poly-acrylamide gels:

1. Gel-staining solution: 0.1% Coomassie brilliant blue R250, 50% methanol, 10% acetic acid.
2. Destaining solution: 7% Acetic acid, 12% methanol.
3. Wash solution: 50% Methanol.
4. Extraction buffer: 100 mM NaHCO$_3$, 8M urea, 3% sodium dodecyl sulfate (SDS), 0.5% Triton X-100 (reduced), 25 mM dithiothreitol (DTT).

3. Methods

3.1. Using a Stirred Cell for Concentration

1. Prefilter or centrifuge any solution containing particulate matter, such as cell debris or precipitates (*see* **Note 9**).
2. Fill a cell of appropriate size (*see* **Subheading 1.2.2.**) with the sample and then secure it on a magnetic stirring table. Connect the inlet line to a regulated gas pressure source providing 2.7–3.4 atm (40–50 psi). Nitrogen is recommended (*see* **Note 8**).

Table 4
Centrifugation Guidelines

Membrane MWCO[a]	Maximum *g*-force	Spin times[b]	
		4°C	25°C
3000	14,000	185	95
10,000	14,000	50	35
30,000	14,000	20	12
50,000	14,000	10	6
100,000	3000[c]	35	15

[a]MWCO in Daltons.
[b]Spine time in minutes; 500-µL samples concentrated to 10 µL
[c]500*g* for DNA/RNA samples.
Source: Courtesy of Amicon, Inc.

3. Pressurize the cell according to instructions. Keep within cell pressure limits. When operating with either hazardous or specially valuable materials, always pressure-check the cell first to assure that all components are properly assembled.
4. Turn on the stirring table and adjust the stirring rate until the vortex created is approximately one-third the depth of the liquid volume.
5. When the run is completed, continue stirring for a few minutes after depressurization to maximize recovery of retained substances. This will resuspend the polarized layer at the membrane surface.

3.2. Protein and Peptide Recovery From Polyacrylamide Gels

The following protocol is recommended by Amicon for its Micropure™ microporous inserts and Microcon® microconcentrators:

1. Stain the gel with staining solution. Remove excess dye by soaking the gel in destaining solution with gentle agitation for approx 2 h or until a clear background is obtained.
2. Cut out the gel piece containing the band of interest with a clean scalpel or razor blade.
3. Place the gel piece in a microcentrifuge tube. Remove dye from the protein by adding 1 mL of wash buffer and sonicating for 5–15 min at 50–60°C or until the gel clears.
4. Remove the wash buffer. Add 50–100 BL of extraction buffer. Incubate for 20–30 min at 65°C.
5. Homogenize the gel using a motor-driven pestle homogenizer.
6. Incubate the tube with homogenized gel in a bath at 50–60°C for 2–3 h or overnight.
7. Place the Micropure-0.22 insert into the Microcon sample reservoir (use Microcon-100 or Microcon-50 for large proteins and Microcon-10 or Microcon-3 for polypeptides). Add 100 BL wash solution. Transfer the gel slurry into the Micropure with a pipet. Rinse the tube with 100 BL of wash solution to remove remaining slurry. Transfer to the Micropure.
8. Spin the assembly until all liquid is removed from the Micropure insert (13,000*g* for 20–30 min in Microcon-10; for others, *see* **Table 4**). Discard the Micropure. Proteins or peptides are retained above the Microcon membrane, depending on selected membrane MWCO.
9. To remove the extraction buffer from the protein, add 400 BL of wash solution to the Microcon units. Spin as indicated in **Table 4**. To complete buffer exchange, add 400 BL of the desired buffer and spin at 13,000*g* for 30 min (for Microcon-10; for others, *see* **Table 4**).
10. To recover the concentrated sample, invert the Microcon into a new vial. Spin at 1000*g* for 3 min.

3.3. Concentration of Antibodies From a Hybridoma

By inserting a microporous filter into a centrifugal UF unit, the sample can be run through two stages of separation at the same time. The insert, with a porosity of 0.22 or 0.45 Bm, retains all particulate matters (such as bacteria, cell fragments, or electrophoretic gel particles) but lets the solution containing the solutes of interest—such as proteins—flow into the UF element where the protein is concentrated. The small microporous insert can also be used by itself, with a standard microcentrifuge vial, to free samples of particles. The following is a protocol offered by Amicon for its Micropure microporous inserts and Microcon microconcentrators.

1. Place the Micropure-0.45 insert into the Microcon sample reservoir (use Microcon-50 for IgG or Microcon-100 for IgM).
2. Pipet up to 350 BL of sample into the Micropure insert.
3. Spin at 500g for 5 min or until there is no more liquid in the Micropure insert.
4. Discard the Micropure. Antibody is retained above the membrane in the Microcon.
5. If the antibody is to be desalted or if buffer is to be exchanged, fill the Microcon sample reservoir with 450 BL of desired buffer. Spin the Microcon-50 at 12,000g for 8 min. For the Microcon-100, spin at 3000g for 15 min.
6. To recover the concentrated antibody, invert the Microcon into a new vial. Spin at 1000g for 3 min.

4. Notes

1. The 90% rejection standard and the significance of molecular shape rather than weight make it very important that the selected membrane cutoff be well below the molecular weight of the solute to be retained. As a rule, it is best to choose a membrane or device with cutoff at about half of the molecular weight of the protein to be concentrated. This provides a balance between high protein recovery and minimal filtration time. For example, if bovine serum albumin ($M_r = 67,000$) were the protein of interest, a 30,000-MWCO (rather than 50,000 MWCO) membrane or device would result in the most efficient concentration and recovery of the protein in the retentate. For the highest possible retention, a membrane cutoff one-tenth of the M_r of the solute should be used (10,000 in the above example). Of course, the tighter the membrane, the slower the filtration rate. When species of low molecular weight are to be exchanged, the membrane cutoff should be substantially above that of the passing solute.
2. Recovery of protein can sometimes be improved by passivation of the membrane. For very dilute protein solutions (in the range of 1 Bg/mL), concentrate recovery is often not quantitative (*see* **Fig. 5**). This loss of protein can be caused by nonspecific binding of the protein to exposed binding sites on the plastic of the concentration device. The extent of nonspecific binding varies with the relative hydrophobicity of individual protein conformations. Pretreating (passivating) the plastic by blocking the available binding sites before concentration can often improve the recovery yield from dilute protein solutions. For passivation procedures, the manufacturers' recommendations should be followed. The procedure is simple, requiring the filling of devices with a prescribed solution, leaving them overnight, and then rinsing thoroughly. Among recommended passivation solutions are 1% IgG in phosphate-buffered saline, Tween-20, or Triton-X in distilled water, bovine serum albumin, and even powdered milk in distilled water.
3. Highly viscous solutions filter slowly, as do solutions containing particulate matter, such as colloids. Where a viscous agent (sucrose, glycerin, etc.) is to be removed, flow can often be increased by predilution.

4. For rapid desalting (diafiltration), the UF cell can be connected to an auxiliary reservoir containing a diafiltrate solution with the desired microsolute concentration, if any. The auxiliary reservoir is then gas pressurized to 2.7–3.4 atm (40–50 psi). The cell's fluid volume, as a well as the macrosolute concentration, remain constant as filtrate is automatically replaced by diafiltrate solution. This technique provides a simple means for rapid microsolute exchange. It is typically used as a substitute for dialysis. A concentration/Dialysis Selector Switch (Amicon) permits simple switching between concentration and diafiltration (*see* **Fig. 3**).

5. Fully membrane-permeating molecules pass at the same rate as salt because their transport through the membrane is independent of their individual molecular sizes. Proteins of $M_r >$ 3000 may be separated from each other and from smaller molecules if they differ by a factor of about 10 in molecular size and are not associated in solution. For example, one cannot diafilter bovine serum albumin ($M_r = 67,000$) from IgG ($M_r = 160,000$), but can easily diafilter biotin ($M_r = 244$) from cytochrome-*c* ($M_r = 12,400$). Small volumes should be prediluted to assist in diafiltration.

6. Ultrafiltration is not primarily a fractionation technique. It can only separate molecules that differ by at least an order of magnitude in size and is best done in a diafiltration arrangement, with multiple washes to separate the mixture of macromolecules. Molecules of similar size cannot be separated by UF. During concentration polarization, the gel layer on the membrane surface superimposes its own rejection characteristics on those of the membrane. Usually, concentration polarization increases rejection of lower-M_r species. A membrane with a nominal 100,000 MWCO may reject 10–20% of albumin ($M_r = 67,000$) in a 0.1% solution of pure albumin. However, in the presence of larger solutes, such as IgG, it may reject 90% of the albumin.

7. For example, the monomer of Triton X-100 ($M_r = 500–650$) should pass readily through a 3000-MWCO membrane. However, at concentrations above its critical micelle concentration of 0.2 mM, Triton X-100 forms micelles composed of approx 140 monomeric units. During UF, the micelles behave like globular proteins of $M_r = 70,000–90,000$. Therefore, above the critical micelle concentration of Triton X-100, a 100,000-MWCO membrane would be required to removed the detergent effectively.

8. Use of compressed air can cause large pH shifts resulting from dissolution of carbon dioxide. With sensitive solutions, oxidation can occur, leading to other potential problems.

9. Because of their unique design, which uses gravitational force to counteract deposition of suspended particles on the membrane, certain concentrators (Centriprep®, Amicon) are especially useful for UF of solutions with high solid content. The feature counteracts membrane fouling and eliminates the need for prefiltration of samples. It can also be obtained with a microporous filter for separation of antibodies from cell culture supernatant or for desalting of bacteria. Another centrifugal UF device uses a low-adsorptive, polypropylene sample reservoir, which minimizes nonspecific adsorption of solutes to the walls of the devices. It also contains a low-adsorptive, hydrophilic membrane. No passivation is required before processing very dilute protein solutions. By employing a final inverted-spin feature, high solute recovery is possible, even from dilute solutions in the 1 Bg/mL range (Centriplus™).

13

Bulk Purification by Fractional Precipitation

Shawn Doonan

1. Introduction

The solubility of a particular protein in aqueous solution depends on the solvent composition and on the pH; hence, variation in these parameters provides a way of purifying proteins by fractional precipitation. The factors that dictate solubility are complex, because the surface of a protein is itself complex, containing ionized residues, polar regions, and hydrophobic patches, all of which will interact with the solvent in ways that are not completely understood. Hence, it is not possible to elaborate a theoretical approach to fractional precipitation (1); rather, the methods are used in an essentially empirical fashion. The most widely used procedures are precipitation by the addition of salt (salting out) and by the addition of organic solvents; these are the focus of this chapter (see **Note 1**).

The addition of high concentrations of salt to a protein solution causes precipitation largely by removing water of solvation from hydrophobic patches on the protein's surface, thus allowing these patches to interact with resulting aggregation. For a pure protein, the relationship between solubility S (in g/kg of water) and the ionic strength I (in mol/kg water) is given by

$$\log S = \beta - K_s[(I/2)] \tag{1}$$

where β and K_s are constants for a particular protein at a particular pH and temperature. The point here is that a protein will precipitate over a range of ionic strength values (determined by the value of K_s) and that different proteins will precipitate over different, but frequently overlapping, ranges. This highlights the fact that it is rarely possible to purify a particular protein from a complex mixture by using fractional precipitation alone (although heroic attempts were made to do this in the early days of protein purification). Rather, the value of the method is that it provides a simple procedure for enrichment of the protein of interest from large volumes of extracts and, at the same time, can be used to concentrate the fraction (see Chapter 10); with large-scale purifications in particular, this is important in reducing the problem to a manageable scale. Hence, fractional precipitation by salt is almost invariably used at an early stage of a purification procedure, often on the clarified extract, to obtain an initial purification and con-

From: *Methods in Molecular Biology, vol. 244: Protein Purification Protocols: Second Edition*
Edited by: P. Cutler © Humana Press Inc., Totowa, NJ

centration (*see* **Note 2**). Further purification is then achieved by chromatographic methods, which have higher resolving power, but generally lower capacity.

A variety of salts has been used in the past for this purpose, including NaCl, Na_2SO_4, KCl, $CaCl_2$, and $MgSO_4$, and these are still sometimes used for particular applications. By far the most frequently used salt is, however, ammonium sulfate ($[NH_4]_2SO_4$). The reasons for this include its high solubility in water (about 4 *M* at saturation), its low heat of solution, the fact that the density of saturated solutions (1.235 g/mL) is less than that of proteins, hence allowing for their collection by centrifugation, and the essentially innocuous nature to proteins of its constituent ions. Concentrations of the salt are traditionally expressed in terms of percentage saturation at a particular temperature; **Table 1** gives the amounts of solid ammonium sulfate required to obtain the required concentration at 0°C *(2)*, the temperature at or near which fractional precipitation is usually carried out (*see* **Note 3**).

The precipitation of proteins by addition of organic solvents is a more complex process. Factors involved probably include the decrease in dielectric constant, which promotes aggregation by charge interaction, as well as sequestration of water of solvation of the protein. The same caveats about the usefulness of the method apply as discussed for salt fractionation, although solvent fractionation has achieved some noteworthy successes, such as the classical Cohn fractionation of plasma proteins (summarized in **ref. 3**). There is with this method, however, the added problem that organic solvents can cause protein denaturation by interaction with hydrophobic residues in the protein's interior. Hence, it is usually essential to carry out solvent precipitation at a low temperature to minimize denaturation. The solvents used need to be miscible with water in all proportions and nontoxic; acetone and ethyl alcohol best meet these requirements (*see* **Note 4**).

Whichever of the two methods is used, there will necessarily be a trade-off between degree of purification achieved and yield. An ammonium sulfate or solvent "cut" over a 10% concentration range might give a yield of 70% of the desired protein with a purification factor of 5. Increasing the concentration range may increase the yield, but with a corresponding decrease in purification factor. For an example of the use of both methods in purification of the same protein, *see* **ref. 4**.

2. Materials

2.1. Fractional Precipitation With Ammonium Sulfate

1. Solid ammonium sulfate (*see* **Note 5**).
2. Magnetic stirrer or electrical paddle stirrer.
3. Ice bath.
4. Refrigerated centrifuge and rotor (e.g., a 6 × 250-mL angle rotor).
5. Screw-top plastic bottles for the rotor to be used.
6. Visking dialysis tubing (14- or 19-mm inflated diameter; 26- or 31-mm flat width).

2.2. Fractional Precipitation With Acetone

1. Analar acetone precooled to −20°C (*see* **Note 6**).
2. Electrical paddle stirrer.
3. Ice-salt cooling bath.
4. Centrifuge, rotor, bottles, and dialysis tubing as in **Subheading 2.1**.

Table 1
Amounts of Solid Ammonium Sulfate Required to Change the Concentration of a Solution From a Given Starting Value to a Desired Target Value at 0°C

Initial percentage saturation at 0°C	Target percentage at 0°C[a]																
	20	25	30	35	40	45	50	55	60	65	70	75	80	85	90	95	100
0	106	134	164	194	226	258	291	326	361	398	436	476	516	559	603	650	697
5	79	108	137	166	197	229	262	296	331	368	405	444	484	526	570	615	662
10	53	81	109	139	169	200	233	266	301	337	374	412	452	493	536	581	627
15	26	54	82	111	141	172	204	237	271	306	343	381	420	460	503	547	592
20		27	55	83	113	143	175	207	241	276	312	349	387	427	469	512	557
25			27	56	84	115	146	179	211	245	280	317	355	395	436	478	522
30				28	56	86	117	148	181	214	249	285	323	362	402	445	488
35					28	57	87	118	151	184	218	254	291	329	369	410	453
40						29	58	89	120	153	187	222	258	296	335	376	418
45							29	59	90	123	156	190	226	263	302	342	383
50								30	60	92	125	159	194	230	268	308	348
55									30	61	93	127	161	197	235	273	313
60										31	62	95	129	164	201	239	279
65											31	63	97	132	168	205	244
70												32	65	99	134	171	209
75													32	66	101	137	174
80														33	67	103	139
85															34	68	105
90																34	70
95																	35

[a]Gram of solid ammonium sulfate per liter of solution.

3. Methods

3.1. Fractional Precipitation With Ammonium Sulfate

It is assumed that a trial experiment has been carried out in order to determine the optimal concentration range of ammonium sulfate for the particular protein sample to be fractionated (*see* **Note 7**). Suppose that the trial showed the best results to be obtained with the fraction precipitated between 35% and 50% saturation. Proceed as follows.

1. Measure the volume of the protein solution to be fractionated and pour it into a glass beaker of capacity about twice the measured volume of solution.
2. Weigh out 0.194 g of solid ammonium sulfate for every 1 mL of protein solution (*see* **Table 1**). If the ammonium sulfate contains lumps, then these should be broken up using a mortar and pestle.
3. Place the beaker containing the protein on ice and stir either with a magnetic stirrer or with a paddle stirrer. A slow rate of stirring should be used to avoid foaming.
4. Add the ammonium sulfate to the protein solution in small batches over a period of several minutes, ensuring that one lot of ammonium sulfate has dissolved before adding the next (*see* **Note 8**).
5. After addition is complete, leave to stand for 10 min to ensure equilibrium, and then remove precipitated protein by centrifugation at about 5000*g* and 4°C for 30 min using screw-top plastic tubes (*see* **Note 9**).
6. Decant off the supernatant solution from each tube into a measuring cylinder and determine the total volume (*see* **Note 10**).
7. Pour the combined supernatants into a beaker in the ice bath and add 0.087 g of ammonium sulfate/mL protein solution (the amount required to take the concentration from 35% to 50% saturation; *see* **Table 1**) using the same precautions as in **steps 2–4**.
8. Recover the precipitated protein by centrifugation as in **step 5**, decant the supernatant solution into a beaker (*see* **Note 10**), and suspend the protein pellets in the minimum volume of water or of an appropriate buffer (*see* **Note 11**).
9. Transfer the protein suspension to a sack of Visking dialysis tubing and dialyze against at least two changes of buffer using 100 times the volume of the sample and allowing 3–4 h for equilibration (*see* **Note 12**).
10. After dialysis, remove any precipitated material by centrifugation.

3.2. Fractional Precipitation With Acetone

Again, a trial experiment must be carried out to determine the optimum precipitation range for the particular protein fraction (*see* **Note 7**). Assuming this to be between 37.5% and 50% (v/v), proceed as follows.

1. Measure the volume of the protein solution, pour it into a beaker (*see* **Note 13**) immersed in an ice-salt bath, and stir until the temperature reaches 0°C.
2. For each 1 mL of protein solution, add 0.60 mL of acetone (precooled to −20°C) dropwise with constant stirring and at such a rate that the temperature does not rise above 0°C (*see* **Notes 8** and **14**). After the addition of acetone is complete, continue stirring for 10 min with constant control of temperature.
3. Remove precipitated protein by centrifugation at 0°C for 10 min at 3000*g* using precooled centrifuge tubes (*see* **Note 15**).
4. Measure the volume and return the combined supernatant solutions to the beaker in the ice-salt bath. Add a further 0.25 mL of acetone/mL protein solution using the precautions described in **steps 1** and **2**.

5. Recover the precipitated protein by centrifugation as in **step 3**. Pour off the supernatant solutions and invert the centrifuge tubes over filter paper to drain; blot off any drops of solution adhering to the walls of the tubes.

6. Resuspend the pellets in water or an appropriate buffer and remove residual acetone by dialysis, membrane filtration, or gel filtration (*see* **Note 16**).

4. Notes

1. An alternative that is sometimes used is precipitation by alteration of the pH. Proteins will generally have minimum solubility at their isoelectric points, where the net charge is zero and there is no electrostatic repulsion. The approach is to incubate the protein fraction at various pH values to see if the species of interest precipitates; obviously, if the isoelectric point of the protein is known, then that pH should be used. A problem with the method is that the protein may not be stable at its isoelectric point if this is far removed from neutrality. In addition, the protein may not precipitate, particularly if its concentration is low, in the absence of added salt; this makes establishing the conditions for pH fractionation quite difficult.

2. There is a further advantage in using ammonium sulfate fractionation immediately after homogenization and clarification (*see* Chapter 2). Some tissues give homogenates that are very difficult to clarify because of the presence of membrane fragments and nucleoprotein complexes, which resist sedimentation. This particulate matter usually aggregates at low concentrations of ammonium sulfate and sediments readily in the first fraction. Hence, unless the protein of interest is contained in this fraction, the protein solution obtained from this procedure will be devoid of suspended matter, as required for use of such techniques as column chromatography. Fractionation with organic solvents confers the same advantage. In this case, it is because of the low density and viscosity of water–solvent mixtures, which facilitate sedimentation of particulate matter.

3. **Reference 3** gives a corresponding table for 25°C, which may be used if it is preferable to carry out fractionation at room temperature. Alternatively, the following formula can be used:

$$g = [533(P_2 - P_1)/(100 - 0.3 P_2)] \qquad (2)$$

where g is the number of grams of ammonium sulfate required to change the concentration of 1 L of solution from P_1% to P_2% saturation at 20°C.

4. An alternative to using organic solvents is provided by fractionation with water-soluble polymers, of which polyethylene glycol (PEG) is the most commonly employed. The mechanism of precipitation seems to be by steric exclusion; that is, the protein is concentrated in the extrapolymer space of the solution until its solubility limit is exceeded and precipitation occurs (*5*). Consistently, larger proteins tend to precipitate earlier than smaller ones, and precipitation is relatively insensitive to pH and ionic strength. The commonly used precipitants are PEG 4000 or PEG 6000 (i.e., PEGs with molecular-weight averages of 4000 or 6000). The great advantage of these precipitants is that they have little tendency to denature proteins. For further details, *see* **ref. 5**.

5. Analar ammonium sulfate is more than twice as expensive as the general-purpose grade (around $150 for 5 kg as compared to about $70 for general purpose reagent [GPR] grade). It is doubtful whether the extra cost is worth it when dealing with large volumes of crude protein mixtures; Aristar grade is certainly not worth using at about $300/kg. The major problem with less expensive grades is the presence of low amounts of heavy metal contaminants; if the protein of interest is metal-sensitive, then EDTA (10 mM) can be added to the solution to remove these contaminants.

6. Use analar grade, which is only marginally more expensive than the general-purpose grade. If carrying out fractionation with ethanol, then either absolute ethanol (>99%) or the more commonly available 96% variety may be used; in the latter case, the volume added to achieve the desired concentration would need to be adjusted to allow for the water content.

7. Optimum in this context means the range of ammonium sulfate concentrations that gives the desired balance between yield and purification. If the protein of interest has been purified previously, then the required range may be available in the literature, but it should be kept in mind that precipitation will depend on pH, on buffer composition, and on the protein concentration and composition of the fraction. Hence, unless these are identical to those in published procedures, it is unwise to assume that the protein will behave identically in your purification.

   Ideally, the trial ammonium sulfate fractionation should be carried out exactly as described in **Subheading 3.1.**, but using a small volume (approx 20 mL) of fraction; that is, ammonium sulfate should be added to concentrations of 0–20%, 20–30%, and so on, in 10% steps at each stage, removing precipitated protein by centrifugation before increasing the salt concentration. The recovered precipitates should be dissolved in buffer and assayed for the protein of interest and for total protein; the latter can be done most simply by measuring the absorbance of a suitably diluted sample at 280 nm and using the approximate relationship that an absorbance of 1 corresponds to a protein concentration of 1 mg/mL (accurate enough for present purposes). If the test for activity of the protein is sensitve to NH_4^+ or SO_4^{2-} ions, then the dissolved pellets will have to be dialyzed before assay. This trial should give the necessary information to proceed to large-scale work, but if, for example, the protein of interest precipitates equally in the ranges of 20–30% and 30–40%, it may be worthwhile checking to see if the range 25–35% gives better results than simply taking a 20–40% cut. If the protein precipitates over a very broad range so that only a low degree of purification can be achieved, then it is probably better to abandon the idea of fractionation and use precipitation only as a means of concentration if a large volume is a problem (*see* Chapter 10).

   A quicker way of carrying out a trial is to take several samples of the protein fraction and add 20% ammonium sulfate to the first, 30% to the second, and so on. Then, after equilibration, recover the precipitated protein by centrifugation, dissolve, and assay as above. The problem with this is that the precipitation behavior of a particular protein will depend on the precise protein composition of the solution, so that the conditions in a trial of this sort will not properly reflect the conditions in the large-scale procedure. The use of this method is not, therefore, recommended. The above considerations apply equally to trial solvent fractionation experiments.

8. *Important*: If the salt or organic solvent is added too quickly, then high local concentrations will develop and proteins will be precipitated that would remain soluble at the target ammonium sulfate or solvent concentration. Such proteins may not readily redissolve, and the result will be decreased purity of the active fraction and possible loss of the protein of interest in lower fractions. With solvents, there is also the increased risk of denaturation. Some authors recommend the use of saturated solutions of ammonium sulfate for fractionation, but this has the disadvantage of leading to large volume increases and should not be necessary if due care is taken when adding the solid salt.

9. Care must be taken with this step to protect both the centrifuge and the operator! The tubes must be balanced to within 0.1–0.2 g across the rotor axis (*see* **Note 3** of Chapter 2), and it is particularly important not to counterbalance a tube full of ammonium sulfate solution with a tube full of water because of the large difference in density between the two. Either the ammonium sulfate/protein suspension should be divided between two tubes or the suspension should be balanced using a solution of ammonium sulfate of the same concentra-

tion. It is also most important to avoid spillage of ammonium sulfate solutions into the centrifuge head. Such solutions are extremely corrosive to the materials of which rotors are constructed and can lead to irreversible damage, rendering the rotors unsafe for use. After centrifugation of ammonium sulfate (or other salt) solutions, rotors should always be removed and washed in warm water.

Note that proteins precipitated in ammonium sulfate are usually stable and, hence, it is often convenient to store the suspension overnight at 4°C in this form before proceeding with the purification schedule.

10. The pellets of precipitated protein from this step may be discarded, but to be on the safe side, it is worth keeping them until it has been established that the protein of interest is indeed obtained in the next fraction. If something has gone wrong, then it is easier to reprocess the 0–35% precipitate than to go back to the beginning of the preparation and start again! Similarly, do not throw the supernatant from the 35–50% cut away until you are sure that your protein has been precipitated.

11. This step is crucial for obtaining a concentrated fraction. Add a very small volume of water or buffer (about 10 mL if using a 250-mL tube) and resuspend the protein pellet using a glass rod. If several centrifuge tubes have been used, then transfer the protein suspension from tube to tube, resuspending the pellet each time; only add more water or buffer if the suspension becomes too thick to transfer readily. After transfer of the final suspension to the dialysis bag, a further small aliquot of water or buffer can be used to rinse out the tubes and the rinsings combined with the suspension. A common error is to attempt to redissolve the protein pellets after centrifugation; because the pellets contain considerable quantities of ammonium sulfate, this takes a large volume of buffer and it is easy to end up with a volume comparable to that of the original fraction. When doing a trial fractionation (*see* **Note 7**), it is acceptable to redissolve the precipitates, as the volume will not generally be important.

12. Visking tubing comes in several sizes. For a given volume of solution, the larger the diameter of the dialysis tubing used, the shorter the piece that will be required; attainment of equilibrium, however, will be slower with short, fat bags than with long, thin ones. The sizes recommended are a compromise between these two factors. For most applications, it is only necessary to soak the dry tubing in water or in the buffer to be used for dialysis, and it is ready to be used. The tubing does, however, contain significant quantities of sulfur compounds and of heavy metal ions; the latter may be a problem if the protein of interest is metal sensitive. They may be removed by boiling the dialysis tubing in 2% (w/v) sodium bicarbonate, 0.05% (w/v) EDTA for about 15 min, washing with distilled water, and then boiling in distilled water twice for 15-min periods. Prepared tubing can be stored indefinitely at 4°C in water or buffer containing 0.1% (w/v) sodium azide. For most applications, Visking tubing is perfectly adequate, but problems will arise if the protein of interest is small. The pores in this type of tubing are of such a size that the nominal molecular-weight cutoff (NMWC) is about 15,000, although larger proteins may still pass through the pores if they have an elongated shape. As a rule of thumb, Visking tubing can be used with confidence if M_r >20,000. For dialysis of smaller proteins, Spectropor tubings with a range of NMWC values starting at 1000 are available (from Serva), and the appropriate tubing should be selected. These tubings are much more expensive than Visking tubing, and the rate of dialysis decreases with pore size; hence, they should only be used if strictly necessary.

To remove ammonium sulfate from the protein suspension, a piece of dialysis tubing with a volume about twice that of the suspension is taken and securely closed with a double knot at one end. The suspension is then poured into the bag using a funnel, air is removed from the top part of the bag by running it between the fingers, and the top secured with a double knot. It is very important to have this space in the bag to allow for expansion, as water will flow in while the internal salt concentration is high. If insufficient space is left, the bag can become very tight owing to this inflow. Bags rarely burst, because the membranes are quite strong, but they are tricky to open in this state; the best way is to insert one end into a measuring cylinder and then prick the bag with a scalpel. Equilibrium will be reached in about 3–4 h, but only if the system is stirred; otherwise, about 6 h should be allowed. Care should be taken when stirring to ensure that the magnetic pellet or stirring paddle does not tear the dialysis bag. The volume of dialysis solution to be used depends, of course, on the sample volume and the final concentration of ammonium sulfate desired. If a ratio of 1:100 (sample:dialysis fluid) is used, then at equilibrium, the ammonium sulfate concentration will have been reduced 100-fold. (This is not strictly true, because it ignores the Donnan effect, but will do as an approximation.) A second dialysis will then result in a total decrease of 10,000-fold, and so on. The buffer to be used for dialysis is usually dictated by the requirements of the next step in the purification schedule. For example, if this is to be ion-exchange chromatography, then column equilibration buffer is the logical choice.

13. Glass is acceptable, but stainless steel is to be preferred because of the more rapid heat transfer. Plastic will not do.

14. The addition of ethanol or acetone to water leads to a volume reduction of about 5%, but this is usually ignored in calculating percentage concentrations. To calculate the volume (v) (in mL) of solvent required to change the concentration of 1 L of solution at $P_1\%$ to $P_2\%$, use the formula:

$$v = [1000(P_2 - P_1)/(100 - P_2)] \tag{3}$$

15. Only short centrifugation times are required because of the low density and viscosity of water–solvent mixtures.

16. Do not attempt to resuspend the precipitates in too small a volume (in distinction to the practice with salt fractionation), because the result would be a high solvent concentration with attendant risk of denaturation. Similarly, if the protein is sensitive to solvents, it is necessary to remove residual solvent quickly, so membrane filtration or gel filtration may be preferable to dialysis.

References

1. Englard, S. and Seifter, S. (1990) Precipitation techniques. *Methods Enzymol.* **182,** 285–300.
2. Dawson, R. M. C., Elliot, D. C., Elliot, W. H., and Jones, K. M. (1986) *Data for Biochemical Research,* 3rd ed., Oxford University Press, Oxford, pp. 537–539.
3. Green, A. A. and Hughes, W. L. (1955) Protein fractionation on the basis of solubility in aqueous solutions of salts and organic solvents. *Methods Enzymol.* **1,** 67–90.
4. Banks, B. E. C., Doonan, S., Lawrence, A. J., and Vernon, C. A. (1968) The molecular weight and other properties of aspartate aminotransferase from pig heart muscle. *Eur. J. Biochem.* **5,** 528–539.
5. Ingham, K. C. (1990) Precipitation of proteins with polyethylene glycol. *Methods Enzymol.* **182,** 301–306.

14

Ion-Exchange Chromatography

Chris Selkirk

1. Introduction

Ion-exchange chromatography is one of the most widely used forms of column chromatography. It is used in research, analysis, and process-scale purification of proteins. Ion exchange is ideal for initial capture of proteins because of its high capacity, relatively low cost, and its ability to survive rigorous cleaning regimes. Ion exchange is also ideal for "polishing" of partially purified material on account of the high-resolution attainable and the high capacity giving the ability to achieve a high concentration of product. Ion-exchange chromatography is widely applicable because the buffer conditions can be adapted to suit a broad range of proteins rather than being applicable to a single functional group of proteins.

Ion-exchange chromatography matrices are available as dry granular material or as preswollen loose beads, but prepacked columns (Bio-Rad, Amersham Bioscience) are now common, particularly for small-scale analytical and method development work. Ion exchange can now also be carried out on monolithic columns (Bio-Rad), on membranes (Pall, Sartorius), and on ion-exchange high-performance liquid chromatography (HPLC) columns. The method is essentially the same whichever of these formats is employed. The method described in this chapter will assume that the column to be used is packed ready for use.

Ion-exchange chromatography relies on the interaction of charged molecules in the mobile phase (buffer + sample) with oppositely charged groups coupled to the stationary phase (column packing matrix). The charged molecules in a buffer solution come from the buffer components (e.g., salts). The charged groups on a protein are provided by the different amino acids in the protein. Lysine, arginine, and histidine have a positive charge at physiological pH, whereas aspartic acid and glutamic acid have a negative charge at physiological pH.

Charges on amino acids at physiological pH:
+ve   lysine, arginine, histidine
−ve   aspartic acid, glutamic acid
Charges on ion-exchange matrices
+ve   DEAE, QAE
−ve   CM, SP, sulfonic acid

From: *Methods in Molecular Biology, vol. 244: Protein Purification Protocols: Second Edition*
Edited by: P. Cutler © Humana Press Inc., Totowa, NJ

Table 1
Ion-Exchange Groups

Ion-exchanger type	Strong exchangers	Weak exchangers
Cation exchangers	SP (sulfopropyl)	CM (carboxymethyl)
	S (Methyl sulfonate)	
Anion exchangers	Q (quaternary ammonium)	DEAE (diethylaminoethyl)
	QAE (quaternary aminoethyl)	

The net charge on a protein molecule will depend on the combination of positively and negatively charged amino acids in the molecule. The charges of the amino acid groups varies depending on the hydrogen ion concentration (acidity) of the solution and, thus, the overall charge on a protein varies according to the pH. The more acidic the solution, the more groups will be positively charged; the more alkaline the solution, the more negatively charged the protein will become. The pH at which the negative charges on a protein balance the positive charges and, therefore, the overall charge on that protein is zero is called the isoelectric point (pI) for that protein. It is useful to know the pI of a protein if it is to be purified by ion-exchange chromatography. This information will assist in deciding on the best starting conditions for optimization of the purification conditions (*see* **Note 1**).

Binding and elution of proteins is based on competition between charged groups on the protein and charged counterions in the buffer for binding to oppositely charged groups on the stationary phase. The higher the concentration of charged salt molecules in the solution, the greater is the competition for binding to the ligands on the matrix, so the greater is the tendency for the protein to dissociate from the ion-exchange matrix.

The protein sample is applied to the ion-exchange column in a solution of low salt concentration. The counterions with which the column has been charged are not permanently bound but are held by electrostatic interaction. Therefore, there is a continual binding and unbinding of counterions. Under low-salt conditions, charged groups on the protein have a greater probability of binding to charged counterions on the ion exchanger and become bound to the ion-exchange column. During elution, the salt concentration is increased, so that when a protein group dissociates from an ionic group on the stationary phase, there is an increased probability that ions in the mobile phase will bind to the charged group on the protein and the ionic group on the stationary phase. Thus, the proteins dissociate from the ion-exchange matrix and are eluted as the salt concentration increases. The more strongly bound the protein, the greater is the salt concentration required to elute it.

Ion-exchange matrices are divided into two major types according to the charge on the ion-exchange ligands (*see* **Table 1**):

Cation Exchange: Cation-exchange resins have negatively charged groups on the surface. These are used to bind proteins that have an overall positive charge. Proteins will have an overall positive charge at a pH below their isoelectric point. Therefore, cation exchange is used at a pH below the isoelectric point of the protein(s) to be bound.

Anion Exchange: Anion-exchange resins have positively charged groups on the surface. These are used to bind proteins that have an overall negative charge. Proteins will have an

Table 2
Buffers: Anion Exchange

Buffer	Anion	pH Range
N-methyl piperazine	Cl^-	4.5–5.0
Piperazine	Cl^-	5.0–6.0
L-Histidine	Cl^-	5.5–6.5
Bis-Tris	Cl^-	5.8–6.8
Bis-Tris propane	Cl^-	6.4–7.3
Triethanolamine	Cl^-	7.3–8.2
Tris	Cl^-	7.5–8.0
Diethanolamine	Cl^-	8.4–9.4

overall negative charge at a pH above their isoelectric point. Therefore, anion exchange is used at a pH above the isoelectric point of the protein(s) to be bound.

Ion exchangers are also divided into strong and weak ion exchangers. Strong ion-exchange ligands maintain their charge characteristics, and therefore ion-exchange capacity, over a wide pH range, whereas weak ion-exchange ligands show a more pronounced change in their exchange capacity with changes in pH *(1)*. DEAE–Sepharose Fast Flow (weak anion) has a working pH range of 2.0–9.0, whereas Q–Sepharose Fast Flow (strong anion) has a working pH range of 2.0–12.0. CM–Sepharose Fast Flow (weak cation) has a working pH range of 6.0–10.0 and SP–Sepharose Fast Flow (strong cation) has a working pH range of 4.0–13.0. If your purification is to be carried out at pH above 9 for anion exchange or below 6 for cation exchange, then it is likely that you will need to use a strong ion exchanger. However, if your purification is to be carried out at a less extreme pH, then the slight differences in selectivity mean that it may be worth comparing the results obtained with both weak and strong ion exchangers to optimize your purification.

2. Materials

1. Binding buffer of appropriate pH and composition for binding of protein to matrix (*see* **Notes 2–4**). **Tables 2** and **3** list some suitable buffers and the pH ranges over which they are useful is included.
2. Elution buffer (often the same as binding buffer but with higher salt concentration) (*see* **Note 5**).
3. Regeneration buffer (e.g., 1 *M* NaCl). *Note*: Buffers and samples should be filtered (0.45 µm) before applying to the column to avoid blockage of the column flow path by particulates. Buffers should be stored sterile or with addition of a bacteriostat (e.g., 0.02% [w/v] sodium azide).
4. Desalting column. These can be purchased ready to use (e.g., Amersham Bioscience Hi-Trap desalting, Bio-Rad Econo-Pac P6 or Perbio D-Salt columns) or can be prepared in the lab.
5. Ion-exchange column. Ion-exchange columns can be purchased ready to use (e.g., Amersham Bioscience or Bio-Rad) or can be prepared in the lab by packing a column with loose ion-exchange beads according to the manufacturer's instructions.
6. Chromatography equipment (*see* **Notes 6** and **7**).
7. Assay methods for analysis of the purified materials will be required to determine the success of the purification.

Table 3
Buffers: Cation Exchange

Buffer	Cation	pH range
Maleic acid	Na$^+$	1.5–2.5
Formic acid	Na$^+$	3.3–4.3
Citric acid	Na$^+$	2.6–6.0
Lactic acid	Na$^+$	3.6–4.3
Acetic acid	Na$^+$	4.3–5.3
MES/NaOH	Na$^+$	5.5–6.7
Phosphate	Na$^+$	6.7–7.7
MOPS	Na$^+$	6.5–7.5
HEPES	Na$^+$	7.5–8.2

3. Methods

3.1. Preparation

1. The starting material must first be equilibrated in the binding buffer before ion exchange can be commenced. If the sample is not already prepared in a suitable buffer for the desired components to bind to the matrix, then the buffer can be replaced either by dialysis or by using a desalting column (*see* Chapter 26). If recovery of the protein of interest is to be measured following ion-exchange chromatography, then it is worthwhile also measuring recovery of the protein after the preparatory desalting. Some proteins precipitate in low-ionic-strength buffers close to the protein p*I* and these are the conditions employed in ion-exchange buffers. Small changes in buffer pH, ionic strength, or buffer composition can make major changes in the recovery of protein both on the desalt step and on the ion-exchange purification.

2. The sample should be filtered (0.45 μm) before applying to the column to reduce the risk of column blockage.

3. Before use, the ion-exchange column should be charged with the counterion. The most commonly used counterions are sodium (Na$^+$) for cation exchange and chloride (Cl$^-$) for anion exchange. Many ion exchangers are supplied charged with Na$^+$ or Cl$^-$; however, if this is not the case or if a different counterion is to be used, then the column will need to be charged with the appropriate counterion. This is most easily done by pumping 1–2 column volumes of high-ionic-strength elution buffer through the column (*see* **Note 8**).

4. Once charged with the counterion, the column needs to be thoroughly washed with the binding buffer (5–10 column volumes) to ensure equilibration in the low-salt buffer prior to application of the protein. Measure pH and conductivity of the buffer eluted from the column and compare this with the pH and conductivity of the binding buffer being applied to the column. Once the column is equilibrated, then the measurements for eluate and binding buffer should be the same.

3.2. Chromatography

1. The sample, in the binding buffer, is applied to the column either by gravity flow or preferably using a pump. The recommended flow rate for the ion-exchange medium should be included in the suppliers instructions. The chromatography steps are often carried out at a lower flow rate than column washing and equilibration because the proteins are larger than the buffer ions so it will take longer to diffuse into the pores of the stationary phase.

2. The eluate from the column should be collected for analysis to confirm that the protein of interest has bound to the column.

3. Once the sample has been applied, the column is washed with several column volumes of binding buffer (around 5 column volumes, depending on sample and on column packing) to ensure that all nonbound proteins are washed out of the column. Monitoring the column eluate with an ultraviolet (UV) detector at 280 nm gives an immediate visual indication of the amount of protein or other UV-absorbing material present in the eluate from the column.

4. Elute bound proteins by washing the column with an increasing salt gradient of 0–500 m*M* NaCl in binding buffer over 10–15 column volumes (*see* **Notes 5 and 9–11**).

5. Collect the eluted protein in fractions for analysis.

6. Analyse both nonbound material and eluted fractions to determine in which fractions the protein of interest has been isolated and whether contaminants have coeluted. Based on this analysis, required modifications to the chromatographic conditions can be planned (*see* **Note 10**).

3.3. Column Regeneration and Storage

1. Ion-exchange columns should be cleaned and regenerated between purifications, otherwise the binding can rapidly be reduced and the column can become blocked by contaminants. Ion-exchange columns can be washed with a high-salt solution as part of the elution gradient or as a separate cleaning step. Use of 1 *M* NaCl will elute most covalently bound contaminants not eluted during the purification (*see* **Note 8**).

2. Ion-exchange columns can be stored packed provided they contain a bacteriostatic solution. The column can be equilibrated in buffer containing 0.02% (w/v) sodium azide or (depending on the bead material) in a buffer containing 20% ethanol. Consult the manufacturer's instructions for the recommended storage options. Ensure that the column tubing is sealed to prevent drying of the column packing during storage. Columns should preferably be stored at 4°C.

3. Before subsequent reuse, any cleaning or storage solution must be washed out and the column must again be fully equilibrated by pumping 5–10 column volumes of the appropriate binding buffer through the column.

4. During long-term storage, there is a likelihood that the column may start to dry out. Therefore for extended storage periods, it is recommended that the column should be unpacked and the ion-exchange matrix stored in a buffer containing a bacteriostatic agent (e.g., 20% ethanol).

4. Notes

1. The isoelectric point (p*I*) can be determined experimentally by isoelectric focusing. Isoelectric points for many proteins can be found in the literature *(2)*. The p*I* of novel proteins can be predicted from the amino acid sequence of a protein if this is known. Software packages are available that will calculate the p*I* and there are sites accessible on the Internet where p*I* and other properties for a protein can be calculated if the amino acid sequence or the base sequence of the DNA coding for the protein is entered (e.g., Swissprot, www.expasy.ch). Once into the Swissprot site, go to "Proteomics Tools" and then click on "Primary structure analysis." There are several options within this part of the site that offer predictions of protein properties, including p*I*.

2. When scouting for the best pH for ion-exchange purification, start by trying a pH around 1–1.5 pH units from the p*I* of the protein being purified: one unit above the p*I* for anion exchange and one unit below the p*I* for cation exchange. Having analyzed the separation achieved at this pH, the buffer can be adjusted slightly in subsequent runs to improve the results.

3. It is often desirable to choose conditions such that the protein of interest elutes early in the elution gradient. Under such conditions, little of the binding capacity of the column will be occupied by weaker binding species, so the column will have a greater capacity for the protein of interest.

4. The ionic strength of the buffer is as important as the pH when carrying out ion-exchange purification. In most cases, a starting buffer of 20–50 mM is suitable. Many proteins tend to aggregate in solutions close to the protein's pI; this aggregation is increased in solutions of lower ionic strength. Hydrophobic interactions between proteins and the chromatography matrix will also increase at low ionic strength and these can affect the separation achieved creating a "multimode" chromatography. It is therefore not advisable to use buffers of less than around 10–20 mM for most applications.

5. Proteins are most commonly eluted from the ion-exchange matrix by increasing the ionic strength of the solution. Usually, this is achieved by increasing the sodium chloride concentration, but it can also be achieved by increasing the molarity of the buffer components. Proteins can also be eluted by a change in buffer pH, raising the pH to elute from cation exchangers and lowering the pH to elute from anion exchangers. However, it is more difficult to control pH gradients on standard ion-exchange columns than to control changes in ionic strength, so pH gradients are only commonly applied with chromatofocusing columns.

6. Ion-exchange chromatography can be carried out in short, wide columns because the bed volume is more important than the bed height. Bed heights of 5–10 cm are frequently used. Once the capacity of the column has been determined for your protein under your conditions, then the column volume can be set to fit the amount of protein to be purified in one run.

7. The equipment used for ion-exchange chromatography can be as basic as a gravity-fed column with the eluate fractions collected manually. However, the use of programmable chromatography control equipment is recommended, as it will make it easier to carry out purification reproducibly and aid in analysis of the chromatography results.

   In addition to the equipment used for other chromatography methods, a conductivity meter is desirable to measure the changes in buffer conductivity during the chromatography process. An in-line conductivity cell and a conductivity meter allow continuous monitoring of the eluate conductivity. This can be recorded alongside the ultraviolet (UV) monitor trace on a two-channel chart recorder, allowing determination of the conductivity at the point each protein peak is eluted.

   A gradient mixer is required for elution of proteins by salt gradient. Gradient mixers allow the formation of a controlled and reproducible salt gradient that is essential for run to run consistency.

8. It is common to charge the ionic groups on the column matrix with the counterion by flushing the column with the high-salt buffer used for protein elution. However, other high-salt solutions can be used. Sodium chloride (1 M) is often used for cleaning ion-exchange columns between purification runs. This will charge the column with chloride (anion exchange) or sodium (cation exchange) at the same time as cleaning. Sodium hydroxide (0.5–1 M) can be used for cleaning and sanitizing of Sepharose ion-exchange columns; on cation-exchange columns, it will also serve to charge the matrix with sodium. Washing the column with NaOH (0.5 M for 30 min) acts as a bacteriocide and will also destroy endotoxin bound to the column. The instructions provided by the ion-exchange resin manufacturer should be consulted for appropriate methods of sanitizing or sterilizing the chromatography medium you are using.

9. If the bound proteins are to be eluted using a continuous gradient, then a low-salt buffer, normally the binding buffer, and a high-salt buffer (binding buffer plus 0.1–1 M sodium chloride) are prepared. The gradient mixer can then be programmed to produce a gradient by mixing a low-salt solution with increasing amounts of a high-salt solution. The volume over which the gradient is run can be altered according to the required resolution. For initial investigations, a gradient of 0–500 mM NaCl over 10–15 column volumes is often suitable.

10. After initial purification runs have been analyzed it may be desired to alter the gradient to improve separation of eluted proteins. The gradient can be altered either by increasing or decreasing the ionic strength of the high-salt buffer or by altering the volume over which the gradient is applied. Applying a more gradual gradient will have the effect of increasing peak separation but will also spread peaks, reducing the concentration at which each protein is collected. Applying a steeper gradient sharpens the eluted peaks, but may cause closely eluting peaks to merge, increasing contamination of the product. By careful analysis of the peaks eluted, the gradient can be fine-tuned to optimize purification of the desired protein. Although ion-exchange columns have a high capacity, the resolution achieved can be improved by loading the column to well below the maximum binding capacity (10–20%).

11. Elution by stepwise increases in sodium chloride can be used if a gradient mixer is not available or once elution conditions have been determined well enough to determine the appropriate salt concentrations for each step. In this case, prepare solutions of (binding) buffer-containing sodium chloride at appropriate concentrations over the desired range. Step elution using appropriate concentrations can achieve the highest elution concentrations for a protein. However, if the wrong concentrations are used, peak splitting or coelution of peaks can result. The solutions are run through the column in turn from the lowest salt concentration to the highest. With each solution, the volume applied should be sufficient to elute all of the proteins that can be eluted at that concentration. This should be monitored using an UV monitor.

References

1. Amersham Pharmacia Biotech (1999) *Ion Exchange Chromatography Principles and Methods*, Amersham Pharmacia Biotech, Uppsala, Sweden.
2. Righetti, P. G. and Carravaggio, T. (1976) Isoelectric points and molecular weights of proteins: a table. *J. Chromatogr.* **127,** 1–28.

15

Hydrophobic Interaction Chromatography

Paul A. O'Farrell

1. Introduction

Hydrophobic interaction chromatography (HIC) is a technique for the separation of biological macromolecules based on their surface hydrophobicity. Although proteins maintain their tertiary structure by burying a hydrophobic core and exposing polar residues to the solvent, it is still the case that they have hydrophobic areas on their surfaces [as much as 50% of the surface area *(1)*]. Indeed, these areas are often critical to a protein's function, as many of them are involved in protein–protein interactions. It is the variety in the extent and character of these hydrophobic patches that is exploited for separation in HIC. It can be a powerful technique, especially because it operates on a principle different from either of the two most commonly used general chromatographic methods, ion exchange and size exclusion, and can thus be used to separate components that these techniques cannot. Careful manipulation of the conditions can enable it to be very sensitive. For example, it is capable of separating proteins that differ by as little as one amino acid residue and of separating native from incorrectly folded forms *(2)*.

The physical mechanism by which hydrophobic interaction occurs is not fully understood. A number of different mechanisms have been put forward, including some based on surface tension, van der Waals forces, the charge-masking effect of small ions, and entropic considerations *(3)*. There is a certain degree of overlap between these explanations and it is likely that a number of different effects are involved. However, it is clear that the properties of water—in particular its ability to form structure—are of central importance. Water is a very polar molecule and forms a net of hydrogen bonds in its liquid form, which in the short range has a great degree of order. When a nonpolar solute is dissolved in liquid water, this structure is disturbed. The water molecules cannot hydrogen-bond to it and are forced to form a shell around the solute that is, in fact, more ordered than the surrounding bulk solvent. In the case of a protein, such shells are formed around hydrophobic patches on its surface. The formation of these ordered shells reduces the overall entropy of the solution and, consequently, is thermodynamically unfavorable. When two nonpolar solutes come into contact or a hydrophobic patch on a protein's surface comes into contact with a hydrophobic ligand,

From: *Methods in Molecular Biology, vol. 244: Protein Purification Protocols: Second Edition*
Edited by: P. Cutler © Humana Press Inc., Totowa, NJ

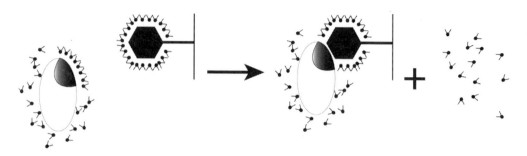

Fig. 1. Protein binding to a hydrophobic ligand. Shells of ordered water surround the hydrophobic patch on the protein's surface and the hydrophobic ligand attached to the column matrix. When the ligand and hydrophobic patch come into contact, the ordered water is displaced to the bulk solvent. The resultant increase in overall entropy favors the interaction.

these ordered water shells can amalgamate, releasing some of the water molecules to a less ordered state. This results in an increase in entropy, conferring a thermodynamic advantage that favors the interaction (*see* **Fig. 1**). This is the classic interpretation of the hydrophobic interaction: It is not an attractive force *per se*, but an interaction that is conferred on nonpolar solutes by the structure of a polar solvent. It follows that the addition of substances that alter the structure of the solvent can have an effect on the interaction. Structure-forming (lyotropic) salts favor hydrophobic interaction by increasing the degree of order in the shell of water surrounding the nonpolar solute and thus increasing the thermodynamic advantage to be gained by releasing those molecules to the bulk solvent.

This explanation of the hydrophobic effect was developed using small molecules, and although these ideas are still applicable, complications arise when we consider hydrophobic interaction chromatography of proteins. Proteins are flexible, labile, and, in some cases, fragile, and in addition to having a hydrophobic character, they also possess polar and ionizable charged groups. These additional factors make it difficult to predict how a particular protein will interact with a hydrophobic matrix. Thus, although an understanding of the theoretical aspects can be very useful, method development for any specific protein purification procedure is essentially empirical (*see* **Note 7**). The simplest method involves application of the sample in a high concentration of a structure-forming salt and elution by reducing the salt concentration. The procedure presented here is not part of a purification scheme: The sample used here consists of two commercially available proteins; but it can be useful for evaluation of a column's performance, and the method can easily be adapted to the reader's requirements.

2. Materials

1. Buffer 1: 50 mM sodium phosphate, pH 7.0; 2 M ammonium sulfate (*see* **Notes 1** and **2**).
2. Buffer 2: 50 mM sodium phosphate, pH 7.0.
3. Sample: 1 mg/mL each chicken egg lysozyme and whale myoglobin dissolved in buffer 1.
4. Chromatography system: AKTA purifier (Amersham Biosciences).
5. Column: Phenyl Sepharose HP, 1-mL column volume (Amersham Biosciences). The matrix consists of beads of crosslinked agarose substituted with a phenyl ligand (*see* **Note 3**).

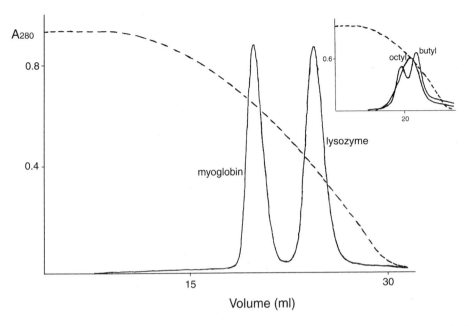

Fig. 2. HIC on phenyl–sepharose HP column. Absorbance at 280 nm is shown by a solid line and the conductivity is represented by a dashed line. The inset shows a similar experiment carried out using chromatography columns bearing butyl and octyl ligands; these more hydrophobic ligands do not result in adequate separation.

3. Method

1. The method is carried out at 4°C (*see* **Note 4**).
2. Filter and degas all buffers before use. The protein sample should also be filtered or centrifuged, as precipitation is a possibility under high-salt conditions.
3. The flow rate used here is 2 mL/min; this is dependent on the pressure limits of the column and the back-pressure generated by the system.
4. The following steps are programmed in the method. Absorbance can be monitored at 280 nm and at 410 nm.
   a. Equilibrate column with 5 column volumes (cv) buffer 1.
   b. Load 1.5 mL sample through loop.
   c. Wash with 4 cv buffer 1.
   d. Elute with gradient of 0–100% buffer 2 over 30 cv; collect 1-mL fractions (*see* **Note 5** and **Fig. 2**).
   e. Wash with 4 cv buffer 2.
5. Analyze fractions. (The myoglobin peak can also be identified by its absorbance at 410 nm.)
6. Cleaning. The column can be further cleaned with more buffer 1, distilled H_2O, or 50% ethanol. Any suitable buffer can be used for short-term storage. Longer-term storage requires a biocidal agent such as 20% ethanol or 0.02% sodium azide (*see* **Note 6**).

4. Notes

1. A number of different structure-forming salts can be used for binding in HIC. The Hofmeister series (*see* **Table 1** and **ref. 4**) lists ions in order of their effectiveness in promoting the

Table 1
The Hofmeister Series

Anions	Cations
PO_4^{3-}	NH_4^+
SO_4^{2-}	Rb^+
CH_3COO^-	K^+
Cl^-	Na^+
Br^-	Cs^+
NO_3^-	Li^+
ClO_4^-	Mg^{2+}
I^-	Ca^{2+}
SCN^-	Ba^{2+}

Note: Ions are listed in the order from those that most favorable hydrophobic interaction at the top, to less favorable ions at the bottom.

hydrophobic effect (*see also* **ref. 3**). Often, the salt of choice is ammonium sulfate. This is because it has high solubility, good lyotropic properties, and low absorbance at 280 nm and is inexpensive. Indeed, these are also the properties that make it the reagent of choice for fractional precipitation by salting out. The choice of a salt need not always be determined simply by its lyotropic properties however. For example, if HIC is to be performed immediately following ion-exchange chromatography, it is economical to use the salt that was used for elution in that procedure. This avoids the need for dialysis or other methods of buffer exchange; the salt concentration simply needs to be increased to a level that ensures binding to the hydrophobic matrix. This kind of economy also often leads to HIC being used immediately after ammonium sulfate precipitation in a purification scheme.

2. As pH only affects the charged residues, it theoretically should not impact on hydrophobic interaction. However, it has been observed that there is a general decrease in the strength of interactions between proteins and hydrophobic matrices with increasing pH. This is presumably caused by increasing hydrophilicity of the protein resulting from the titration of charged groups. This general trend is supplemented by the fact that pH effects are different for different proteins. Thus, it is possible to modify elution profiles and improve separation by carrying out the procedure at various pH values. It is important, however, to work at pH values at which the protein of interest is stable and to bear in mind that proteins may precipitate near their isoelectric points (p*I*s), resulting in lower recovery.

3. Many different hydrophobic matrices are available commercially. Matrices are generally substituted with phenyl or alkyl ligands of varying chain lengths. Some products also combine hydrophobicity with other effects, such as charge. The choice of which ligand to use is empirical and should be determined by small-scale pilot experiments. Prior knowledge about the hydrophobicity of the protein of interest can guide these experiments. For example, membrane proteins will likely be very hydrophobic and may irreversibly bind to a strongly hydrophobic ligand, leading to sample loss. A weakly hydrophobic ligand such as ether may be the initial choice in such a case. In fact, with an uncharacterized protein, it is generally advisable to begin with a weakly hydrophobic ligand, particularly if the amount of available material is low and high recovery is required.

   The matrix used to support the ligand needs two characteristics. First, it should be inert; that is, there should be no functional groups (e.g., charged groups) that would provide a binding surface for the protein, as we want to control the activity only with our choice of ligand. Second, it must have sufficient mechanical strength to withstand the pressures to be used. Manufacturers of chromatography systems generally sell prepacked columns with suitable matrices that are available with a variety of hydrophobic ligands.

4. Because hydrophobic interaction is primarily driven by entropy changes, it is apparent that temperature will have an effect on binding *(5)*. In general the strength of the interaction increases with increasing temperature. Although, theoretically, it should be possible to bind a protein to a hydrophobic matrix at high temperature and to elute it by reducing the temperature, such temperature shifts are not used in practice. Temperature can affect protein stability and conformation, and these properties, in turn, impact on a protein's interaction with the hydrophobic matrix. As these effects are different for different proteins, it is possible to alter chromatographic profiles by carrying out the procedure at different temperatures and this may result in better separation. These effects are difficult to predict however, and it is often simpler just to work at a temperature where the protein of interest is stable. In many laboratories, it is not possible to change the temperature (e.g., the chromatography system may be in the cold room). In cases where the temperature can be changed, its effects should be remembered during method development, which should be carried out at the same temperature at which it is intended to carry out the purification.

5. Elution is generally effected by reducing the salt concentration. This can be achieved either via a gradient or in a stepwise fashion. Although a gradient elution should give better separation, peaks can sometimes be broad, resulting in larger elution volumes and decreased resolution. Stepwise elution often gives lower elution volumes, which may be useful if the protein is unstable in dilute solution or if HIC is to be followed by size-exclusion chromatography. In some cases, simply reducing the salt concentration may not be sufficient to release the protein from the column. More stringent elution conditions include the use of alcohols (e.g., 0–80% ethylene glycol and up to 30% isopropanol) or detergents (e.g., Triton-X-100, 1% [w/v]). Care must be taken with such conditions, as these agents can have deleterious effects on protein structure. In addition, detergents can be difficult to remove from hydrophobic matrices. The need for such measures may sometimes be avoided by the use of a less hydrophobic matrix. It is possible to use a ternary gradient, increasing the concentration of a chaotropic agent (e.g., alcohol) at the same time as decreasing the salt concentration. This technique has been reported to give increased resolution *(6)*. Some workers have also found that sharper peaks can be achieved by including a low level of solvent (0.1–5% ethanol) in all buffers.

6. It is generally recommended to clean the column regularly to prevent slow buildup of contaminants. More stringent cleaning can be achieved by following the manufacturer's instruction for column regeneration. These methods generally involve using a strong base such as 1 *M* sodium hydroxide followed by copious flushing with water. In addition to the column, the chromatography system itself must be cleaned. Many modern fast protein liquid chromatography (FPLC) systems have very narrow tubing, both to reduce mixing and to reduce dead volume. The high salt concentrations used in HIC can quickly result in the formation of salt crystals because of evaporation, which can block this tubing and are very difficult to remove. Thus, it is important to flush the system thoroughly after chromatography. In addition, chloride ions can be very corrosive to stainless-steel components and should be flushed from vulnerable systems.

7. The following provides a short description of possible steps in method development for HIC:
   a. Sample: Know your protein of interest (POI) with respect to its stability in various conditions (e.g., pH and temperature). Work within a suitable range. What is the p*I*? Does the POI require additives for stability? Is it soluble only in a limited range of ionic

strength? Are detergents necessary? Detergents can be difficult to remove from HIC columns and may argue against using this chromatographic mode; a limited ionic strength range will affect the experimenter's ability to effect binding and elution.

   b. Sequencing: Where in the overall purification scheme will the HIC step lie? Carrying out HIC immediately after ion-exchange chromatography or ammonium sulfate precipitation is convenient, allowing the experimenter to take advantage of the salt already present in the sample. Following HIC with a size-exclusion step will allow the salt to be removed without resorting to dialysis.

   c. Salt concentration: Perform a salting-out experiment with the salt to be used for binding. At what concentration does the POI precipitate? A concentration just below this can be used for binding. Do contaminating proteins precipitate before or after the POI? If many contaminating proteins precipitate before the POI, it may be worthwhile to do a fractional precipitation step—both as purification in itself and to avoid binding too many contaminants to the column. If most contaminants precipitate after the POI, it is likely that they will not bind to the column under conditions that will allow binding of the POI.

   d. Ligand: If amounts of the POI are limited, it is best to begin with a ligand of low hydrophobicity to avoid sample loss? However, if amounts are plentiful, experiments can begin with a ligand of intermediate hydrophobicity. Ideally, we want to find a ligand that allows elution of the POI in the middle of the gradient. If the POI does not bind, move to a more hydrophobic ligand. If it binds but cannot be eluted by simply reducing the salt concentration, move to a less hydrophobic ligand. If the POI will not bind to a HIC column, perhaps HIC can be used as a "negative" chromatographic step—removing contaminants by allowing them to bind to the column while the POI flows through.

   e. Elution: Elute by decreasing the salt concentration via a gradient. The slope of the gradient can be adjusted to balance the desired resolution with the desired elution volume. If the POI is sufficiently resolved, a step gradient can be used to reduce the elution volume. Changing the pH at this stage may allow the resolution to be improved. Other means of improving resolution can also be attempted at this stage, such as the use of a ternary gradient or the addition of small amounts of solvent to all buffers, as mentioned in **Note 5**.

References

1. Lee, B. and Richards, F. M. (1971) The interpretation of protein structures: estimation of static accessibility. *J. Mol. Biol.* **55(3)**, 379–400.
2. Jing, G. (1994) Resolution of proteins on a phenyl-Superose HR5/5 column and its application to examining the conformation homogeneity of refolded recombinant staphylococcal nuclease. *J. Chromatogr.* **685(1)**, 31–37.
3. Melander, W. (1977) Salt effect on hydrophobic interactions in precipitation and chromatography of proteins: an interpretation of the lyotropic series. *Arch. Biochem. Biophys.* **183(1)**, 200–215.
4. Hofmeister, F. (1888) Zur lohre von der wirkung der salze. Zweite mittheilung. *Arch. Exp. Pathol. Pharmakol.* **24**, 247–260.
5. Haidacher, D., Vailaya, A., and Horvath, C. (1996) Temperature effects in hydrophobic interaction chromatography. *Proc. Natl Acad. Sci. USA* **93(6)**, 2290–2295.
6. El Rassi, Z., DeCampo, L. F., and Bacolod, M. D. (1990) Binary and ternary salt gradients in hydrophobic interaction. *J. Chromatogr.* **(499)**, 141–152.

16

Affinity Chromatography

Paul Cutler

1. Introduction

1.1. General Principles

Affinity chromatography is a method of selectively and reversibly binding proteins to a solid support matrix based on the exploitation of known biological affinities between molecules. The interaction between the target protein and the matrix is not based on general properties such as the isoelectric point (pI) or hydrophobicity, which are more commonly used in adsorption chromatography, but on individual structural properties such as the interaction of antibodies with antigens, enzymes with substrate analogues, nucleic acid with binding proteins, and hormones with receptors. In order to exploit the interaction for the purposes of purification, one of the components, the *ligand*, must be immobilized onto a solid matrix in a manner that renders it stable and active. Once produced, the matrix can be used to purify its specific target from a suitable biological extract *(1)*.

Affinity chromatography is a particularly powerful technique because it offers the potential of purifying target proteins that exist in very low titer from complex mixtures with a very high degree of purification in a single step. Because the aim is to purify material on the basis of biological function, it is even possible to selectively separate active and inactive forms of the same material.

Affinity matrices can be formed from ligands which are either mono-specific or group-specific. Mono-specific ligands recognize a single form of a protein such as a receptor with affinity for a specific hormone or an enzyme recognizing an inhibitor. These matrices are often "tailor-made" for particular separations. Group-specific ligands include enzyme cofactors, plant lectins, and protein A from *Staphylococcus aureus* (*see* **Table 1**). Because of their generic nature, group-specific matrices are frequently available commercially from a range of suppliers, often in prepacked columns.

Originally, in order to purify a protein by affinity chromatography, it was necessary to find a naturally occurring ligand. In recent years, the definition of affinity chromatography has expanded to include the separation of proteins by specific interactions other than purely biological interactions. Included in this category are dye affinity ligands such as cibacron blue *(2)* and immobilized metal ion affinity matrices *(3)*. The orig-

From: *Methods in Molecular Biology, vol. 244: Protein Purification Protocols: Second Edition*
Edited by: P. Cutler © Humana Press Inc., Totowa, NJ

Table 1
Some Commonly Employed Group-Specific Affinity Ligands

Ligand	Target protein
5' AMP, ATP	Dehydrogenases
NAD, NADP	Dehydrogenases
Protein A	Antibodies
Protein G	Antibodies
Lectins	Polysaccharides, glycoproteins
Histones	DNA
Heparin	Lipoproteins, DNA, RNA
Gelatin	Fibronectin
Lysine	rRNA, dsDNA, plasminogen
Arginine	Fibronectin
Benzamidine	Serine proteases
Polymyxin	Endotoxins
Calmodulin	Kinases
Cibacron blue	Kinases, phosphatases, dehydrogenases, albumin

inal dye-based affinity adsorbents, such as those based on triazine dyes, lacked specifity. However, with the advent of ligands designed around natural ligands usign computers, biomimetic dyes are expected to have an important role as affinity chromatography tools *(4)*. A similar *in silico* approach can be taken with finding peptide affinity reagents or via combinatorial chemistry *(5,6)*. Such approaches include the use of customized affinity matrices derived from phage display. The peptides or proteins can be expressed on the surface of a bacteriophage via fusing the sequence of the protein to that of a surface coat protein of the phage *(7)*.

With the advent of monoclonal antibodies and recombinant protein technology, it is possible to manipulate a biological affinity. By using hybridoma technology, it is possible to produce economically viable amounts of monoclonal antibodies where the target protein is the antigen. Immobilizing the antibody on a solid support creates an *immunopurification* matrix *(8)*.

It is also possible with recombinant DNA technology to produce a target protein in cell culture that has a specific affinity region, the *tag*, engineered into the product *(9,10)*. The tag is used to purify the protein on a relatively inexpensive and well-defined affinity matrix. Enzymic cleavage points may also be inserted between the protein and the tag sequences so that once purified from the culture, the tag can be removed. An example of this is a fusion containing the target protein with a hexahistidine sequence that shows preferential binding to immobilized metal affinity matrices *(11)*. Immobilized metal affinity matrices are based on the interaction between an immobilized transition metal (nickel, copper, zinc, etc.) and specific amino acid side chains. The residue exhibiting the strongest affinity is histidine because of the ability to form coordination bonds between the electrons in histidine and the metal ion. Therefore, by engineering six consecutive histidine residues into the protein, the expressed protein can be efficiently removed from complex mixtures *(12)*.

A further example is the engineering a streptavidin binding sequence via a fusion protein, which, although occupying the same pocket as biotin, is a nine-amino-acid peptide

(13). This has the advantage over biotin of being expressed as part of the protein and facilitating elution under relatively mild conditions while still permitting detection via standard streptavidin based Western blot and enzyme-linked immunosorbent assay (ELISA) systems.

As our understanding of cellular processes and molecular biology increases, the potential exists to use other biologically based molecules to isolate certain proteins, the use of DNA-based affinity chromatography to isolate transcription factors was recently reviewed *(14)*. Following the development of methods for increased protein characterization including posttranslational modifications, the affinity isolation of proteins according to glyosylation *(15)* or phosphorylation *(16)* is now possible.

Strategies for affinity chromatography vary depending on the target protein, stability, scale of operation, and so on forth *(17,18)*. The use of double affinity purification, using either endogneous of genetically engineered tags, can offer advantages in terms of selectivity *(17,19)*. This has been particularly effective for the isolation and characterization of complexes *(20)*.

Two schools of thought exist regarding where an affinity chromatography step should lie in a purification scheme. One suggests that because of the expense of producing affinity matrices, the material should be partially purified prior to the affinity step to protect the matrix from proteases and so forth. Conversely, the affinity matrix can first rapidly purify what may be an unstable protein to near homogenity in a single high yielding step while dramatically reducing the volume of material to be processed. With the increasing availability of affinity matrices manufactured using recombinant technologies to produce less costly ligands, the latter school of thought appears to be gaining popularity. However, the actual position of the affinity chromatography step within a purification scheme must be decided on a case-by-case basis.

1.2. Affinity Matrices

1.2.1. Ligand

The immobilized affinity ligand must be able to form a reversible complex with the target protein. The binding constant (the inverse of the dissocciation constant) should be high enough to enable stable complex formation ideally at physiological conditions. However, the affinity must be sufficiently weak to facilitate the elution of the protein under relatively mild conditions, thereby avoiding denaturation of either the ligand or target protein. Binding constants of 10^5–10^{11} *M* are usually considered a good working range (*see* **Note 1**).

A detailed description of the immobilization of the ligand to generate an affinity support is beyond the scope of this chapter, however, reviews exist describing immobilization techniques (*see* **ref. 21** and **Note 2**). In addition to immobilization in a form that retains biological function of the ligand, good chemical stability must be maintained in order withstand the harsh elution and cleaning regimes used in affinity chromatography process. Leaching of the ligand from the matrix must be minimized to avoid contamination of the purified protein.

1.2.2. Matrix

The matrix of choice should show the physical and chemical properties required for adsorption chromatography techniques *(22,23)*. The matrix typically constitutes macroporous beads (50–400 μm) to produce a large surface area that maximizes the capacity of the activated matrix and prevents any size-exclusion effects. It must be hydrophillic but uncharged to prevent nonspecific ionic binding. It must demonstrate good mechanical rigidity and chemical stability under the conditions used for activiation, derivitization, elution, and regeneration. Although the most commonly used matrix support is agarose, a range of other matrix supports, including silica and polyacrylamide, are available. Selection of the matrix should be made based on the physicochemical properties required.

1.2.3. Spacers

One of the common problems encountered when derivitizing a matrix for affinity chromatography is the steric hindrance observed when the ligand is too close to the solid phase. To minimize this, affinity matrices are often activated with a spacer arm, commonly a six-carbon hydrophilic chain to distance the ligand from the solid phase. The length of the spacer arm should be optimized for each individual matrix. Although longer spacer arms may allow greater ligand availability, they can lead to steric hindrance and confer unwanted hydrophobicity onto the support.

1.2.4. Activation of the Matrix

The protein ligand is almost always covalently attached to the solid support. The matrix is activated with a reagent such as cyanogen bromide, which reacts with the epsilon amino groups of lysine residues in protein ligands resulting in the protein being covalently attached. Other chemistries enable immobilization of proteins via carboxyl, hydroxyls, and thiol functionalities (*see* **ref. *21*** and **Note 2**).

1.3. Practical Aspects

Purification by affinity chromatography is a relatively straightforward technique because of the selective binding of the target protein (*see* **Fig. 1**). The crude extract is passed through the column, the target material binds, and the matrix is then washed to remove the nonbound fraction. This may involve more than one wash step if nonspecific interactions or selective binding of impurities to the bound target protein are suspected. The conditions of the mobile phase are then modified usually by changes in pH or ionic strength or by inclusion of either a general chaotropic agent or a competing soluble ligand to facilitate elution of the protein. The eluted protein is collected and the column regenerated under suitable conditions.

1.3.1. Sample Preparation

Efficient binding is facilitated by ensuring that the sample and the column are equilibrated in a buffer that is optimal for binding, this requires a defined pH, ionic strength, and so forth. As most biological affinities in mammalian systems occur at physiological pH, buffers such as PBS (phosphate-buffered saline) are common. The buffer should contain any elements such as cofactors required to maintain activity of the target protein. The sample can be conditioned for binding to the matrix by methods such as dialysis, desalting, ultrafiltration, or simply by pH adjustment. If crude material is to be loaded, then some form of pretreatment may be required, such as filtration or centrifugation. It may be necessary to include protease inhibitors such as leupeptin or phenylmethylsulfonyl fluoride (PMSF) to prevent proteolytic degradation of the target (*see* **Note 3**).

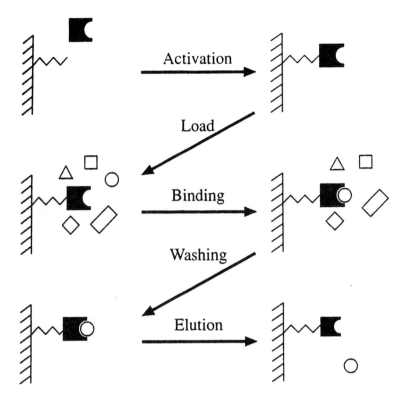

Fig. 1. Principle of affinity chromatography. The ligand is immobilized onto a solid support matrix. The crude extract is passed through the column. The target molecule for which the ligand possesses affinity is retained, whereas all other material is eluted. The bound target protein is eluted by alteration of the mobile-phase conditions.

1.3.2. Column

The column size and the degree of ligand substitution determines the capacity of the column for the target protein. The affinity constant of the ligand for the target protein under the operating conditions will also dictate column performance. This can be determined empirically by overloading the column or, in a more sophisticated manner, by producing a breakthrough curve to calculate the adsorption isotherm (*see* **Fig. 2**). Capacities of 1–20 mg protein bound/mL matrix are common. In addition to standard chromatographic columns, other affinity separation methodologies exist, including activated filters *(24)* and stirred tanks or expanded beds (*see* **ref. *25*** and **Note *4***).

Under normal conditions, short, wide columns are used to facilitate rapid separations. However, where the affinity is low and the material is only retarded, separation is enhanced by use of a thinner, longer column *(26)*.

Columns should be packed in accordance with the standard procedures described by the manfacturers. Briefly, the column should be packed in a glass or acrylic column with flow adaptors to ensure a minimum of dead space. For small-scale separations, less expensive disposable columns can be used under gravimetric flow.

The flow rate used for adsorption depends on the physical properties of the matrix support and the binding constant of the ligand for the target protein. Maximum adsorp-

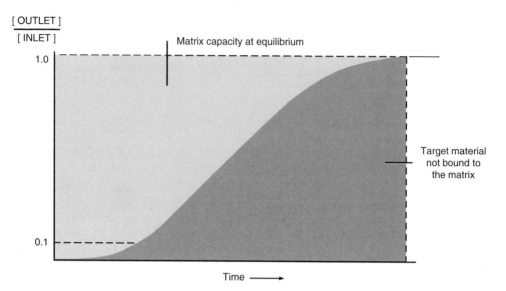

Fig. 2. Breakthrough curves can be used to establish the effective capacity of an affinity matrix. By specifically detecting or assaying the target protein appearing in the flowthrough from the column, the breakthrough of material can be plotted. Eventually, as the matrix reaches equilibrium (theoretical maximum capacity), the inlet concentration and outlet concentration are equivalent. The effective, or dynamic, capacity of the matrix is reached sooner and is often taken when the ratio of outlet to inlet is 0.1.

tion usually occurs at a low flow rate, although flow rates tend to be compromised between optimal binding and process time, which affects cost and target protein stability. In practice, linear flow rates of 30–50 cm/h are used for low-pressure preparative separations.

1.3.3. Loading

Loading volume is generally not critical so long as the total amount of target protein does not exceed the capacity of the column. The passage of proteins through the column is monitored most commonly using an ultraviolet (UV) detector. Following loading, the column is washed with equilibration buffer to remove any unbound material. Once baseline has been achieved on the detector, a more stringent wash can be made if necessary to remove any weakly binding material that may have bound nonspecifically. This wash is designed so as not to cause elution of the target protein. Often, the wash buffers can contain low levels of detergents or moderate concentrations of salt.

1.3.4. Elution

The protein of interest is eluted by weakening the ligand–protein interaction. This can be done either nonspecifically (e.g., by using changes in pH, ionic strength, etc.) or by the addition of a specific solute in the eluant to selectively remove the target protein (e.g., a competing ligand). In extreme circumstances, chaotropic agents may be used although these significantly increase the likelihood of denaturation. Because of the selectivity of binding, it is common for affinity elutions to be performed in a stepwise mode (*see* **Note 1**). However, in certain circumstances, particularly where group-specific ligands are used, a gradient elution may be used. This has been used successfully to separate antibodies of differing subclasses on protein A affinity matrices.

The conditions for elution are one of the primary reasons why affinity chromatography fails to yield active protein (*see* **Note 5**). The need to use harsh conditions to maximize recovery must be balanced against potential denaturation and inactivation of the target protein. A knowledge of the physiochemical properties of the target protein is invaluable when designing an elution buffer. In general, the pH of the buffer should not be near the p*I* of the protein. It is advisable to avoid buffering systems with chelation effects (e.g., citrate) for purification of metalloproteins.

The purified protein should be placed in an environment that promotes its stability as soon as possible after elution. Frequently, in the case of pH elution, this involves titration to near neutrality. The material may be placed in a suitable buffer by using buffer-exchange techniques such as desalting, ultrafiltration, or dialysis. Treatment of the eluated protein will be largely influenced by subsequent purification steps and the use for which the protein is designed (*see* **Note 6**).

1.3.5. Purity Analysis

The performance of the affinity chromatography is usually determined by comparison of the purity before and after purification by polyacrylamide gel electrophoresis (PAGE). Other techniques for assessing purity include gel filtration and reverse-phase high-performance liquid chromatography (HPLC). Where possible, specific assays should be used to detect recovery (% yield) such as Western blotting techniques or, in the case of enzymes, calculating the activity per mass of total protein using techniques such as ELISA.

It is important to analyze the flowthrough to ensure maximum recovery. Several methods exist for analyzing the flowthrough material, such as rechromatography to establish that the capacity of the column has not been exceeded. This, however, will not reflect the inability to bind resulting from denaturation of the ligand or inappropriate binding conditions.

1.3.6. Matrix Regeneration and Storage

It is often the case that the harshest conditions that a solid-phase chromatography matrix is exposed to are those used for regeneration and sanitization. Where a labile protein is used as a ligand, the affinity matrix cannot withstand treatment with stringent agents such as sodium hydroxide. Commonly used regeneration regimes involve the use of chaotropic agents such as 6 *M* guanidine hydrochloride or 3 *M* sodium thiocyanate.

Affinity matrices should be stored under conditions that prevent bacterial and fungal growth. Matrices are typically stored in the presence of antimicrobial agents such as 20% (v/v) ethanol, 0.02% (w/v) Thimerasol, or 0.01% (w/v) sodium azide at standard cold room/refrigerator temperatures (<8°C).

2. Materials

1. Protein A–Sepharose 4B (Amersham Biosciences, Uppsala, Sweden).
2. Buffer A: 100 m*M* Tris-HCl, pH 7.5, 3 *M* NaCl.
3. Buffer B: 50 m*M* Tris-HCl, pH 7.5, 1.5 *M* NaCl.
4. Buffer C: 0.1 *M* citrate, pH 3.0.
5. Buffer D: 0.1 *M* citrate, pH 2.0.

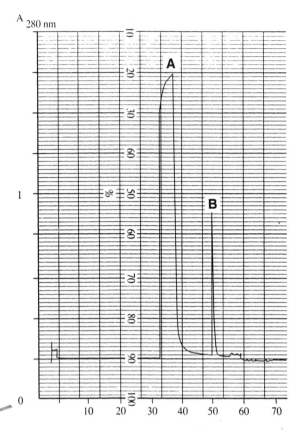

Fig. 3. Purification of goat immunoglobin from serum by protein A affinity chromatography. The majority of the protein elutes in the flowthrough fraction (*peak A*). The bound immunoglobin is eluted by lowering the pH of the mobile phase (*peak B*).

3. Methods

Immunoglobulin Gs (IgGs) bind to Protein A from *S. aureus* via the immunoglobulins' Fc region. The affinity of the antibody for protein A is both species and subclass dependent. A range of protein A affinity matrices can be bought commercially. The crude antibody extract can be prepared by dilution or by dialysis into the binding buffer. The matrix is equilibrated with the binding buffer before the extract is loaded. The unbound material is washed away with the buffer prior to elution with a low pH buffer. An example purification strategy is as follows (*see* **ref. 27** and **Fig. 3A**).

1. Dilute 5 mL of goat polyclonal antiserum raised against human albumin with 5 mL of buffer A and retitrate to pH 7.5.
2. Pack a 5-mL (1.0 cm × 6.4 cm) protein A–Sepharose 4B (Pharmacia) column and equilibrate in buffer B. Apply the diluted serum to the column at a flow rate of 50 cm/h. Wash the column with five column volumes (25 mL) of buffer B.
3. Program the fraction collector to collect 1 mL fractions. Exchangre the inlet and outlet tubing of the column to reverse the flowthrough in the column.
4. Elute the antibody with 5 column volumes of buffer C. Wash the column with 5 column volumes of buffer D, followed by 2 column volumes of 6 *M* guanidine hydrochloride.

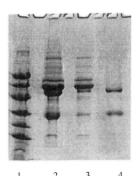

Fig. 4. Analysis of the purification of immunoglobins from goat serum by SDS-PAGE. The crude serum (*lane 2*) was purified by protein A affinity chromatography. The unbound material (*lane 3*) was eluted in the column flowthrough. The purified antibody (*lane 4*) was eluted from the column by lowering the pH of the mobile phase. The SDS-PAGE is run under reducing conditions resulting in the antibody appearing as its constituent "heavy" chains (approx 50 kDa) and "light" chains (approx 25 kDA). The molecular weights are compared to standard marker proteins (*lane 1*).

5. Pool the fractions containing antibody and adjust the pH to pH 7.5 with 1 *M* untitrated Tris base. Analyze the antibody for purity by sodium dodecyl sulfate (SDS)-PAGE (*see* **Fig 4**).

4. Notes

1. It is a commonly held belief that the higher the capacity of a column the better. Although a high capacity is clearly advantageous, it should be remembered that on adsorption, the protein of interest is being removed from a dilute solution and concentrating on the solid phase. On elution, the protein is often released in a relatively high concentration form in a dissociating buffer. This can lead to insolubility of the protein and dramatic losses. This concentration can be partly modulated by the elution flow rate. In general, low flow rates promote elution in a concentrated form, whereas higher flow rates produce more dilute solutions, which may be advantageous in certain cases. Where the binding constant is too high to facilitate elution of the protein in an active form, it may be necessary to attenuate the ligand affinity via chemical modification. However, this is a relatively complicated process requiring extensive knowledge of the interaction between ligand and target molecule.

2. Knowledge of the nature of the interaction between ligand and target protein is invaluable when selecting the chemistry by which a ligand is to be immobilized. A common immobilization strategy is to immobilize proteins via lysine residues using cyanogen bromide-activated supports. If it is known that a lysine residue on the ligand is involved in the interaction that is being exploited, then it is clear that the matrix may show poor performance. A recent advance is site-directed immobilization (*28*). Many ligands are glycoproteins with defined regions of oligosaccharide commonly linked to asparagine residues. Use of hydrazine chemistry that reacts specifically with the cis diols of the sugar moities allows the ligand to be selectively immobilized. This has been used successfully in immunopurification matrices where antibodies are immobilized via their oligosaccharide-containing Fc regions, leaving their antigen-binding regions free to interact with their target protein. Ligands have also been engineered to include defined biotinylated residues that can then be immobilized on streptavidin activated matrices in an orientated manner.

3. An important factor in protein purification is the stability of the target protein. The target protein may be susceptible to inactivation or degradation because of the physicochemical conditions employed during purification, proteolytic degradation by enzymes that may be inadvertently activated during purification, and the harsh conditions that may be used for elution of the bound material. All of these must be considered when planning a purification protocol.

4. Affinity chromatography has evolved to fit into the format of general protein purification and, as such, has been predominantly performed in columns. As methodologies are required for preliminary purification of crude extracts, the use of the matrix in "batch mode" has begun to be developed. The advantage of the batch mode is that the matrix can be applied to very crude cell cultures or tissue extracts with only minimal need for any form of clarification (removal of cell debris, etc.) At the lab scale, matrices such as silica are popular because of their density allowing rapid recovery of the matrix from the extract by centrifugation. At the analytical scale, the methodology can be applied to separations in microcentrifuge tubes. Once the target protein has bound to the ligand, the matrix can be washed and either eluted in batch mode or packed into a column format. On a larger scale, the method has been performed with purpose designed equipment as "fluidized beds" (e.g., StreamLine® [Pharmacia]).

5. If a highly concentrated eluate is required, then reverse elution can often be advantageous. When a column is loaded to under capacity, the protein occupies ligands at the top of the column preferentially. By eluting in the reverse mode, the material elutes in a tighter band. This is also advantageous if there is a heterogeneous pool of proteins with differing affinities, such as an extract of serum continuing polyclonal antibodies with differing affinities for the immunogen that is acting as a ligand. The antibodies with the highest avidity for the epitope rests at the top of the column, so reverse elution aids recovery.

6. Purification is always a compromise between purity and recovery. The more steps used in the purification, the purer the end product but the lower the recovery. Losses are incurred during the chromatography and in the manipulation of the protein between chromatography steps. The eluted protein fractions commonly undergo buffer exchange and concentration prior to subsequent steps. By selecting elution conditions that are consistent with subsequent chromatography steps such as high salt for hydrophobic chromatography or defined pH changes for ion-exchange chromatography, the target may be eluted in a form requiring minimal conditioning prior to further purification and thereby maximizing the recovery.

References

1. Ostrove, S. (1990) Affinity chromatography: general methods, in *Methods in Enzymology* (Deutscher, M. P., ed.), Academic Press, London, pp. 357–371.
2. Clonis, Y. D. (1988) The application of reactive dyes in enzyme and protein downstream processing. *Crit. Rev. Biotechnol.* **7(4)**, 263–280.
3. Sulkowski, E. (1989) The saga of IMAC and MIT. *Bioessays* **10(5)**, 170–175
4. Clonis, Y. D., Labrou, N. E., Kotsira, V. Ph., Mazitsos, C., Melissis, S., and Gogolas, G. (2000) Biomimetic dyes as affinity chromatography tools in enzyme purification. *J. Chromatogr. A* **891**, 33–44.
5. Lowe, C. R. (2001) Combinatorial approaches to affinity chromatography. *Curr. Opin. Chem. Biol.* **5**, 248–256.
6. Hammond, D. J. (1998) Idenitifcation of affinity ligands from peptide libraries and their applications. *Chromatographia* **46(7/8)**, 475–476.
7. Larrrson, L-J. (2001) Customised ligands optimise affinity chromatography procedures. *BioPharmacology* **14**, 42–44.

8. Desai, M. A. (1990) Immunoaffinity adsorption: process scale isolation of therapeutic-grade biochemicals. *J. Chem. Tech. Biotechnol.* **48**, 105–126

9. Sherwood, R. (1991) Protein fusions: bioseparation and application. *TIBTECH* **9**, 1–3.

10. Sassenfield, H. M. (1990) Engineering proteins for purification. *TIBTECH* **8**, 88–93.

11. Schmitt, J., Hess, H., and Stunnenberg, H. G. (1993) Affinity purification of histidine tagged proteins. *Mol. Biol. Rep.* **(18)**, 223–230.

12. Bornhorst, J. A. and Falke, J. J. (2000) Purification of proteins using polyhistidine affinity tags. *Methods Enzymol.* **326**, 245–254.

13. Skerra, A. and Schmidt, T. G. M. (1999) Applications of a peptide ligand for streptavidin: the Strep-tag. *Biomol. Eng.* **16** 79–86.

14. Gadgil, H., Jurado, L. A., and Jarrett, H. W. (2001) DNA affinity chromatography of transcription factors. *Anal. Biochem.* **290**, 147–178.

15. Caron, M., Seve, A-P., Bladier, D., and Joubert-Caron, R. (1998) Glycoaffinity chromatography and biological recognition. *J. Chromatogr. B* **715**,153–161.

16. Holmes, L. D. and Schiller, M. R. (1997) Immobilised iron(III) metal affinity chromatography for the separation of phosphorylated macromolecules: ligands and applications. *J. Liq. Chromatogr. Related Technol.* **20** 123–142.

17. Porath, J. (2001) Strategy for differential protein affinity chromatography. *Int. J. Bio-Chromatogr.* **6(1)**, 51–78.

18. Burton, S. J. (1996) Affinity chromatography in *Downstream Processing of Natural Products* (Verall, M. S., eds.),Wiley, New York, pp. 193–207.

19. Hage, D. S. (1999) Affinity chromatography, a review of clinical applications. *Clin. Chem.* **45**, 593–615.

20. Puig, O., Caspary, F., Rigaut, G., et al. (2001) The tandem affinity purification (TAP) mehod: a general procedure of protein complx purification. *Methods* **24**, 218–229.

21. Dean, P. D. G., Johnson, W. S., and Middle, F. A. (1985) *Affinity Chromatography*, IRL Press, Oxford, pp. 31–59.

22. Groman, E. V. and Wilchek, M. (1987) Recent developments in affinity chromatography supports. *TIBTECH* **5**, 220–224.

23. Narayanan, S. U. and Crane, L. J. (1990) Affinity chromatography supports: a look at performance requirements. *TIBTECH* **8**, 12–16.

24. Bamford, C. H., Al-Lamee, K. A., Purbrick, M. D., and Wear, T. J. (1992) Studies of a novel membrane for affinity separations. *J. Chromatogr.* **606**, 19–31.

25. Chase, H. A. (1994) Purification of proteins by adsorption chromatography in expanded beds *TIBTECH* **12**, 296–303.

26. Ohlson, S., Lundblad, A., and Zopf, D. (1988) Novel approach to affinity chromatography using 'weak' monoclonal antibodies. *Anal. Biochem.* **169**, 204–208.

27. Perry, M. and Kirby, H. (1990) Monoclonal antibodies and their fragments, in *Protein Purification Applications* (Harris, E. L. V. and Angal, S., eds.), IRL Press, Oxford, pp. 147–164.

28. O'Shannessy, D. J. and Quarles, R. H. (1985) Specific conjugation reactions of the oligosaccharide moieties of immunoglobulins. *J. Appl. Biochem.* **7**, 347–355.

17

Dye-Ligand Affinity Chromatography

Anne F. McGettrick and D. Margaret Worrall

1. Introduction

Dye-ligand affinity is based on the ability of the reactive dyes to bind proteins in a selective and reversible manner *(1,2)*. The dyes are generally either monochlorotriazine compounds (two example structures are shown in **Fig. 1**) and were originally developed in the textile industry. The reactive chloro group allows easy immobilization of the triazine dye to a support matrix, such as Sepharose or agarose, and, more recently, to nylon membranes.

The initial discovery of the ability of these dyes to bind proteins came from the observation that blue dextran (a conjugate of cibacron blue FG-3A), used as a void volume marker on gel filtration columns, could retard the elution of certain proteins *(3)*. A number of studies have been carried out on the specificity of the dyes for particular proteins, mostly using the prototype cibacron blue dye. The dyes appear to be most effective at binding proteins and enzymes that utilize nucleotide cofactors, such as kinases and dehydrogenises, although other proteins such as serum albumin also bind tightly. It has been proposed that the aromatic triazine dye structure resembles the nucleotide structure of nicotinamide adenine dinucleotide (NAD) and that the dye interacts with the dinucleotide fold in these proteins *(4)*. In many cases, bound proteins can be eluted from the columns by a substrate or nucleotide cofactor in a competitive fashion *(5)*, and dyes have been shown to compete for substrate-binding sites in free solution *(6)*. It seems likely that these dyes can bind proteins by electrostatic and hydrophobic interactions and by more specific "pseudoaffinity" interactions with ligand-binding sites. Enhancing the specificity of dye ligands by modification to further resemble ligands (biomimetic dyes) has been successful in the purification of a number of dehydrogenases and proteases *(7)*.

The degree of purification achieved with unmodified dye-ligand chromatography is generally better that that obtained with less specific techniques such as ion-exchange or gel filtration chromatography. A further advantage is that the reactive dyes are relatively inert and unaffected by enzymes in crude cellular extracts.

Individual dyes show differences in binding profiles, and it is generally useful for carrying out a screening procedure to determine the best dye ligand for a given purification protocol. It has been reported that NAD-utilizing enzymes bind cibacron blue, whereas

From: *Methods in Molecular Biology, vol. 244: Protein Purification Protocols: Second Edition*
Edited by: P. Cutler © Humana Press Inc., Totowa, NJ

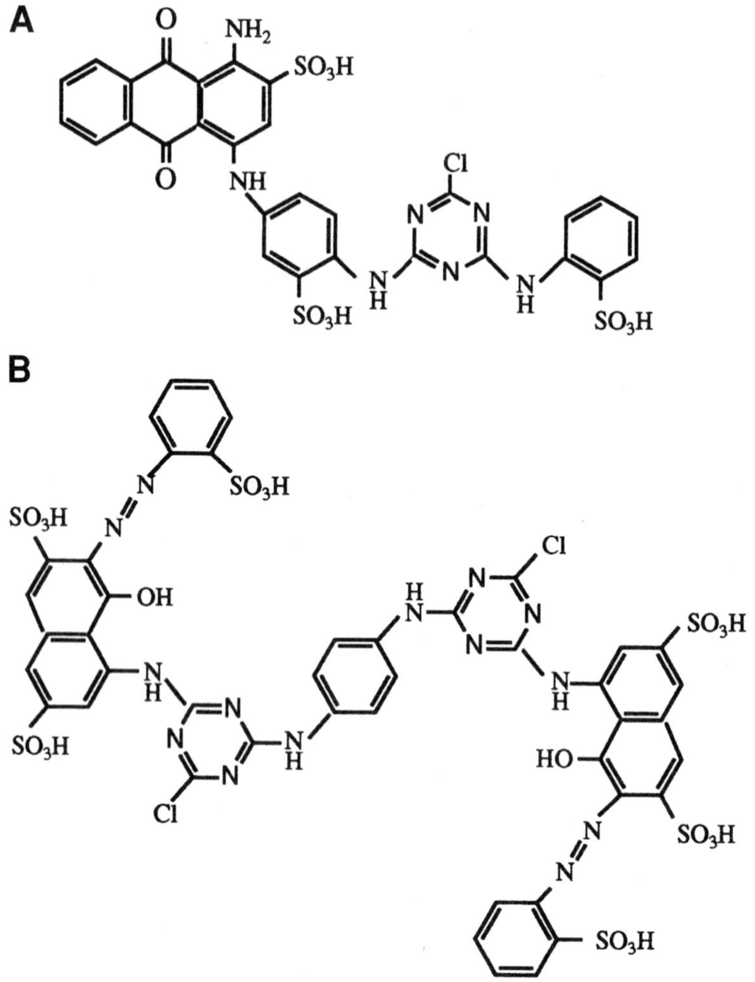

Fig. 1. (A) Structure of cibracon blue F3G-A (procion blue H-B, Reactive Blue 2) and **(B)** structure of procion red HE-3B (reactive red 120).

NADP-utilizing enzymes preferentially bind to procion red HE-3B *(8)*. In general, however, it is difficult to predict which ligand will give the best purification, and a detailed method for screening is given in **Subheading 3.1.**

Elution of the bound protein is usually carried out by increasing the ionic strength or by using a competing ligand, such as the substrate or cofactor. The optimized conditions for purification of a mammalian bifunctional enzyme CoA synthase, using both of these methods, are described.

2. Materials

1. Dye screening kits are available from Sigma Chemical Co. and Millipore (formerly Amicon dyes matrices) containing a range of immobilized dyes as described in **Table 1**. Both

Table 1
Some of the More Commonly Used Commercially Available Dyes and Resins

Procion dye	Millipore	Sigma
Blue H-B (Cibacron Blue)	DyeMatrex Gel Blue A	Reactive Blue 2
Blue MX-R		Reactive Blue 4
Red HE-3B	DyeMatrex Gel Red A	Reactive Red 120
Yellow H-A	DyeMatrex Gel Orange A	Reactive Yellow 3
Yellow MX-3R	DyeMatrex Gel Orange B	Reactive Yellow 86
Green H-4G		Reactive Green 5
Green H-E4BD	DyeMatrex Gel Green A	Reactive Green 19
Brown MX-5BR		Reactive Brown

    of these suppliers use agarose as the support matrix. Different trade names and company numbers assigned to the various dyes and matrices can make direct comparisons difficult. The Sigma-Aldrich website (www.sigma-aldrich.com) gives a full list, with the equivalent cibracron or procion dye information alongside the reactive dye number. Dyes can be immobilized in-house (*see* **Note 1**).

2. Blue Sepharose CL-6B (cibacron blue F3G-A linked to Sepharose CL-6B) and red Sepharose (procion red HE-3B–Sepharose CL-6B) are available from Pharmacia.
3. Equilibration buffer: 20 mM Tris-HCl, pH 8.0, 0.5 mM dithiothreitol (DTT) (*see* **Notes 2** and **3**).
4. Elution buffer: 20 mM Tris-HCl, pH 8.0, 1 M KCl, 0.5 mM DTT.
5. Blue Sepharose affinity elution buffer: 0.1 mM coenzyme A, 0.1 M KCl in equilibration buffer.

    All buffers contain analytical-grade reagents. DTT is added freshly before use.

3. Methods

3.1. Screening Dye Ligands

1. Prepacked columns are available to screen for binding of the protein of interest. Otherwise, pour individual columns of 1-mL packed bed volume of a range of absorbants, such as those listed in **Table 1**. Wash the columns with 10 mL of equilibration buffer.
2. Adjust the pH of the protein sample to be applied to the columns to that of the equilibration buffer. The ionic strength of the protein sample should also be close to that of the equilibration buffer and preferably not exceeding a total of 0.05 M. This can be achieved by simple dilution. The protein concentration should not exceed 10–20 mg/mL, and, if necessary, the sample should be centrifuged or filtered to remove any particular matter.
3. Load 1 mL of the sample to the columns under gravity and wash through with 5 mL of the equilibration buffer, collecting the unbound proteins in one fraction. Elute the bound protein with 5 mL elution buffer and collect in a fresh tube.
4. Assay both fractions from each column for total protein using a quantitative assay, such as the Bradford assay *(9)*. Assay for the protein of interest using the most quantitative method available. Calculate the purification factor achieved in each case.
5. The best binding dye will be the one that effectively binds all of the target protein but allows much of the contaminants to pass through. Further development of this step will involve optimization of the elution conditions to remove contaminants by step or gradient elution. A dye that does not bind the protein of interest but absorbs some of the contaminating proteins can also be useful as a negative binding purification step (*see* **Note 3**).

3.2. Optimization of Purification

Further small-scale columns are recommended for optimization of purification conditions prior to scaling up.

1. For a negative binding step, vary the composition of the equilibration buffer with respect to both ionic strength and pH. Lowering of the ionic strength and lowering the pH will generally increase the amount of total protein binding to the immobilized dyes. As in the dye screening step, collect the unbound material and assay for total protein and protein of interest. Calculate the specific activity and the recovery of the protein of interest. The best buffer conditions will give the highest specific activity in the unbound fraction, provided a good yield of the protein is also achieved.

2. For a positive binding purification step, the equilibration buffer conditions should similarly be varied in order to maximize binding of the protein of interest and to minimize binding of contaminants proteins. Temperature should be kept constant (*see* **Note 5**).

3. The capacity of the dye for the protein of interest can be determined by frontal analysis. This is achieved by continuous loading of the sample solution onto the column until the protein of interest is detected in the eluate. This occurs when the target protein is displaced by proteins with higher affinities for the immobilized dye. Frontal analysis can be useful for examining the relative affinities of more than one protein in a mixture of proteins for a particular matrix. Optimum loading is 80% of the sample volume required for frontal detection.

4. Elution conditions can now be optimized. Wash the column with 5 vol of equilibration buffer to remove nonbinding proteins fully. Apply increasing salt concentrations to the column up to 1 M NaCl (or KCl), either in a stepwise fashion with increases of 200 mM/step (2 column volumes/step) or with a salt gradient.

5. The ability of specific ligands, such as a substrate, cofactor, or inhibitor, to elute the protein of interest can also be screened for, and this "affinity elution" can often provide a more powerful purification step.

    As earlier, a small-scale column is loaded with sample and washed with equilibration buffer. Having already assessed the ionic-strength conditions necessary for elution as described in the previous step, this information is used to optimize affinity elution conditions. First, wash the column with buffer of an ionic strength just below that required to elute the protein in order to remove contaminants (try 20 mM < eluting buffer). Check that the eluate does not contain the protein of interest; if so, lower the salt concentration of the buffer further (by 5–10 mM, depending on the ionic strength of the specific ligand) and spike this solution with the specific ligand. In cases where the specific ligand will elute many proteins (e.g., ATP), better purification may be attained by applying a concentration gradient of the affinity ligand.

6. When scaling up of the purification step, keep the proportion of sample to column volume constant, although the column size should be increased with larger diameters rather than larger column lengths. The linear flow rate should remain constant so that the sample protein will have the same contact time with the dye.

7. Regeneration of dye-ligand absorbants can be carried out using either high-salt (2 M NaCl) and/or high-pH (1 M NaOH) solutions, and this is generally quite effective. However, if the column capacity is noticeably decreasing with continued use, then it may be improved by using chaotropic agents, such as 8 M urea or 6 M guanidinium hydrochloride, to denature and so remove residual protein.

3.3. Example Purification: Isolation of Coenzyme A Synthase

The bifunctional enzyme CoA synthase (phosphopantetheine adenylyl transferase and dephospho-CoA kinase) from pig liver was purified to near homogeneity from a partially purified extract using two types of immobilized dyes (*10*).

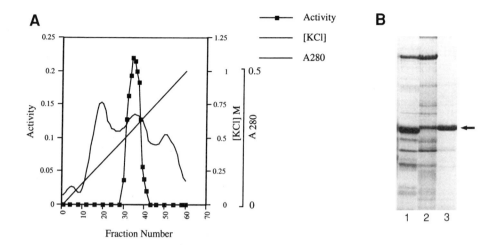

Fig. 2. Purification of CoA synthase on procion red HE-3B Sepharose and affinity elution from blue Sepharose (F3G-A) **(A)** Elution profile of CoA synthase from red Sepharose using a 0–1 *M* KCl gradient. Fractions 30–40 were pooled, diluted with 2 vol of 0.5 m*M* DTT and applied to the blue Sepharose column for affinity elution with 0.1 m*M* CoA. **(B)** Sodium dodecyl sulfate–polyacrylamide gel electrophoresis of purification steps; *lane 1*, partially purified extract (CoA synthase 60 kDa band not visible); *lane 2*, pooled red Sepharose fractions; *lane 3*, blue-Sepharose affinity-eluted protein.

The crude protein solution was loaded onto a red Sepharose column equilibrated with 10 vol of equilibration buffer. The column was further washed with 2 column volumes of equilibration buffer, and a linear gradient of 0–1 *M* KCl in 20 m*M* Tris-HCl, pH 8.0, was applied to elute the protein. The fractions containing CoA synthase (*see* **Fig. 2A**) were pooled and diluted with 2 vol of 0.5 *M* DTT in order to reduce the ionic strength and allow binding to blue Sepharose. This material was loaded onto a blue Sepharose CL-6B column equilibrated with 10 m*M* Tris-HCl, pH 8.0. The column was then washed with 0.12 *M* KCl, in 10 m*M* Tris-HCl, pH 8.0, which does not elute the CoA synthase and removes many of the contaminating proteins. The enzyme was then eluted with a solution of 0.11 *M* KCl, 10 m*M* Tris-HCl, pH 8.0, containing 0.1 m*M* coenzyme A. As can be seen in **Fig. 2B**, the combination of these two steps can purify this protein from a crude extract where it is not visible on sodium dodecyl sulfate–polyacrylamide gel electrophoresis to near homogeneity.

Essentially the same dye-ligand chromatography steps were more recently used for purifying monofunctional phosphopantetheine adenylyltransferase from *Escherichia coli* and for isolating recombinant human CoA synthase (*11,12*).

4. Notes

1. Immobilization of dyes onto a support matrix can be readily carried out in the laboratory, and this may be desirable for cost-saving measures, particularly for large-scale column purification. The reactive dyes can be purchased relatively inexpensively from Sigma and Polysciences. Detailed methods for coupling the dyes to the support matrix are given else-

where *(13)*. The major drawback for homemade columns is usually leakage of the dye from the matrix even after extensive washing. Novel purpose-made dye ligands have been synthesized to reduce leakage problems and increase protein-binding capacity *(14)*.

2. The composition of the equilibration and elution buffer can be varied greatly. The examples given use Tris buffers at pH 8.0, which is relatively high, and so will reduce the amount of total protein binding. If binding of the protein of interest is low or weak in these conditions, it may be desirable to decrease the pH, in which case, phosphate buffers may be more suitable.

3. The addition of divalent and trivalent metal ions (e.g., Mg^{2+}, Ca^{2+}, Zn^{2+}, Cu^{2+}, Mn^{2+}, Fe^{3+}) to the equilibration buffer can promote binding of certain proteins *(15)* through the formation of ternary complexes. Similarly, the addition of EDTA to elution buffers can increase desorption of proteins.

4. Negative binding steps have proven to be very effective in the purification of proteins from plasma and from cell culture extracts, which contain large amounts of fetal calf serum. This is the result of the high affinity of serum albumin, which is the major contaminating protein in such preparations, to cibacron blue and procion red HE-3B dye *(16)*.

5. Temperature can effect binding and so should be kept constant throughout the purification procedure. The necessity for cold-room conditions will depend on the protein of interest, but as with other agarose-based supports, moving columns to warmer conditions will generate air bubbles and should be avoided.

High-performance liquid chromatography-grade support matrices for dye-ligand chromatography have been used for purification of a number of proteins by immobilization of dyes onto silica-based matrices *(17)*. Prepacked columns of procion blue MX-R, procion red HE-3B, and cibacron blue immobilized to silica (300 and 500 Å) are now commercially available for Serva Fine Chemicals.

7. Flow rates are also limited by the support matrix and generally 10–30 cm/h is appropriate. The use of nylon-based membrane supports for immobilized dyes has been investigated by a number of groups *(1,18)*. These have the advantage of large surface areas and rapid flow rates, but precoated dye-affinity membranes are not yet available commercially.

References

1. Denizli, A. and Piskin, E. (2001) Dye-ligand affinity systems. *J. Biochem. Biophys. Methods* **49**, 391–416.

2. Stellwagen, E. (1990) Chromatography on immobilised reactive dyes. *Methods Enzymol.* **182**, 343–357.

3. Kopperschlager, G., Freyer., R., Diezel, W., and Hofmann, E. (1968) Some kinetic and molecular properties of yeast phosphofructokinase. *FEBS Lett.* **1**, 137–141.

4. Thompson, S. T., Cass, K. H., and Stellwagen, E. (1975) Blue dextran–Sepharose: an affinity column for the dinucleotide fold in proteins. *Proc. Natl. Acad. Sci. USA* **72**, 669–672.

5. Baird, J. K., Sherwood, R. F., Carr, R. J., and Atkinson, A. (1976) Enzyme purification by substrate elution chromatography from procion dye–polysaccharide matrices. *FEBS Lett.* **70**, 61–66.

6. Reuter, R., Metz, P., Lorenz, G., and Kopperschlager, G. (1990) Interaction of bacterial glucose-6-phosphate dehydrogenase with triazine dyes: a study by means of affinity partitioning and kinetic analysis. *Biomed. Biochim. Acta.* **49**, 151–160.

7. Clonis, Y. D., Labrou, N. E., Kotsira, V. P., Mazitsos, C., Melissis, S., and Gogolas, G. (2000) Biomimetic dyes as affinity chromatography tools in enzyme purification. *J. Chromatogr. A* **891**, 33–44.

8. Watson, D. H., Harvey, M. J., and Dean, P. D. (1978) The selective retardation of NADP+-dependent dehydrogenases by immobilized procion red HE-3B. *Biochem. J.* **173**, 591–596.

9. Bradford, M. M. (1976) A rapid and sensitive method for the quantitation of microgram quantities of protein utilizing the principle of protein-dye binding. *Anal. Biochem.* **72**, 248–254.

10. Worrall, D. M. and Tubbs, P. K. (1983) A bifunctional enzyme complex in coenzyme A biosynthesis: purification of pantetheine phosphate adenylyltransferase and dephospho-CoA kinase. *Biochem. J.* **215**, 153–157.

11. Geerlof, A., Lewendon, A., and Shaw, W. V. (1999) Purification and characterization of phosphopantetheine adenylyltransferase from *Escherichia coli.* *J. Biol. Chem.* **274**, 27, 105–27,111.

12. Aghajanian, S. and Worrall, D. M. (2002) Identification and characterization of the gene encoding the human phosphopantetheine adenylyltransferase and dephospho-CoA kinase bifunctional enzyme (CoA synthase). *Biochem. J.* **365**, 13–18.

13. Lowe, C. R. and Pearson, J. C. (1984) Affinity chromatography on immobilized dyes. *Methods Enzymol.* **104**, 97–113.

14. Lowe, C. R., Burton, S. J., Burton, N., et al. (1990) New developments in affinity chromatography. *J. Mol. Recogn.* **3**, 117–122.

15. Hughes, P. (1989) Metal enhanced interactions of proteins with triazine dyes and applications to downstream processing, in *Protein–dye Interactions: Developments and Applications* (Vijayalakshima, M. A. and Bertrand, O., eds.), Elsevier Applied Science, London, pp. 207–215.

16. Travis, J., Bowen, J., Tewksbury, D., Johnson, D., and Pannell, R. (1976) Isolation of albumin from whole human plasma and fractionation of albumin-depleted plasma. *Biochem. J.* **157**, 301–306.

17. Groman, E. V. and Wilchek, M. (1987) Recent developments in affinity chromatography supports. *Trends Biotechnol.* **5**, 220–224.

18. Weissenborn, M., Hutter, B., Singh, M., Beeskow, T. C., and Anspach, F. B. (1997) A study of combined filtration and adsorption on nylon-based dye-affinity membranes: separation of recombinant L-alanine dehydrogenase from crude fermentation broth. *Biotechnol. Appl. Biochem.* **25**, 159–168.

LIVERPOOL
JOHN MOORES UNIVERSITY
AVRIL ROBARTS LRC
TEL. 0151 231 4022

18

Lectin Affinity Chromatography

Iris West and Owen Goldring

1. Introduction

Lectins are glycoproteins or proteins that have a selective affinity for a carbohydrate, or a group of carbohydrates. Many purified lectins are readily available, and these may be immobilized to a variety of chromatography supports.

Immobilized lectins are powerful tools that can be used to separate and isolate glycoconjugates *(1–7)*, released glycans *(8)*, glycopeptides *(9)*, polysaccharides, soluble cell components, and cells *(10)* that contain specific carbohydrate structures. Glycopeptide mixtures can be fractionated by sequential chromatography on different lectins *(11,12)*. Immobilized lectins can also be used to remove glycoprotein contaminants from partially purified proteins. In addition, immobilized lectins can be used to probe the composition and structure of surface carbohydrates *(13)*. Lectin chromatography has also been utilized in the analysis of protein complexes *(14)* and to isolate glycosylated forms on proteins from their nonglycoylated forms *(15)*. Subtle changes in glycoprotein structure owing to disease may be reflected in altered affinity for lectins *(16–18)*. The predominant lectins used for affinity chromatography are listed in **Table 1**.

In principle, the mixture of glycoconjugates to be resolved is applied to the immobilized lectin column. The glycoconjugates are selectively adsorbed to the lectin, and components without affinity for the lectin are then washed away. The adsorbed carbohydrate-containing components are dissociated from the lectin by competitive elution with the hapten sugar (i.e., the best saccharide inhibitor). This may be a simple monosaccharide or an oligosaccharide *(35,36)*.

There are many matrices suitable for coupling to lectins, and a number of reliable coupling methodologies have been developed. Many supports are now supplied preactivated. These are simple to use. **Table 2** lists a few of the most popular activated matrices (*see* **Note 1**).

The lectin must be chosen for its affinity for a carbohydrate or sequence of sugars. Many of the lectins are not specific for a single sugar and may react with many different glycoproteins. Concanavalin A (Con A) has a very broad specificity and will bind many serum glycoproteins, lysosomal hydrolases, and polysaccharides. The binding efficiency of Con A is reduced by the presence of the detergents 0.1% (w/v) sodium dodecyl sulfate

From: *Methods in Molecular Biology, vol. 244: Protein Purification Protocols: Second Edition*
Edited by: P. Cutler © Humana Press Inc., Totowa, NJ

Table 1
Common Lectins Used in Affinity Chromatography

Lectin	Specificity	Useful eluants	Uses	Refs.
Concanavalin A	α-D-Mannopyranosyl with free hydroxyl groups at C3, C4, and C6	0.01–0.5 *M* Methyl α-D-mannoside	Separation of glycoproteins	**19**
		D-Mannose	Purification of glycoprotein enzymes	**20**
		D-Glucose	Partial purification of IgM	**21**
			Separation of lipoproteins	**22**
Lens culinaris	α-D-Glucopyranosyl residues	Methyl-α-D-glucoside	Purification of gonadotrophins	**23**
	α-D-Mannopyranosyl	0.1 *M* Na borate, pH 6.5	Purification of HeLa cells	**24**
		0.15 *M* Methyl α-D-mannoside	Isolation of mouse H antigens	**25**
	Binds less strongly than Con A		Purification of detergent-solubized glyco-proteins	**26**
			Schistosoma mansoni: surface membrane isolation	**27**
Tritium vulgaris	*N*-Acetyl-D-glucosamine	0.1 *M N*-Acetyl-glucosamine	Biochemical characterization of H4G4 antigen from HOON pre-B leukemic cell line	**28**
			Purification and analysis of RNA polymerase transcription factors	**29**
Ricinus communis	α-D-Galactopyranosyl residues	0.15 *M* D-Galactose	Fractionation of glycopeptide-binding proteins	**30**
Jacalin	D-Galactopyranosyl residues	0.1 *M* Melibiose in PBS	Separating IgA1 and IgA2	**31,32**
			Purification of C1 inhibitor	**33**
Bandeira simplicifolia	α-D-Galactopyranosyl and *N*-acetyl-D-galactosamyl	PBS	Resolving mixtures of nucleotide sugars	**34**

Note: PBS = phosphate-buffered saline.

Table 2
Commonly Available Activated Matrices

Commercial name	Supplier[a]
Affi 10, Affi 15	Bio-Rad
Affi Prep 10 and 15 *(37)*	Bio-Rad
CNBr agarose	Sigma
Epoxy-activated agarose	Sigma
CDI agarose	Sigma
Reacti 6X	Pierce
Polyacrylhydrazide agarose	Sigma

[a]Locations: Bio-Rad (Hercules, CA); Sigma (St. Louis, MO); Pierce (Rockford, IL).

(SDS), 0.1% (w/v) sodium deoxycholate, and 0.1% (w/v) cetyltrimethyl-ammonium bromide (CTAB) *(38)*. *Lens culinaris* lectin has similar carbohydrate specificities to Con A, but binds less strongly. The lectin retains its binding efficiency in the presence of sodium deoxycholate. Therefore, it is useful for the purification of solubilized membrane glycoconjugates. Wheat germ lectin can interact with mucins that contain *N*-acetylglucosamine residues. It retains its binding efficiency in the presence of 1% (w/v) sodium deoxycholate. Jacalin is a lectin isolated from the seeds of *Artocarpus integrifolia*, which has an affinity for D-galactose residues *(33)*. It is reported to separate IgA_1 from IgA_2 subclass antibodies. Therefore, it is useful for removing contaminating IgA from preparations of purified IgG.

Two simple procedures (**Subheadings 3.1.** and **3.2.**) for the immobilization of lectins are given, one for Con A onto carbonyldiimidazole (CDI)-activated agarose, and the other for Con A onto Affigel 15. Once the Con A has been immobilized to the matrices, the purification and elution procedures are common. A single protocol (**Subheading 3.3.**) is given detailing this procedure.

2. Materials

2.1. Immobilization of Con A on CDI-Activated Agarose

1. CDI-activated agarose (Sigma). The activated gel is stable in the acetone slurry for 1–2 y at 4°C.
2. Affinity-purified Con A.
3. Protective sugar: Methyl-α-D-mannoside.
4. Coupling buffer: 0.1 *M* Na_2CO_3, pH 9.5.
5. 0.1 *M* Tris (hydroxymethyl) methylamine, pH 9.5.
6. Phosphate-buffered saline (PBS).
7. 0.1 *M* Sodium acetate, pH 4.0.
8. 0.1 *M* $NaHCO_3$, pH 8.0.

2.2. Immobilization of Con A on Affigel 15

1. Affigel 15.
2. Affinity-purified Con A.
3. Protective sugar: Methyl-α-D-mannoside.
4. Coupling buffer (PBS): 8.0 g NaCl, 0.2 g KH_2PO_4, 2.9 g $Na_2HPO_4 \cdot 12H_2O$, 0.2 g KCl made up to 1 L with distilled water.

5. Equilibration buffer: 5 mM Sodium acetate buffer containing 0.1 M NaCl, 1 mM MnCl$_2$, 1 mM CaCl$_2$, 1 mM MgCl$_2$, and 0.02% (w/v) sodium azide (*see* **Note 2**).

2.3. Purification of Glycoproteins on Immobilized Con A

1. Equilibration buffer: 5 mM Sodium acetate buffer, pH 5.5, containing 0.1 M NaCl, 1 mM CaCl$_2$, 0.02% (w/v) sodium azide.
2. 0.01 M Methyl-α-mannoside in acetate buffer.
3. 0.3 M Methyl-α-mannoside in acetate buffer.
4. 5 mM Sodium acetate buffer containing 1 M NaCl.

3. Methods
3.1. Immobilization of Con A (see Note 3) on CDI-Activated Agarose

1. Place the gel in a sintered glass funnel. Apply gentle suction (a water pump is adequate).
2. Wash the gel free of acetone with 10 vol of ice-cold distilled water (*see* **Note 4**). Do not dry the cake.
3. Dissolve Con A at 20 mg/mL in 0.1 M Na$_2$CO$_3$ buffer, pH 9.5, containing the protective sugar 0.5 M methyl-α-D-mannoside (*see* **Note 3**). Keep at 4°C.
4. Gently break up the gel cake; then, add the cake directly to the coupling buffer containing the ligand, 1 vol of gel to 1 vol of buffer.
5. Incubate for 48 h at 4°C with gentle mixing.
6. Transfer the gel to a column (*see* **Note 5**) and then wash the gel with 0.1 M Tris to block any remaining unreacted sites.
7. Wash the gel with 3 vol of 0.1 M sodium acetate buffer, pH 4.0, and then with 3 vol of 0.1 M NaHCO$_3$, pH 8.0.
8. Repeat **step 7**.
9. Wash the gel with PBS containing 0.02% (w/v) sodium azide and then store the gel at 4°C in buffer containing 0.02% (w/v) sodium azide.

3.2. Immobilization of Con A (see Note 3) on Affigel 15 (see Notes 6–8)

1. Place the gel in a sintered glass funnel. Apply gentle suction to remove the isopropyl alcohol. Do not let the gel dry out.
2. Wash the gel with 3 vol of cold distilled water (*see* **Note 4**).
3. Drain off excess water.
4. Dissolve Con A at 10 mg/mL in PBS containing 0.3 M methyl-α-D-mannoside (*see* **Note 3**).
5. Transfer the gel to the coupling buffer, 1 vol gel:1 vol coupling buffer (*see* **Note 9**).
6. Mix gently for 2 h at 4°C.
7. Transfer the gel to a column (*see* **Note 5**) and wash with 5 vol PBS.
8. Wash with equilibration buffer and then store the column at 4°C in the presence of 0.2% (w/v) sodium azide.

3.3. Purification of Glycoproteins on Immobilized Con A

1. Equilibrate the column with 5 mM sodium acetate buffer containing 0.1 M NaCl, 1 mM MnCl$_2$, 1 mM CaCl$_2$, 1 mM MgCl$_2$, 0.02% (w/v) sodium azide (*see* **Note 10**); flow rate 0.1 mL/min.
2. Either dissolve solid glycoprotein in a little equilibration buffer, or if the glycoprotein is a solution, dilute 1:1 with equilibration buffer. Remove any particulate matter.
3. Apply to the column and wash with equilibration buffer to remove nonbound components. Monitor effluent at 280 nm until the A_{280} is 0 (*see* **Note 11**).

Table 3
Common Lectins and Their Protective Saccharides

Lectin	Hapten saccharide	Ref.
Con A	Methyl-α-D-mannoside	*39*
L. culinaris	Methyl-α-D-mannoside	—
T. vulgaris	Chitin oligosaccharides	*40*
R. communis	Methyl-β-galactoside	*29*
Jacalin	D-Galactose	*33*

4. Elute with 0.01 M methyl-α-D-mannoside in acetate buffer (5 vol). Collect 1-mL fractions and store on ice.
5. Elute with 0.3 M methyl-α-D-mannoside in acetate buffer (*see* **Notes 12** and **13**). Collect 1-mL fractions and store on ice.
6. Wash column with acetate buffer containing 1 M NaCl and then regenerate the column by extensive washing with equilibration buffer.
7. Pooled bound fractions can either be concentrated by ultrafiltration or dialyzed against a buffer to remove the hapten sugar.

4. Notes

1. When using CNBr-activated matrices, charged groups may be introduced during the coupling step. This may lead to nonspecific adsorption because of ion-exchange effects.
2. Con A requires Mn^{2+} and Ca^{2+} to preserve the activity of the binding site.
3. Coupling methods are similar for different lectins; however, the lectin-binding site must be protected by the presence of the hapten monosaccharide. **Table 3** lists the common lectins and the saccharides used to protect the active site during immobilization.
4. Adequate washing of the gel prior to coupling is required. Many proteins are sensitive to the acetone or alcohols used to preserve the activated gels.
5. The size of a column depends on the amount of sample to be purified. Preliminary experiments can easily be carried out using Pasteur pipets, with supports of washed glass wool. Alternatively, the barrels of disposable plastic syringes can be used. Long, thin columns will give the best resolution of several components. However, where the requirement is to remove only contaminants from the protein, a short, fat column will elute the proteins in a smaller volume.
6. Affigel 10 and 15 are *N*-hydroxysuccinimide esters of crosslinked agarose. Both are supplied as a slurry in isopropyl alcohol. The gels can be stored without loss of activity for up to a year at $-20°C$.
7. Affigel 10 couples neutral or basic proteins optimally. The coupling buffers used should be at a pH at, or below, the pI of the ligand. To couple acidic proteins, coupling buffers should contain 80 mM $CaCl_2$. Affigel 15 couples acidic proteins optimally. The pH of coupling buffers should be above or near the pI of the ligand. To couple basic proteins, include 0.3 M NaCl in the coupling buffer.
8. **Steps 1–4** should be completed in 20 min to ensure optimum coupling.
9. Buffers containing amino groups should not be used for coupling. Suitable buffers include HEPES, PBS, and bicarbonate.
10. Equilibration buffers used will depend on the stability of the glycoconjugate being purified, and where protease activity is suspected, the appropriate protease inhibitors should be included. Some glycoconjugates are sensitive to acetate *(29)*. Phosphate buffers will precip-

itate out any Ca²⁺ added as calcium phosphate. The optimal pH for the equilibration buffer should be established to ensure maximum adsorption to the affinity column. Binding may be enhanced in certain cases by the inclusion of NaCl in the buffer *(41)*.

11. Sparingly reactive glycoconjugates may be retarded rather than bound by the column *(42)*. Washing should continue for at least five bed volumes. The nonbound fractions can be retained and reapplied to the column on a subsequent occasion, when it is suspected that the load applied to the column is saturating.

12. Elution can be achieved either as a "step" (i.e., a change to a buffer containing the hapten sugar, or using a gradient of the hapten sugar. Different inhibiting sugars can be used for elution (e.g., D-glucose, α-D-mannose, methyl-α-D-glucoside).

13. Occasionally, glycoconjugates may bind very tightly to the matrix and may not elute even with 0.5 *M* hapten sugar. Sodium borate buffers offer an alternative: These may elute components with high affinity for the lectin without denaturing it. The eluant temperature can be raised above 4°C, dependent on the protein stability. Detergents, e.g., 0–5% (w/v) SDS together with 6 *M* urea will often strip components from the lectin, but may irreversibly denature the lectin. Occasionally, it may be expedient to use stringent denaturing conditions to desorb bound fractions with a high yield. Glycoproteins have been eluted from Con A–Sepharose by heating in media containing 5% (w/v) SDS and 8 *M* urea *(43)*. 1 *M* NH₄OH has also been used as an eluant *(24)* with high yield, but irreversibly denaturing the lectin.

References

1. Kennedy, J. F. and Rosevear, A. (1973) An assessment of the fractionation of carbohydrates on Concanavalin A Sepharose by affinity chromatography. *J. Chem. Soc. Perkin Trans.* **19,** 2041–2046.
2. Fidler, M. B., Ben-Yoseph, Y., and Nadler, H. L. (1979) Binding of human liver hydrolases by immobilised lectins. *Biochem. J.* **177,** 175.
3. Kawai, Y. and Spiro, R. G. (1980) Isolation of major glycoprotein from fat cell plasma membranes. *Arch. Biochem. Biophys.* **199,** 84–91.
4. Center, R. J., Schuck, P., Leapman, R. D., et al. (2001) Oligomeric structure of virion-associated and soluble forms of the simian immunodeficiency virus envelope protein in the perfusion activated conformation. *Proc. Nat. Acad. Sci.* **98,** 14877–14882.
5. Ambudkar, S. V., Lelong, I. H., Zhang, J. and Cardarelli, C. (1998) Purification and reconstitution of human P-glycoprotein. *Methods Enzymol.* **292,** 492–504.
6. Simoni, M., Peters, J., Behre, H. M., Kliesch, S., Leifke, E., and Nieschlag, E. (1996) Effects of gonadotropin-releasing hormone on bioactivity of follicle-stimulating hormone (FSH) and microstructure of FSH, luteinizing hormone and sex hormone-binding globulin in a testosterone-based contraceptive trial: evaluation of responders and non-responders. *Eur. J. Endocrinol.* **135,** 433–439.
7. Lee, K. Y. and Do, S. I. (2001) Differential and cell-type specific microheterogeneity of high mannose-type Asn-linked oligosaccharides of human transferrin receptor. *Mol. Cells* **12,** 239–243.
8. Tuisiani, D. R. (2000) Structural analysis of the asparagine-linked glycan units of the ZP2 and ZP3 glycoproteins from mouse zone pellucida. *Arch. Biochem. Biophys.* **382,** 275–283.
9. Schuette, C. G., Weisgerber, J., and Sandhoff K. (2001) Complete analysis of the glycosylation and disulfide bond pattern of human beta-hexosaminidase. *Glycobiology* **11,** 549–556.
10. Sharma, S. K. and Mahendroo, P. P. (1980) Affinity chromatography of cells and cell membranes. *J. Chromatogr.* **184,** 471–499.
11. Cummings, R. D. and Kornfeld, S. (1982) Fractionation of asparagine linked oligosaccharides by serial lectin affinity chromatography. *J. Biol. Chem.* **257,** 11,235–11,240.

12. Aeed, P. A., Geng, J. G., Asa, D., Raycroft, L., Ma, L., and Elhammer, A. P. (1998) Characterization of the O-linked oligosaccharide structures on P-selectin glycoprotein ligand-1 (PSGL-1). *Glycoconjugate J.* **15**, 975–985.

13. Boyle, F. A. and Peters, T. J. (1988) Characterisation of galactosyltransferase isoforms by ion exchange and lectin affinity chromatography. *Clin. Chim. Acta.* **178**, 289–296.

14. Borchers, C., Boer, R., Figala, V., et al. (2002) Characterization of the dexniguldipine binding site in the multidrug resistance-related protein P-glycoprotein by photoaffinity labelling and mass spectrometry. *Mol. Pharmacol.* **61**, 1366–1376.

15. Sun, S., Albright, C. F., Fish, B. H., et al. (2002) Expression, purification and kinetic characterization of full-length human fibroblast activation protein. *Protein Express. and Purif.* **24**, 274–281.

16. Pos, O., Drechoe, A., Durand, G., Bierhuizen, M. F. A., Van der Stelt, M. E., and Van Dijk, W. (1989) Con A affinity of rat α-1-acid glycoprotein (rAGP): changes during inflammation, dexamethasone or phenobarbital immunoelectrophoresis (CAIE) are not only a reflection of biantennary glycan content. *Clin. Chim. Acta* **184**, 121–132.

17. Winograd, E. and Rojas, A. P. (1999) Identification of two nonglycosylated polypeptides of Taenia solium recognised by immunoglobulins from patients with neurocysticercosis. *Parasitol. Res.* **85**, 513–517.

18. Tertov, V. V., Sobenin, I. A., Kaplun, V. V., and Orkhov, A. N. (1998) Antioxidant content in low density lipoprotein and lipoprotein oxidation in vivo and in vitro. *Free Radical Res.* **29**, 165–173.

19. Nikawa, T., Towatari, T., and Katunuma, N. (1992) Purification and characterisation of Cathepsin J from rat liver. *Eur. J. Biochem.* **204(1)**, 381–393.

20. Esmann, M. (1980) Concanavalin A Sepharose purification of soluble Na,K-ATPase from rectal glands of the spiny dogfish. *Anal. Biochem.* **108**, 83–85.

21. Weinstein, Y., Givol, D., and Strausbauch, D. (1972) The fractionation of immunoglobulins with insolubilised concanavalin A. *J. Immunol.* **109**, 1402–1404.

22. Yamaguchi, N., Kawai, K., and Ashihara, T. (1986) Discrimination of gamma-glutamyltranspeptidase from normal and carcinomatous pancreas. *Clin. Chim. Acta* **154**, 133–140.

23. Kinzel, V., Kÿbler, D., Richards, J., and St?hr, M. (1976) *Lens culinaris* lectin immobilised on Sepharose: binding and sugar specific release of intact tissue culture cells. *Science* **192**, 487–489.

24. Matsuura, S. and Chen, H. C. (1980) Simple and effective solvent system for elution of gonadotropins from concanavalin A affinity chromatography. *Anal. Biochem.* **106**, 402–410.

25. Kvist, S., Sandberg-Tragardh, L., and Ostberg, L. (1977) Isolation and partial characterisation of papain solubilised murine H-2 antigens. *Biochemistry* **16**, 4415–4420.

26. Dawson, J. R., Silver, J., Shepherd, L. B., and Amos, B. D. (1974) The purification of detergent solubilised HL-A antigen by affinity chromatography with the haemagglutin from *Lens culinaris*. *J. Immunol.* **112**, 1190–1193.

27. Pujol, F. H. and Cesari, I. M. (1993) *Schistosoma Mansoni:* surface membrane isolation with lectin coated beads. *Membr. Biochem.* **10(3)**, 155–161.

28. Gougos, A. and Letarte, M. (1988) Biochemical characterisation of the 44g4 antigen from the HOON pre-B leukemic cell line. *J. Immunol.* **141(6)**, 1934–1940.

29. Dulaney, J. T. (1979) Eractors affecting the binding of glycoproteins to concanavalin A and *Ricinus communis* agglutin and the dissociation of their complexes. *Anal. Biochem.* **99**, 254.

30. Gregory, R. L., Rundergren, J., and Arnold, R. R. (1987) Separation of human IgA$_1$ and IgA$_2$ using Jacalin–agarose chromatography. *J. Immunol. Methods* **99**, 101–106.

31. Jackson, S. F. and Tjian, R. (1989) Purification and analysis of RNA polymerase 11 transcription factors by using wheat germ agglutinin affinity chromatography. *Proc. Natl. Acad. Sci. USA* **86(6)**, 1781–1785.

32. Pilatte, Y., Hammer, C. H., and Frank, M. M. (1983) A new simplified procedure for C1 inhibitor purification. *J. Immunol. Methods* **120,** 37–43.

33. Roque-Barreira, M. C. and Campos-Neto, A. (1985) Jacalin: an IgG-binding lectin. *J. Immunol.* **134,** 1740.

34. Blake, D. A. and Goldstein, I. J. (1982) Resolution of carbohydrates by lectin affinity chromatography. *Methods Enzymol.* **83,** 127–132.

35. Debray, H., Pierce-Cretel, A., Spik, G., and Montreuil, J. (1983) Affinity of ten insolubilised lectins towards various glycoproteins with the *N*-glycosylamine linkage and related oligosaccharides, in *Lectins in Biology, Biochemistry, Clinical Biochemistry*, vol. 3 (Bog-Hanson, T. C. and Spengler, G. A., eds.), de Gruyter, Berlin, p. 335.

36. Debray, H., Descout, D., Strecker, G., Spik, G., and Montreuil, J. (1981) Specificity of twelve lectins towards oligosaccharides and glycopeptides related to *N*-glycosylglycoproteins. *Eur. J. Biochem.* **117,** 41–55.

37. Matson, R. S. and Siebert, C. J. (1988) Evaluation of a new *N*-hydroxysuccinimide activated support for fast flow immunoaffinity chromatography. *Prep. Chromatogr.* **1,** 67–91.

38. Lotan, R. and Nicolson, G. L. (1979) Purification of cell membrane glycoproteins by lectin affinity chromatography. *Biochem. Biophys. Acta* **559,** 329–376.

39. Cook, J. H. (1984) Turnover and orientation of the major neural retina cell surface protein protected from tryptic cleavage. *Biochemistry* **23,** 899–904.

40. Mintz, G. and Glaser, L. (1979) Glycoprotein purification on a high capacity wheat germ lectin affinity column. *Anal. Biochem.* **97,** 423–427.

41. Lotan, R., Beattie, G., Hubbell, W., and Nicolson, G. L. (1977) Activities of lectins and their immobilised derivatives in detergent solutions. Implications on the use of lectin affinity chromatography for the purification of membrane glycoproteins. *Biochemistry* **16(9),** 1787.

42. Narashimhan, S., Wilson, J. R., Martin, E., and Schacter, H. (1979) A structural basis for four distinct elution profiles on concanavalin A Sepharose affinity chromatography of glycopeptides. *Can. J. Biochem.* **57,** 83–97.

43. Poliquin, L. and Shore, G. C. (1980) A method for efficient and selective recovery of membrane glycoproteins from concanavalin A Sepharose using media containing sodium dodecyl sulphate and urea. *Anal. Biochem.* **109,** 460–465.

19

Immunoaffinity Chromatography

Paul Cutler

1. Introduction

1.1. General Principles

The principle of *immunoaffinity* or *immunoadsorption chromatography* is based on the highly specific interaction of an antigen with its antibody *(1)*. Immunoaffinity chromatography is a specialized form of affinity chromatography and, as such, utilizes an antibody or antibody fragment as a ligand immobilized onto a solid support matrix in a manner that retains its binding capacity. Although the technique includes the separation of antibodies using immobilized antigens *(2–4)*, it is more commonly performed for the identification, quantification, or purification of antigens (*see* **Note 1** and **refs. 5** and **6**). The crude extract is pumped through the column and the unbound material washed clear prior to elution of the retained antigen by alterations to the mobile-phase conditions that weaken the antibody–antigen interaction.

As with other affinity chromatography methods, the power of immunoaffinity chromatography lies in the ability to separate a target biomolecule from a crude mixture such as a cell extract or culture medium with a high degree of specificity and in high yield. The target can be concentrated from a dilute solution to a high degree of purity, typically 10^3- to 10^4-fold purification in a single chromatographic step. The most common format for both analytical and preparative separations is column chromatography *(7)*.

Antibodies are produced by B-lymphocytes primarily as part of the adaptive immune system. They are produced in response to specific agents (antigens) and, as such, display selective affinity in the antigen binding (Fab) region (*see* **Fig. 1**). The constant region (Fc) may mobilize the immune system by activation of complement and binding to the Fc receptors on host tissue cells such as phagocytes. Antibodies are raised to specific regions (epitopes) on antigens; hence, antibodies can be induced in the laboratory by the introduction of a specific antigen (immunogen) into an animal, giving rise to polyclonal antibodies of differing affinities and selectivities (i.e., to different epitopes) (*see* **ref. 8** and **Note 2**). An individual clone from a lymphocyte can be made immortal by fusion with an appropriate myeloma cell line to produce a hybridoma; hence, via cell culture, a stable supply of monoclonal antibodies (mAbs) *(9)*. Once the genetic code is known for the specific mAb or an active fragment of antibody (e.g., a Fab region), the molecule can be gen-

From: *Methods in Molecular Biology, vol. 244: Protein Purification Protocols: Second Edition*
Edited by: P. Cutler © Humana Press Inc., Totowa, NJ

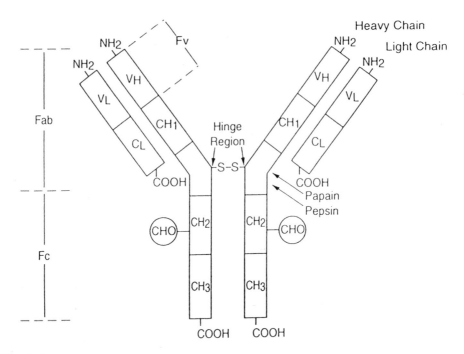

Fig. 1. Immunoglobulin G (IgG) structure. IgG molecules are tetramers of heavy and light chains. The Fab (antigen-binding fragment) region mediates selective target isolation. The Fc (crystallizable fragment) region controls receptor binding and complement activation. The Fc region also plays a role in binding to protein A and protein-G. *N*-Linked oligosaccharide moieties are found on Asn-297 of the heavy chain in the CH2 region but may be found at additional sites.

erated by recombinant technology. The antibody must be capable of stable binding at neutral pH and yet be readily dissociated at either mild acid or alkali conditions (*see* **Note 3**). An alternative to both polyclonal antibodies and monoclonal antibodies are recombinantly expressed antibodies and antibody fragments. Immobilized protein A and protein G have been used extensively for the affinity purification of antibodies (*see* **ref. 10** and **Note 4**).

1.2. Practical Aspects

1.2.1. Matrices

The range of solid-phase supports for affinity chromatography is now extremely large. However, the attributes of matrices that determine their suitability for activation and immobilization of the antibody are universal:

Uniformly sized particles with a macroporous structure facilitating a high area-to-volume ratio and without size-exclusion effects even for large protein molecules.

- Hydrophilic and inert material with low nonspecific absorption.
- Readily derivatised with a stable linkage between matrix and ligand.
- Rigidity to withstand high flow rates.
- Stability to the extremes of pH and dissociating agents used in the separation.

The most commonly used matrices are based around spherical agarose beads that can be bought commercially and chemically crosslinked to improve chemical stability and rigidity. Acrylamide has also been used; however, it is often supplied as a copolymer with agarose to improve its physical stability. Controlled pore glass has been used to improve physical stability and thereby improve flow rates. As with all affinity purifications, the technique is based on selective adsorption and desorption and not continuous partitioning. As such, size exclusion is not desired and the commercially available matrices should show no size-exclusion effects. If large proteins or complexes are to be purified, then the pore size of the matrix may have to be addressed.

1.2.2. Activation and Immobilization

The relative expense of mAbs production has resulted in a need to establish matrices, which will retain a high capacity for the target over many cycles of reuse and without significant mAb leakage. Various functional groups are introduced onto matrices for the immobilization of ligands.

The types of functional group exploited for the immobilization of immunoglobulins are directed toward the amino acid side-chain residues (amines, sulfhydryls, carboxylic acids), the termini (amino and carboxylic acid), and sugars of the oligosaccharide chains. In addition, matrices exist that are tailored for derivitized proteins such as biotinylated immunoglobulins. The major reactive groups are listed in **Table 1**. The first example of immunoaffinity chromatography utilized diazotized aminobenzylcellulose to immobilize an antigen for subsequent antibody purification *(11)*. The usual method for immobilization is via primary amino groups using agents such as cyanogen bromide (CNBr) (*see* **Fig. 2**) or hydroxysuccinimide (*see* **Fig. 3**).These methods have the advantage of being well established and relatively easy to perform. Use of CNBr activation is complicated by the toxicity of CNBr. This may be overcome by the use of commercially available preactivated matrices.

Multipoint attachment of the mAb can also induce steric hindrance with respect to antigen binding. At high levels of activation, the antibodies themselves can sterically hinder binding. Several methods have been devised to overcome these issues by utilizing site-directed immobilization. Most are based on the principle of immobilizing the immunoglobulin at the Fc region, resulting in a more efficient use of the ligand (*see* **Note 5**).

Activation of agarose gels by CNBr is based on the conversion of the hydroxy groups of the agarose gel to a cyanate ester group. This "activated matrix" is then capable of reacting with the ε-amino groups (lysines) of immunoglobulins and other proteins (*see* **Note 6**).

N-Hydroxysuccinimide (NHS) is an alternative to CNBr for the immobilization of antibodies (and other proteins) via their primary amine groups. The linkage generated by NHS is extremely stable under normal operating and storage conditions of immunoadsorbents. Coupling is relatively fast and efficient, with a high degree of selectivity for primary amines, although sulfhydryls will also react (*see* **Notes 7** and **8**).

1.2.3. Determining Coupling Efficiency

Before using the prepared affinity matrix, it is prudent to ensure that ligand immobilization has succeeded. This can be done by the determination of the ligand concentra-

Table 1
Reactive Groups for Derivitization

Functional group	Structure on matrix	Target moiety on ligand
Cyanogen Bromide		-NH$_2$
Epoxide / Oxarine		-NH$_2$, -OH, -SH, Sugars
Carbonyldiimidazole		-NH$_2$
N-Hydroxysuccinimide		-NH$_2$
Vinyl Sulphone		-NH$_2$, -SH, -OH
Thiol		-SH
Tosyl Chloride		-NH$_2$, -SH
Tresyl Chloride		-NH$_2$, -SH
Azlactone		-NH$_2$, -SH, -OH
Hydrazine		Sugars
Avidin	Protein[a]	Biotinylated proteins, etc.
Protein A/protein-G	Protein[b]	Immunoglobulin via Fc region

[a]Avidin is a tetrameric protein from egg white with a subunit molecular weight of 15,000.

[b]Protein A (molecular weight 42,000) is a cell-wall protein from *Staphylococcus aureus*. Protein-G is a cell-wall protein from *Streptococci* (molecular weight 35,000)

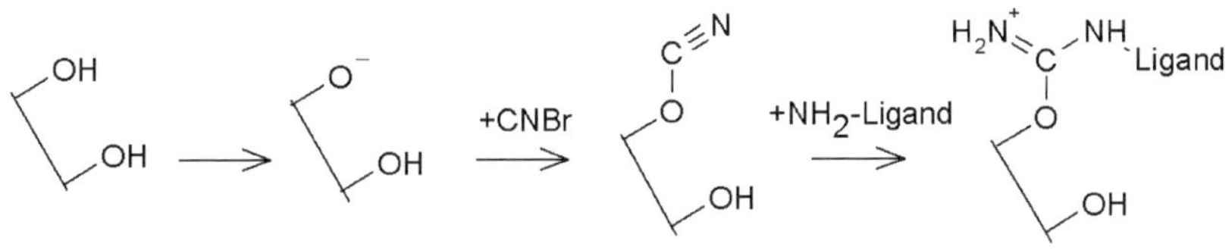

Fig. 2. Activation and immobilization via cyanogen bromide.

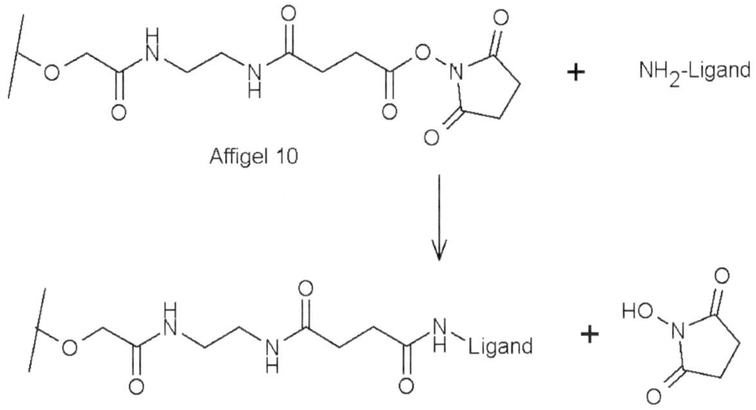

Fig. 3. Immobilization of a ligand via an *N*-hydroxysuccinimide linkage.

tion in terms of micromoles (or milligrams) ligand per milliliter of gel. By reference to the starting level of immunoglobulin, the coupling efficiency can be determined. Several methods are available for achieving ligand density. The simplest is the "indirect method" calculated by subtraction of the unbound antibody present in the solution after coupling from that present prior to coupling (*see* **Note 9**).

1.3. Practical Aspects

1.3.1. Immunoaffinity Purification

Immunoaffinity columns are usually packed in to glass columns for low-pressure chromatography or stainless-steel columns for high performance immunoaffinity chromatography (HPIAC). The equipment required includes a pump, detector, recorder, and, if possible, a fraction collector. High-performance liquid chromatography (HPLC) systems can be purchased from several companies including Agilent, Waters, and Dionex. Complete low-pressure systems are readily available from several suppliers, including Amersham Biosciences, Bio-Rad, and Applied Biosystems. Alternatives include stirred tanks, which allow the isolation from crude cell homogenates.

Whichever method is selected, the reagents used must be of the highest grade possible and all solutions must be sterile-filtered prior to application to the column. These precautions may significantly prolong the lifetime of the matrix and prevent column fouling.

The type of chromatography employed and scale of operation will depend on the aim of the study. It is worth considering running the chromatography using the actual antibody and comparing the output with that from a control column preferably containing an unrelated antibody from the same species and subclass. This will allow the elimination of proteins binding via nonspecific adsorption. A column containing deliberately inactivated antibody may serve the same purpose.

1.3.2. Sample Preparation

Samples are often supplied as crude extracts which may contain cell debris and other insoluble matter. This necessitates some form of pretreatment of the sample to prevent column fouling and/or unwanted nonspecific binding. Suitable methods include centrifugation and or filtration through a 0.4-μm nitocellulose filter (or finer).

1.3.3. Binding

An important part of successful immunopurification is the choice of buffer solutions used to elicit binding of the target protein and, equally important, the conditions used to facilitate elution of the bound target. The buffer solution chosen is largely dependent on the sample to be processed. The most commonly used binding buffers are pH 7–8 (e.g., phosphate or borate) and often contain 0.1–0.5 M NaCl and/or a small amount of surfactant to reduce nonspecific binding. The actual conditions will depend on the affinity constant (K_a) and the nature of the impurities present. High levels of NaCl may induce binding of impurities by promoting hydrophobic interactions.

Surfactants include the nonionic detergents Triton X-100, Tween, and Lubrol PX. Lubrol has the advantage of being nonabsorbing at 280 nm, a common wavelength for monitoring chromatography. More stringent ionic detergents such as sodium deoxycholate may be needed for solubilization of membrane proteins. When using surfactants, consideration should be given to the ultimate purpose of the purified reagent, as their removal is often not a trivial exercise. Analysis of the immunoadsorbent characteristics (e.g., sodium dodecyl sulfate-polyacrylamide gel electrophoresis [SDS-PAGE]) as an analytical technique may aid optimization of binding and elution conditions *(12)*.

1.3.4. Flow Rates

The sample is introduced onto the matrix in a manner limited by the mechanical ridigity of the support and the diffusion of the material into the pores of the matrix. For optimal flow rates, it is always advisable to consult the manufacturers' recommendations. A flow of approx 30–50 cm/h is suitable for low-pressure chromatography and 300–500 cm/h for HPIAC.

1.3.5. Pre-elution Washing

Following loading, the column is washed with the binding buffer to remove any unbound material. The amount of nonspecific binding will depend on the original binding conditions. Use of 0.5 M NaCl in the initial buffer will result in relatively low-impurity binding and reduced requirement for washing. If hydrophobic interaction of impurities is suspected, low-ionic-strength washing may be effective. Extended washing may remove some nonspecific binding material, but a balance between yield and purity has to be obtained. In general, washing should be continued until a stable ultraviolet (UV) baseline is achieved.

1.3.6. Elution

Effective elution requires rapid desorption of concentrated functional product. The selection of the correct elution conditions is as important a part of the protocol design as any other aspect and probably the most likely to require tailoring to the particular tar-

Table 2
Mobile Phases for Elution From Immunoaffinity Supports

Class of eluent	Principle
0.1 M Glycine–NaOH, pH 10.0 50 mM Diethylamine pH 11.5 1 M NH$_4$OH	High pH
0.1 M Glycine–HCl, pH 2.0 20 mM HCl 0.1 M Sodium citrate, pH 2.5 1 M Propionic acid	Low pH
50% (v/v) Ethylene glycol Dimethyl sulfoxide (DMSO) Acetonitrile 10% (v/v) Dioxane	Organic solvent
0.1 M Tris-HCl, pH 8.0, + 2 M NaCl	High ionic strength
Deionized water	Low ionic strength
1 M Ammonium thiocyanate 3 M Potassium chloride 5 M Potassium iodide 4 M Magnesium chloride	Chaotropes
6 M Guanidine HCl, pH 3.0 6–8 M Urea	Denaturant
1% (w/v) SDS	Surfactant
1–10 mM EDTA or EGTA Sodium citrate	Metal-ion chelator

get molecule *(13)*. The elution conditions should facilitate the release of the bound substance in a manner that retains the functional and structural integrity of both the target and the immobilized ligand. The most common way of releasing the target protein is to alter the mobile phase (*see* **Table 2**, **Notes 10** and **11**, and **refs.** *13* and *14*).

2. Materials

2.1. Immobilization of Immunoglobulins Via N-Hydroxysuccinimide

1. Agarose gel, nonactivated (e.g., Sepharose CL-6B [Amersham Biotech]) or preactivated (e.g., Affi-Gel 10 and 15 [Bio-Rad]).
2. Antibody solution in coupling buffer at approx 1–5 mg/mL gel.
3. *N-N'*-Disuccinimidylcarbonate (DSC).
4. Dry acetone.
5. Dry pyridine.
6. Coupling buffer: 0.1 M Sodium phosphate, pH 7.5.
7. Blocking buffer: 50 mM Ethanolamine, pH 8.0.
8. Storage buffer: Phosphate-buffered saline (PBS) containing 0.1% (w/v) sodium azide.

2.2. Immunopurification

1. Empty chromatography column of appropriate dimensions.
2. Suitable immunoaffinity matrix.

3. Suitable binding buffer (e.g., PBS).
4. Crude antigen-containing solution/extract.
5. Wash buffer: 0.01 mM sodium phosphate, pH 7.4, for example.
6. Elution buffer: 0.1 M sodium acetate, pH 4.0, for example.
7. Regeneration buffer: 0.1 M Sodium actetate, pH 4.0, containing 8 M urea.
8. Storage buffer: PBS containing 0.1% (w/v) sodium azide.

3. Methods

3.1. Activation

1. Wash the agarose gel in a Buchner funnel with 10 vol of deionized water under low vacuum.
2. Wash sequentially with 4–10 vol of 30% (v/v) aqueous acetone, 50% (v/v) aqueous acetone, 70% (v/v) aqueous acetone, and, finally, dry acetone.
3. Resuspend the gel in dry acetone and make the suspension a 50% gel slurry containing 80 g/L (0.3 M) DSC.
4. While mixing, in a fume hood add 1 gel volume of dry pyridine containing 7.5% (v/v) anhydrous triethylamine dropwise over 30–60 min. Continue mixing for a further 60 min.
5. Wash the gel sequentially with 4–10 vol of dry acetone and 4–10 vol of dry isopropanol.
6. Store the activated gel at 4°C as a 50% (v/v) slurry in isopropanol.

3.2. Immobilization

The amount of gel prepared will depend on the amount of immunoglobulin available and the scale of operation desired. The recommended level of immobilization is 4–5 mg antibody/mL gel.

1. Dissolve, desalt, or dialysis the antibody to be immobilized in the coupling buffer and cool to 4°C.
2. Wash the 50% (v/v) suspension of activated gel/Affi-Gel 10 with 10 vol of ice-cold deionized water in a sintered glass funnel to remove the storage buffer (isopropanol).
3. Resuspend the gel to a 50% (v/v) slurry in ice-cold coupling buffer and transfer the slurry to a flask containing the immunoglobulin in coupling buffer.
4. Stir the flask gently and allow to mix for 4–16 h at 4°C.
5. Coupling can be monitored by following the $A_{280\,nm}$ of the coupling solution. As antibody binds, the absorbance at $A_{280\,nm}$ falls.
6. Wash the gel and resuspend in 1 vol of blocking buffer and allow to mix for 1 h.
7. Transfer the slurry to storage buffer using the Buchner funnel and store at 4°C prior to use.

3.3. Immunoaffinity Purification of Target Protein From Crude Extract

1. Pack the column in accordance with the manufacturer's instructions.
2. Wash the column with deionized water and equilibrate in binding buffer.
3. Equilibrate the starting material in the binding buffer via dialysis, desalting, or ultrafiltration and centrifuge to remove insoluble material immediately prior to loading on to the column.
4. Load the starting material onto the column at 30 cm/h and monitor the effluent by UV absorbance.
5. Once loaded, wash the unbound material clear with binding buffer until the UV trace returns to baseline.
6. Wash the column with 2 column volumes of low ionic strength wash buffer (e.g. binding buffer minus the sodium chloride).
7. If possible, reverse the direction of the flow through the column.

8. Elute the bound material with elution buffer and collect into a suitable container. The eluted protein is neutralized or dialyzed as appropriate to remove the dissociating agent.
9. Adjust the eluent to neutral pH.
10. Wash the column sequentially with 4–10 column volumes deionized water, regeneration buffer, and, finally, storage buffer.

4. Notes

1. Whereas affinity chromatography exploits known physiological interactions such as receptors binding hormones, enzymes binding inhibitors, and so forth, antibodies are now available for immunoaffinity purification of target proteins (e.g., structural proteins), for which biological ligands were previously not available. The technique can be applied to any molecule capable of eliciting an antibody response. Although it was a long-stated axiom that any molecule of a molecular weight greater than 5 kDa is theoretically capable of such a response, synthetic peptides as short as hexamers have been used to raise antibodies. In certain cases, the technique has been applied to small molecules by raising antibodies to epitopes linked to larger molecules. This has been exploited in the generation of immunoaffinity methodology for highly sensitive drug residue analysis *(15–17)*. The specificity of the matrices also means that columns of differing selectivities can be used in series to analyze or prepare a number of different solutes in one experiment *(18)*.
2. The selection of the antibody for use in immunoaffinity purification is critical. Initial immunopurification studies were performed using polyclonal antisera. This makes the selection of affinities suitable for binding and elution difficult. With the advent of mAbs, these problems were to a large extent eliminated. Despite the initial workup of a hybridoma cell line for the production of mAbs, the increased control and supply of material from mAbs has resulted in a rapid growth in the use of immunopurification reagents. The methodology is now being applied to process scale isolations *(17)* and to the isolation of therapeutic-grade proteins *(18)*.
3. The value of the dissociation constant (K) can vary within a range of 10^{-3} to $10^{-14}\,M^{-1}$. The ideal range for an immunoaffinity reagent must be determined empirically, but if $K > 10^{-4}\,M^{-1}$, then the antigen may not be retained sufficiently to be practicable. When $K < 10^{-8}\,M^{-1}$, the affinity may be too strong and harsh elution conditions may be necessary to elicit dissociation. A value in the range 10^{-7} to $10^{-8}\,M^{-1}$ is often considered optimal.
4. The antibodies used for the immunoaffinity purification must be of high purity and are often themselves purified by affinity techniques such as protein A or protein-G affinity chromatography (as described earlier) or using anti-isotype or anti-idiotype antibodies for immunopurification of the immunopurification reagent itself. This may appear to be a circular argument, but general antibodies raised against the constant region of a antibody from a particular species are often readily available commerically. Polyclonal antibodies by virtue of their source often require more cleanup before use.
5. In theory, an immobilized IgG antibody should be capable of binding 2 mol antigen/mol antibody, although a binding efficiency of around 10% is more likely to be the norm. Explanations for this poor performance include incorrect orientation of the antibody such that the Fab region has been utilized in the matrix-coupling reaction. In the methodology described in **Figs. 2** and **3**, immobilization is random with different points of the molecule forming the linkage with the solid matrix. Purely on the basis of probability, a proportion of the antibodies will be bound via the Fab region, leading to functional inactivation and poor column capacity.
6. Cyanogen bromide activation has been used most extensively because of its almost universal applicability to agarose, polysaccharide, and hydroxyl group containing synthetic polymers. They are also useful for coupling small molecules as well as large macromoleu-

cles such as antibodies. The immobilization procedures are simple, reproducible, and quantitative. An important factor is the ability of the immobilization to operate under mild conditions with a range of pH's that is possible, an important consideration for immunoglobulin immobilization. CNBr-based immobilization does, however, have its disadvantages. The instability of the isourea bond leads to a low but constant leakage of ligand under operating conditions resulting from nucleophilic attack. The isourea moiety also acts as a weak anion-exchange matrix at neutral pH. In addition to ensuring protein purity to reduce nonspecific adsorption, when using CNBr matrices it is necessary to consider potential DNA because of the ability of CNBr-activated matrices to immobilize nucleic acids. Despite these disadvantages, CNBr remains a popular method of immobilization.

7. As with CNBr-activated matrices, the supports can either be activated in the laboratory or bought as activated matrices (Affi-Gel 10 and Affi-Gel 15 [Bio-Rad], NHS-activated Superose, and HiTrap NHS [Amersham Biosciences]). The Affi-Gel matrices are available as loose column packing; the Amersham Biosciences matrices are also available in prepacked columns with methodology for immobilization *in situ*. Differences in the commercially available matrices reflect the spacer used. Affi-Gel 10 contains a noncharged 10-atom spacer and is used for the immobilization of proteins with pIs' of 6.5–11. Affi-Gel 15 has a positively charged 15-atom spacer arm and is useful for proteins with a pI below 6.5. An advantage of Affi-Gel is that coupling can occur over a wide pH range (pH 3–10). However, efficiency will vary markedly with protein immobilized and pH used (consult manufacturer's recommendations).

8. One advantage of the NHS-activated supports is their ability to immobilize ligands in either aqueous or nonaqueous solutions, although immobilization of antibodies usually proceeds in aqueous buffer solutions. (Competing amines such as Tris or glycine should be avoided.) For immobilization, an uncharged amine is required and efficient immobilization is usually performed at pH 7.5–9.0, because at elevated pH's, the hydrolysis of the activated matrix will accelerate, leading to poor ligand immobilization. Affi-Gel 10 and Affi-Gel 15 are supplied in organic phases (isopropanol) to stabilize the matrices. As soon as exchange into aqueous buffer takes place the activated groups will begin to hydrolyze. The half-life of hydrolysis at pH 8.0 is approx 20 min at 4°C. This requires careful but rapid buffer transfer and ligand introduction in order to maximize the reproducibility of the immobilization.

9. Care must be taken in the method employed for determining the antibody concentration, as common methods such as measurements of $A_{280\,nm}$ and the Lowry assay can give erroneous results. The UV absorbance at $A_{280\,nm}$ can be complicated by the presence of other $A_{280\,nm}$ absorbing agents such as the surfactant Triton X-100, DTT, or high levels of Tris. The Lowry assay and A_{280} can be inappropriate for NHS-activated matrices as the NHS interferes with the assay at neutral or basic pH. The solution should therefore be made acidic by the addition of 10 mM HCl prior to analysis. Amino acid analysis is a more accurate but expensive option for determining protein concentration. As amino acid analysis is used following extended hydrolysis, this method can be used to determine the protein attached to the matrix. Other methods for determining coupling efficiency include fluorescence and radioimmunoassay.

10. Chaotropes affect elution by altering the structure of the water associated with the interaction and by disrupting the hydrophobic interactions, which may contribute to the overall binding reaction. As protein folding can be largely affected by intermolecular and intramolecular hydrophobic interactions, these agents can also lead to protein denaturation. Therefore, it is necessary to remove the chaotropic agent promptly after elution using dialysis or desalting column chromatography. For this reason, selection of the correct chaotrope at an appropriate concentration is critical. Chaotropes can be listed in order of stringency, as in the *Hofmeister* (also known as the lyotropic) series:

NH4$^+$<K$^+$<Na$^+$<Cs$^+$<Li$^+$<Mg^{2+}<Ca^{2+}<Ba^{2+}<PO$_4^{2-}$<SO$_4^{2-}$<CH$_3$COO$^-$<Cl$^-$<Br$^-$<NO3$^-$<ClO$_4^-$<I$^-$<SCN$^-$

11. Denaturants such as guanidine, urea, and ethylene glycol act by a combination of modifying protein-folding (mild denturation) and chaotropic activity. Some antibody–antigen interactions are cation dependent and use of chelators such as EDTA can elicit elution. It should be noted that when EDTA elution is used, it should be operated at pH's above neutrality to maintain solubility.

References

1. Phillips, T. M. (1989). High-performance immunoaffinity chromatography. *Adv. Chromatogr.* **29,** 133–173.
2. Weliky, N., Weetall, H. H., Gilden, R. V., and Campbell, D. H. (1964) Synthesis and use of some insoluble immunologically specific adsorbents. *Immunochemistry* **1,** 219–229.
3. Kristiansen, T. (1976). Matrix-bound antigens and antibodies. *Scand. J. Immunol.* (**Suppl.**) **3,** 19–27
4. Santucci, A., Soldani, D., Lozzi, L., et al. (1988) High performance liquid chromatography immunoaffinity purification of antibodies and antibody fragments. *J. Immunol. Methods* **114,** 181–185.
5. Muronetz, V. I., Sholukh, M., and Korpela, T. (2001) Use of protein–protein interactions in affinity chromatography. *J. Biochem. Biophys. Methods* **49,** 29–47.
6. Hage, D. S. (1998) Survey of recent advances in analytical applications of immunoaffinity chromatography. *J. Chromatogr.* **715,** 3–28
7. Hill, C. R., Kenney, A. C., and Goulding, L. (1987) The design, development and use of immunopurification reagents. *Biotech. Forum* **4,** 168–171.
8. Dunbar, B. S. and Schwoebel, E. D. (1990) Preparation of polyclonal antibodies, in *Methods in Enzymology* (Deutscher, M. P., ed.), Academic, London, vol. 182, pp. 663–670.
9. Kohler, G. and Milstein, C. (1975) Continuous cultures of fused cells secreting antibody of predefined specificity. *Nature* **256,** 495–497
10. Eliasson, M., Andersson, R., Olsson, A., Wigzell, H., and Uhlen, M. (1989) Differential IgG-binding characteristics of staphylococcal protein A, streptococcal protein G, and a chimeric protein AG. *J. Biol. Chem.* **142,** 575–581.
11. Campbell, D. H., Luescher, E., and Lerman, L. S. (1951) Immunologic adsorbents. I. Isolation of antibody by means of a cellulose-protein antigen. *Proc. Natl. Acad. Sci. USA* **37,** 575–578.
12. Janatova, J. and Gobel, R. J. (1984) Rapid optimization of immunoadsorbent characteristics. *Biochem. J.* **221,** 113–120.
13. Yarmush, M. L., Antonsen, K. P., Sundaram, S., and Yarmush, D. M. (1992) Immunoadsorption: strategies for antigen elution and production of reusable adsorbents. *Biotechnol. Prog.* **8,** 168–178.
14. Yan, S. B. (1996) Review of conformation-specific affinity purification methods for plasma vitamin K-dependent proteins. *J. Mol. Recogn.* **9,** 211–218.
15. Katz, S. E. and Brady, M. S. (1990) High-performance immunoaffinity chromatography for drug residue analysis. *Anal. Chem.* **73,** 557–560.
16. Katz, S. E. and Siewierski, M. (1992) Drug residue analysis using immunoaffinity chromatography. *J. Chromatogr.* **624,** 403–409.
17. VanGinkel, L. A. (1991) Immunoaffinity chromatography, its applicability and limitations in multi-residue analysis of anabolizing and doping agents. *J. Chromatogr.* **564,** 363–384.
18. Jack, G. W. and Wade, H. E. (1987) Immunoaffinity chromatography of clinical products. *TIBTECH* **5,** 91–95.

20

Immobilized Metal-Ion Affinity Chromatography

Tai-Tung Yip and T. William Hutchens

1. Introduction

Immobilized metal-ion affinity chromatography (IMAC) *(1–4)* is also referred to as metal chelate chromatography, metal-ion interaction chromatography, and ligand-exchange chromatography. We view this affinity-separation technique as an intermediate between highly specific, high-affinity bioaffinity separation methods, and wider-spectrum, low-specificity adsorption methods, such as ion exchange. The IMAC stationary phases are designed to chelate certain metal ions that have selectivity for specific groups (e.g., His residues) in peptides (e.g., **refs. 5–9**) and on protein surfaces *(10–15)*. The number of stationary phases that can be synthesized for efficient chelation of metal ions is unlimited, but the critical consideration is that there must be enough exposure of the metal ion to interact with the proteins, preferably in a biospecific manner. Several examples are presented in **Fig. 1**. The challenge to produce new immobilized chelating groups, including protein surface metal-binding domains *(17,18)* is being explored continuously *(19)*. A common fusion protein is the hexahistidine tag for purification *(20)*. **Table 1** presents a list of published procedures for the synthesis and use of stationary phases with immobilized chelating groups. This is by no means exhaustive and is intended only to give an idea of the scope and versatility of IMAC.

The number and spectrum of different proteins (*see* **Fig. 2**) and peptides (*see* **Fig. 3**) characterized or purified by use of immobilized metal ions are increasing at an incredible rate. The three reviews listed *(10,23,24)* barely present the full scope of activities in this field. Beyond the use of immobilized metal ions for protein purification are several analytical applications, including mapping proteolytic digestion products *(7)*, analyses of peptide amino acid composition (e.g., **refs. 7** and **8**), evaluation of protein surface structure (e.g., **refs. 10–12**), monitoring ligand-dependent alterations in protein surface structure (*see* **Fig. 4**) *(14,19)*, and metal-ion exchange or transfer (e.g., **refs. 17** and **25**). Recently, Fe(III) has been used for purification of proteins, especially phosphoprotiens *(26,27)*.

The versatility of IMAC is one of its greatest assets. However, this feature is also confusing to the uninitiated. The choice of stationary phases and the metal ion to be immobilized is actually not complicated. If there is no information on the behavior of the

From: *Methods in Molecular Biology, vol. 244: Protein Purification Protocols: Second Edition*
Edited by: P. Cutler © Humana Press Inc., Totowa, NJ

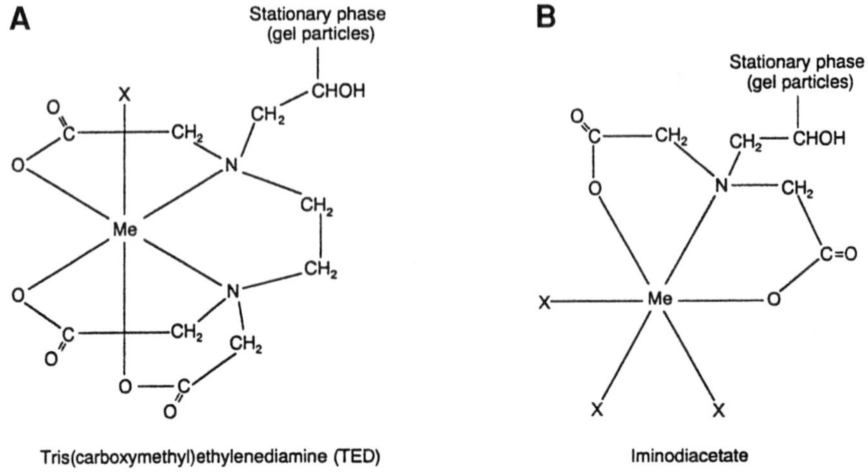

A

Tris(carboxymethyl)ethylenediamine (TED)

B

Iminodiacetate

Fig. 1. Schematic illustration of several types of immobilized metal-chelating group, including iminodiacetate (IDA), tris(carboxymethyl) ethylenediamine (TED), and the metal-binding peptides (GHHPH)$_n$G (where n = 1, 2, 3, and 5. (From **refs. 17** and **18**.)

Table 1
Immobilized Chelating Groups and Metal Ions Used
for Immobilized Metal-Ion Affinity Chromatography

Chelating group	Suitable metal ions	Ref.	Commercial source[a]
IDA	Transitional	*(1,2)*	Pharmacia LKB Pierce Sigma Boehringer-Mannheim TosoHaas
2-Hydroxy-3[N-(2-pyridylmethyl)glycine]propyl	Transitional	*5*	Not available
α-Alkyl nitrilotriacetic acid	Transitional	*6*	Not available
Carboxymethylated aspartic acid	Ca(II)	*15*	Not available
TED	Transitional	*2*	Not available
(GHHPH)$_n$G[b]	Transitional	*17,18*	Not available

[a]Locations: Pharmacia LKB, Uppsala, Sweden; Pierce, Rockford, IL; Sigma, St. Louis, MO; Boehringer-Mannheim, Mannheim, Germany; TosoHaas, Philadelphia, PA.

[b]Letters represent standard one-letter amino acid codes (G = glycine; H = histidine; P = proline). The number of internal repeat units is given by n (n = 1, 2, 3, and 5).

particular protein or peptide on IMAC in the literature, use a commercially available stationary phase (immobilized iminodiacetate) and pick the relatively stronger affinity transitional metal ion, Cu(II), to immobilize. If the interaction with the sample is found to be too strong, try other metal ions in the series, such as Ni(II) or Zn(II), or try an immobilized metal-chelating group with a lower affinity for proteins *(2,28)*. An important contribution to the correct use of IMAC for protein purification is a simplified presentation of the various sample elution procedures. This is especially important to the first-time user. There are many ways to decrease the interaction between an immobilized

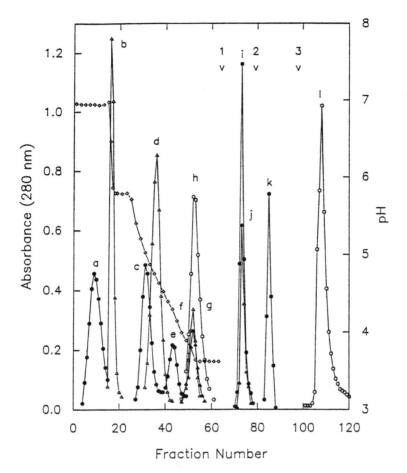

Fig. 2. Protein elution from immobilized (IDA) Cu(II) ions as function of decreasing pH and increasing imidazole concentration. Proteins were eluted in the following order: chymotrypsinogen (a), chymotrypsin (b), cytochrome-*c* (c), lysozyme (d), ribonuclease A (e), ovalbumin (f), soybean trypsin inhibitor (g), human lactoferrin (h), bovine serum albumin (i), porcine serum albumin (j), myoglobin (k), and transformed (DNA-binding) estrogen receptor (l). Open triangles represent pH values of collected fractions. Arrows 1–3 mark the introduction of 20, 50, and 100 m*M* imidazole, respectively, to elute high-affinity proteins resistant to elution by decreasing pH. Protein elution was evaluated by absorbance at 28 nm. In the case of the [3H]estradiol-receptor complex, receptor protein elution was determined by liquid scintillation counting. Except for the estrogen receptor (l), protein recovery exceeded 90%. Only 50–60% of the DNA-binding estrogen receptor protein applied at pH 7.0 was eluted with 100–200 m*M* imidazole. (Reproduced with permission from **ref. *21*.**)

metal ion and the adsorbed protein. Two of these methods are efficient and easily controlled; they will be presented in detail in this chapter. Interpretation of IMAC results for purposes other than separation (i.e., analysis of surface topography and metal-ion transfer) has been discussed elsewhere and is beyond the scope of this chapter.

2. Materials

The following list of materials and reagents is only representative. Other stationary phases, metal ions, affinity reagents, and mobile-phase modifiers are used routinely.

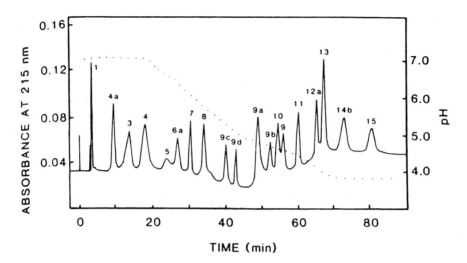

Fig. 3. Seperation of bioactive peptide hormones by IMAC. The pH-dependent separation of a synthetic peptide hormone mixture (19 peptides) was accomplished using a TSK chelate-5PW column (8 × 75 mm, 10-μm particle diameter) loaded with Cu(II). A 20-μL sample (1–4 μg of each peptide) ws applied to the column equilibrated in 20 m*M* sodium phosphate containing 0.5 *M* NaCl, pH 7.0. After 10 min of isocratic elution, pH-dependent elution was initiated with a 50-min gradient to pH 3.8 using o.1 *M* sodium phosphate containing 0.5 *M* NaCl at a flow rate of 1 mL/min. Peptide detection during elution was by UV absorance at 215 nm (0.32 AUFS). The pH profile of effluent is indicated by the dotted line. Sample elution peaks were identified as follows: 1, neurotensin; 4a, sulfated [leu[5]] enkephalin; 3, oxytocin; 4, [leu[5]] enkephalin; 5, mastoparan; 6a, tyr-bradykinin; 7, substance P; 8, somatostatin; 9c, [Asu[1.7]] eel calcitonin; 9d, eel calcitonin (11–32); 9a, Asu[1.7]] human calcitonin; 9b, human calcitonin (17–32); 10, bombesin; 9, human calcitonin; 11, angiotensin II; 12a, [Trp (for)25,26] human GIP (21–42); 13, LH–RH; 14b, human PTH (13–34); 15, angiotensin I. (Reproduced with permission from **ref. 22.**)

2.1. Stationary Phase for IMAC

2.1.1. Conventional Open-Column Stationary Phases (Agarose)

One example is Chelating Sepharose Fast Flow (Pharmacia), which uses the iminodiacetate (IDA) chelating group. Another example is Tris(carboxymethyl)ethylenediamine (TED) (*see* **Table 1** and **Fig. 1**). This stationary phase is used for proteins whose affinity for IDA–metal groups is too high *(2,28)*.

2.1.2. High-Performance Stationary Phases (Rigid Polymer)

One example is TSK Chelate 5PW (TosoHaas) (IDA chelating group).

2.1.3. Immobilized Synthetic Peptides as Biospecific Stationary Phases

Stationary phases of this type are designed based on the known sequence of protein surface metal-binding domains *(17,18)*. The metal-binding sequence of amino acids is first identified (e.g., **refs.** *17* and *29*). The synthetic protein surface metal-binding domain is then prepared by solid-phase methods of peptide synthesis *(17)* and verified to have metal-binding properties in solution *(17,30)*. Finally, the peptides are immobilized

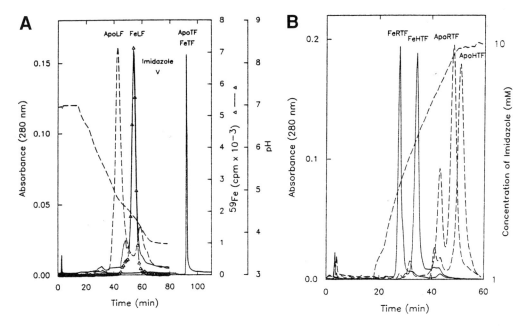

Fig. 4. (A) Separation of apolactoferrin (dashed line) and iron-saturated (solid line) human lactoferrins on high-performance IDA–Cu(II) columns with a phosphate-buffered gradient of decreasing pH in the presence of 3 M urea. Both apotransferrin (ApoTF) and iron-saturated human serum transferrins (FeTF) were eluted only on introduction of 20 mM imidazole (imidazole-labeled arrow). UV absorbance was monitored at 280 nm. The elution of iron-saturated human lactoferrin was determined by measuring protein-bound ^{59}Fe radioactivity (open triangles). **(B)** Separation of iron-free (dashed line) and iron-saturated (solid line) transferrins on a high-performance IDA–Cu(II) affinity column using an imidazole elution gradient protocol. The apo and holo forms of human serum transferrin (hTF) and rabbit serum transferrin (rTF) are shown. The Fe and Apo prefixes to these abbreviations designate iron-saturated and metal-free transferrins, respectively. (Reproduced with permission from **ref. *14*.**)

(e.g., to agarose) using chemical coupling procedures consistent with the retention of metal-binding properties *(18)* (*see* **Table 1** and **Fig. 1**).

The procedures outlined in this chapter emphasize the specific use of agarose-immobilized IDA metal-chelating groups. In general, however, these procedures are all acceptable for use with a wide variety of different immobilized metal-chelate affinity adsorbents.

2.2. Metal-Ion Solutions in Water

1. 50 mM CuSO$_4$.
2. 50 mM ZnSO$_4$.
3. 50 mM NiSO$_4$.

2.3. Buffers

1. Buffer 1: 20 mM Sodium phosphate (7.8 mM NaH$_2$PO$_4$, 12.2 mM Na$_2$HPO$_4$), 0.5 M sodium chloride, pH 7.0.
2. Buffer 2: 0.1 M Sodium acetate, 0.5 M sodium chloride, pH 5.8. Use 0.1 M acetic acid and adjust the pH with 2–5 M sodium hydroxide.

3. Buffer 3: 0.1 *M* Sodium acetate, 0.5 *M* sodium chloride, pH 3.8. Use 0.1 *M* acetic acid and adjust the pH with 2–5 *M* sodium hydroxide.
4. Buffer 4: 50 m*M* Sodium dihydrogen phosphate, 0.5 *M* sodium chloride. Add concentrated HCl until pH is 4.0.
5. Buffer 5: 20 m*M* Imidazole (use the purest grade or pretreat with charcoal), 20 m*M* sodium dihydrogen phosphate (or HEPES), 0.5 *M* sodium chloride. Adjust to pH 7.0 with HCl.
6. Buffer 6: 2 m*M* Imidazole, 20 m*M* sodium dihydrogen phosphate (or HEPES), 0.5 *M* sodium chloride. Adjust to pH 7.0 with HCl.
7. Buffer 7: Buffer 1 containing 50 m*M* EDTA.
8. Buffer 8: 20 m*M* Sodium phosphate (3.2 m*M* NaH_2PO_4, 16.8 m*M* Ha_2HPO_4), 0.5 *M* sodium chloride, 3 *M* urea, pH 7.5.
9. Buffer 9: 50 m*M* Sodium dihydrogen phosphate, 0.5 *M* sodium chloride, 3 *M* urea. Add concentrated HCl until pH is 3.8.
10. Milli-Q (Millipore) water or glass-distilled, deionized water.

Urea (1–3 *M)* and ethylene glycol (up to 50%) have been found useful as additives to the above-mentioned buffers (*see* **Notes 1** and **2**).

2.4. Columns and Equipment

1. 1-cm-Inner-diameter (id) columns, 5–10 cm long for analytical and micropreparative scale procedures.
2. 5 to 10-cm-Inner-diameter columns, 10–50 cm long for preparative scale procedures.
3. Peristaltic pump.
4. Simple gradient forming device to hold a 10–20X column bed volume (if stepwise elution is unsuitable).
5. Flowthrough ultraviolet (UV) detector (280 nm) and pH monitor.
6. Fraction collector.

3. Methods

3.1. Loading the Immobilized Metal-Ion Affinity Gel With Metal Ions and Column-Packing Procedures (Agarose-Based IDA Chelating Gel)

1. Suspend the IDA gel slurry well in the bottle supplied. Pour an adequate portion into a sintered glass funnel. Wash with 10 bed volumes of water to remove the alcohol preservative.
2. Add 2–3 bed volumes of 50 m*M* metal-ion solution in water. Mix well.
3. Wash with 3 bed volumes of water (use buffer 3 for IDA–Cu^{2+}) to remove excess metal ions.
4. Equilibrate gel with 5 bed volumes of buffer 1.
5. Suspend the gel and transfer to a suction flask. Degas the gel slurry.
6. Add the gel slurry to column with the column outlet closed. Allow the gel to settle for several minutes; then, open the outlet to begin flow.
7. When the desired volume of gel has been packed, insert the column adaptor. Pump buffer through the column at twice the desired end flow rate for several minutes. Readjust the column adapter until it just touches the settled gel bed. Re-equilibrate the column at a linear flow rate (volumetric flow rate/cross-sectional area of column) of approx 30 cm/h.

3.2. Elution of Adsorbed Proteins

3.2.1. pH Gradient Elution (Discontinuous Buffer System)

1. After sample application, elute with 5 bed volumes of buffer 1.
2. Change buffer to buffer 2, and elute with 5 bed volumes (*see* **Note 3**).

3. Elute with a linear gradient of buffer 2 to buffer 3. Total gradient vol should be equal to 10–20 bed volumes.
4. Finally, elute with additional buffer 3 until column effluent pH is stable and all protein has eluted (recovery should exceed 90%).

3.2.2. pH Gradient Elution (Continuous Buffer System)

1. After sample application, elute with 2.5 bed volumes of buffer 1.
2. Start a linear pH gradient of buffer 1 to buffer 4. Total gradient volumes should be equal to approx 15 bed volumes (*see* **Note 4**).
3. Elute with additional buffer 4 until the column effluent pH is stable.
4. If the total quantity of added protein is not completely recovered, elute with a small volumes (<5 bed volumes) of buffer 4 adjusted to pH 3.5.

3.2.3. Affinity Gradient Elution with Imidazole

1. Equilibrate the column first with 5 bed volumes of buffer 5. Now, equilibrate the column with 5–10 bed volumes of buffer 6 (*see* **Note 5**).
2. After sample application, elute with 2.5 bed volumes of buffer 6.
3. Elute with a linear gradient of buffer 6 to buffer 5. Total imidazole gradient volume should equal 15 bed volumes (*see* **Notes 6** and **7**).

3.3. Evaluation of Metal-Ion Exchange or Transfer From the Stationary Phase to the Eluted Peptide/Protein

1. Use trace quantities of radioactive metal ions (e.g., ^{65}Zn) to label the stationary phase (i.e., immobilized) metal-ion pool (*see* **Subheading 3.1.**). After elution of adsorbed proteins (*see* **Subheading 3.2.**), determine the total quantity of radioactive metal ions transferred to eluted proteins from the stationary phase (by use of a gamma counter).
2. To avoid the use of radioactive metal ions, the transfer of metal ions from the stationary phase to apo (metal-free) peptides present initially in the starting sample may be determined by either of two methods of soft ionization mass spectrometry. Both electrospray ionization mass spectrometry *(31)* and matrix-assisted UV laser desorption time-of-flight mass spectrometry *(17,29–31)* have been used to detect peptide–metal ion complexes (*see* **Fig. 5**). Both techniques are rapid (<10 min), sensitive (picomoles), and are able to address metal-binding stoichiometry.

3.4. Column Regeneration

1. Wash with 5 bed volumes of buffer 7.
2. Wash with 10 bed volumes of water. The column is now ready for reloading with metal ions.

3.5. High-Performance IMAC

For example, use a TSK chelate 5PW column (7.5 mm inner diameter × 750 mm) 10-μm particle size.

1. High-performance liquid chromatography (HPLC) pump system status.
   a. Flow rate: 1 mL/min.
   b. Upper pressure limit: 250 psi.
2. Metal-ion loading.
   a. Wash the column with 5 bed volumes of water.
   b. Inject 1 mL of 0.2 *M* metal sulfate in water.
   c. Wash away excess metal ion with 3 bed volumes of water; for Cu(II), wash with 3 bed volumes of buffer 3.

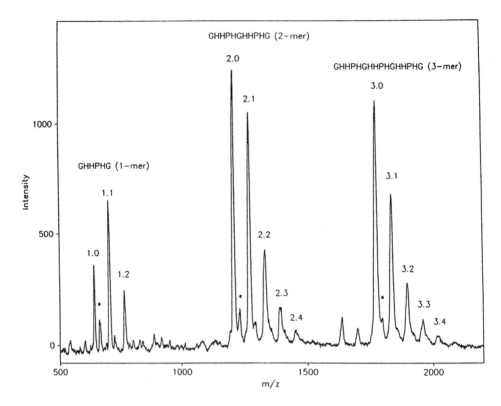

Fig. 5. Matrix-assisted UV laser desorption time-of-flight mass spectrometry (LDTOF) of a mixture of three synthetic metal-binding peptides (1-, 2-, and 3-mer) *after* elution from a column of immobilized GHHPHGHHPHG (2-mer) loaded with Cu(II) ions *(17)*. The synthetic peptide–metal ion affinity column used for metal-ion transfer was prepared by coupling GHH-PHGHHPHG (2-mer) to Affi-10 (Bio-Rad, Hercules, CA). Cu(II) ions were loaded as described in **Subheading 3.1.** The column was equilibrated with 20 m*M* sodium phosphate buffer (pH 7.0) with 0.5 *M* NaCl. An equimolar mixture of the three different synthetic peptides (free of bound metal ions) was passed through the column unretained. Flowthrough peptide fractions were analyzed directly by LDTOF *(23–25)*. The metal-ion-free peptides GHHPHG (1-mer peak 1.0), GHHPHGHHPHG (2-mer peak 2.0), and GHHPHGHHPHGHHPHG (3-mer peak 3.0) are observed along with peptides with one, two, three, or four bound Cu(II) ions (e.g., 3.1., 3.2., 3.3., 3.4). The small peaks marked by an asterisk indicate the presence of a peptide–sodium adduct ion. A detailed description of these results is provided in **ref. *17*.** (Reproduced with permission.)

3. pH Gradient elution: phosphate buffers pH 7.0 and pH 4.0.
   a. 100% Buffer 1, duration 5–10 min.
   b. 0–80% Buffer 4, duration 25 min.
   c. 80–100% Buffer 4, duration 20 min.
   d. 100% Buffer 4 until column effluent pH is constant or until all proteins have been eluted.
4. pH Gradient elution in 3 *M* urea: phosphate buffers pH 7.5–3.8.
   a. 100% Buffer 8, duration 5 min.
   b. 0–10% Buffer 9, duration 10 min.
   c. 10–80% Buffer 9, duration 18 min.
   d. 80–100% Buffer 9, duration 25 min.
   e. 100% Buffer 9 until eluent pH is constant or all proteins have been eluted.

5. Imidazole gradient elution: 1 mM imidazole (5% buffer 5 in buffer 1) to 20 mM imidazole (100% buffer 5) (*see* **Notes 7** and **8**).
   a. 5–10% Buffer 5, immediately after sample injection, duration 10 min.
   b. 10–100% Buffer 5, duration 30 min.
   c. 100% Buffer 5 until all samples have been eluted.

4. Notes

1. To facilitate the elution of some proteins, mobile-phase modifiers (additives), such as urea, ethylene glycol, detergents, and alcohols, can be included in the column equilibration and elution buffers *(12,21–32)*.
2. To ensure reproducible column performance for several runs without a complete column regeneration in between, low concentrations of free metal ions can be included in the buffers to maintain a fully metal-charged stationary phase. This will not affect the resolution and elution position of the proteins *(9,21,33)*. On the other hand, if free metal ions are not desired in the protein eluent, a separate metal-ion scavenger column (e.g., a blank or metal-free IDA–gel or TED–gel column) may be connected in series.
3. The discontinuous buffer pH gradient is ideal for the pH 6.0–3.5 range. We have observed that acetate is also a stronger eluent than phosphate.
4. The phosphate buffer pH gradient is good for the pH 7.0–4.5 range. It has the advantage of UV transparency and is particularly suitable for peptide analysis *(22)*.
5. HEPES can be used instead of phosphate for both discontinuous buffer pH gradients and in the imidazole gradient elution mode. HEPES is a weaker metal-ion "stripping" buffer than phosphate. HEPES is also good for preserving the metal-binding capacity of some carrier proteins such as transferrin *(14)*.
6. For well-characterized proteins, a stepwise gradient of either pH or imidazole can be used to eliminate the need for a gradient-forming device.
7. For the affinity elution method with imidazole, the imidazole gradient actually formed must be monitored by, for example, absorbance at 230 nm or by chemical assay, if reproducible results are desired. Even when the column is presaturated with concentrated imidazole and then equilibrated with buffers containing a substantial amount (2 mM) of imidazole the immobilized metal ions can still bind additional imidazole when the affinity elution gradient is introduced. As a result, when simple (nonprogrammable) gradient-forming devices (typical for open-column chromatography) are used, a linear imidazole gradient is not produced; a small imidazole elution front (peak) at the beginning of the gradient will cause some proteins to elute "prematurely." The multistep gradient described for use with the HPLC systems is designed to overcome this problem. However, we emphasize that this particular program is custom designed only for the high-performance immobilized metal-ion column of given dimensions and capacity. The program must be adjusted for other column types.
8. The imidazole gradient of up to 20 mM is only an example. Quite often, much higher concentrations (up to 100 mM) are required to elute higher-affinity proteins *(11,12,21)* (*see* **Fig. 2**).

Acknowledgment

This work was supported, in part, by the US Department of Agriculture, Agricultural Research Service Agreement No. 58-6250-1-003. The contents of this publication do not necessarily reflect the views or policies of the US Department of Agriculture, nor does mention of trade names, commercial products, or organizations imply endorsement by the US government.

References

1. Porath, J., Carlsson, J., Olsson, I., and Belfrage, G. (1975) Metal chelate affinity chromatography, a new approach to protein fractionation. *Nature* **258,** 598–599.
2. Porath, J. and Olin, B. (1983) Immobilized metal ion affinity adsorption and immobilized metal ion affinity chromatography of biomaterials. Serum protein affinities for gel-immobilized iron and nickel ions. *Biochemistry* **22,** 1621–1630.
3. Garberc-Porekar, V. and Menart V. (2001) Perspectives of immobilized-metal affinity chromatography. *J. Biochem. Biophys. Methods* **49**, 335–360.
4. Chaga, G. S. (2001) Twenty-five years of immobilized metal ion affinity chromatography: past, present and future. *J. Biochem. Biophys. Methods* **49**, 313–334.
5. Monjon, B. and Solms, J. (1987) Group separation of peptides by ligand-exchange chromatography with a Sephadex containing *N*-(2-pyridyl-methyl)glycine. *Anal. Biochem.* **160,** 88–97.
6. Hochuli, E., Dobeli, H., and Schacher, A. (1987) New metal chelate adsorbent selective for proteins and peptides containing neighbouring histidine residues. *J. Chromatogr.* **411,** 177–184.
7. Yip, T.-T. and Hutchens T. W. (1989) Development of high-performance immobilized metal affinity chromatography for the separation of synthetic peptides and proteolytic digestion products, in *Protein Recognition of Immobilized Ligands* (Hutchens, T. W., ed.), UCLA Symposia on Molecular and Cellular Biology Vol. 80, Alan R. Liss, New York, pp. 45–56.
8. Yip, T. T., Nakagawa, Y., and Porath, J. (1989) Evaluation of the interaction of peptides with Cu(II), Ni(II), and Zn(II) by high-performance immobilized metal ion affinity chromatography. *Anal. Biochem.* **183,** 159–171.
9. Hutchens, T. W. and Yip, T. T. (1990) Differential interaction of peptides and protein surface structures with free metal ions and surface-immobilized metal ions. *J. Chromatogr.* **500,** 531–542.
10. Sulkowski, E. (1985) Purification of proteins by IMAC. *Trends Biotechnol* **3**, 1–7.
11. Hutchens, T. W. and Li, C. M. (1988) Estrogen receptor interaction with immobilized metals: differential molecular recognition of Zn^{2+}, Cu^{2+}, and Ni^{2+} and separation of receptor isoforms. *J. Mol. Recogn.* **1,** 80–92.
12. Hutchens, T. W., Li, C. M., Sato, Y., and Yip, T.-T. (1989) Multiple DNA-binding estrogen receptor forms resolved by interaction with immobilized metal ions. Identification of a metal-binding domain. *J. Biol. Chem.* **264**, 17,206–17,212.
13. Hemdan, E. S., Zhao, Y.-J., Sulkowski, E., and Porath, J. (1989) Surface topography of histidine residues: a facile probe by immobilized metal ion affinity chromatography. *Proc. Natl. Acad. Sci. USA* **86,** 1811–1815.
14. Hutchens T. W. and Yip, T.-T. (1991) Metal ligand-induced alterations in the surface structures of lactoferrin and transferrin probed by interaction with immobilized Cu(II) ions. *J. Chromatogr.* **536,** 1–15.
15. Mantovaara-Jonsson, T., Pertoft, H., and Porath, J. (1989) Purification of human serum amyloid ccmponent (SAP) by calcium affinity chromatography. *Biotechnol. Appl. Biochem.* **11,** 564–571.
16. Fiedler, M. and Skerra, A. (2001) Purification and characterization of His-tagged antibody fragments. in *Antibody Engineering* (Kontermann R., ed) Springer-Verlag, Berlin, pp. 243–256.

17. Hutchens, T. W., Nelson, R. W., Li, C. M., and Yip, T.-T. (1992) Synthetic metal binding protein surface domains for metal ion-dependent interaction chromatography. I. Analysis of bound metal ions by matrix-assisted UV laser desorption time-of-flight mass spectrometry. *J. Chromatogr.* **604,** 125–132.

18. Hutchens, T. W. and Yip, T.-T. (1992) Synthetic metal binding protein surface domains for metal ion-dependent interaction chromatography. II. Immobilization of synthetic metal-binding peptides from metal-ion transport proteins as model bioactive protein surface domains. *J. Chromatogr.* **604,** 133–141.

19. Hutchens, T. W. and Yip, T.-T. (1990) Model protein surface domains for the investigation of metal ion-dependent macromolecular interactions and metal ion transfer. *Methods* **4,** 79–96.

20. Bornhorst, J. A. and Falke, J. J. (2000) Purification of proteins using polyhistidine affinity tags. *Methods Enzymol.* **326,** 245–254.

21. Hutchens, T. W. and Yip, T.-T. (1990) Protein interactions with immobilized transition metal ions: quantitative evaluations of variations in affinity and binding capacity. *Anal. Biochem.* **191,** 160–168.

22. Nakagawa, Y., Yip, T. T., Belew, M., and Porath, J. (1988) High performance immobilized metal ion affinity chromatography of peptides: analytical separation of biologically active synthetic peptides. *Anal. Biochem.* **168,** 75–81.

23. Fatiadi A. J. (1987) Affinity chromatography and metal chelate affinity chromatography. *CRC Crit. Rev. Anal. Chem.* **18,** 1–44.

24. Kagedal, L. (1989) Immobilized metal ion affinity chromatography, in *High Resolution Protein Purification* (Ryden, L. and Jansson, J.-C., eds.), Verlag Chemie, Deerfield Beach, FL, pp. 227–251.

25. Muszynska, G., Zheo., Y.-J., and Porath, J. (1986) Carboxypeptidase A: a model for studying the interaction of proteins with immobilized metal ions. *J. Inorg. Biochem.* **26,** 127–135.

26. Holmes, L. D. and Schiller, M. R. (1997) Immobilized iron (III) metal affinity chromatography for the separation ogf phophorylated macromolecules: ligands and applications. *J. Liquid Chromatogr.* **20,** 123–142.

27. Andersson, L. (1996) The use of immobilized Fe3+ and other hard metal ions in chromatography of peptides and proteins. *Int. J. Biochromatogr.* **2,** 25–36.

28. Yip, T.-T. and Hutchens, T. W. (1991) Metal ion affinity adsorption of a ZN(II)-transport protein present in maternal plasma during lactation: structural characterization and identification as histidine-rich glycoprotein. *Protein Express. Purif.* **2,** 355–362.

29. Hutchens, T. W., Nelson, R. W., and Yip, T.-T. (1992) Recognition of transition metal ions by peptides: identification of specific metal-binding peptides in proteolytic digest maps by UV laster desorption time-of-flight spectrometry. *FEBS Lett.* **296,** 99–102.

30. Hutchens, T. W., Nelson, R. W., and Yip, T.-T. (1991) Evaluation of peptide-metal ion interactions by UV laser desorption time-of-flight mass spectrometry. *J. Mol. Recogn.* **4,** 151–153.

31. Hutchens, T. W., Nelson, R. W., Allen, M. H., Li, C. M., and Yip, T.-T. (1992) Peptide metal ion interactions in solution: detection by laser desorption time-of-flight mass spectrometry and electrospray ionization mass spectrometry. *Biol. Mass Spectrom.* **21,** 151–159.

32. Hutchens, T. W. and Yip, T.-T. (1991) Protein interactions with surface-immobilized metal ions: structure-dependent variations in affinity and binding capacity constant with temperature and urea concentration. *J. Inorg. Biochem.* **42,** 105–118.

33. Figueoroa, A., Corradini, C., Feibush, B., and Karger, B. L. (1986) High-performance immobilized metal ion affinity chromatography of proteins on iminodiacetic acid silica-based bonded phases. *J. Chromatogr.* .**371,** 335–352.
34. Hutchens, T. W., Yip, T.-T., and Porath, J. (1988) Protein interaction with immobilized ligands. Quantitative analysis of equilibrium partition data and comparison with analytical affinity chromatographic data using immobilized metal ion adsorbents. *Anal. Biochem.* **170,** 168–182.
35. Hutchens, T. W. and Li, C. M. (1990) Ligand-binding properties of estrogen receptor proteins after interaction with surface-immobilized Zn(II) ions: evidence for localized surface interactions and minimal conformational changes. *J. Mol. Recogn.* **3,** 174–179.

21

Chromatography on Hydroxyapatite

Shawn Doonan

1. Introduction

Hydroxyapatite (HT/HTP) has achieved only limited popularity as a chromatographic material for the purification of proteins. This is for a variety of reasons, including difficulties in predicting its chromatographic behavior, its relatively low capacity, and the fact that its handling properties are not ideal. For most applications, other materials with superior chromatographic properties are to be preferred. That being said, because its modes of protein adsorption and desorption are different from those of techniques such as ion-exchange chromatography (*see* next paragraph), it is sometimes possible to achieve fractionation using HT/HTP when other techniques have failed; the example described in **Subheading 3.** is a case in point. Aficionados of HT/HTP may consider this relegation of the material to "last chance" status as unjustified, but it is nonetheless the case that most practitioners of protein purification will try something else first.

Hydroxyapatite is a crystalline form of calcium phosphate with the molecular formula $Ca_{10}(PO_4)_6(OH)_2$. The mechanism of interaction of proteins with the material has been investigated and reviewed by Gorbunoff (*1*). When the HT/HTP is equilibrated with phosphate buffer (the most common mode of use), it then appears that positively charged proteins interact nonspecifically with the general negative charge on the column produced by immobilized phosphate ions. The proteins can then be eluted by increasing the phosphate concentration, by adding a salt, such as sodium chloride, or by using Ca^{2+} or Mg^{2+}, which complex with phosphate ions on the column and effectively neutralize it. In the case of negatively charged proteins, interaction is a balance between electrostatic repulsion by the negative charge on the column and specific complexation between protein carboxylic acid groups and column calcium sites; the latter effect will depend on the disposition of carboxylate side chains in the protein rather than simply on their number. Elution is effected by ions that complex more strongly with calcium than do carboxylic acids (e.g., phosphate or fluoride); Cl^- is not effective because it does not complex with calcium.

These considerations have led to the formulation of set of guidelines for the use of HT/HTP to fractionate an unknown mixture of proteins (*1*). The column should be equilibrated with dilute phosphate buffer (e.g., 1 m*M*, pH 6.8) if the objective is to retain

From: *Methods in Molecular Biology, vol. 244: Protein Purification Protocols: Second Edition*
Edited by: P. Cutler © Humana Press Inc., Totowa, NJ

basic proteins and some of the acidic ones from a mixture; alternatively, equilibration can be in an unbuffered NaCl solution (1 mM) if the protein of interest is acidic and is to be retained on the column. Elution should then carried out by using an initial wash with 5 mM MgCl$_2$ (or a gradient of 1–5 mM) to remove basic proteins, a wash with 1 M NaCl (or a gradient of 0.01–1 M) to elute proteins with isoelectric points around neutrality, and, finally, a gradient of 0.1–0.3 M phosphate to elute acidic proteins. Clearly, for particular protein mixtures, it may be possible to use a simpler elution schedule depending on the composition, but the above-outlined procedures provide a basis for initial examination of an unknown protein sample from which simpler protocols can be derived. Some typical uses of HT/HTP chromatography are given in **refs. *1–5*.**

The particular application described in **Subheading 3.** concerns the separation of the cytosolic and mitochondrial isoenzymes of fumarase *(5)*. These two proteins are products of the same gene and differ only in a small number (so far undefined) of amino acid residues at the N-terminus. They are not separable by conventional ion-exchange chromatography, but can be easily resolved by chromatography on HT/HTP.

2. Materials

1. Hydroxyapatite (HT/HTP) from Bio-Rad (*see* **Note 1**).
2. Buffer A: 20 mM Potassium phosphate, pH 7.0, containing 10 μM phenylmethanesulfonyl fluoride (PMSF) (*see* **Note 2**).
3. Buffer B: 350 mM Potassium phosphate, pH 7.0, containing 10 μM PMSF.

3. Method

1. Suspend about 25 g of HT/HTP in 100 mL of buffer A. Stir gently, allow to settle, and then remove fines by aspiration. Repeat the last step (*see* **Note 3**).
2. Pack the HT/HTP into a column (2.0 × 10 cm), and wash with 5 bed volumes of buffer A at a flow rate of 30 mL/h (*see* **Note 4**).
3. Apply the protein sample previously dialyzed in buffer A and continue washing the column with the same buffer until no more protein is eluted as judged from A$_{280}$ measurements.
4. Elute bound protein by application of a linear gradient formed from 400 mL each of buffers A and B.
5. Regenerate the column by washing with 2 column volumes of buffer B, followed by 5 column volumes of buffer A. Store in buffer A containing 0.02% sodium azide or in buffer A saturated with toluene.

The elution profile obtained is shown in **Fig. 1.** The source of the partially purified mixture of fumarases is described in **Note 5.**

4. Notes

1. Hydroxyapatite suitable for chromatography can be prepared in the laboratory. The procedure involves mixing calcium chloride and sodium phosphate to form the amorphous compound brushite, and then converting this to crystalline hyroxyapatite by boiling in sodium hydroxide or ammonia. A detailed procedure is given by Bernardi *(6)*. Although this product may be superior to commercial HT/HTPs *(7)*, it is likely that the occasional user will prefer to purchase the material.

   Other products on the market include solid supports with surface coatings of HT/HTP. For example, CHT from Bio-Rad consists of coated ceramic beads. This material is available as Bio-Scale prepacked columns.

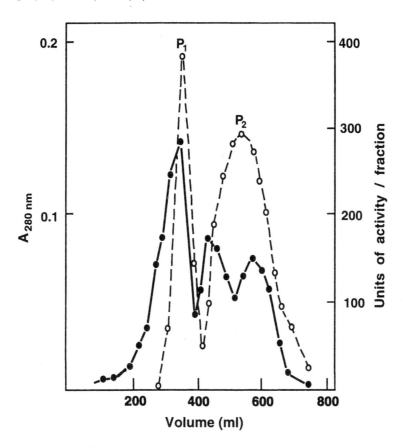

Fig. 1. Separation of cytosolic and mitochondrial fumarases from baker's yeast by HT/HTP chromatography. Solid and dashed traces show absorbance at 280 nm and enzyme activity, respectively. (Reprinted from **ref. 5** with kind permission from Elsevier Science, Ltd.)

2. The PMSF was included in this particular application because it was vital to the objective of the investigation to preclude the possibility of proteolysis *(5)*. The necessity to add protease inhibitors and which inhibitors to add will depend on the problem at hand *(see* Chapter 9).

    Stock solutions of PMSF (50–100 m*M*) should be made in dry propan-2-ol, methanol, or ethanol; in this form, they are stable for some months at 4°C. The half-life in aqueous solutions is short and, hence, fresh PMSF should be added to protein solutions or to buffers daily.

3. Crystalline HT/HTPs are not very stable mechanically, and it is important to avoid fragmentation resulting from rough handling and to be sure to remove fines before packing columns. Failure to do so will result in high back-pressures and low flow rates. Similarly, the flow rates of columns of HT/HTP decrease with repeated use, and when this occurs, it is necessary to repack them. Absorption of carbon dioxide and consequent formation of a plug at the top of the column is one reason for the development of back-pressure; this can be avoided by boiling buffers before use to remove dissolved CO_2 or, more simply, by replacement of the top 1–2 cm of the column bed if plugging occurs.

4. Use of high flow rates, particularly with crystalline HT/HTP, will lead to compaction of the material and necessitate repacking of the column.

5. The fumarase isoenzymes were extracted from 1 kg of baker's yeast and then partially purified by precipitation with 60% saturation ammonium sulfate, followed by ion-exchange

chromatography on carboxymethyl cellulose CM-23 at pH 5.8. It was this fraction that was subjected to chromatography on HT/HTP, with the results shown in **Fig. 1**. Peak 1 eluted from the column at about 120 mM phosphate buffer and contained mainly the cytosolic fumarase *(8)*. Peak 2 eluted at about 250 mM phosphate buffer and contained mainly the mitochondrial form. The separation was not complete, so fractions from the front of peak 1 and from the back of peak 2 were combined to give the respective pools for the two isoenzymes; the central fractions were rejected. Final purification of the separated isoenzymes was achieved by affinity chromatography using pyromellitic acid–Sepharose 4B *(9)*. Apart from HT/HTP chromatography, the only other technique that has been found to be effective in separating fumarase isoenzymes is chromatofocusing, which was used for the isoenzymes from rat liver *(10)*.

References

1. Gorbunoff, M. J. (1990) Protein chromatography on hydroxyapatite columns. *Methods Enzymol.* **182,** 329–339.
2. Gorbunoff, M. J. (1980) Purification of ovomucoid by hydroxyapatite chromatography. *J. Chromatogr.* **187,** 224–228.
3. Mizuuchi, K., O'Dea, M. H., and Gellert, M. (1978) DNA gyrase: subunit structure and ATPase activity of the purified enzyme. *Proc. Natl. Acad Sci. USA* **75,** 5960–5963.
4. Peng, H. and Marians, K. J. (1993) *Escherichia coli* topoisomerase IV: purification, characterisation, subunit structure and subunit interactions. *J. Biol. Chem.* **268,** 24,481–24,490.
5. Boonyarat, D. and Doonan, S. (1988) Purification and structural comparisons of the cytosolic and mitochondrial fumarases from baker's yeast. *Int. J. Biochem.* **20,** 1125–1132.
6. Bernardi, G. (1971) Chromatography of proteins on hydroxyapatite. *Methods Enzymol.* **22,** 325–339.
7. Roe, S. (1989) Separation based on structure, in *Protein Purification Methods: A Practical Approach* (Harris, E. L. V. and Angal, S., eds.), IRL, Oxford, pp. 238–242.
8. Hiraga, K., Inoue, I., Manaka, H., and Tuboi, S. (1984) Chromatographic differentiation of the mitochondrial and cytosolic fumarases of rat liver and baker's yeast and differential induction of two fumarases of baker's yeast. *Biochem. Int.* **9,** 455–461.
9. Beeckmans, S. and Kanarek, L. (1977) A new purification procedure for fumarase based on affinity chromatography: isolation and characterisation of pig-liver fumarase. *Eur. J. Biochem.* **78,** 437–444.
10. O'Hare, M. C. and Doonan, S. (1985) Purification and structural comparisons of the cytosolic and mitochondrial fumarases from rat liver. *Biochim. Biophys. Acta* **827,** 127–134.

22

Thiophilic Affinity Chromatography and Related Methods

Paul Matejtschuk

1. Introduction

Thiophilic affinity chromatography (TAC) is a powerful and simple method for the purification of proteins—in particular, immunoglobulins. It relies upon the affinity of a sulfur-containing ligand, immobilized on a suitable matrix, for immunoglobulins in the presence of high concentrations of lyotropic (water-forming) salts, and their elution is mediated by removal of these salts. Although appearing similar at first consideration to hydrophobic interaction chromatography (HIC) there are clear differences both in the specificity and chemistry of the interactions that occur. Indeed, proteins that bind tightly to HIC resins, such as albumin, are not necessarily retained on TAC *(1)*. TAC has broad specificity for immunoglobulins of different classes and subclasses from a wide range of species and has capacity similar to that of other affinity methods. Its advantages are the simplicity of the sample preparation, applicability with a range of source materials and protein concentration ranges, and gentle elution conditions, unlike the popular protein A/protein-G separation methods that require pH extremes for elution to which the antibodies may not be sufficiently stable. It is also inexpensive compared to protein A/protein-G and its range of specificity gives it a marked advantage over the protein A/protein-G techniques, which are often limited in terms of their specificity for both subclasses and species. Its main disadvantage is in the need to pretreat the starting material with high levels of lyotropic salts. However, recently, a new sulfur-containing affinity matrix has been developed (for what is described as hydrophobic charge induction chromatography) that does not require the high-salt loadings and offers a very attractive widely applicable option for the purification of antibodies.

Thiophilic affinity chromatography (TAC) was introduced by Porath in 1985 *(2)*. The homemade resin was synthesised by the reaction of β-mercaptoethanol with divinyl sulfone-activated agarose. The resin was termed T-gel (*see* **Fig. 1**) and has been compared and contrasted with HIC *(1,4,5)*. The binding profile and salt concentration required differed between the two methods, and whereas HIC suffered from residual "tightly bound" protein that remained after cleaning, the thiophilic gels appeared to be capable of being cleaned readily. The marked specificity for immunoglobulins was recognized from the outset and its application within this area of protein purification filled a gap

From: *Methods in Molecular Biology, vol. 244: Protein Purification Protocols: Second Edition*
Edited by: P. Cutler © Humana Press Inc., Totowa, NJ

1)   Matrix– O – CH$_2$–CH$_2$–SO$_2$–CH$_2$–CH$_2$–S–CH$_2$–CH$_2$–OH

2)   Matrix–O –CH$_2$– CH$_2$–SO$_2$ – CH$_2$ –CH$_2$–S–pyridine

3)   Matrix– CH$_2$ –O –CH$_2$–CH$_2$–CH$_2$ –S–CH$_2$–CH$_2$–pyridine

Fig. 1. Sulfur-containing ligands used for immunoglobulin purification: Key: (1) T-gel, (2) Py-S-sulfone (**3**), MEP Hypercel.

between resins such as protein A with a restricted binding profile and the more general techniques such as immobilized metal affinity chromatography (IMAC) and HIC, which often fail to yield pure immunoglobulin from a single chromatographic pass.

The TAC technique binds IgG from human *(6)*, bovine *(7)*, murine *(8)*, and other species *(9)* and from a range of start materials, including ascites, cell culture harvest, serum, and whey. Thiophilic purification of IgA *(10)* and IgM *(11)* have also been reported. Although several groups published the successful synthesis of T-gel *(12,13)* real interest developed when resins became commercially available. TAC has been applied to the purification of IgG fragments such as Fab$_2$ *(14)*, Fab *(6,15)* with or without affinity tags, and Fc *(14)*. Recombinant scFv antibodies *(17)*, diabodies *(16)*, and IgY from egg yolk *(18)* have also been purified.

Other proteins that are bound by T-gel include α-2-macroglobulin *(2)*, and T-gel has also been used as a purification stage in the isolation of a number of other diverse proteins, including wheat serpins *(19)*, yeast acetolactate synthetase *(20)*, collagen peptides *(21)*, hydrophobic prion peptide markers, prostate-specific antigen *(22)*, and certain bacterial antigens *(23)* presumably acting on hydrophobic motifs similar to those found in immunoglobulins.

Ammonium, sodium, and potassium sulfate have all been used as the lyotropic salt. Although the presence of one or more sulfur atoms in the ligand is typical it has been shown that other ligands can bind IgG, providing both electron-donating and electron-accepting groups are present *(24–26)*.

Using IgG purification as the criterion, it was discovered that certain modified ligands showed no specificity, whereas others, particularly pyridine rings, gave good binding specificity *(25)*. Several studies in the late 1990s demonstrated that certain ligands bound IgG in the presence of lyotropic salts and then retained them when the salts were removed. In particular, Scholz et al. *(3,27)* showed that certain mercaptoheterocyclic ligands could bind IgG without the use of high salts.

In 2000, Biosepra/Life Technologies launched MEP Hypercel which it claimed operated by hydrophobic charge induction chromatography (HCIC). This relied on specific hydrophobic interaction between the IgG and the thiol-containing pyridyl ligand to bind antibody from medium without the need for lyotropic salts. Elution was mediated by charge induction in the pyridyl ring at pH below 5. This offered great advantage over the traditional T-gel resins *(28)*.

Thiophilic affinity chromatography relies on the interaction between thiol containing ligand and aromatic regions on the protein, principally phenylalanine and tryptophan residues *(1)*. It differs from HIC, and although the mechanism of action is still not entirely clear, it seems to be by either electron donor-acceptor, or charge transfer *(1,26)*. The mechanism of action of HCIC resin is particularly elegant, allowing the binding of immunoglobulin by the hydrophobic spacer arm and sulfur-containing ligand from start material without pretreatment. The immunoglobulin is then eluted by pH manipulation, which induces a charge on the pyridyl ring of the ligand of similar nature to that on the immunoglobulin at low pH (both positive).

2. Materials

2.1. Ligands

The range of ligands that have been used for thiophilic chromatography have varied since its first introduction. The simplest ligands were produced by the action of β-mercaptoethanol on divinyl sulfone-activated agarose. A simple strategy for the manufacture of a "homemade" thiophilic gel has been described *(13)* and can be summarized as follows (all steps should be performed in a fume cupboard, as the reagents divinyl sulfone and β-mercaptoethanol are toxic):

1. Resuspend 1 mL of settled agarose chromatography matrix in 1 mL of 0.5 *M* sodium carbonate buffer (pH 10).
2. Add 0.05 mL of divinyl sulfone and mix for 3 h at ambient temperature (on an end-over-end mixer).
3. Wash the gel with excess deionized water.
4. Resuspend the gel in 1 mL of 0.5 *M* sodium carbonate, pH 10, and add 0.1 mL of β-mercaptoethanol and incubate overnight with shaking.
5. Wash the gel exhaustively (at least three times) with 15–20 volumes of deionized water, aspirating the supernatant on each occasion.
6. The gel is now ready for use or may be stored at 2–8°C until use with 0.02% sodium azide as the bacteriostatic agent.

However, it is far more convenient to use one of the thiophilic gels commercially available.

2.2. Commercially Available Matrices

The initial thiophilic gels were prepared from derivatized agarose (as earlier). However, since their introduction, many other matrices have been used. A list of currently available matrices is given in **Table 1**. Crosslinked agarose has a predominant position as the matrix of choice in all forms of affinity chromatography, because of its high level of substitution, excellent flow properties, and broad pH stability. However, for high-performance liquid chromatography (HPLC) applications, a control pore glass or silica

Table 1
Some of the Commercially Available Thiophilic and HCIC Resins

Resin name	Supplier[a]	Matrix	Quoted capacity (IgG/mL)	Ligand	Comments
Affi-T	Kem-en-Tec	Agarose	10–15 mg/mL	Thioether (not stated)	
Thiosorb	Millipore	Control pore glass	10–18 mg/mL	Not stated	
Thiosorb M	Millipore	Control pore glass	2 mg/mL IgM	Not stated	Marketed for IgM purification
T-Gel	Affiland	Sepharose CL-4B Sepharose 4 FF StreamLine®	10–22 mg/mL	Thioether	
Hi-Trap IgM	Amersham Biosciences	Highly crosslinked agarose	5 mg/mL IgM	2-Mercapto pyridine	Marketed for IgM purification
HiTrap IgY	Amersham Biosciences	Highly crosslinked agarose	20 mg/mL IgY	2-Mercapto pyridine	Marketed for IgY purification
Fractogel EMD TA	Merck	Crosslinked polymethacrylate resin	30 mg/mL	Sulfone thioether	
Thiophilic Superflow Resin	Clontech, Becton Dickinson	Crosslinked agarose	25 mg/mL	Sulfone thioether	
Thiophilic Uniflow Resin	Clontech Becton Dickinson	Crosslinked agarose	20 mg/mL	Sulfone thioether	
MEP Hypercel	Biosepra	Agarose	>20 mg/mL	4-Mercaptoethyl pyridine	

[a]Manufacturers/suppliers: Affiland, 304 rue de l'Yser, B-4430 Ans-Liege, Belgium; fax: (+32) 4246 1506; website: www.affiland.com; Amersham Biosciences UK Ltd, Amersham Place, Little Chalfont, Bucks, HP7 9NA, UK; +fax (+44) 1494 542179; website: www.amershambiosciences.com; Biosepra S.A. (a process division of Ciphergen), 48 Avenues de Genottes, 95800 Cergy Saint Christophe, France; fax: (+33) 13420787878; website: www.biosepra.com; Clontech BD Biosciences, Between Towns Road, Cowley, Oxford, OX4 3LY, UK; fax: (+44) 1865 748844; website: www.clontech.com; Kem-en-tec A/S, Lersø Parkallé 42, DK-2100 Copenhagen, Denmark; fax: (+45) 39200178; website: www.kem-en-tec.com; Merck KGaA, Merck EuroLab Ltd, Poole Dorset BH15 1TD UK; fax (+44) 1202 665599; website www.merck.de; Millipore(UK) Ltd, Unit 3 and 5 The Courtyards, Hatters Lane, Watford, Hert, WD18 8YH, UK; fax: (+44) 1923 818297; website: www.millipore.com.

resin may be more appropriate. For industrial use, in particular with difficult feedstocks, expanded bed resins such as StreamLine (Amersham Biosciences) may offer the preferred solution. In industrial settings, sanitization of the resin may be of key importance and so a matrix compatible with sodium hydroxide should be chosen. Manufacturers of gels with poor stability at high pH have suggested alternative cleaning regimens with denaturants or alcohols, but these may not be acceptable in manufacturing applications aimed at therapeutics production.

2.3. Chromatographic Equipment

Thiophilic affinity chromatography, as with all affinity methods, can be readily performed with basic chromatographic equipment, at its simplest being merely a column to hold the matrix and an ultraviolet (UV) detector to monitor protein elution. However, gradient pump facilities will be required especially if a linear gradient is used to trial elution conditions. Medium pressure systems such as Akta or fast protein liquid chromatography (FPLC) (Amersham Biosciences) or their equivalents are ideally suited to scouting elution conditions for a given application and, indeed, for scale-up to modest levels. Large-scale systems are beyond the scope of this chapter, but there are no special requirements resulting from the separation methodology. In order to calculate protein recovery, an UV/vis spectrophotometer and a simple protein assay system such as the Bradford method would be required.

3. Methods

3.1. Thiophilic Chromatography

Thiophilic chromatography relies upon a lyotropic binding buffer containing high molarity salts and a nonlyotropic eluant, typically neutral pH-buffered solution. For scouting purposes, a linear gradient between the two is formed and the elution point of the immunoglobulin determined. Chromatography can be split into several stages: (1) Sample and matrix preparation; (2) Sample loading; (3) Product elution; and (4) Column cleanup.

3.1.1. Sample and Matrix Preparation

3.1.1.1. SAMPLE PREPARATION

The sample must be high in lyotropic salts for the immunoglobulin to bind. The concentration of the immunoglobulin in the sample to be loaded (feedstock) is not of itself a limitation to the thiophilic chromatography. However, for practical purposes, it may be expensive and time-consuming to load large volumes of an immunoglobulin-containing medium to a high concentration with ammonium sulfate. Hence, some sample concentration may be advisable. The sample should be loaded up with solid ammonium sulfate to 0.5–1.5 *M* final concentration and up to 1 *M* NaCl may also be added. The maximum concentration of lyotrope that does not induce precipitation of the immunoglobulin should be determined empirically and then used for the TAC method (*see* **Note 1**). If working with plasma or serum, a clarification step such as filtration or centrifugation will reduce the risk of blocking the column with lipid or particulates. As a general precaution, 0.2-μm filtration of the feedstock is recommended.

3.1.1.2. MATRIX PREPARATION

The matrix is packed according to the manufacturer's particular instructions, but, typically, the required volume of matrix is removed and allowed to settle and the storage solution aspirated. The gel is then suspended in an equilibration buffer and the column poured in a single step, allowing the bed to pack under gravity. The column is then removed to the chromatographic rig and packed to constant bed height at a flow rate significantly higher than the operating pressure but within the operating maximum pressure of the matrix. At least 10–20 vol of buffer is passed through the column, and any adjustment of the top adaptor made to minimize the head space above the column. The eluate should be monitored to ensure that there is a stable baseline with neglible UV absorbance.

3.1.2. Sample Loading

The sample is applied at the flow rate recommended by the manufacturer. For example, for Prosep Thiosorb, a flow rate of 50–1000 cm/h can be used. The eluate should be collected and checked to ensure that none of the immunoglobulin remains uncaptured, by specific assay (such as enzyme-linked immunosorbent assay [ELISA], immunodiffusion or nephelometry) or using an analytical method such as sodium dodecyl sulfate-polyacrylamide gel electrophoresis (SDS-PAGE), size-exclusion chromatography, or agarose electrophoresis. As the medium will almost always contain other proteins monitoring of the optical density (OD) of the eluate alone is not informative. Although a maximum flow rate is usually quoted by the manufacturer, the capacity of the resin often falls off near its maximum operating rate.

Once loading is complete, the column should be washed with equilibration buffer having the same concentration of lyotropic salt until the OD of the eluate is reduced to a negligible level. The column is then ready for elution of the specifically bound immunoglobulin.

3.1.3. Product Elution

In principle, immunoglobulin binds to thiophilic gel in the presence of lyotropic salts and elutes when they are removed. Hence, all eluants have lower concentrations of the lyotropic salts than the equilibrating buffer. Although buffer alone can be used, it may well be that immunoglobulin will elute at a reduced concentration of lyotropic salt, and for maximum recoveries, the concentration of lyotrope that yields pure immunoglobulin should be scouted for each individual application (*see* **Note 2**). The addition of 1 *M* NaCl to the wash and elution buffers may result in greater recoveries of IgG.

The immunoglobulin-containing peak is collected and can be analyzed by SDS-PAGE or size-exclusion chromatography (which, in addition, allows assessment of antibody aggregation) as appropriate. Yields may be calculated by estimating total protein and immunoglobulin concentrations by means of protein assays and immunoglobulin-specific assays as described earlier.

3.1.4. Column Cleanup

It is important to clean the column of any material that remains tightly bound to the resin and not eluted even after prolonged application of eluting buffer. This may be achieved with 60% ethylene glycol, a chaotrope such as 6 *M* urea, or high ethanol concentrations (70% [v/v]). Some T-gel resins are resistant to sodium hydroxide and so may be cleaned with this solution.

Once thoroughly cleaned of bound protein material, the resin may be stored as in

Subheading 3.1.5. or re-equilibrated with buffer as described earlier, in readiness for a further purification cycle.

3.1.5. Column Storage

Follow the manufacturer's instructions, but typically store the resin in 20% ethanol (with or without buffer) or 0.02% sodium azide as a bacteriostatic agent. Some manufacturers recommend a high-salt buffer solution for storage. No chromatographic gel should be allowed to freeze.

3.2. Optimization

Several steps are important in the optimization of TAC, including capacity determination, flow rate optimization, and selection of suitable wash buffers prior to elution.

3.2.1. Capacity

Load a packed column of T-gel (of determined volume) with feedstock until immunoglobulin is detected in the eluate to a predetermined breakthrough level (e.g., 5–10% of the loading concentration). This will give the dynamic capacity of the resin. Alternatively, static capacity can be determined simply by incubation of a known amount of gel with varying ratios of feedstock and measuring the amount of immunoglobulin remaining in the feedstock supernatant after sedimenting the gel by gentle (300 *g*, 5 min) centrifugation. It is important when using thiophilic chromatography to also elute the bound immunglobulin with the chosen eluant, following thorough washing of the gel to ensure removal of entrapped but unbound immunoglobulin from the feedstock. This will indicate the likely recoverable yield of immunoglobulin, as it is not uncommon for a proportion of the antibody to remain tightly bound and only to be eluted by the cleaning solution.

3.2.2. Elution Flow Rate

Use a low flow rate initally (linear flow 30–60 cm/h), and if successful purification results, increase the flow rate within the range recommended for the matrix by the manufacturer, at each step measuring the capacity and purity as previously described. Select a flow rate dependent upon feedstock volume and the required purity/yield criteria.

3.2.3. Purity

If the purity of the eluted immunoglobulin is inadequate, it may be necessary to develop more appropriate elution conditions. This can be done in the static mode outlined earlier or more conveniently in a dynamic mode by application of a linear gradient or sequential step gradient with buffers containing decreasing concentrations of lyotropic salt. Each protein peak eluted should be assayed and the purity of the immunoglobulin fraction assessed. The eluant yielding the highest purity and best recovery of immunoglobulin can then be selected.

3.3. Hydrophobic Charge Induction Chromatography

Hydrophobic charge induction chromatography differs from thiophilic chromatography in that it does not require the presence of lyotropic salt for immunoglobulin to bind. This makes sample preparation much easier—no pretreatment other than pH adjustment to neutrality (pH 6.5–8.0) if necessary and 0.2 μm filtration is required. This is a distinct

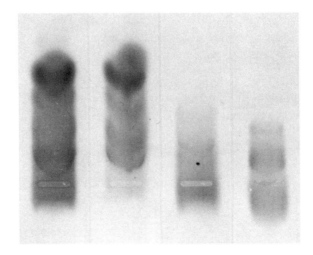

Fig. 2. Agarose electrophoresis of samples from thiophilic purification of plasma on Thiosorb column. *Lane 1*: plasma supernatant for loading post ammonium sulfate addition; *lane 2*: column flowthrough; *lane 3*: bound IgG eluted with 0.5 *M* NaCl and 20 m*M* HEPES, pH 7.5; *lane 4*: material stripped from column by 6 *M* urea. Electrophoresis at 200 V constant voltage (Cellogel, Helena Biosciences). Fixation in 20% (w/v) sulfosalicylic acid followed by brief water rinse, staining with 1% (w/v) aqueous Coomassie blue R250 with destaining in deionized water. Anode at top.

advantage. Wash down (*see* **Note 3**), elution, re-equilibration, and storage are, in principle, as for thiophilic chromatography. Elution is achieved by application of a pH 4.0 buffered solution (such as acetate or citrate) and the immunoglobulin collected.

3.4. Example Purification

Thiophilic purification of immunoglobulin from human plasma on Prosep Thiosorb (Millipore).

Plasma was adjusted to 0.5 *M* NaCl, 7.5% (w/v) ammonium sulfate, 20 m*M* HEPES, pH 7.5 (final concentrations) by dilution one-to-one with 2X concentrated buffer. Any precipitate was spun off and tested for the presence of antibody. A column of Thiosorb was packed in HR5/5 FPLC column (Amersham Biosciences) and equilibrated using 20 m*M* HEPES, 0.5 *M* NaCl, and 7.5% ammonium sulfate, pH 7.5. Plasma sample was applied at a linear flow rate of 150 cm/h and the column washed with the same equilibrating buffer. Elution of antibody was achieved by application of 20 m*M* HEPES, 0.5 *M* NaCl, pH 7.5, and the column cleaned using 6 *M* urea. The resultant immunoglobulin obtained was essentially free of albumin and α-globulins although some other bands (possibly other immunoglobulin classes) remained on analysis by agarose electrophoresis (*see* **Fig. 2**).

4. Notes

1. Phenyl red sometimes used in tissue culture medium binds avidly to some thiophilic matrices *(25)*, resulting in marked loss of apparent capacity. This dye should be omitted from feedstocks to be used with thiophilic chromatography.

2. Immunoglobulin fragments (Fab, Fc) can elute at differing concentrations of lyotropic salts than do intact immunoglobulin and so a gradient elution of decreasing lyotrope should always be used when a mixture of immunoglobulin forms are expected. No overall differential selection of immunoglobulin subclasses by TAC has been reported by some workers *(13)*, whereas others have reported some preferential binding *(8)*. Again, when working with more than one isotype, the composition before and after purification should be checked if this is of concern for the intended application.

3. The use of a wash step of buffer containing 25 m*M* caprylate (octanoate) is advocated to minimize albumin contamination of antibodies isolated from albumin-containing medium by HCIC. However, in our experience, this is not always necessary and can be omitted if it does not influence the purity of the eluted immunoglobulin.

Acknowledgments

Thanks to colleagues John More, Mike Harvey, and Deborah Wong at Bio Products Laboratory, Elstree, Hertfordshire, UK, where the experimental work was performed.

References

1. Porath, J. and Belew, M. (1987) Thiophilic interaction and the selective adsorption of proteins. *Trends Biotechnol.* **5**, 225–229.
2. Porath, J., Maisano, F., and Belew, B. (1985) Thiophilic adsorption—a new method for protein fractionation. *FEBS Lett.* **185**, 306–310.
3. Scholz, G. H., Wippich, P., Leistner, S., and Huse, K. (1998) Salt-independent binding of antibodies from human serum to thiophilic heterocyclic ligands. *J. Chromatogr. B* **709**, 189–196.
4. Oscarsson, S., Angulo-Tatis, D., Chaga, G., and Porath, J. (1995) Ampiphilic agarose-based adsorbents for chromatography. *J. Chromatogr. A* **689**, 3–12.
5. Hutchens, T. W. and Porath, J. (1987) Thiophilic adsorption: a comparison of model protein behaviour. *Biochemistry* **26**, 7199–7204.
6. Behazid, M., Ambler, D., Matejtschuk, P., and Lowe, D. (1993) A universal method for the purification of immunoglobulin classes and their derivatives using a new affinity matrix. *Int. Biotech. Lab.* July, p. 30.
7. Konecny, P., Brown, R. J., and Scouten, W. H. (1994) Chromatographic purification of immunoglobulin G from bovine milk whey. *J. Chromatogr. A* **673**, 45–53.
8. Finger, U. B., Brummer, W., Thommes, J., and Kula, M-R. (1996) Investigation on the specificity of thiophilic interaction for monoclonal antibodies of different subclasses. *J. Chromatogr. B* **675,** 197–204.
9. Lihme, A. and Heegaard, P. M. H. (1991) Thiophilic adsorption chromatography: the separation of serum proteins. *Anal. Biochem.* **192**, 64–69.
10. Leibl, H., Tomasits, R., and Mannhalter, J. W. (1995) Isolation of human serum IgA using thiophilic adsorption chromatography. *Protein Express. Purif.* **6**,408–410.
11. Rapoport, E. M., Zhigris, L. S., Vlasova, E. V., Piskarev, V. E., Bovin, N. V., and Zubov, V. P. (1995) Purification of monoclonal antibodies to Le^y and Le^d carbohydrate antigens by ion-exchange and thiophilic adsorption chromatography. *Bioseparation* **6**, 165–184.
12. Sulk, B., Birkenmeier, G., and Kopperschlager, G. (1992) Application of phase partitioning and thiophilic adsorption chromatography to the purification of monoclonal antibodies from cell culture fluid. *J. Immunol. Methods* **149**, 165–171.
13. Bridonneau, P. and Lederer, F. (1993) Behaviour of human immunoglobulin G subclasses

on thiophilic gels: comparison with hydrophobic interaction chromatography. *J. Chromatogr.* **616**, 197–204.

14. Yurov, G. K., Neugodova, G. L., Vekhovsky, O. A., and Naroditsky, B. S.(1994) Thiophilic adsorption: rapid purification of F(ab)$_2$ and Fc fragments of IgG1 antibodies from murine ascitic fluid. *J. Immunol. Methods* **177**, 29–33.

15. Fiedler, M. and Skerra, A. (1999) Use of thiophilic adsorption chromatography for the one step purification of a bacterially produced antibody Fab fragment without the need for an affinity tag. *Protein Express. Purif.* **17**, 421–427.

16. Schulze, R. A., Kontermann, R. E., Queitsch, I., Dubel, S., and Bautz, E. K. F. (1994) Thiophilic adsorption chromatography of recombinant single-chain antibody fragments. *Anal. Biochem.* **220**, 212–214.

17. Muller, K. M., Arndt, K. M., and Pluckthun, A. (1998) A dimeric bispecific miniantibody combines two specificities with avidity. *FEBS Lett.* **432**, 45–49

18. Hansen, P., Scoble, J. A., Hanson, B., and Hoogenraad, N. J. (1998) Isolation and purification of immunoglobulins from chicken eggs using thiophilic interaction chromatography. *J. Immunol. Methods* **215**, 1–7.

19. Rosenkrands, I., Hejgaard, J., Rasmussen, S. K., and Bjorn, S. E. (1994) Serpins from wheat grain. *FEBS Lett.* **343**, 75–80.

20. Poulsen, C. and Stougaard, P. (1989) Purification and properties of *Saccharomyces cerevisiae* acetolactate synthase from recombinant *Escherichia coli*. *Eur. J. Biochem.* **185**, 433–439.

21. Pedersen, B. J. and Bonde, M. (1994) Purification of human procollagen type I carboxy-terminal propeptide cleaved as in vivo from procollagen. *Clin. Chem.* **40**, 811–816.

22. Chadha, K. C., Kaminski, E., and Sulkowski, E. (2001) Thiophilic interaction chromatography of prostate-specific antigen *J. Chromatogr. B* **754**, 521–525.

23. Rosenkrands, I., Rasmussen, P. B., Carnio, M., Jacobsen, S., Theisen, M., and Andersen, P. (1998) Identification & characterization of a 29kD Protein from *Mycobacterium tuberculosis* culture filtrate. *Infect. Immun.* **66**, 2728–2735.

24. Scoble, J. A., and Scopes, R. K. (1997) Ligand structure of the divinyl sulphone-based T-gel. *J. Chromatogr. A* **787**, 47–54.

25. Schwarz, A., Kohen, F., and Wilchek, M. (1995) Novel heterocyclic ligands for thiophilic purification. *J. Chromatogr. B* **664**, 83–88.

26. Berna, P. P., Berna, N., Porath, J., and Oscarsson, S. (1998) Comparison of the protein adsorption selectivity of salt-promoted agarose based adsorbents *J. Chromatogr. B* **800**, 151–159.

27. Scholz, G. H., Vieweg, S., Leistner, S., Seissler, J., Scherbaum, W. A., and Huse, K. (1998). A simplified procedure for the isolation of immunoglobulins from human serum using a novel type of thiophilic gel at low salt concentration. *J. Immunol. Methods* **219**, 109–118.

28. Guerrier, L., Girot, P., Schwartz, W., and Boschetti, E. (2000) New methods for the selective capture of antibodies under physiological conditions. *Bioseparation* **9**, 211–221.

23

Affinity Precipitation Methods

Jane A. Irwin and Keith F. Tipton

1. Introduction

1.1. Overview

Affinity chromatography (*see* Chapter 16) is a powerful protein purification technique that exploits the specific interaction between a biological ligand (e.g., a substrate, coenzyme, hormone, antibody, or nucleic acid) or its synthetic analog and its complementary binding site on a protein. One of the variations on this technique (*see* **refs.** *1* and *2* for reviews) was that of affinity precipitation. As in affinity chromatography, the protein binds to a specific ligand, but the latter is free in solution, rather than bound to an insoluble support. Ligand binding results in the precipitation of the protein, which may then be separated by centrifugation. The pellet contains the protein of interest and the ligand, whereas the other components of the mixture remain in the supernatant, allowing easy separation.

There are two main approaches to affinity precipitation. The first of these is called the "bis-ligand" or "homobifunctional ligand" approach. The ligand is bifunctional, bearing two identical ligands connected by a spacer arm. If the spacer is long enough, each ligand can bind to a ligand-binding site on a different protein molecule. Oligomeric proteins can bind two or more bis-ligands, with the consequent formation of crosslinked lattices. When the lattice becomes large enough, it will precipitate from solution.

The second approach to affinity precipitation involves "heterobifunctional" ligands. It differs from the bis-ligand approach in that the affinity ligand has two functions, one of which binds the target protein and a second that promotes the precipitation of the aggregate. It is usually prepared by binding an affinity ligand to a polymer that can be reversibly rendered soluble and insoluble. These polymers can be either of natural origin (e.g., chitosan, alginate) or synthetic (e.g., Eudragit S-100 or polymers of *N*-isopropylacrylamide). The polymer–affinity ligand–target protein complex is precipitated by altering the conditions to change the solubility of the polymer. Changes in pH, temperature, and ionic strength as well as the addition of metal ions, water-soluble organic solvents, or oppositely charged electrolytes may precipitate the polymer. The target protein can then be dissociated using methods similar to those used to elute proteins from columns in affinity chromatography *(2)*. This technique is referred to in the literature as

From: *Methods in Molecular Biology, vol. 244: Protein Purification Protocols: Second Edition*
Edited by: P. Cutler © Humana Press Inc., Totowa, NJ

affinity precipitation, but, in contrast to the first approach, the precipitation of the complex does not occur as a direct consequence of the affinity interaction between the protein and its ligand, but by changing the conditions to effect precipitation. An alternative method involves converting an insoluble hydrophilic polymer (e.g., dextran) to an affinity gel and hydrolyzing this to water-soluble affinity polymeric ligands that can be recovered and purified *(3)*. This general approach has also been employed for sequence-specific affinity precipitation of oligonucleotides *(4)*.

This chapter will be confined to a discussion of the bis-ligand form of affinity precipitation, which can be described as "true" affinity precipitation, occurring solely as a result of the direct interaction between the ligand and the protein and not as a result of a change in conditions.

1.2. Bis-Coenzymes and Their Applications

Mosbach's group was the first to synthesize a bifunctional affinity precipitation reagent *(5)*. This affinity precipitation reagent was N_2,N_2'-adipodihydrazido-bis-(N^6-carbonylmethyl)NAD$^+$, abbreviated to bis-NAD$^+$. It consisted of two molecules of the NAD$^+$ derivative N^6-carboxymethyl-NAD$^+$, linked by an adipic acid dihydrazide spacer arm (*see* **Fig. 1**). This compound was used for the affinity precipitation of purified lactate dehydrogenase from bovine heart, with a recovery of up to 90%. Bovine liver glutamate dehydrogenase and yeast alcohol dehydrogenase were also precipitated by this technique *(6–8)*. The latter enzyme also required high ionic strengths for precipitation to occur.

An alternative synthesis of a range of bis-NAD$^+$ affinity reagents has been developed *(9)*. It involves the carbodiimide-mediated condensation of two molecules of N^6-(2-aminoethyl)NAD$^+$ with different dicarboxylic acids (*see* **Fig. 2**). The synthesis of bis-coenzyme derivatives has not been limited only to NAD$^+$; a synthesis of a bis-ATP derivative, N_2,N_2'-adipodihydrazido-bis-(N^6-carbonylmethyl)ATP *(10)* is also described here.

1.3. Structural and Kinetic Requirements for Affinity Precipitation With Bis-Coenzymes

The technique of bis-ligand affinity precipitation with bis-NAD$^+$ only works under certain conditions *(11)*, summarized as follows:

1. The enzyme of interest has to contain more than one coenzyme-binding site. An enzyme with two coenzyme-binding sites can form linear crosslinked polymers, whereas enzymes with more binding sites can form lattices involving many crosslinked molecules (*see* **Fig. 3**).

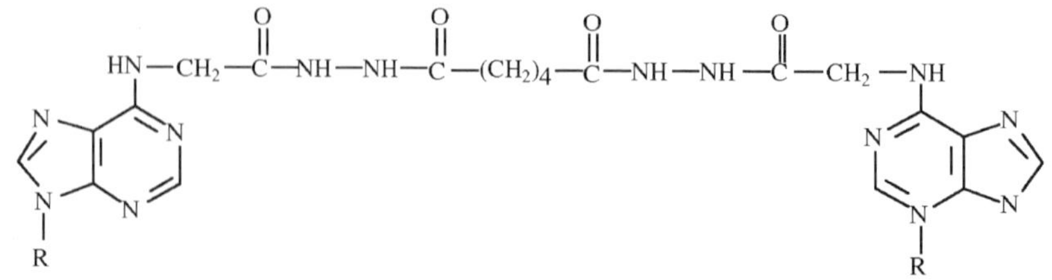

Fig. 1. Structure of N_2,N_1'-adipodihydrazido-bis-(N^6-carbonylmethyl)-NAD$^+$, first synthesized by Larsson and Mosbach *(5)* and used as an affinity ligand for several dehydrogenases. R represents nicotinamide mononucleotide phosphoribose.

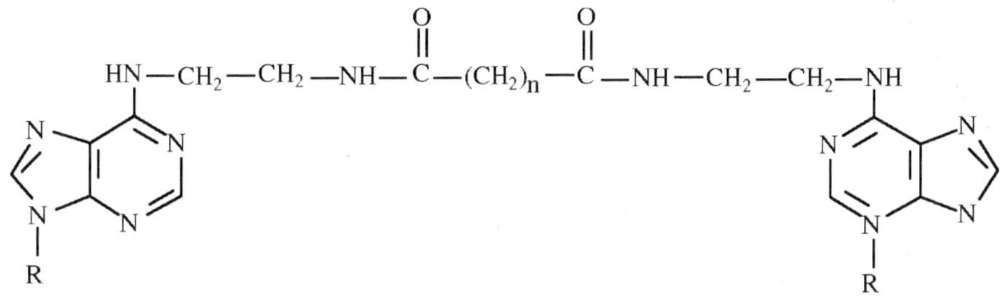

$$HN-CH_2-CH_2-NH-\overset{\overset{\displaystyle O}{\|}}{C}-(CH_2)_n-\overset{\overset{\displaystyle O}{\|}}{C}-NH-CH_2-CH_2-NH$$

Fig. 2. General structure of bis-NAD$^+$ derivatives, based on N^6-(2-aminoethyl)-NAD$^+$ as a starting meaterial. These include N,N'-bis-(N^6-ethylene-NAD$^+$) glutaramide ($n = 3$), N,N'-bis-(N^6-ethylene-NAD$^+$) adipamide ($n = 4$), and N,N'-bis-(N^6-ethylene-NAD$^+$)pimelamide ($n = 5$). R represents nicotinamide mononucleotide phosphoribose.

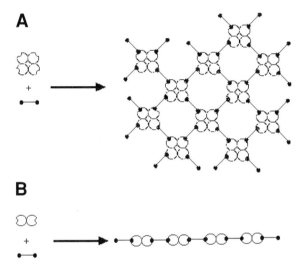

Fig 3. Crosslinking of oligomeric enzymes by bis-NAD$^+$. **(A)** A tetramer (e.g., mammalian lactate dehydrogenase) can form crosslinks to four other tetramers. **(B)** A dimer (e.g., horse liver alcohol dehydrogenase) can be linked to two other dimers, forming a linear polymer.

2. The bis-ligand has to have a strong affinity for the enzyme.
3. The spacer connecting the two ligands has to be long enough to bridge the distance between two ligand-binding sites on two different enzyme molecules. In the case of N_2,N_2'-adipodihydrazido-bis-(N^6-carbonylmethyl)NAD$^+$, the spacer length was approx 1.7 nm, which permitted easy simultaneous access for a molecule of bis-NAD$^+$ to two different molecules of a dehydrogenase.

An important factor for successful affinity precipitation is the ratio of coenzyme derivative to enzyme subunit. If this is low, a lattice will not form, because there are not enough crosslinks; if it is too high, each dehydrogenase subunit can be occupied by a separate molecule of bis-NAD$^+$ and no crosslinks will form. Maximum crosslinking occurs at an optimum ratio of NAD$^+$ equivalents per enzyme subunit (assuming two NAD$^+$ equivalents per bis-NAD$^+$), which, in the case of tetramers, has been found to

LIVERPOOL JOHN MOORES UNIVERSITY
LEARNING SERVICES

be approximately unity. This can vary; for example, the hexameric mammalian glutamate dehydrogenase (GDH) precipitated over a broad range of NAD^+ equivalents to active-site ratios (0.3–10), with up to 70% precipitation occurring at a ratio as low as 0.16, although this probably involves the existence of higher GDH polymers *(6)*. The ratio of approximately unity for tetramers is similar to the behaviour observed in immunoprecipitation, in which two antigen molecules per antibody results in the optimum precipitation of immune complexes *(12)*.

Adding bis-NAD^+ alone to a crude extract containing many different dehydrogenases will not in itself cause specific affinity precipitation, because it is a "general ligand" that will bind to form binary (E–NAD^+) complexes with most NAD^+-dependent dehydrogenases. The addition of a substrate analog (X) that is a competitive inhibitor, relative to the enzyme's second substrate, strengthens the binding interaction by forming an E–NAD^+–X ternary complex. Because NADH binds, on average, one order of magnitude more tightly to the active sites of many dehydrogenases than NAD^+, it displaces bis-NAD^+ and is commonly used to dissolve crosslinked aggregates.

The specificity of bis-NAD^+ of affinity precipitation for any given dehydrogenase is conferred by a property described by O'Carra as the "locking-on" effect *(13)*. This was originally developed to increase the strength of enzyme binding to an immobilized ligand by adding an analog of the specific substrate of the enzyme under study. In the case of lactate dehydrogenase, for example, the strength of adsorbtion to an immobilized NAD^+ derivative was increased by adding oxalate, a structural analog of lactate, to the irrigating buffer. The enzyme was subsequently eluted by simply leaving out the oxalate.

This "locking-on" property does not occur with all enzymes. It is confined to those with sequential mechanisms (*see* **refs. *1*** and ***13*** for more details). Most coenzyme-dependent enzymes, including the majority of dehydrogenases, have ordered sequential kinetic mechanisms, in which the coenzyme binds before the second substrate. If this is the case, the "locking-on" effect occurs. The addition of an unreactive, competitive inhibitor of the second substrate (X) will displace the coenzyme-binding equilibrium by ternary complex formation (sometimes described as an "abortive" complex) and thereby increase the strength of binding.

$$E \overset{NAD^+}{\rightleftharpoons} E\text{–}NAD^+ \overset{X}{\rightleftharpoons} E\text{–}NAD^+\text{–}X$$

In affinity precipitation, the bis-coenzyme is "locked" into place by the substrate analog, provided it is saturating. Some enhancement of bis-ligand binding can occur for an enzyme with a random sequential mechanism, provided that the equilibrium of the reaction under the conditions employed in affinity precipitation is such that the binding of the substrate analog then favors coenzyme binding through ternary complex formation.

The identity of the substrate analog will determine which dehydrogenase is recipitated from a crude extract. For example, adding bis-NAD^+ and glutarate will lead to the precipitation of glutamate dehydrogenase from a crude extract: the addition of pyrazole or oxalate would favor alcohol dehydrogenase or lactate dehydrogenase precipitation, respectively. This is the key to making the technique specific for one enzyme.

Reduction of bis-NAD^+ to bis-NADH by chemical or enzymatic means gave rise to a bis-ligand that precipitated yeast alcohol dehydrogenase, glutamate dehydrogenase, and lactate dehydrogenase in the absence of substrate analog *(14)*. This is because NADH has a much higher affinity for these three dehydrogenases than bis-NAD$^+$, resulting in bis-NADH being unsuitable for the selective precipitation of a single dehydrogenase from a crude mixture, as it is too nonspecific.

In the case of the only reported bis-ATP derivative, which was used for affinity precipitation of bovine heart phosphofructokinase, precipitation was not accomplished with the aid of a second substrate analog, but by the use of the allosteric inhibitor citrate. ATP is both a substrate and an allosteric inhibitor of this enzyme, and this inhibition is potentiated by citrate *(10)*.

1.4. Affinity Precipitation With Other Bifunctional Ligands

Examples of bis-ligand affinity precipitation in the literature are few, and the technique has not gained widespread use. This may be because it has only been found to work with a relatively small number of proteins. Furthermore, the benefit of simplicity that may result from this approach may not outweigh the work necessary to synthesize the appropriate bifunctional ligand. In order to extend the applicability of this approach, other bis-ligands have been synthesized, but these have been limited so far to triazine dyes and immobilized metal ions. Triazine dyes have been widely used as pseudoaffinity ligands (*see* Chapter 17 for a discussion of dye-ligand chromatography). A bis-derivative of Cibacron blue F3GA was found to precipitate bovine serum albumin and lactate dehydrogenase *(15,16)* and a monofunctional synthetic analog of Procion blue H-B was found to precipitate rabbit muscle lactate dehydrogenase selectively *(17,18)*. The synthesis of this analog is reported in **ref. 17** and the method for large-scale lactate dehydrogenase (LDH) purification is described in **ref. 18**. Only two examples of metal-ion affinity precipitation have been reported to date, one of which consisted of the precipitation of recombinant galactose dehydrogenase containing a pentahistidine affinity tail by a (Zn)$_2$–EGTA chelate *(19)*. In the second case, myoglobin and hemoglobin were precipitated by an EGTA–Cu(II) chelate or, alternatively, by a bis-ligand consisting of Cu(II) cations chelated by molecules of iminodiacetic acid immobilized on each end of a molecule of polyethylene glycol *(20)*.

Table 1 gives a list of examples of both bis-ligand and heterobifunctional ligand *(21–32)* affinity precipitation. Further applications of bis-NAD$^+$ are also given in **Table 1**, and some limitations on its use are described in **Note 1**.

2. Materials

2.1. Synthesis of Coenzyme Derivatives

2.1.1. Synthesis of N$_2$, N$_2$-Adipodihydrazido-bis-(N^6-carbonylmethyl)NAD$^+$

This compound was formerly sold by Sigma-Aldrich, but is no longer available. The starting material is N^6-carboxymethyl-NAD$^+$, and the following reagents are required for its synthesis.

2.1.1.1. N^6-CARBOXYMETHYL-NAD$^+$

1. NAD$^+$ (98%, free acid).
2. Iodoacetic acid.
3. 2 *M* LiOH.

Table 1
Published Examples of Protein Purification by Affinity Precipitation

Protein	Bis-ligand	Ref.
Bis-ligands[a]		
Lactate dehydrogenase	Bis-NAD+	5–9
Glutamate dehydrogenase	Bis-NAD+	6–8, 33
Yeast alcohol dehydrogenase	Bis-NAD+	6, 8
Isocitrate dehydrogenase	Bis-NAD+	8
Phosphofructokinase	Bis-ATP	10
Lactate dehydrogenase (rabbit)	Bis-Cibacron blue F3G-A	15, 16
Bovine serum albumin	Bis-Cibacron blue F3G-A	15, 16
Lactate dehydrogenase (rabbit)	methoxylated *p*-sulfonated isomer of Procion blue H-B	17, 18
Galactose dehydrogenase (recombinant)	EGTA (Zn)$_2$	19
Human hemoglobin, sperm whale myoglobin	Cu(II)$_2$EGTA, Cu(II)$_2$ polyethylene glycol-(iminodiacetic acid)$_2$	20

Protein	Ligand	Carrier	Precipitant	Ref.
Hetero-bifunctional ligands[b]				
Trypsin	*p*-Aminobenzamidine	*N*-Acryloyl-aminobenzoic acid	Low pH	21
Trypsin	Soybean trypsin inhibitor	Chitosan	High pH	22
Wheat germ agglutinin	*N*-Acetyl-D-glucosamine	Chitosan	High pH	23
Recombinant protein A	IgG	Eudragit S100	Low pH	24
IgG	Protein A	Galactomannan	Potassium borate	25
Lactate dehydrogenase	Cibacron blue	Dextran	Concanavalin A	26
Trypsin	Soybean trypsin inhibitor	Alginate	Ca^{2+} ions	27
Endo-polygalacturonase	Alginate	Alginate	Ca^{2+}, low pH	28
IgG	Protein A	*N*-Isopropylacrylamide polymer	Temperature increase	29
Subtilisin Carlsberg	*m*-Aminophenylboric acid	*N*-Diethylacrylamide polymer	Ethylene glycol	30
Avidin	Iminobiotin	*N*-Isopropylacrylamide polymer	Temperature increase	31
Protein inhibitors	Cu(II)ions	*N*-Isopropylacrylamide polymer	Temperature increase	32

[a]These do not require a second component to effect affinity precipitation.
[b]These require a ligand carrier and a third component, to promote the precipitation of the ligand–protein complex.

4. 96% (w/v) Ethanol.
5. Sodium dithionite.
6. 0.24 M NaHCO$_3$.
7. Yeast alcohol dehydrogenase (crystalline, Sigma-Aldrich or Boehringer).
8. A sintered funnel (fairly large, at least 10 cm in diameter).
9. AG 1X2 anion-exchange resin (200–400 mesh, Cl⁻ form, Bio-Rad).
10. CaCl$_2$.

2.1.1.2. Bis-NAD⁺

1. Adipic acid dihydrazide dichloride (Sigma-Aldrich).
2. 2 M NH$_4$OH.
3. NH$_4$HCO$_3$.
4. DEAE-cellulose (Whatman DE-52, equilibrated with 10 column volumes of 1 M NH$_4$HCO$_3$, and washed with water).

2.1.2. N$_2$, N$_2'$-Adipodihydrazido-bis-(N^6-carbonylmethyl-ATP)

As for N^6-carboxymethyl-NAD⁺, except that NAD⁺ is replaced by ATP and LiCl is used in place of CaCl$_2$. Neither this ATP derivative nor N^6-carboxymethyl-ATP are commercially available.

2.1.3. Other Bis-NAD⁺ Derivatives

1. NAD⁺ (Sigma-Aldrich, Roche Molecular Biochemicals; 98% free acid).
2. Ethyleneimine (obtainable from Serva, Heidelberg, Germany) *Caution*: This is toxic and carcinogenic.
3. 70% HClO$_4$.
4. 96% Ethanol.
5. LiCl.
6. Bio-Rex 70 cation-exchange resin (100–200 mesh, Na⁺ form, Bio-Rad).
7. 1 M LiOH.
8. 1 M HCl.
9. 1-Ethyl-3-(3-dimethylaminopropyl)-carbodiimide hydrochloride (EDC).
10. 5 M NaOH.
11. 0.5 M Ammonium acetate, pH 7.0.
12. 0.05 M Ammonium acetate, pH 7.0.
13. DE-52 cellulose.
14. TLC solvent system A: (NH$_4$)$_2$SO$_4$: 0.1 M potassium phosphate, pH 6.8: 1-propanol (60:100:2, w/v/v).
15. TLC solvent system B: Isobutyric acid: 1 M aqueous NH$_3$ (5:3, v/v), saturated with Na$_2$-EDTA.
16. TLC solvent system C:Isobutyric acid:water: 25% aqueous NH$_3$ (66:33:1, v/v/v).
17. μBondapak C-18 column (Waters or the equivalent from an alternative source; 3.9 × 300 mm)
18. 0.1 M Potassium dihydrogen phosphate, pH 6.0, containing 10% (v/v) methanol for separating monosubstituted NAD⁺ derivatives.
19. 0.1 M Potassium dihydrogen phosphate, pH 6.0, containing 20% (v/v) methanol for separation of bis-NAD⁺ derivatives.

Bio-Rex 70 can be converted to the H⁺ form by washing it exhaustively with 0.5–1.0 M HCl. The absence of Na⁺ ions can be tested by flame photometry. The column is then washed with 1 m*M* HCl, pH 3.0, and equilibration is checked by pH and conductivity measurements.

The DE-52 cellulose is equilibrated with 0.5 *M* ammonium acetate, pH 7.0, to convert it to the acetate form; this is then equilibrated for chromatography by washing it with 10 column volumes of 0.05 *M* ammonium acetate, pH 7.0.

For Bis-NAD$^+$ synthesis glutaric, adipic and pimelic acid (all available from Sigma-Aldrich) have been found to serve as satisfactory, water-soluble spacers.

In addition to this, a rotary evaporator and chromatographic columns (dimensions variable) are needed.

The plates have the following qualities: aluminium backed, silica gel 60, fluorescent indicator F_{254}, layer thickness 0.2 mm (Merck). A source of fluorescent light (wavelength 254 nm) can be used to visualize the spots, which appear purple on a green background.

The reversed phase, μBondapak C-18 is equilibrated with 0.1 *M* potassium dihydrogen phosphate, pH 6.0, containing 10% (v/v) methanol for separating monosubstituted NAD$^+$ derivatives and 20% methanol for separation of bis-NAD$^+$ derivatives. All solutions used in HPLC must be degassed using a vacuum pump and filtered with a 0.22-μm filter, to exclude particulate matter and avoid air bubbles. In addition, all samples must be centrifuged for 2 min in a minifuge to pellet particles, before application to the column. The absorbance should be monitored at 254 nm.

2.2. Pilot Affinity Precipitation Studies

1. A source of the protein of interest (e.g., a crude tissue extract supernatant) or a commercially available preparation of the protein, to ensure the reagent is effective.
2. Bis-ligand.
3. Stock solutions of a substrate analog; for example, 560 m*M* oxalate for lactate dehydrogenase, 560 m*M* pyrazole for alcohol dehydrogenase, or 700 m*M* glutarate for glutamate dehydrogenase.
4. 0.4 *M* Potassium phosphate buffer (62.4 g/L NaH$_2$PO$_4$, titrated to pH 7.4 with 5 *M* KOH), for some applications.
5. 1.5-mL Polypropylene minifuge tubes
6. Assay reagents for determining enzyme activities (*see* **ref. 34**).

2.3. Enzyme Purification

Apart from the reagents and apparatus mentioned in **Subheading 2.2.**, reagents and apparatus for electrophoresis (e.g., PAGE/SDS-PAGE [sodium dodecyl sulfate-polyacrylamide gel electrophoresis]) are needed, to check the purity of the affinity precipitated protein (*35*). In addition, reagents for a protein assay (e.g., by the Lowry or Bradford methods) are required to determine the protein concentration.

For the separation of LDH isoenzymes (*see* **Subheading 3.3.3.**), materials for starch gel electrophoresis are appropriate. These include hydrolyzed potato starch (Sigma), electrophoresis-grade Trizma base to buffer the gel, and grade III NAD$^+$ (Sigma). The activity stain for LDH contains 1.21 g Tris, 7.72 mL 70% Na DL-lactate (8 g), 50 mg nitroblue tetrazolium, 50 mg grade-III NAD$^+$, and 4 mg phenazine methosulfate. (*Caution:* Nitroblue tetrazolium and phenazine methosulfate are toxic. The latter compound is light sensitive.) Make the volume up to 200 mL with water and adjust the pH to 7.1 with 6 *M* HCl.

3. Methods

3.1. Synthesis of Coenzyme Derivatives

3.1.1. N_2,N_2-Adipodihydrazido-bis-(N⁶-carbonylmethyl)NAD⁺

The synthesis of N^6-carboxymethyl–NAD⁺ is described in **ref. 36**, but some modifications are given here. The synthesis of bis–NAD⁺ is a modified and more detailed version of that described in **ref. 5**.

1. Dissolve 9 g of fresh iodoacetic acid in approx 10 mL of water, neutralize the solution with 2 M LiOH, and add 3 g of NAD⁺. Adjust the pH to 6.5 with 2 M HCl (the total volume should be about 30 mL) and leave the solution in darkness for 7 d at room temperature (approx 20°C), or at 37°C for 2 d. The pH should be checked daily (or every 4–6 h at the higher temperature) and readjusted to 6.5 with 2 M HCl as required. The progress of the reaction should also be followed by thin-layer chromatography (TLC) and/or high-performance liquid chromatography (HPLC).

2. When the reaction is complete, adjust the pH to 3.0 with 6 M HCl and add 2 volumes of 96% (v/v) ethanol. This gives a milky, pink-tinged suspension, which precipitates on the addition of a further 10 vol of cold 96% ethanol. (The water/ethanol ratio is important; using a wet vessel can give rise to the formation of a brown, sticky substance that is water soluble. This also applies to the corresponding ATP derivative.) Filter the crude $N(1)$-carboxymethyl-NAD⁺ on a sintered funnel, wash with ethanol and diethyl ether, and dry under vacuum (average yield 2.9 g). Store the product at −20°C under vacuum.

3. Dissolve the crude $N(1)$-carboxymethyl-NAD⁺ in 0.24 M NaHCO₃ (90 mL), which gives a pale orange solution, and adjust the pH to 8.5 with 1 M NaOH. Deoxygenate the solution by bubbling N₂ gas through it for 2 min, add 1.5 g of sodium dithionite, and leave the solution in the dark until maximum reduction is achieved. This depends on the dithionite; monitor the reaction by taking samples, diluting them 1:50 or 1:100 and measuring the increase in A_{340}.

4. Terminate the reaction by stirring vigorously for 10 min to oxygenate the solution and then bubble N₂ gas through for 2 min. Adjust the pH to 11.5 with 1 M NaOH and leave in a water bath at 75°C to allow the Dimroth rearrangement from the $N(1)$- to the N^6-substituted derivative to occur (*see* **ref. 37** for a reaction mechanism). Monitor the rearrangement by HPLC or TLC (*see* **Subheading 2.1.3.**). Cool the reaction mixture to room temperature and add 6 mL of 2 M Tris and 1.5 mL of redistilled acetaldehyde.

5. Adjust the pH to 7.5 with 1 M HCl and add 8–24 mg of yeast alcohol dehydrogenase (2500–7500 units; *see* **Note 2**). Monitor the reaction at 340 nm as in **step 3**. When a minimum A_{340} is reached, add 1 vol of 96% ethanol and pour the milky flocculent precipitate into 10 vol of vigorously stirred 96% ethanol. Leave for 30 min (or overnight, if desired) and collect the precipitate by filtration. The crude N^6-carboxymethyl-NAD⁺ can be stored at −20°C under vacuum for up to 1 mo (yield approx 2.6 g).

6. The product from **step 5** can be purified by dissolving the crude powder (2.6 g) in 30 mL water, adjusting the pH to 8.0 with 1 M LiOH and applying this solution to a column of Dowex AG 1X2 (200–400 mesh, Cl⁻ form, 4 × 10 cm). Wash the column beforehand with 1 L of 3 M HCl, followed by exhaustive washing with at least 20 L of water until neutral pH is reached. After applying the coenzyme solution, wash the column with 0.5 L of water, followed by 1 L of 5 mM CaCl₂, until the pH of the effluent is 2.8. Apply a linear gradient (2 × 1 L) from 5 mM CaCl₂, pH 2.7, to 50 mM CaCl₂, pH 2.0.

7. Collect 10- to 20-mL fractions and monitor the absorbance at 260 nm. The composition of the fractions can be monitored by TLC. R_f values: system A—NAD⁺, 0.41; $N(1)$-car-

boxymethyl-NAD$^+$, 0.31; N^6-carboxymethyl NAD$^+$, 0.22; system B—NAD$^+$, 0.44; $N(1)$-carboxymethyl-NAD$^+$, 0.27; N^6-carboxymethyl NAD$^+$, 0.22. The values can vary slightly with changes in the solvent composition over time (*see* **Note 3**). Pool the fractions containing the N^6 derivative, adjust the pH to 7.0 with 2 M Ca(OH)$_2$, and concentrate to 5–10 mL by rotary evaporation at 40°C. Precipitate with 96% ethanol as in **step 5** and dry under vacuum. This gives a pale yellow compound (yield 0.82 g, 25%). $\varepsilon = 19{,}300\ M^{-1}\ cm^{-1}$ at 266 nm.

8. To synthesize bis-NAD$^+$, dissolve 0.82 g of purified N^6-carboxymethyl-NAD$^+$ and 105 mg of adipic acid dihydrazide dihydrochloride in 20 mL water, to give a brownish solution. Adjust the pH to 4.6 with 1 M HCl and then add 0.5 M EDC at 0°C in 15 100-μL aliquots over a period of 35 min. Monitor the pH and readjust it to 4.6 before each addition. Add water (2 L) to dilute the solution 100-fold, adjust the pH to 8.0 with 2 M NH$_4$HCO$_3$, and apply the solution to a DE-52 column (2.5 × 30 cm) equilibrated with 1 M NH$_4$HCO$_3$, pH 8.0, and then wash with water. Wash the column with water until the A$_{260}$ is less than 0.1 and apply a linear gradient (2X 1 L) from 0 to 0.25 M NH$_4$HCO$_3$, pH 8.0. Monitor the R_f values of the fractions by TLC (R_f of bis-NAD$^+$: system A, 0.09; system B, 0.05; *see* **Note 3**). The yield from pure N^6-carboxymethyl-NAD$^+$ is approx 14%. $\varepsilon = 42{,}800\ M^{-1}\ cm^{-1}$ at 266 nm.

3.1.2. N$_2$, N$_2'$-Adipodihydrazido-bis-(N^6-carbonylmethyl)ATP (see **refs. 38** and **39**)

1. Dissolve 3.4 g of fresh iodoacetic acid in 1 mL of water and adjust the pH to 7.0 with 2 M LiOH. Dissolve 1.16 g of ATP in the solution and leave it at pH 6.5 in the dark for 4 d at 30°C, adjusting the pH to 6.5 every day. Monitor the reaction by HPLC and/or TLC, or follow the decrease in absorbance at 640 nm. When the reaction is complete, as judged by TLC, HPLC, or absorbance change, add 1 vol of cold 96% ethanol to give a milky suspension and add this to 6–8 vol of stirring absolute ethanol at 0°C. Filter and wash as for the corresponding NAD$^+$ derivative (typical yield 82–87%, approx 1.4 g).

2. Dissolve 1.4 g of product in 30 mL of water to give a reddish solution. Adjust the pH to 8.5 with 1 M LiOH and incubate at 90°C for 100 min, with pH adjustment every 20 min. Monitor the reaction over this time by TLC or HPLC. The solution can be cooled on ice and stored overnight at 4°C overnight, if required. HPLC retention times are as follows: ATP, 4.5 min; $N(1)$-carboxymethyl-ATP, 3.1 min; N^6-carboxymethyl-ATP, 4.1 min.

3. Adjust the pH of the solution at 20°C to 2.75 with 6 M HCl and apply it to an AG 1 × 2 column (200–400 mesh, Cl$^-$ form, 4 × 9 cm), prepared as described in **Subheading 3.1.1.** Wash the column with 1 L of 0.3 M LiCl, pH 2.75, until the A$_{260}$ is <1.0 and then apply a 2X 1-L linear gradient, 0.3 M LiCl, pH 2.75–0.5 M LiCl, pH 2.0. Monitor the R_f values of the fractions by TLC as previously. (R_f values: system A—ATP, 0.54; $N(1)$-carboxymethyl-ATP, 0.73; N^6-carboxymethyl-ATP, 0.54; system B—ATP, 0.35; $N(1)$-carboxymethyl-ATP, 0.27; N^6-carboxymethyl-ATP, 0.21; *see* **Note 3**). Pool the fractions containing the N^6 derivative and adjust the pH with 2 M LiOH. Yield: 32–38% from ATP; $\varepsilon = 17{,}300\ M^{-1}\ cm^{-1}$ at 266 nm. This procedure should be carried out as quickly as possible, preferably within 1 d, as ATP is unstable under acidic conditions.

4. Bis-ATP. Dissolve 105 mg of adipic acid dihydrazide dihydrochloride in a solution containing 0.42 g of N^6-carboxymethyl-ATP in 100 mL. Adjust the pH to 4.0 with 1 M HCl and add three 1-mL aliquots of 1 M EDC over a period of 45 min while stirring continuously and monitoring the pH. Dilute to 2 L with water to stop the reaction and adjust the pH to 8.0 with 2 M NH$_4$OH. This solution can be stored at −20°C for about 1 wk.

5. The compound can be purified by applying the dilute solution to a DE-52 column (2.5 × 25 cm), equilibrated with 2–4 L of 1 M NH$_4$HCO$_3$ and then washed with 8 L water. Elute the bis-ATP with a 0- to 0.4-M NH$_4$HCO$_3$ gradient, pH 8 (2 × 1 L). Collect 10- to 20-mL fractions and pool and freeze-dry the fractions containing bis-ATP, as judged by HPLC or

by affinity precipitation of phosphofructokinase (*see* **Subheading 3.3.4.**). Ensure that the compound is completely dry before storage at $-20°C$, as it is very hygroscopic. Yield: 0.06 g from 0.42 g N^6-carboxymethyl-ATP (yield approx 2% from ATP). $\varepsilon = 39,600\ M^{-1}\ cm^{-1}$ at 266 nm (*see* **Note 4**).

3.1.3. Other Bis-NAD$^+$ Derivatives

1. The synthesis of N^6-(2-aminoethyl)-NAD$^+$ is carried out essentially as described in **ref. 40**. However, $N(1)$-(2-aminoethyl)NAD$^+$ can be rearranged to the N^6-substituted derivative without removal of the unreacted NAD$^+$ by ion-exchange chromatography (**41**) and the following procedure is a modification to take account of that. Dissolve crude $N(1)$-(2-aminoethyl)-NAD$^+$ [contains approx 70% $N(1)$-(2-aminoethyl)NAD$^+$; 3.05 g, 4.3 mmol] in 850 mL of water. Adjust the pH to 6.5 with 1 M LiOH. Place the solution in a water bath at 50°C and allow the rearrangement to proceed for 6–7 h. Monitor the reaction by HPLC/TLC/change in absorbance maximum over this time. The wavelength of maximum absorbance should change from 260 nm to at least 264 nm, as the λ_{max} for N^6-(2-aminoethyl)-NAD$^+$ is 266 nm. Terminate the reaction by cooling the solution to room temperature and adjust the pH to 5.5 with 1 M HCl. Concentrate the solution by rotary evaporation at 35°C to approx 20 mL. The compound is stable in concentrated solution for at least 1 wk at 4°C.

2. Apply the concentrated reaction mixture to a Bio-Rex 70 cation-exchange column (H$^+$ form, 2.7 × 73 cm), pre-equilibrated with 1 mM HCl, pH 3.0. Wash the column with this solution and collect fractions of 10–20 mL, in a total volume of approx 4 L. Monitor the A_{260} as an indicator of coenzyme concentration. The last peak to elute contains the N^6-(2-aminoethyl)-NAD$^+$. Some overlap may occur with the previous peak, containing the tricyclic NAD$^+$ derivative 1,N^6-ethanoadenine-NAD$^+$. Pool and rotary evaporate the fractions containing the desired compound to approx 5 mL. Lyophilization gives a fluffy pale yellow compound, stable for at least 12 mo at $-20°C$ in an airtight container. $\varepsilon = 21,600\ M^{-1}\ cm^{-1}$ at 266 nm. HPLC retention times: NAD$^+$, 6.3 min; $N(1)$-(2-aminoethyl)NAD$^+$, 4.4 min; N^6-(2-aminoethyl)-NAD$^+$, 5.7 min; 1,N^6-ethanoadenine-NAD$^+$, 4.7 min. TLC R_f values: system C—NAD$^+$, 0.28; $N(1)$-(2-aminoethyl)NAD$^+$, 0.10; N^6-(2-aminoethyl)-NAD$^+$, 0.16; 1,N^6-ethanoadenine-NAD$^+$, 0.08.

3. Dissolve the spacer (glutaric, adipic, or pimelic acid) in water to give a 50-mM solution. Adjust the pH to 7.0 with 5 M NaOH and add 700 μL of this solution to 50 mg (67 μmol) of N^6-(2-aminoethyl)NAD$^+$ to give a 2:1 molar ratio of coenzyme to spacer arm. Adjust the pH from approx 3.0 to 5.4 with 5 M NaOH. Add solid EDC (34 mg, 335 μmol) to the solution in six to seven portions over 20 min while monitoring the reaction by TLC/HPLC. The pH tends to rise; readjust it to 5.5–6.0 when it reaches a value of 6.9–7.0. Stop the reaction after 40 min by freezing the reaction mixture at $-20°C$. TLC indicates that at least four side products are formed in addition to bis-NAD$^+$.

4. To purify the bis-NAD$^+$, dilute the sample twofold with 50 mM ammonium acetate, pH 7.0, and apply it to a Whatman DE-52 anion-exchange column (2.3 × 16 cm), equilibrated as described in **Subheading 2.1.3**. Wash the column with this buffer. The last major peak contains the bis-NAD$^+$ derivative. This can be checked by TLC, HPLC, or by the ability of the fractions to cause affinity precipitation of glutamate dehydrogenase. Pool and rotary evaporate the fractions containing bis-NAD$^+$ to 3–5 mL and freeze-dry these (stable for at least 4 mo at 4°C or $-20°C$). The compound can also be stored for approx 1 mo in solution at 4°C, although bacterial contamination may result. TLC R_f values (system C) for bis-NAD$^+$ derivatives: N,N'-bis (N^6-ethylene-NAD$^+$)glutaramide, 0.09; N,N'-bis(N^6-ethylene-NAD$^+$) adipamide, 0.12; N,N'-bis-(N^6-ethylene-NAD$^+$)pimelamide, 0.15. $\varepsilon = 44,200 \pm 1400\ M^{-1}\ cm^{-1}$. HPLC retention times (in 0.1 M potassium phosphate buffer, pH 6, containing 20% methanol) are 7.8, 11.2, and 13 min for the above compounds, respectively.

3.2. Pilot Precipitation Study

1. Take a source of the enzyme of interest (e.g., a crude supernatant) and assay it for enzyme activity (*see* **ref. 32** for a wide range of assays) to determine the specific activity.
2. Calculate the approximate concentration of enzyme, based on the specific activity and sub-umit relative molecular mass.
3. Add 100–200 µL of the enzyme solution to a series of minifuge tubes. If the study is to be carried out with purified enzyme, dilute a stock solution of 400 mM potassium phosphate buffer pH 7.4, to give a final concentration of 20 mM. Set up all samples in duplicate, including two control tubes to which water has been added in place of the coenzyme derivative. Set up additional duplicate controls, which contain no bis-ligand or substrate analogue. This enables the percentage inhibition resulting from the presence of the substrate analog to be calculated.
4. Calculate the concentration of a stock solution of bis-coenzyme spectrophotometrically, using the extinction coefficients given in **Subheading 3.1**.
5. Add bis-NAD$^+$ (or bis-ATP) to give a range of coenzyme equivalents/ enzyme subunit from approx 0.1 to 20, calculating the concentration from the extinction coefficient.
6. Add the appropriate substrate analog and leave at 4°C for at least 30 min (or overnight, if required).
7. Centrifuge in a minifuge for 5 min and assay the supernatant. Calculate the precipitation as a percentage of the activity remaining in the control sample. The tube giving the minimal residual activity shows the maximum affinity precipitation. This indicates the appropriate concentration of bis-coenzyme to add to obtain maximum affinity precipitation.

3.3. Enzyme Purification Protocols

The following are a selection of protocols for purifying specific enzymes with bis-ligands.

3.3.1. Purification of Yeast Alcohol Dehydrogenase (E.C. 1.1.1.1; YADH)

1. The yeast lysis is carried out according to a modification of the procedure in **ref. 42**. Crumble 40 g of fresh baker's yeast into 21 mL of toluene, preheated to 45°C in a water bath. (*Caution:* Keep the vessel covered or incubate in a fume cupboard.) Allow the mixture to liquefy over a period of 90 min, with occasional stirring (*see* **Note 5**).
2. Leave the mixture at room temperature for 3 h and add 42 mL of 1 mM Na$_2$EDTA as a protease inhibitor. Stir for 2 h at 4°C and leave the mixture at that temperature overnight.
3. Centrifuge the lysate at 47,800g for 10 min, remove the top fatty layer by aspiration, and retain the supernatant. Add solid (NH$_4$)$_2$SO$_4$ slowly, until 60% saturation is reached (7.22 g to 20 mL) and centrifuge for 10 min at 47,800g. The ammonium precipitation step removes some proteins that would otherwise precipitate on addition of bis-NAD$^+$ and gives a purification of approximately threefold to fourfold. Dissolve the pellet in approx 3 mL of 20 mM potassium phosphate buffer, pH 7.4, containing 1 mM Na$_2$EDTA and dialyze against two changes of 2 L of this buffer to remove the salt (alternatively, gel-filter on a column of Sephadex G-25, at least 1 × 50 cm long).
4. Dilute this dialysate 1:4 with buffer for a pilot precipitation study. Add pyrazole and (NH$_4$)$_2$SO$_4$ to give a final concentrations of 28 mM and of 0.5 M, respectively, and different quantities of bis-NAD$^+$. A suitable range of concentrations of NAD$^+$ equivalents is 0–20 µM. Allow precipitation to occur for at least 3 h (preferably overnight) at 4°C. The presence of 0.5 M (NH$_4$)$_2$SO$_4$ is required to promote precipitation. The affinity precipitation can also be performed without the dialysis to remove (NH$_4$)$_2$SO$_4$ (*see* **step 3**). In that case, it is unnecessary to add additional (NH$_4$)$_2$SO$_4$.

5. Assay the supernatants as described in **ref. *34***, diluting the enzyme solution if necessary, and determine which concentration of bis-NAD$^+$ gives maximum precipitation. Add the appropriate concentration of bis-NAD$^+$ to whatever volume of supernatant is required, along with $(NH_4)_2SO_4$ and pyrazole. Allow precipitation to occur as described in **Subhead 3.2.** and centrifuge the precipitate for 10 min at 12,000g to pellet the crosslinked enzyme aggregate.

6. Resolubilize the pellet by adding 0.6 mM NADH in 20 mM potassium phosphate buffer, pH 7.4. (The volume can be varied, depending on what final enzyme concentration is required.) Allow resolubilization to proceed for at least 3 h, preferably 12 h, at 4°C on a rocking tray or with occasional stirring (*see* **Note 5**). The enzyme should be essentially homogeneous, as shown by SDS-PAGE on a 12.5% gel.

3.3.2. Purification of Glutamate Dehydrogenase (E.C. 1.4.1.3; GDH) From Beef and Rat Liver (see **ref. *33***)

1. If beef liver is used, transport this from the abattoir on ice.
2. Homogenize the tissue and carry out ammonium sulfate precipitation and DEAE-cellulose chromatography as described in **ref. *43***.
3. Pool the fractions from the DE-52 column that contains GDH activity (*34*) and concentrate them to a volume of approx 10 mL by ultrafiltration through an Amicon XM-50 membrane. Dialyze the solution overnight at 4°C against 200 mL of sodium or potassium phosphate buffer, pH 7.4, with at least one change of buffer.
4. Carry out a pilot precipitation as described in **Subheading 3.2**. Incubate 100-µL samples of the dialyzed solution in the presence of 12.7 µL of 0.7 M glutarate, pH 7.0, and bis-NAD$^+$ for 15 min. Assume a subunit M_r for GDH of 56,700. The optimum ratio of NAD$^+$ equivalents/enzyme subunit may vary with the preparation used; for example, with a preparation of beef liver enzyme that had a specific activity of 2.7 units/mg (*see* **Note 2**), about half of the activity was precipitated at a ratio of 2 NAD$^+$ equivalents/GDH subunit. However, a preparation with the lower specific activity of 0.8 units/mg required approx 8 NAD$^+$ equivalents/GDH subunit to precipitate half the enzyme activity. The precipitation yield also depends on the protein concentration.
5. Take the dialyzed solution and add glutarate to a final concentration of 79 mM. Add the appropriate amount of bis-NAD$^+$ to achieve maximum precipitation. Keep the mixture on ice overnight and centrifuge at 10,000g for 15 min. Redissolve the pellet by stirring for 6 h in 1 mL of 20 mM sodium or potassium phosphate buffer, pH 7.4, containing 0.6 mM NADH. Dialyze the redissolved pellet against 200 mL of the same buffer as in **step 3**. A purification summary is given in **Table 2**. This method appears to avoid some of the proteolysis encountered in more conventional methods of purifying this enzyme (*see* **ref. *43***). Although affinity precipitation was obtained with preparations that had not been subjected to chromatography on DEAE-cellulose, the yield was much lower and the precipitated material was not completely pure.

3.3.3. Separation of Lactate Dehydrogenase (EC 1.1.1.27; LDH) Isoenzymes

This technique is based on the principle that both M and H isoenzyme subunits of LDH from abortive complexes with NAD$^+$ and oxalate, whereas the H form gives rise to abortive ternary complexes with NAD$^+$ and oxamate. Thus, in a mixture of H and M subunits containing bis-NAD$^+$ and oxamate, ternary complexes and thus crosslinks will form only with the H subunits, and tetramers with predominantly M subunits will tend not to be precipitated.

Table 2
Purification of Beef and Rat Liver GDH

Step	Volume (mL)	Total protein (mg)	Total activity (units)	Specific activity (units/mg)	Purification (factor)	Yield (%)
Homogenate	370 (90)	8580 (2720)	1920 (760)	0.2 (0.3)	—	100 (100)
$(NH_4)_2SO_4$	75 (35)	2010 (980)	1640 (505)	0.3 (0.5)	4 (2)	86 (67)
DEAE-cellulose chromatography	8.4 (9.8)	139 (130)	408 (195)	2.9 (1.5)	13 (5)	21 (26)
Affinity precipitation	1.2 (1.0)	9.4 (5.1)	376 (190)	40 (37)	180 (140)	20 (25)

Note: The first three steps were carried out as described by McCarthy et al. *(43)*. The purification of the enzyme from beef liver used 46 g and that from rat (values in brackets) used 10 g of liver.
Source: From **ref. 33**.

1. Select a source of LDH isoenzymes (e.g. a crude supernatant, or Sigma Type X bovine muscle LDH that contains varying proportions of all five isoenzymes). Determine the specific activity and hence estimate the concentration of the enzyme *(34)*.

2. Carry out a pilot precipitation as described in **Subheading 3.2**. The precipitation solution should contain 20 m*M* potassium phosphate buffer, pH 7.4, and potassium oxalate to a final concentration of 20 m*M*. Optimum affinity precipitation should be obtained with a ratio of approx 1 NAD^+ equivalent/LDH subunit. Allow precipitation to occur overnight at 4°C. Centrifuge for 10 min and retain the supernatant for further analysis.

3. Resolubilize the precipitated enzyme by adding 50 μL of 0.6 m*M* NADH in the above buffer and incubating overnight. Centrifuge the samples for 10 min at 12,000*g* to pellet any remaining crosslinked enzyme.

4. The isoenzymic composition of the pellet and supernatant can be analyzed by starch-gel electrophoresis. This is carried out by the method described in **ref. 44**. Make an 11% starch gel. Apply 10-μL samples of the enzyme solution to the gel on small squares of filter paper (about 0.5 cm × 0.5 cm) and carefully insert these, using a forceps, in wells cut across the middle of the gel. Place a piece of polyethylene wrap over the gel and put a cooling plate on top. Run the gel for 4 h at 550 V and 60 mA in a cold room at 4°C. After electrophoresis, stain the gel with an activity stain for LDH (*see* **Subheading 2.3.**) until the enzyme-activity bands appear. Wash the gel exhaustively with water (otherwise overstaining can easily occur) and treat the gel gently, as starch gels are fragile. **Figure 4** shows a starch gel, on which the separation pattern can be seen.

3.3.4. Purification of Beef Heart Phosphofructokinase (EC 2.7.1.11; PFK)

1. Obtain a fresh portion of beef heart from a freshly slaughtered animal and keep it on ice. Remove the ventricular muscles and adipose tissue, dice the remaining tissue, and wash it in distilled water. Homogenize it in a precooled blender for 10 s and take 500 g of mince and add 10 m*M* Tris-HCl buffer containing 2 m*M* EDTA, pH 8.2, to a total volume of 1.25 L. Homogenize this for 90 s at high speed and then centrifuge the homogenate at 2600*g* for 10 min. Discard the supernatant.

2. Resuspend the pellet in 10 m*M* Tris-HCl containing 0.5 m*M* ATP, 50 m*M* $MgSO_4$, and 5 m*M* of 2-mercaptoethanol, pH 8.0, to a total volume of 1 L. Prewarm the buffer to 37°C. Stir the suspension at 37°C for 30 min. This extraction procedure gives rise to a decrease in pH of 0.6–0.8 pH units. Remove solid material by centrifugation for 15 min at 8000*g*,

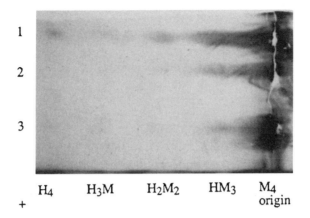

Fig. 4. Starch gel electrophoresis of LDH isoenzymes. The gel was stained with an affinity stain for LDH, as described in the text. *Lane 1* consists of Sigma Type X bovine muscle LDH (1.4 units) containing all isoenzymes of LDH with M and H subunits. *Lane 2* contains the supernatant after affinity precipitation in the presence of bis-NAD^+ (1 NAD^+ equivalent/LDH subunit) and 20 mM oxamate. *Lane 3* contains the resolubilized pellet from the same sample as *Lane 2*. The pellet was resolubilized in 20 mM potassium phosphate, pH 7.4, containing 0.6 mM NADH. Affinity precipitation and resolubilization were carried out in a final volume of 100 µL.

    followed by a further centrifugation at 12,000g. The supernatant contains about 90% of the PFK activity.

3. Set up a pilot precipitation (*see* **Subheading 3.2.**) to determine the concentration of bis-ATP giving optimum precipitation. The following volumes are appropriate; 100 µL enzyme solution, 30 µL of 30 mM citrate, pH 7.0, 60 µL 0.16 M $MgSO_4$. Add bis-ATP to give a range of ATP equivalent/ PFK subunit ratios (0–8 is suitable, at a enzyme concentration of 0.38 mg/mL PFK). Make the volume up to 500 µL with extraction buffer. A ratio of 4–6 is usually found to be optimum for precipitation (*see* **Fig. 5**). A similar pilot experiment varying the enzyme concentration can also be set up, at a fixed concentration of bis-coenzyme. A sample purification table is given (*see* **Table 3**).

4. The volume for precipitation can be scaled up as required. The enzyme can be recovered by dialysis, which causes resolubilization of the pellet.

4. Notes

1. Bis-NAD^+ has been used as a ligand with several NAD^+-linked dehydrogenases, but only for three enzymes (YADH, LDH, GDH) has it successfully been applied for purification purposes. A partial purification of isocitrate dehydrogenase giving rise to a yield of 18% has been described *(8)*. Horse liver alcohol dehydrogenase, which is a dimer, has also been observed to form aggregates with bis-NAD^+ and pyrazole *(6,45)* but it does not affinity precipitate, despite the fact that it may form larger linear polymers. Other enzymes that do not successfully affinity precipitate include malate, alanine and glyceraldehyde-3-phosphate dehydrogenases. Malate dehydrogenase is a dimer, and the other two enzymes, although hexameric and tetrameric, respectively, may fail to form crosslinked ternary complexes, possibly owing to steric or kinetic factors.

2. One unit of enzyme activity is defined as the amount catalyzing the production of 1 µmol of product (or the disappearance of 1 µmol of substrate) in 1 min under stated conditions.

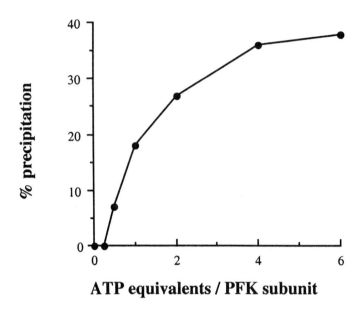

ATP equivalents / PFK subunit

Fig. 5. Affinity precipitation of phosphofructokinase. A pilot precipitation was set up as described in the text. The precipitation mixture contained 0.1 mL enzyme supernatant, 20 µL of 0.5 M citrate, pH 7.0, bis-ATP (0.25–6 ATP equivalents/ATP subunit), and 20 µL 0.5 M MgSO$_4$. The tubes were left overnight at 0°C and centrifuged for 10 min to pellet the enzyme.

Table 3
Purification of PFK From Bovine Heart (200 g Tissue)

Step	Volume (mL)	Total protein (mg)	Total activity (units)	Specific activity (units/mg)	Purification (factor)	Yield (%)
Crude extract						
Resuspended precipitate	490	3675	970	0.26	1	100
Supernatant after extraction	548	376	883	3.2	12.3	91
Affinity precipitation	3	7.1	214	30.1	115.7	22

Note: The method used is that given in **Subheading 3.3.4.**
Source: From **ref. 38**.

3. The Rf values given for TLC of coenzyme derivatives are not exact. They can vary by up to ±0.05. This may result from slight changes in the solvent composition over time, possibly as a result of evaporation.
4. To date, bis-ATP has only been used in the purification of phosphofructokinase. The yields obtained for this compound are low and it is unstable, possibly owing to hydrolysis of the spacer arm, catalyzed intramolecularly by a phosphate group. It should be used within 4 d of synthesis. Because of this, the reagent must be regarded as being of limited practical use, and further studies on the synthesis of a more stable bis-ATP derivative are necessary.
5. Yeast alcohol dehydrogenase has essential thiol groups and if these are oxidized the enzyme loses activity. They should be kept reduced by the addition of 1 mM dithiothreitol or 1 mM of 2-mercaptoethanol. The breaking of the yeast cells may also be carried out by grinding with glass beads.

Acknowledgments

Support from Trinity College, Dublin and EOLAS (Forbairt/Enterprise Ireland) is gratefully acknowledged.

References

1. Irwin, J. A. and Tipton, K. F. (1995) Affinity precipitation: a novel approach to protein purification. *Essays Biochem.* **29**, 137–156.
2. Gupta, M. N., Kaul, R., Guoqiang, D., Dissing, U., and Mattiasson, B. (1996) Affinity precipitation of proteins. *J. Mol. Recogn.* **9**, 356–359.
3. Chaga, G. S., Guzman, R., and Porath, J. O. (1997) A new method of synthesizing biopolymeric affinity ligands. *Biotechnol. Appl. Biochem.* **26**, 7–14.
4. Mori, T., Umeno, D., and Maeda, M. (2001) Sequence-specific affinity precipitation of oligonucleotide using poly (*N*-isopropylacrylamide)-oligonucleotide conjugate. *Biotech. Bioeng.* **72**, 261–268.
5. Larsson, P.-O., and Mosbach, K. (1979). Affinity precipitation of enzymes. *FEBS Lett.* **98**, 333–338.
6. Flygare, S., Griffin, T., Larsson, P.-O., and Mosbach, K. (1983) Affinity precipitation of dehydrogenases. *Anal. Biochem.* **133**, 409–416.
7. Larsson, P.-O., Flygare, S., and Mosbach, K. (1984). Affinity precipitation of dehydrogenases. *Methods Enzymol.* **104**, 364–369
8. Beattie, R. E., Graham, L. D., Griffin, T. O., and Tipton, K. F. (1985). Purification of NAD$^+$-dependent dehydrogenases by affinity precipitation with adipo-N_2,N_2'-dihydrazido bis-(N^6-carboxymethyl-NAD$^+$) (bis-NAD$^+$) *Biochem. Soc. Trans.* **12**, 433.
9. Irwin, J. A. and Tipton, K. F. (1995) Resolution of lactate dehydrogenase isoforms by affinity precipitation. *Biochem. Soc. Trans.* **23**, 365S.
10. Beattie, R. E., Buchanan, M., and Tipton, K. F. (1987). The synthesis of N_2,N_2'-adipodihydrazido-bis-(N^6-carboxymethyl-ATP) and its use in the purification of phosphofructokinase. *Biochem. Soc. Trans.* **15**, 1043–1044.
11. Larsson, P.-O. and Mosbach, K. (1981). Novel affinity techniques. *Biochem. Soc. Trans.* **9**, 285–287.
12. Feinstein, A. and Rowe, A. J. (1965) Molecular mechanism of formation of an antigen–antibody complex. *Nature* **205**, 147–149.
13. O'Carra, P. (1978). Theory and practice of affinity chromatography, in *Chromatography of Synthetic and Biological Polymers,* vol. 2. (Epton, R., ed.), Ellis Horwood, for the Chemical Society, London, pp. 131–158.
14. Irwin, J. A. and Tipton K. F. (1996) Affinity precipitation of dehydrogenases with bis-NADH derivatives. *Biochem. Soc. Trans.* **24**, 11S.
15. Hayet, M and Vijayalakshmi, M. A. (1986). Affinity precipitation of proteins using bis-dyes. *J. Chromatogr.* **376,** 157–161.
16. Lowe, C.R. and Pearson, J.C. (1983). Bio-mimetic dyes, in *Affinity Chromatography and Biological Recognition* (Chaiken, I. M., Wilchek, M., and Parikh, I., eds.), Academic, London, pp. 421–432.
17. Pearson, J. C., Burton, S. J., and Lowe, C. R. (1986). Affinity precipitation of lactate dehydrogenase with a triazine dye derivative: selective precipitation of rabbit muscle lactate dehydrogenase with a Procion Blue H-B analog. *Anal. Biochem.* **158**, 382–389.
18. Pearson, J. C., Clonis, Y. D., and Lowe, C. R. (1989) Preparative affinity preparation of L-lactate dehydrogenase. *J. Biotechnol.* **11**, 267–274.
19 Lilius, G., Persson, M., Bülow, L., and Mosbach, K. (1991) Metal affinityprecipitation of proteins carrying genetically attached polyhistidine affinity tails. *Eur. J. Biochem.* **198**, 499–504.

20. Van Dam, M. E., Wuenschell, G. E., and Arnold, F. H. (1989). Metal affinity precipitation of proteins. *Biotechnol. Appl. Biochem.* **11**, 492–502.
21. Schneider, M., Guillot, C., and Lamy, B. (1981) The affinity precipitation technique. Application to the isolation and purification of trypsin from bovine pancreas. *Ann. NY Acad. Sci.* **369**, 257–263.
22. Senstad, C. and Mattiasson, B. (1989) Affinity-precipitation using chitosan as ligand carrier. *Biotechnol. Bioeng.* **33**, 216–220.
23. Senstad, C. and Mattiasson, B. (1989) Purification of wheat germ agglutinin using affinity flocculation with chitosan and a subsequent centrifugation or flotation step. *Biotechnol. Bioeng.* **34**, 387–393.
24. Kamihira, M., Kaul, R., and Mattiasson, B. (1992). Purification of recombinant protein A by aqueous two-phase extraction integrated with affinity precipitation. *Biotechnol. Bioeng.* **40**, 1381–1387.
25. Bradshaw, A. P. and Sturgeon, R. J. (1990) The synthesis of soluble polymer–ligand complexes for affinity precipitation studies. *Biotechnol. Techn.* **4**, 67–71
26. Senstad, C. and Mattiasson, B. (1989). Preparation of soluble affinity complexes by a second affinity interaction: a model study. *Biotechnol. Appl. Biochem.* **11**, 41–48.
27. Linné, E., Garg, N., Kaul, R., and Mattiasson, B. (1992). Evaluation of alginate as a ligand carrier in affinity precipitation. *Biotechnol. Appl. Biochem.* **16**, 48–56.
28. Gupta, M. N., Dong, G. Q., and Matiasson, B. (1993). Purification of endo-polygalacturonase by affinity precipitation using alginate. *Biotechnol. Appl. Biochem.* **18**, 321–328.
29. Chen, J. P. and Hoffman, A. S. (1990). Polymer–protein conjugates. II. Affinity precipitation separation of human immunogammaglobulin by a poly (*N*-isopropylacrylamide)–protein A conjugate. *Biomaterials* **11**, 631–634.
30. Eggert, M., Baltes, T., Garret-Flaudy F., and Freitag, R. (1998) Affinity precipitation—an alternative to fluidized bed adsorption? *J. Chromatogr. A* **827**, 269–280.
31. Garret-Flaudy, F. and Freitag, R. (2001) Use of the avidin (imino)biotin system as a general approach to affinity precipitation. *Biotechnol. Bioeng.* **71**, 223–234.
32. Kumar, A., Galaev, I. Y., and Mattiasson, B. (1998) Metal chelate affinity precipitation: a new approach to protein purification. *Bioseparation* **7**, 185–194.
33. Graham, L. D., Griffin, T. O., Beatty, R. E., McCarthy, A. D., and Tipton, K. F. (1985). Purification of liver glutamate dehydrogenase by affinity precipitation and studies on its denaturation. *Biochim. Biophys. Acta.* **828**, 266–269.
34. Bergmeyer, H. U., Graß, M., and Walter, H.-E. (1983), in *Methods of Enzymatic Analysis*, Vol. 2, 3rd ed. (Bergmeyer, H. U., Bergmeyer, J., and Graß, M., eds.), Verlag Chemie, Weinheim, pp. 126–328.
35. LaemmLi, U. K (1970) Cleavage of structural proteins during the assembly of the head of bacteriophage T4. *Nature* **227**, 680–685.
36. Mosbach, K., Larsson, P.-O., and Lowe, C. (1976) Immobilized coenzymes. *Methods Enzymol.* **44**, 859–887.
37. Engel, J. D. (1975) Mechanism of the Dimroth rearrangement in adenine. *Biochem. Biophys. Res. Commun.* **64**, 581–585.
38. Buchanan, M. (1988) The synthesis of N_2,N_2'-adipodihydrazido-bis-(N^6-carboxymethyl-NAD$^+$) and N_2,N_2'-adipodihydrazido-bis-(N^6-carboxymethyl-ATP) and subsequent affinity precipitation of enzymes. M.Sc. thesis, University of Dublin.
39. Beattie, R. E. (1984) The synthesis of N_2,N_2'-adipodihydrazido-bis-(N^6-carboxymethyl-NAD$^+$) and its use in the purification of dehydrogenases. M.Sc. thesis, University of Dublin.
40. Bückmann, A. F. (1987) A new synthesis of coenzymically active water-soluble macromolecular NAD and NADP derivatives. *Biocatalysis* **1**, 173–186.

41. Bückmann, A. F. and Wray, V. (1992) A simplified procedure for the synthesis and purification of N^6-(2-aminoethyl)-NAD and tricyclic 1,N^6-ethanoadenine NAD. *Biotechnol. Appl. Biochem.* **15**, 303–310.

42. Butler, P. J. G. and Thelwall Jones, G. M. (1970) The preparation of alcohol dehydrogenase and glyceraldehyde-3-phosphate dehydrogenase from baker's yeast. *Biochem. J.* **118**, 375–378.

43. McCarthy, A. D., Walker, J. M., and Tipton, K. F. (1980) Purification of glutamate dehydrogenase from ox brain and liver. Evidence that commercially available preparations of the enzyme have suffered proteolytic cleavage. *Biochem. J.* **191**, 605–611.

44. Phelps, C. (1984), in *Techniques in the Life Sciences*: Volume B1/1 Supplement, BS 104, Protein and Enzyme Biochemistry (Tipton, K. F., ed.), Elsevier, Dublin, pp 1–16.

45. Buchanan, M., O'Dea, C. D., Griffin, T. O., and Tipton, K. F. (1989) Reversible crosslinking of alcohol and lactate dehydrogenases with the bifunctional reagent N_2, N_2'-adipodihydrazido-bis-(N^6-carboxymethyl-NAD$^+$). *Biochem. Soc. Trans.* **17**, 422.

24

Isoelectric Focusing

Reiner Westermeier

1. Introduction

Isoelectric focusing (IEF) is performed in a pH gradient in an electric field. The charged proteins migrate toward the anode or the cathode—according to the sign of their net charge—until they reach the position in the pH gradient where their net charges are zero. This pH value is the isoelectric point (pI) of the substance, an exactly defined physicochemical constant. Because the molecule is no longer charged, it stays there; the electric field does not have any influence on it. Should the protein diffuse away, it will gain a net charge and the applied electric field will cause it to migrate back to its pI. This concentrating effect leads to the name "focusing" and makes the method very useful for purification purposes.

The principle of isoelectric focusing is employed in varying technical approaches. The most complete information about these methods is found in the works of Righetti *(1,2)* and Andrews *(3)*.

The procedure described here is performed in a horizontal flat bed of a granulated dextran gel. Large sample volumes can be mixed with the original gel slurry from which the gel bed is prepared. Labile samples are applied at a defined zone of the gradient. The separation is run at a constant power of 8 W for 14–16 h at controlled temperature. pH measurements and prints on filter paper can be made directly on the gel surface. The sample fractions are collected by sectioning the gel and then eluting the proteins from these sections. The technique, introduced by Radola in 1969 *(4)*, has been selected for several reasons:

1. It can be performed with standard horizontal electrophoresis equipment, which is also used for analytical methods.
2. The method has a high loading capacity (up to gram quantities).
3. A large number of different applications and references are available for this technique; some examples are listed in **refs. 5–15**.
4. The technique is much less sensitive to precipitation of proteins at their isoelectric points compared to preparative isoelectric focusing methods in a free liquid because the precipitate is trapped within the gel bed.
5. The recovery of the protein fractions from a granulated dextran gel is much easier and gives a higher yield compared to compact gel media like agarose and polyacrylamide.

From: *Methods in Molecular Biology, vol. 244: Protein Purification Protocols: Second Edition*
Edited by: P. Cutler © Humana Press Inc., Totowa, NJ

The original method has been refined and optimized by Winter et al. *(16)*.

2. Materials

2.1. Equipment

In addition to standard laboratory equipment, the following are needed:

1. Horizontal electrophoresis chamber (Multiphor II) and Preparative IEF kit (Pharmacia Biotech, Uppsala, Sweden). The Preparative IEF Kit contains a tray with a 5-mm silicone rim, a fractionating grid frame with 20 fractionation blades, a sample applicator, sample elution columns, IEF electrode strips, and print papers.
2. Thermostatic circulator at 10°C.
3. Constant power supply (>1 kV).

2.2. Consumables, Chemicals, and Solutions

1. Desalting columns for sample preparation: PD-10 columns prepacked with Sephadex G-25 (Pharmacia Biotech, product code 17-0851-01).
2. IEF electrode strips (clean filter paper strips approx 5 × 2 mm).
3. Print paper (clean filter paper approx 110 × 250 cm).
4. Granulated gel media with a very low level of charged contaminants: Ultrodex® (Pharmacia Biotech, product code 80-1130-01) (*see* **Note 1**).
5. Carrier ampholytes of wide pH range (e.g., pH 3.5–9.5) and narrow pH ranges (2–3 pH units) depending on the p*I*'s of the proteins to be purified: Ampholine® and Pharmalytes® (Pharmacia Biotech).
6. Detergent solution (0.1% [v/v] Triton X-100): 1 mL Triton X-100 made up to 1 L with distilled water.
7. Anode solution (1 *M* phosphoric acid): 2.8 mL Concentrated H_3PO_4 made up to 50 mL with distilled water.
8. Cathode solution (1 *M* sodium hydroxide): 2 g NaOH made up to 50 mL with distilled water.
9. Fixing solution (10% [w/v] trichloroacetic acid): 100 g Trichloroacetic acid made up to 1 L with distilled water.
10. Destaining solution (10% [v/v] methanol/20% [v/v] acetic acid): Mix 100 mL methanol with 200 mL acetic acid and make up to 1 L with distilled water.
11. Staining solution (0.2% [w/v] Coomassie blue): Dissolve 0.6 g Coomassie blue R 250 in 300 mL destaining solution.

3. Methods

3.1. Gel Preparation

When large volumes of stable sample solution have to be separated, the sample is included in the initial gel slurry. Labile samples are applied in a certain zone of the gel bed with a sample applicator.

3.1.1. Preparing the Carrier Ampholyte Solution

Prepare a total of 110 mL of a 2% (w/v) carrier ampholyte solution (100 mL for the gel bed and 10 mL for the electrode strips) (*see* **Note 2**). The ampholytes are diluted in distilled water or with sample solution if large volumes of sample are to be separated (*see* **Subheading 3.1.**). The pH range of the carrier ampholytes is selected so that the p*I*(s) of the protein(s) to be purified fall(s) approximately in the middle of the range. This pH value can be identified by a previous analytical experiment *(17)* or from information in the literature. A large number of proteins are listed with their isoelectric points

Table 1
Volumes for the Carrier Ampholyte Solution

Ampholyte	pH	pH range (mL)				
		2.5–5.0	2.5–5.0	5.0–7.5	6.0–8.5	7.0–10.0
Pharmalyte	2.5–5.0	5.5				
Ampholine	3.5–5.0		2.75			
Ampholine	4.0–6.0		2.75			
Ampholine	5.0–8.0			5.5		
Ampholine	6.0–8.0				2.75	
Ampholine	7.0–9.0				2.75	2.75
Ampholine	9.0–11.0					2.75
Distilled water (+ sample)		104.5	104.5	104.5	104.5	104.5
Total volume		110	110	110	110	110

in two tables that have been published by Righetti et al. *(18,19)*. Some carrier ampholyte solution recipes are listed in **Table 1**.

3.1.2. Preparing the Gel Bed (see **Note 3**)

1. Clean the glass plate of the tray thoroughly before use in order to avoid uneven spreading of the gel layer.
2. Cut six IEF electrode strips to 10.5 cm and soak them in 10 mL of the carrier ampholyte solution. Place three layers of strips at each end of the tray, weigh the tray, and place it on a horizontal table.
3. Weigh out 4 g of Ultrodex and sprinkle it in small portions over 100 mL of carrier ampholyte solution in a 200-mL beaker. Let the gel sink down into the liquid before adding the next portion.
4. Weigh the beaker and its contents. Homogenize the suspension by gently stirring with a spatula (do not use a magnetic stirrer). When a basic gradient range is selected, degas the slurry with a water jet pump or comparable equipment, in order to remove the atmospheric carbon dioxide.
5. Pour the suspension into the tray (*see* **Fig. 1**). Gently tap against the ends of the tray, so that the suspension spreads evenly.
6. Weigh the beaker and the remainder of the slurry.

3.1.3. Evaporating the Gel Bed to the Correct Water Content

1. Mount a small fan 70 cm above the tray and evaporate excess water with a light stream of air. The speed and distance of the fan must be adjusted so that no ripples are formed on the surface.
2. Control the weight of the tray and its content from time to time. After 1.5–2 h, it should have reached the final weight, which was calculated from the weight difference of the beaker, the weight of the tray with the strips, and the evaporation limit (*see* **Notes 4** and **5**).

3.1.4. Arranging the Tray on the Cooling Plate

1. Apply 2 mL of a 0.1% (v/v) Triton X-100 solution on the cooling plate of the Multiphor electrophoresis chamber, which has already been cooled to 10°C.
2. Transfer the tray onto the cooling plate avoiding air bubbles.
3. Place an electrode strip soaked in anode solution at the anodal side and another strip soaked in cathode solution at the cathodal side, each on top of the strips already in the tray. Cathodal and anodal sides are marked on the cooling plate.

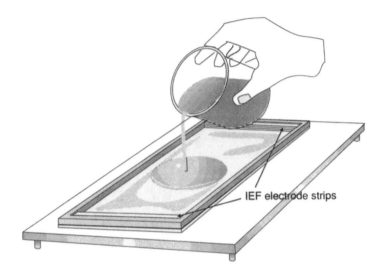

Fig. 1. Pouring the gel slurry into the tray.

4. Cut off the protruding parts so that the electrode strips fit exactly into the tray. No gaps must exist between the ends of the strips and the silicone rubber of the tray.

3.2. Sample Application

3.2.1. Preparing the Sample

1. The sample should not contain any particles or grease. For isoelectric focusing, salt and buffer concentrations over 50 mM should be avoided. Ideally, it should not be higher than 10 mM. The easiest and quickest way to desalt a sample is to perform gel filtration with a PD-10 column or another column packed with Sephadex G-25.
2. The maximum loading capacity is 5–10 mg of protein mixture/mL gel bed (2–4 mg/mL of one single protein) when a narrow pH range is employed (e.g., pH 4.0–6.0). The wider the pH range, the lower the loading capacity.

3.2.2. Loading the Sample

The sample can be already mixed with the initial slurry as described in **Subheading 3.1**. Labile samples are applied with the sample applicator as a narrow zone in the gel bed as follows (*see* **Note 6**):

1. The sample solution should have a volume of 3 mL. Otherwise, dilute it with the remainder of the carrier ampholyte solution, which was used for soaking the electrode strips.
2. Press the sample applicator through the gel bed at the desired position (*see* **Fig. 2**). Scrape off the gel in the applicator, and mix it with the sample in a small beaker.
3. Pour the sample into the applicator, remove the applicator, and allow the the bed to equilibrate for a few minutes.

3.3. Isoelectric Focusing

The resulting separation pattern is dependent on the temperature; conventionally, 10°C is used (*see* **Notes 7** and **8**).

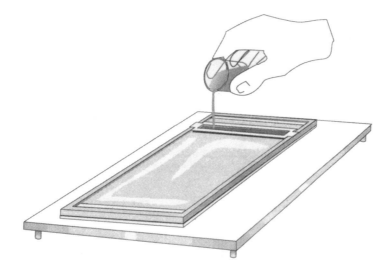

Fig. 2. Pouring the sample into the applicator.

1. Place the platinum electrodes on the respective electrode strips.
2. Perform electrofocusing at a constant current of 8 W for 14–16 h (overnight). The maximal current is 25 mA, and the maximal voltage is 1.5 kV.

3.4. Detection of Protein Zones

1. Mark the anode (+) on a sheet of print paper.
2. Carefully roll the paper onto the gel surface avoiding air bubbles. Leave it for 1 min.
3. Carefully remove the print. Apply the electric field again to the gel during the staining of the print in order to avoid diffusion of the zones.
4. Treat the print as follows:
   a. Three times washing out of the carrier ampholytes in fixing solution for 15 min.
   b. Staining for 10 min in staining solution (*see* **Note 9**).
   c. Destaining for 1 h in destaining solution (*see* **Notes 10–12**).
   d. Dry the print with hot air.

If colored protein fractions are separated, no staining is necessary. **Figure 3** shows the stained paper print of a hemoglobin separation from a pH gradient of 5.0–7.5.

3.5. Measurement of the pH Gradient

1. Press the fractionating grid onto the gel, which is still on the cooling plate at 10°C (*see* **Fig. 4**).
2. Measure the pH gradient by inserting a surface electrode into every second compartment of the 20 fractions of the grid. Wait until the reading has stabilized. For exact results, the electrode has to be calibrated at 10°C. Always start at the cathodal side of the gradient because carbon dioxide from the air will slowly diffuse into the gel surface and the HCO_3^- ions will shift the measured pH to a lower value.

3.6. Collection of the Zones

1. Place the print near the tray to locate the zones of interest.
2. Scrape out relevant bands and transfer to an elution column with a spatula.

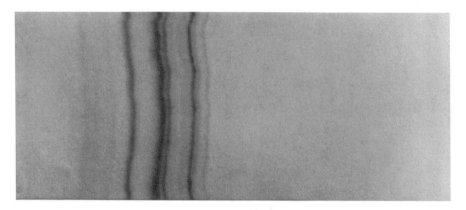

Fig. 3. Paper print of focused hemoglobin bands from a preparative bed of granulated gel with a pH gradient of 6.9–8.5. (Reproduced with permission from Pharmacia Biotech AB.)

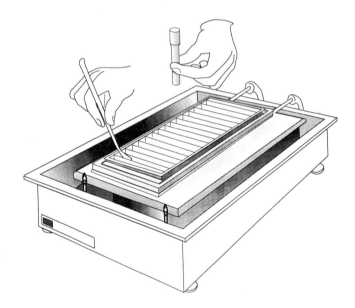

Fig. 4. Scraping the protein sections from the gel using the fractionation grid.

3. Resuspend the gel in the column with a suitable buffer and leave to settle until all buffer has entered the gel.
4. Add one gel volume of buffer and elute the protein from the gel. Alternatively, all fractions can be collected with the help of the fractionating grid (*see* **Fig. 4**) and then transferred to the elution columns.

3.7. Separation of the Protein From the Carrier Ampholytes

In most cases, it is not necessary to remove the carrier ampholytes from the protein fractions for the following reasons:

1. They do not interfere with activity measurements and binding studies.
2. They are not toxic.
3. The fraction can be directly injected for antibody production.

If necessary, several methods can be employed for separating the protein from the carrier ampholytes, such as gel filtration (Chapter 11), ultrafiltration (Chapter 12), dialysis (Chapter 11), ammonium sulfate precipitation (Chapter 10), and electrophoresis (Chapters 34 and 35).

4. Notes

1. Ultrodex is a specially prepared dextran gel with very weak electroendosmotic effects. Untreated Sephadex G-75 or G-100 superfine gels may contain varying amounts of charged, water-soluble contaminants that interfere with the formation of the pH gradient. These gels have to be washed with 10 vol of distilled water and filtered under gentle vacuum.
2. Do not use < 2% (w/v) carrier ampholytes. Otherwise distortions and uneven bands appear in the gel during focusing. Using a higher concentration of carrier ampholytes generally does not improve the separation or the loading capacity.
3. The water content of the final gel bed is very important. If it is too low, the gel cracks. If it is too high, the focused proteins sediment to the lower part of the gel bed. The optimal consistency of the gel bed is controlled by weighing the initial slurry and repeated weighing during an evaporation until a certain percentage of the original weight is reached. This percentage is calculated using the defined evaporation limit, which is indicated on each Ultrodex bottle. For example, a typical evaporation limit might be 34%. Suppose that the weights of the beaker plus slurry before and after filling the tray were 162.6 and 60.3 g, respectively (i.e., 102.3 g of slurry were added to the tray). The evaporation limit is 34% of this (i.e., 34.8 g) and the target weight of slurry is 67.5 g. If the initial weight of the tray and strips was 145.1 g, then the desired final weight after evaporation would be 145.1 + 67.5 = 212.6 g.
4. Alternative to the weighing procedure, the water content of the gel bed can be adjusted in the following way. Cover the mouth of the bottle of Ultrodex with a fine gauze net and fix it with an elastic rubber ring. Sprinkle the dry material onto the surface until the correct consistency has been achieved. This is controlled by tilting the tray to an angle of 45°; when the gel surface does not move, the water content is correct.
5. If the gel cracks during the run in spite of all the precautions described, check the cooling efficiency of the system; the water flow should be 4–12 L/min at 4–10°C.
6. In the case of very labile proteins, sample application should be performed after a prefocusing step of 30 min at a constant 8 W. In this way, the sample can be loaded at a defined optimal pH value that can be predetermined from a previous analytical experiment. Often, it is very helpful to check the literature to find out the optimal place of sample application.
7. When oxygen-sensitive enzymes or very basic proteins are being separated, flush the chamber with a light stream of nitrogen during the focusing experiment.
8. Some enzymes have to be separated at a lower temperature to maintain their biological activity (e.g., at 4°C). pH measurement is then almost impossible, because the electrode would respond extremely slowly at this low temperature. The pH values must always be measured at the focusing temperature because the pK_a values of the carrier ampholytes and of the proteins are highly dependent on temperature.
9. Alternatively or additionally to general protein staining of the paper print, specific staining for glycoproteins or zymogram techniques can be employed. The print can be cut into several strips and stained using different methods.
10. If skew bands are produced, then either the salt content of the sample could be too high or there could be too much electrolyte in the IEF electrode strips. Desalt the sample and/or blot the IEF electrode strips on a filter paper before placing them into the tray.

11. Irregular bands can also appear owing to overloading in a zone; reduce sample concentration or choose a narrower pH range.
12. In some cases, band distortions can be cured by adding nonionic detergents (e.g., 0.5% [v/v] Triton X-100), 10% [v/v] glycerol, 7% [v/v] monoethylene glycol, or 3–7 M urea) to the the gel and the sample.

References

1. Righetti, P. G. (1983) *Isoelectric Focusing: Theory, Methodology and Applications.* Elsevier Biomedical, Amsterdam.
2. Righetti, P. G. (1990) Immobilized pH Gradients: Theory and Methodology. Elsevier, Amsterdam.
3. Andrews, A. T. (1986) Electrophoresis: Theory, Techniques and Biochemical and Clinical Applications. 2nd ed., Clarendon, Oxford.
4. Radola, B. J. (1969) Thin-layer isoelectric focusing of proteins. *Biochim. Biophys. Acta* **194,** 335–338.
5. Delincee, H. and Radola, B. J. (1970) Thin-layer isoelectric focusing on Sephadex layers of horseradish peroxidase. *Biochim. Biophys. Acta* **200,** 404–407.
6. Delincee, H. and Radola, B. J. (1970) Some size and charge properties of tomato pectin methylesterase. *Biochim. Biophys. Acta* **214,** 178–189.
7. Radola, B. J. (1970) Thin-layer isoelectric focusing of sarcoplasmic proteins from irradiated meat. *Rep. Eur.* **4695,** 73.
8. Gordon, B. J. and Dykes, P. J. (1972) Alpha₁-acute-phase globulins of rats: microheterogeneity after isoelectric focusing. *Biochem. J.* **130,** 95.
9. La Gow, J. and Parkhurst, L. J. (1972) Kinetics of carbon monoxide and oxygen binding for eight electrophoretic components of sperm-whale myoglobin. *Biochemistry* **24,** 4520–4525.
10. Coffer, A. J. and King, J. B. (1974) Isoelectric focusing of ³H oestradiol-17s receptor in flat beds of Sephadex. *Biochem. Soc. Trans.* **2,** 1269–1272.
11. Thorpe, R. and Robinson, D. (1975) Isoelectric focusing of isoenzymes of human liver alpha-L-fucosidase. *FEBS Lett.* **54,** 89–92.
12. Parkhurst, L. J. and La Gow, J. (1975) Kinetic and equilibrium studies of the ligand binding reactions of eight electrophoretic components of sperm-whale ferrimyoglobin. *Biochemistry* **14,** 1200–1205.
13. Steinmeier, R. C. and Parkhurst, L. J. (1975) Kinetic studies on the five principal components of normal adult human hemoglobin. *Biochemistry* **14,** 1564–1572.
14. Vacca, C. V., Hall, P. W., III, and Crowley, A. Q. (1979) Preparative electrofocusing in the isolation and purification of human betas microglobulin, in *Electrofocus 78* (Haglund, H., Westerfeld, J., and Ball, J., Jr., eds.), Elsevier/North-Holland, Amsterdam, pp. 49–51.
15. Bours, J., Garbers, M., and Hockwin, O. (1980) Preparative flat-bed isoelectric focusing of LDH, MDH, SDH, and G-6-PDH from rabbit lens, in *Electrophoresis 79* (Radola, B. J., ed.), de Gruyter, Berlin, pp. 539–544.
16. Winter, A., Perlmutter, H., and Davies, H. (1980) Preparative flat-bed electrofocusing in a granulated gel with the LKB 2117 Multiphor. *LKB Application Note* 198. Amersham Biosciences, Upsalla, Sweden.
17. Westermeier, R. (1993) *Electrophoresis in Practice.* VCH, Weinheim, Germany.
18. Righetti, P. G. and Caravaggio, T. (1976) Isoelectric points and molecular weights of proteins. A table. *J. Chromatogr.* **127,** 1–28.
19. Righetti, P. G., Tudor, G., and Ek, K. (1981) Isoelectric points and molecular weights of proteins. A new table. *J. Chromatogr.* **220,** 115–194.

25

Chromatofocusing

Timothy J. Mantle and Patricia Noone

1. Introduction

Chromatofocusing *(1)* is essentially ion-exchange chromatography (conventionally using an anion exchanger) where the elution conditions are obtained by dropping the pH so that the proteins elute in order of their isoelectric points (p*I*). Although in principle this can be achieved by equilibrating an ion exchanger at a relatively alkaline pH and then titrating the resin with a buffer at a lower, more acidic pH, in practice it is more usual to utilize amphoteric buffers titrated to the lower pH to generate a more linear pH gradient. Because the pH gradient is generated gradually as the eluting buffer moves down the column, it is possible to add large volumes to the column or to add a second batch during the run and, in either case, single identical components with a particular p*I* will focus and elute as a single peak.

In practice, this method is extremely straightforward because no gradient makers are required and the separation is normally superior to ion-exchange chromatography.

The protocol described here allows the purification of five major forms of pig liver glutathione-*S*-transferase (GST) by chromatofocusing on polybuffer exchanger PBE 94 (*see* **Note 1**) following the preparation of a GST affinity pool using *S*-hexylglutathione–Sepharose. It is typical of the method used for proteins with p*I* values in the range 6.5–8.5 (*see* **Note 2** for variations in the protocol for proteins with p*I* values outside this range).

2. Materials

2.1. Buffers

Chromatofocusing buffers are available commercially as polybuffers (Pharmacia LKB, Uppsala, Sweden). This protocol uses Polybuffer 96 (*see* **Note 1**), which has a characteristic low absorbance at 280 nm but can be detected at 254 nm. For this reason eluent from a chromatofocusing column must be measured at 280 nm. All other reagents are made up in distilled deionized water at 20°C, unless otherwise indicated.

1. Buffer A: 10% (v/v) Polybuffer 96, pH 6.0 (adjusted with glacial acetic acid).
2. Buffer B: 0.025 *M* ethanolamine, pH 9.4 (adjusted with glacial acetic acid).
3. Buffer C: 0.1 *M* sodium phosphate, pH 6.5.

From: *Methods in Molecular Biology, vol. 244: Protein Purification Protocols: Second Edition*
Edited by: P. Cutler © Humana Press Inc., Totowa, NJ

4. Buffer D: 0.01 *M* Sodium phosphate, pH 7.2.
5. Buffer E: 0.5% (w/v) sodium dodecyl sulfate (SDS), 0.006% (w/v) bromophenol blue, 50% (v/v) glycerol, 63 m*M* Tris-HCl, pH 6.8, 0.5% (v/v) of 2-mercaptoethanol added immediately prior to use.

2.2. Resins

Pack a 25 × 1.5-cm-diameter column (total bed volume 44.3 mL, total capacity 3.5 meq/100 mL of resin) with PBE 94 (*see* **Note 2**).

2.3. Electrophoresis

For SDS polyacrylamide gel electrophoresis (PAGE), Analar-grade chemicals are used with the buffer system of Laemmli *(2)*. A 12% resolving gel was used in the present case.

3. Method

1. Pre-equilibrate the gel to be used (PBE 94) with start buffer (buffer B) and pack into the column.
2. Wash the column with buffer B until the pH of the eluent is the same as that of buffer B (approx 10 column volumes).
3. Precede the application of the sample by the application of 5 mL of the eluent buffer; this ensures the protein sample is never exposed to an extreme of pH.
4. Apply the sample at a flow rate of 1 mL/min (*see* **Note 3**).
5. Elute by the application of the eluent buffer to the column. The volume of the elution buffer depends on the strength of the eluent solution. For pH 6.0–9.0, an elution buffer volume of 10.5 column volumes is recommended (*see* **Note 4**).
6. Assay the fractions for the protein of interest (in this case GST; *see* **Note 5**).
7. Monitor the protein concentration by measuring the absorbance at 280 nm. It is assumed that an absorbance reading of 1 is equivalent to a protein concentration of 1 mg/mL *(3)*. A typical elution profile is shown in **Fig. 1**.
8. Prepare the samples for electrophoresis by adding buffer D to give a final protein concentration of 0.1 mg/mL (*see* **Note 6**).
9. Add buffer E (5 μL) to 20 μL of each sample, mix, and boil on a heating block for 5 min to ensure complete inactivation of proteases and total protein denaturation by SDS.
10. Load the samples (10 μL) on the gel using a Hamilton syringe. Run the gel at 30 mA until the dye front has run off the gel. An example is shown in **Fig. 2**.
11. Remove polybuffer, if required, by extensive dialysis against buffer D or by gel filtration on Sephadex G-25 (*see* **Note 7** and Chapter 11).

4. Notes

1. For most work, all that is required is the PBE 94 and the two amphoteric buffers, Polybuffer 96 and Polybuffer 74, which cover the pH ranges 9.0–6.0 and 7.0–4.0, respectively. If the protein of interest has a more basic p*I*, then it will be necesssary to invest additionally in PBE 118 and Pharmalyte, pH 8.0–10.5. All of these materials can be obtained from Pharmacia. Sophisticated columns are available commercially together with a range of peristaltic pumps and on-line monitors for A_{280} and pH. However, we have always used successfully a long straight glass column with a sinter that allows a good flow rate and gravity.
2. For most separations, a bed volume of approx 20 mL is sufficient for samples containing up to 200 mg of protein; however, this is only a guide and may require adjusting. If the p*I* of the target protein is known (from isoelectric focusing), then **Table 1** should be consulted

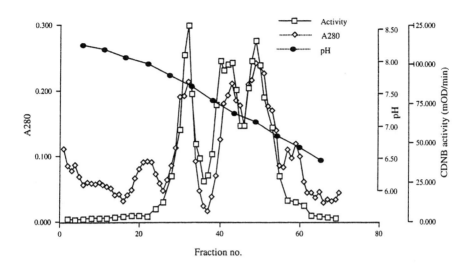

Fig. 1. Chromatofocusing of hepatic porcine GSTs, pH 6.0–9.0.

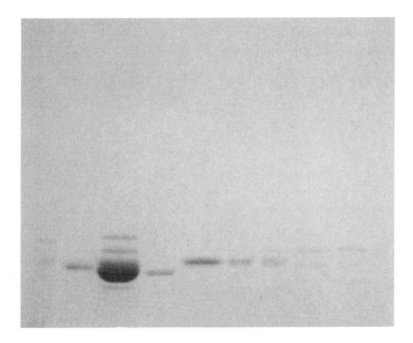

Fig. 2. Chromatofocusing of hepatic porcine GSTs, pH 6.0–9.0 analyzed by SDS-PAGE. Cytosolic GSTs were applied to a chromatofocusing column as described in **Subheading 3**. The peaks from the column were applied to a 12% SDS-PAGE gel and run as described: *Lane 1*, molecular-weight markers; *lane 2*, purified pig GST pi; *lane 3*, mixture of cytosolic GSTs as purified on *S*-hexylglutathione–Sepharose; *lane 4*, flowthrough from the chromatofocusing column; *lanes 5–9*, protein and activity peaks from the chromatofocusing column.

Table 1
Buffers Used in Chromatofocusing[a]

pH Range	Start buffer	Eluent	Dilution factor
10.5–8.0[b]	pH 11.0, 0.25 *M* triethylamine/HCl	pH 8.0, Pharmalyte pH 8.0–10.5/HCl	1:45
9.0–6.0	pH 9.4, 0.025 *M* ethanolamine/HCl	pH 6.0, Polybuffer 96/acetic acid	1:10
7.0–4.0	pH 7.4, 0.025 *M* imadazole/HCl	pH 4.0, Polybuffer 74/HCl	1:8

[a]The gradient volume should be 10 times the column volume in all cases.
[b]The exchanger used in this case is PBE 118; in the other cases, it is PBE 94 (*see* **Note 1**).

for a suitable start buffer. If there is no information on the pI of the target protein, then it is probably sensible to start with PBE 94 and use the pH range 7.0–4.0 (Polybuffer 74). If the target protein does not bind, then try using the pH range 9.0–7.0 (Polybuffer 96). By using these two sets of conditions (pH range 9.0–4.0), the pI of at least 85% of all known proteins (*4*) will have been covered. If the target protein is still not binding to the resin, then it is a very basic protein and PBE 118 should be used. When the choice of exchanger and start buffer has been made, equilibration with start buffer should be commenced. This can take place in the column or, more rapidly, in a sintered glass funnel. In either case, equilibration usually requires 10–15 bed volumes of start buffer. Equilibration should always be checked so that the pH and conductivity of the eluent is identical to that of the start buffer.

3. The sample volume should normally be within 0.5 bed volumes, although the sample volume is unimportant as long as all of the sample is applied to the column before the component of interest has eluted. The sample should not contain large amounts of salt and the ionic strength should be <0.05 *M*. If the buffer concentration of the sample is low, then the pH of the sample is unimportant; however, it is normal to load the sample in either start buffer or eluent buffer, whichever maintains the activity of the component of interest. The loading and elution sequence is as follows:
   a. 5 mL of eluent buffer
   b. Sample (in eluent or start buffer)
   c. Eluent buffer

4. Elution is achieved simply by running 10 column volumes of eluent buffer through the column. The flow rate can be variable and may be regulated by a peristaltic pump or by gravity. Protein should be monitored at 280 nm and the pH of each fraction measured immediately. There is a dead volume of 1.5–2 column volumes, so, in practice, it is usual to use 12 column volumes.

5. The assay conditions for the activity of GST have been described previously (*3*). Reactions are monitored by following a change in absorbance when substrate is conjugated to glutathione.

   Assays are carried out in a final volume of 2 mL in a quartz cuvet at 30°C. The absorbance change at 340 nm is recorded with a Philips (Eindhoven, Holland) PU8625 UV/VIS spectrophotometer attached to a Philips PM 8261 Xt recorder. Glutathione (GSH; 40 m*M* stock solution) is prepared freshly in buffer C. 1-Chloro-2,4-dinitrobenzene (CDNB; 40 m*M*), the electrophilic substrate, is dissolved in ethanol. Buffer C (1.8 mL) incubated at 30°C is placed in the cuvet, glutathione (50 µL) and CDNB (50 µL) are then added. A blank rate is observed before eluent fraction (100 µL) is added.

   Alternatively the assays are carried out in a microwell plate (Nunclon, Kamstrup, Denmark) with a final volume of 200 µL. The absorbance change is monitored at 340 nm (30°C) using a Molecular Devices Thermomax plate reader attached to a Macintosh SE 1/40. The results are analyzed using the Softmax® software. Buffer C (140 µL) is applied to each well followed by protein (10 µL) except for the blanks where buffer C (10 µL) is added. CDNB (0.5 mL) and GSH (0.5 mL) are added to buffer C (9 mL). The reaction mix (50 µL) is added to each well and the absorbance change is observed for 2 min at 6-s intervals. This second method is particularly useful if only small quantities of protein are obtained.

6. If the protein concentration is <0.1 mg/mL, the sample can be concentrated using trichloro-acetic acid (TCA). The desired amount of protein solution is mixed with an equal volume of ice-cold TCA (25% [w/v]). The samples are incubated on ice for 20 min and spun in a microcentrifuge at 12,000g for 3 min. The supernatant is discarded and the pellet is washed twice in ice-cold acetone (1 mL) and allowed to dry. The pellet is resuspended in buffer D (20 µL).

7. There are a number of ways that removal of polybuffer from the sample can be achieved:
   a. Precipitation with ammonium sulfate. Add sufficient solid ammonium sulfate to obtain 85% saturation (608 g/L). After 1 h, precipate the suspension using centrifugation, wash twice with saturated ammonium sulfate, and then store as an ammonium sulfate suspension or dialyse/gel filter to remove the ammonium sulfate. All of these steps should be conducted at 4°C.
   b. Polybuffer can be removed from most proteins by gel filtration providing that the protein of interest has an M_r >15,000.
   c. If the protein of interest binds to an affinity matrix, then this is the most ideal way of obtaining rapid and complete separation.

8. The following cautions should be noted:
   a. In the absence of any data on the pI of the protein of interest, it is well worth a number of preliminary experiments to optimize the binding and eluting conditions.
   b. Care should be taken with regard to the buffer counterion that should have a pK_a at least 2.0 pH units below the lowest point of the chosen gradient. For this reason, chloride is the counterion of choice.
   c. Atmospheric carbon dioxide may casuse a plateau in the region pH 5.5–6.5 and if this is a problem all buffers should be degassed.

References

1. Sluyterman, L. A. and Widnes, J. (1977) Chromatofocussing: isoelectric focussing on ion exchangers in the absence of an externally applied potential, in *Proceedings of the International Symposium Electrofocusing and Isotachophoresis* (Radola, B. J. and Graesslin, D., eds.), deGruyter, Berlin, pp. 463–466.
2. Laemmli, U. K. (1970) Cleavage of structural proteins during the assembly of the head of bacteriophage T4. *Nature* **227,** 680–685.
3. Habig W. H., Pabst, M. J., and Jakoby, W. B. (1974) Glutathione *S*-transferase: the first enzymatic step in mercapturic acid formation. *J. Biol. Chem.* **249,** 7130–7139.
4. Gianazza, E. and Righetti, P. G. (1980) Size and charge distribution of macromolecules in living systems. *J. Chromatogr.* **193,** 1–8.

26

Size-Exclusion Chromatography

Paul Cutler

1. Introduction

Size-exclusion chromatography (also known as gel filtration chromatography) is a technique for separating proteins and other biological macromolecules on the basis of molecular size. Originally developed in the 1950s, the technique was developed using crosslinked dextran *(1,2)*. Some confusion over nomenclature has been created by the term gel permeation, used to describe separation by the same principle in organic mobile phases using synthetic matrices *(3)*. It is now generally agreed that the terms gel filtration and gel permeation do not accurately reflect the nature of the separation. Size-exclusion chromatography (SEC) has been widely accepted as a universal description of the technique and in line with the IUPAC nomenclature, this term will be adopted *(4)*. Size-exclusion chromatography is a commonly used technique, because of the diversity of the molecular weights of proteins in biological tissues and extracts. SEC has been employed for many roles, including buffer exchange (desalting), removal of nonprotein contaminants (DNA, viruses) *(5)*, protein aggregate removal *(6)*, the study of biological interactions *(7–9)*, and protein folding *(10–12)*. It also has the important advantage of being compatible with physiological conditions *(13,14)*.

The solid-phase matrix consists of porous beads (100–250 μm) that are packed into a column with a mobile liquid phase flowing through the column (*see* **Fig. 1**). The mobile phase has access to both the volume inside the pores and the volume external to the beads. The high porosity typically leads to a total liquid volume of >95% of the packed column.

Unlike many other chromatographic procedures, size exclusion is not an adsorption technique. Separation can be visualized as reversible partitioning into the two liquid volumes *(15)*. The elution time is dependent on an individual protein's ability to access the pores of the matrix. Large molecules remain in the volume external to the beads, as they are unable to enter the pores. The resulting shorter flow path means that they pass through the column relatively rapidly emerging early. Proteins that are excluded from the pores completely elute in what is designated the void volume, V_0. This is often determined experimentally by the use of a high-molecular-weight component such as Blue Dextran or calf thymus DNA. Small molecules which can access the liquid within the

From: *Methods in Molecular Biology, vol. 244: Protein Purification Protocols: Second Edition*
Edited by: P. Cutler © Humana Press Inc., Totowa, NJ

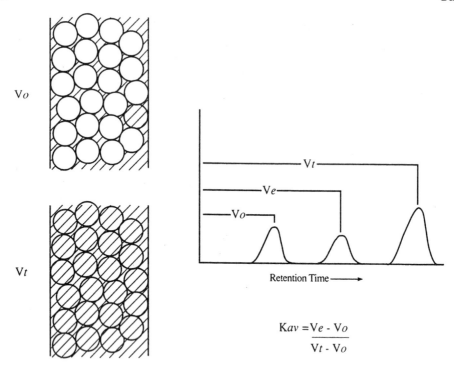

Fig. 1. The basic principle of size exclusion. Solutes are separated according to their molecular size. Large molecules are eluted in the void volume (V_0) and small molecules are eluted in the total volume (V_t). Solutes within the separation range of the matrix are fractionally excluded with a characteristic elution volume (V_e).

pores of the beads are retained longer and therefore pass more slowly through the column. The elution volume for material included in the pores is designated the total volume, V_t. This represents the total liquid volume of the column and is often determined by small molecules such as vitamin B_{12}.

The elution volume for a given protein will lie between V_0 and V_t and is designated the elution volume, V_e (*see* **Note 1**). Intermediate-sized proteins will be fractionally excluded with a characteristic value for V_e. A partition coefficient can be determined for each protein as K_{av}. K_{av} is not a true partition coefficient (K_d) because $V_t - V_0$ includes the volume occupied by the solid matrix component, which is inaccessible to all solutes. However, for any given matrix, the relationship between K_{av} (which is readily determined) and K_d is constant. The K_{av} of a protein does not directly relate to the molecular weight (M_r), but to the hydrodynamic volume or radius of gyration (R). Although there is a relationship between K_{av} and the log of the molecular weight *(16)*, the value of R is also dependent on the tertiary structure of the protein.

Size exclusion tends to be used at the end of a purification scheme when impurities are low in number and the target protein has been purified and concentrated by earlier chromatography steps. An exception to this is membrane proteins, where gel filtration may be used first. In this case concentration techniques are not readily used and the material will be progressively diluted during the purification scheme, making the application of size exclusion early in the scheme advantageous *(17)*.

A range of different preparative and analytical matrices are available commercially. High-performance columns are used analytically for studying protein purity, protein folding, protein–protein interactions, and so forth *(18)*. Preparative separations performed on low-pressure matrices are used to resolve proteins from proteins of different molecular weight, proteins from other biological macromolecules, and for the separation of aggregated proteins from monomers *(19)*. Size exclusion is particularly suited for the resolution of protein aggregates from monomers (*see* **Note 2**).

Several parameters are important in size-exclusion chromatography. The pore diameter controlling the separation is selected for the relative size of proteins to be separated. Many types of matrix are available. Some are used for desalting techniques, where proteins are separated from buffer salts (*see* **ref. *20*** and **Note 3**).

Some matrices offer a wide range of molecular-weight separations and others are high-resolution matrices with a narrow range of operation (*see* **Table 1**).

2. Materials

2.1. Equipment

The preparative separation of proteins by size exclusion is suited to commercially available standard low-pressure chromatography systems. Systems require a column packed with a matrix offering a suitable fractionation range, a method for mobile-phase delivery, a detector to monitor the eluting proteins, a chart recorder for viewing the detector response, and a fraction collector for recovery of eluted proteins (*see* **Fig. 2**). The system should be plumbed with capillary tubing with a minimum hold up volume.

Early systems were less sophisticated with a gravimetric feed of the mobile phase from a suspended reservoir, whereas the most modern systems now have computers to control operating parameters and to collect and store data. The principle of separation, however, remains the same and high resolution is attainable with relatively simple equipment.

2.1.1. Pumps

An important factor in size exclusion is a reproducible and accurate flow rate. The most commonly used pumps are peristaltic pumps, which are relatively effective at low flow rates, inexpensive, and sanitizable. Peristaltic pumps do, however, create a pulsed flow and often a bubble trap is incorporated to both prevent air entering the system and to dampen the pulsing effect. More expensive yet more accurate pumps such as those incorporated in systems such as the Akta purification systems (Amersham Biosciences) and high-performance liquid chromatography (HPLC) systems offering low pulsation. Reciprocating piston pumps are accurate at low flow rates but are prone to pulsation.

2.1.2. Column

Size exclusion, unlike some commonly used adsorption methods of protein separation, is a true chromatography method based on continuous partitioning; hence, resolution is dependent on column length. Column length is a balance between resolution and run time. Preparative columns tend toward being long and thin, typically 70–100 cm long (*see* **Note 4**), although desalting columns tend to be shorter.

The tubing connecting the columns should be as narrow and as short as possible to avoid zone spreading. The column must be able to withstand the moderate pressures generated during operation and be resistant to the mobile phase. The use of columns with flow adapters is recommended to allow the packing volume to be varied and provide a finished support with the required minimum of dead space.

The column must be able to withstand the pressures generated during operation and

Table 1
Mobile Phases Used for Separation of Proteins[a]

Type	Mobile phase	pH Range	Comment
Aqueous			
Nonvolatile		7.2	The ionic strength should be kept within the range equivalent to 0.1–0.5 M NaCl.
	"Good" buffers[b]		Sulfates can lead to salting-out effects. Borates can complex with glucosidic functions on some matrices.
	50 mM Sodium phosphate, 0.15 M NaCl		Physiological.
	0.1 M Tris-HCl	7–9	Good solubility, particularly for RNA and DNA
Volatile	0.1 M Ammonium acetate	7–10	Volatile buffers are used for freeze-drying. They are also relatively compatible with on-line mass spectrometric techniques, but because of the low salt, it can lead to nonideal size-exclusion effects.
	0.1 M Ammonium bicarbonate (untitrated)	7.9	Good for proteins and DNA; should be made immediately prior to use.
	20% (v/v) Acetonitrile in 0.1% (v/v) trifluoracetic acid	2.0	Commonly used for Reverse phase HPLC, therefore compatible with further processing
Detergents	0.1% (w/v) SDS		Often as additions to the aqueous buffers above; can give erroneous M_r estimations.
			Strongly anionic; good solubilizing agent but denaturing; difficult to remove.
	0.2% (w/v) TritonX100		Nonionic; can be used to solubilize hydrophobic proteins while retaining activity.
Denaturing	6 M Guanidine HCl, 50 mM Tris-HCl	8.6	Good ultraviolet properties at 280 nm; proteins can often be renatured on dialysis; used to study folding.
	6–8 M Urea	<7.0	Urea is a good solubilizing agent but may lead to carbamylation of proteins/peptides.
	1 M Propionic acid	1.0	Good solubilizing agents for proteins and peptides.
	70% (v/v) Formic acid	1.4	Good for solubilizing peptides.
	50 mM HCl		
Organic	FACE (formic acid, acetic acid, chloroform, ethanol (1:1:2:1))		High level of glycerol, ethylene glycol, and so forth, can cause severe band spreading.

[a]It is important to consult manufacturer's details with regard to the chemical resistance of individual matrices.
[b]"Good" buffers are a range of zwitterionic buffers specifically developed for biological work and include MES, MOPS, HEPES, and CAPS.

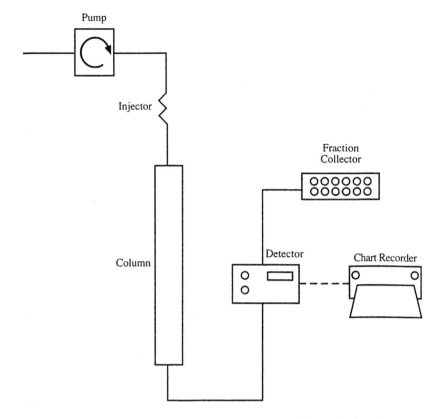

Fig. 2. Schematic diagram of the equipment for preparative size-exclusion chromatography.

be resistant to the mobile phase. Stability is an issue when operating with organic mobile phases, high salt, or denaturants such as guanidine hydrochloride. Stability to organics is generally in the order plastic < polypropylene < glass; however, particular attention should be given to seals and frits and manufacturers' guidelines should be observed. The use of columns with flow adapters is recommended to allow the packing volume to be varied and provide a finished support with the required minimum of dead space. Before use, all column parts should be cleaned in an appropriate solvent and the integrity of the support nets and seals confirmed.

Because of the increased efficiency of HPLC packings, analytical HPLC columns tend, typically, to be shorter than their preparative counterparts *(21)*. The advantage is excellent resolution in a short run time (10–15 min). Longer HPLC columns are, however, available for many matrices, and in our laboratory, we have achieved good results by placing two HPLC columns in series. HPLC columns are supplied prepacked and are best operated with a guard column (2–10 cm) to prevent column fouling.

2.1.3. Detectors

Protein elution is most often monitored by absorbance in the ultraviolet range, either at 280 nm, which is suitable for proteins with aromatic amino acids, or at 206 nm, which detects the peptide bond. Detection at lower wavelengths may be complicated by the absorbance characteristics of certain mobile phases. The advent of diode array detectors has enabled continuous detection at multiple wavelengths enabling characterization of the elutes via analysis of spectral data. The use of SEC in conjunction with on-line mass spectrometry has largely been restricted to synthetic polymers, in part because of the incompatability of common mobile phases with the mass spectrometers. The widespread introduction of matrix-assisted laser desorption time of flight (MALDI-TOF) mass spectrometry has facilitated accurate off-line mass determinations in a range of mobile phases.

Fluorescence detection either by direct detection of fluorescent tryptophan and tyrosine residues or after chemical derivitization (e.g., by fluorescein) have been used as have refractive index, radiochemical, electrochemical, and molecular size (by laser light scattering) *(9)*. In addition to these nonspecific on-line monitoring systems, it is quite common, particularly when purifying enzymes to make use of specific assays for individual target molecules

2.1.4. Fraction Collectors

A key factor in preparative protein purification is the ability to collect accurate fractions. No matter how efficiently the column may have separated the proteins, the accurate collection of fractions is critical. For the detector to reflect as near as possible in real time the fraction collector, the volume between the detector and the fraction collector should be minimal. Systems are available that can collect directly into a 96-well microtiter plate, which can prove highly convenient for assay screening.

2.2. Buffers

Size-exclusion matrices tend to be compatible with most aqueous buffer systems even in the presence of surfactants, reducing agents, or denaturing agents. Size-exclusion matrices are extremely stable, with effective pH ranges of approx 2–12 (*see* **Table 2**). An important exception to this are the silica-based matrices, which offer good mechanical rigidity but low chemical stability at alkaline pH's. Some silica matrices have been coated with dextran and so forth to increase the chemical stability and increase hydrophilicity.

The choice of mobile phase is most often dependent on protein stability necessitating considerations of appropriate pH, solvent composition as well as the presence or absence of cofactors, protease inhibitors, and so forth, which may be essential to maintain the structural and functional integrity of the target molecule. All buffers used in size exclusion should ideally be filtered through a 0.2-µm filter and degassed by low vacuum or sparging with an inert gas such as helium (*see* **Note 5**).

Many matrices retain a residual charge resulting, for example, from sulfate groups in agarose or carboxyl residues in dextran. The ionic strength of the buffer should be kept at 0.15–2.0 *M* to avoid electrostatic or van der Waals interactions, which can lead to nonideal size exclusion *(22,23)*. Crosslinking agents such as those used in polyacrylamide may reduce the hydrophilicity of the matrix, leading to the retention of some small proteins, particularly those rich in aromatic amino acid residues. These interactions have been exploited effectively to enhance purification in some cases but are generally best avoided (*see* **Note 6**).

2.3. Selection of Matrix

The beads used for size exclusion have a closely controlled pore size, with a high chemical and physical stability. They are hydrophilic and inert, to minimize chemical

Table 2
Commonly Used Preparative Size-Exclusion Matrices

Supplier	Matrix	Composition	Size Range[a]		Stability	
			Minimum	Maximum	pH	Organics
Low/medium pressure						
Pharmacia	Sephadex	Dextran	1,000–5,000	5,000–600,000	2.0–10.0	
	Sephacryl	Dextran/bis-acrylamide	1,000–100,000	500,000–>100,000,000	3.0–11.0	
	Sepharose CL	Agarose	10,000–4,000,000	70,000–40,000,000	3.0–13.0	
	Superose	Agarose	1,000–300,000	5,000–5,000,000	3.0–12.0	
	Superdex	Agarose/dextran	100–7,000	10,000–600,000	3.0–12.0	
BioRad	Bio-Gel P	Polyacrylamide	100–1,800	5,000–100,000	2.0–10.0	
	Bio-Gel A	Agarose	10,000–500,000	100,000–50,000,000	4.0–13.0	
Toso Haas	Toyopearl HW	Methacrylate	100–10,000	400,000–30,000,000	1.0–14.0	Yes
Amicon	Cellufine	Cellulose	100–3,000	10,000–3,000,000	1.0–14.0	
Biosepra	Ultrogel AcA	Acrylamide/agarose	1,000–15,000	100,000–1,200,000	3.0–10.0	
	Trisacryl GF	Acrylamide	300–7,500	10,000–15,000,000	1.0–11.0	
Merck	Fractogel TSK	Polymeric	100–10,000	500,000–50,000,000	1.0–14.0	
HPLC						
Toyo Soda	TSK-SW	Silica	5,000–150,000	20,000–10,000,000	2.0–7.0	
	TSK-PW	Polymeric	100–2,000	10,000–200,000	2.0–12.0	Yes?
BioRad	Bio-Sil SEC	Silica	5,000–100,000	20,000–1,000,000	2.0–8.0	Yes
Dupont	BioSeries GF	Zorbax®[c]	10,000–250,000	25,000–900,000	3.0–8.5	
Polymer Laboratories	PL-Gel	Polystyrene/divinylbenzene	0–2000	600,000–10,000,000	?	Yes
	PL-Aquagel-OH	Polymeric	100–30,000	200,000–10,000,000	2.0–10.0	
Phenomenex	BioSep-SEC-S	Bonded silica	1,000–200,000	20,000–3,000,000	2.5–7.5	Yes
	Polysep-GFC-P	Polymeric	100–2,000	10,000–10,000,000	3.0–12.0	
Synchrom	Synchropak GPC	Bonded silica	300–30,000	25,000–50,000,000	?	
Shodex	OHpak	Polyhydroxymethacrylate	100–1000	10,000–20,000,000	?	
	Asahipak	Polyinyl alcohol	100–3,000	10,000–10,000,000	2.0–9.0	Yes
Waters	Ultrahydrogel	Polyhydroxymethacrylate	100–5000	100,000–7,000,000	2.0–12.0	Yes
Macherey-Nagel	Nucleogel SFC	Polymeric	100–100,000	100,000–20,000,000	1.0–13.0	Yes

[a]Size range represents the maximum and minimum separation range for the class of matrices.
[b]Range is given in general as estimates for globular proteins in aqueous buffers as recommended by the manufacturer.
[c]Zorbax® is a trademark of Dupont.

Table 3
Molecular Weight Standards for Size-Exclusion Chromatography

Standard	Molecular weight
Vitamin B_{12}	1350
Glucagon	3550
Aprotinin	6500
Ribonuclease A	13,700
Myoglobin	17,000
Chymotrypsinogen A	25,000
Carbonic anhydrase	29,000
β-Lactoglobulin	37,000
Ovalbumin	43,000
Bovine serum albumin	67,000
Alcohol dehydrogenase	150,000
Bovine gamma globulin	160,000
β-Amylase	200,000
Apoferritin	443,000
Thyroglobulin	669,000
Blue Dextran 2000	approx 2,000,000

interactions between the solutes (proteins) and the matrix itself. The performance and resolution of the technique has been enhanced by the development of newer matrices with improved properties *(24–27)*. Historically, gels were based on starch, although these were superseded by the crosslinked dextran gels (e.g., Sephadex). In addition, polystyrene-based matrices were developed for the use of size exclusion in nonaqueous solutions. Polyacrylamide gels (e.g., Biogel P series) are particularly suited to separation at the lower molecular-weight range because of their microreticular structure. More recently, composite matrices have been developed, such as the Superdex gel (Pharmacia), in which dextran chains have been chemically bonded to a highly crosslinked agarose for high-speed size exclusion (*see* **Table 2**).

3. Methods

3.1. Flow Rates

In chromatography, flow rates should be standardized for columns of different dimensions by stating linear flow rate (cm/h). This is defined as the volumetric flow rate (cm^3/h) per unit cross-sectional area (cm^2) of a given column. As the principle of size exclusion is based on partitioning, success of the technique is particularly susceptible to variations in flow rates. Conventional low-pressure size-exclusion matrices tend to operate at linear flow rates of 5–15 cm/h. Too high a flow rate leads to incomplete partitioning and band spreading. Conversely, very low flow rates may lead to diffusion and band spreading.

3.2. Preparation and Packing

3.2.1. Preparation

Gel matrices are supplied as either preswollen gels or as dry powder. If the gel is supplied as a dry powder, it should be swollen in excess mobile phase as directed by the manufacturers. Swollen gels must be transferred to the appropriate mobile phase. This can be achieved by washing in a sintered glass funnel under low vacuum. During preparation, the gel should not be allowed to dry. The equilibrated gel should be decanted in to a Buchner flask, allowed to settle, and then fines removed from the top by decanting. The equilibrated gel (approx 75% slurry) is then degassed under low vacuum.

3.2.2. Packing

A good column packing is an essential prerequisite for efficient resolution in size-exclusion chromatography. The column should be held vertically in a retort stand, avoiding adverse draughts, direct sunlight, or changes in temperature. The gel should be equilibrated and packed at the final operating temperature. With the bottom frit or flow adaptor in place, degassed buffer (5–10% of the bed volume) is poured down the column side to remove any air from the system. In one manipulation, the degassed gel slurry is poured in to the column using a glass rod to direct the gel down the side of the column, avoiding air entrapment. If available, a packing reservoir should be fitted to columns to facilitate easier packing. The column can be packed under gravity, although a more efficient method is to use a pump to push buffer through the packing matrix. The flow rate during packing should be approx 50% higher than the operating flow rate (e.g., 15 cm/h for a 10-cm/h final flow rate). The bed height can be monitored by careful inspection of the column, as it is packing. Once packed, a clear layer of buffer will appear above the bed and the level of the gel will remain constant. The flow should then be stopped and excess buffer removed from the top of the column, leaving approx 2 cm of the buffer above the gel. The outlet from the column should be closed and the top flow adaptor carefully placed on top of the gel avoiding trapped air or disturbing the gel bed.

3.3. Equilibration

The packed column should be equilibrated by passing the final buffer through the column at the packing flow rate for at least 1 column volume. The pump should always be connected so as to pump the eluent on to the column under positive pressure. Drawing buffer through the column under negative pressure may lead to bubbles forming as a result of the suction. The effluent of the column should be sampled and tested for pH and conductivity in order to establish equilibration in the desired buffer.

3.4. Sample Application

Several methods exist for sample application. It is critical to deliver the sample to the top of the column as a narrow sample zone. This can be achieved by manually loading via a syringe directly on to the column, although this requires skill and practice. The material may be applied through a peristaltic pump, although this will inevitably lead to band spreading because of sample dilution. The sample should never be loaded through a pump with a large holdup volume such as a syringe pump or upstream of a bubble trap. Arguably, the best method of applying the sample is via a sample loop in conjunction with a switching valve, allowing the sample to be manually or electronically diverted through the loop and directly onto the top of the column.

3.5. Evaluation of Column Packing

The partitioning process occurs as the bulk flow of liquid moves down the column. As the sample is loaded, it forms a sample band on the column. In considering the effi-

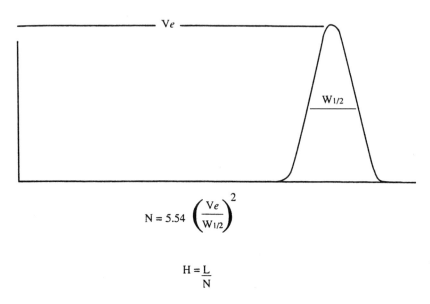

$$N = 5.54 \left(\frac{V_e}{W_{1/2}}\right)^2$$

$$H = \frac{L}{N}$$

Fig. 3. Calculation of the theoretical plate height as a measure of column performance. The number of theoretical plates (N) is related to the peak width at half-height ($W_{1/2}$) and the elution volume (V_e). The height of the theoretical plate (H) is related to the column length (L).

ciency of the column, partitioning can be perceived as occurring in discrete zones along the axis of the columns length. Each zone is referred to as a theoretical plate and the length of the zone is termed the theoretical plate height (H). The value of H is a function of the physical properties of the column, the exclusion limit, and the operating conditions such as flow rate and so forth. Column efficiency is defined by the number of theoretical plates (N). The highest resolving matrices have high values for N and correspondingly low values for H. In practice, the value of N can be measured using a suitable ultraviolet (UV) absorbing material such as 1% (v/v) acetone monitored at 280 nm. Other aromatic compounds can be used, such as tyrosine or benzyl alcohol (*see* **Fig. 3**). Resolution and, hence, the number of theoretical plates is enhanced by increasing column length (*see* **Note 7**).

3.6. Standards and Calibration

Calibration is obtained by use of standard proteins and plotting retention time (V_e) against log molecular weight. Successful calibration requires accurate flow rates. The resultant plot gives a sigmoidal curve approaching linearity in the effective separation range of the gel.

It should be noted that it is not only the molecular weight that is important in size exclusion but the hydrodynamic volume or the Stokes radius of the molecule. Globular proteins appear to have a lower molecular size than proteins with a similar molecular weight, which are in an α-helical form. These, in turn, appear smaller than proteins in random coil form. It is critical to use the appropriate standards for size exclusion where proteins have a similar shape. This has been useful in some cases for studying protein folding and unfolding. Commonly used molecular-weight standards are given in **Table 3**.

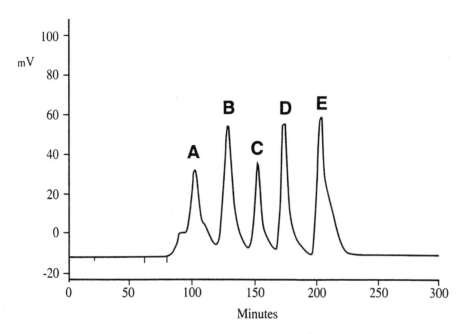

Fig. 4. Size exclusion of five molecular-weight markers on Superdex 200 (Pharmacia) column (1.6 × 75 cm). Thyroglubin, 670,000 (**A**); γ-globulin, 158,000 (**B**); ovalbumin, 44,000 (**C**); myoglobin, 17,000 (**D**), and vitamin B$_{12}$, 1350 (**D**).

3.7. Separation of Proteins

The optimum load of size-exclusion columns is restricted to <5% (typically 2%) of the column volume in order to maximize resolution. Gel filtration columns are often loaded at relatively high concentrations of protein such as 2–20 mg/mL. The concentration is limited by solubility of the protein and the potential for increased viscosity, which begins to have a detrimental effect on resolution. This becomes evident around 50 mg/mL. It is important to remove any insoluble matter prior to loading by either centrifugation or filtration. Because of the limitation on loading, it is often wise to consider the ability of the method for scale-up when optimizing the operating parameters (*see* **ref. 28** and **Fig. 4**).

3.8. Column Cleaning and Storage

Size-exclusion matrices can be cleaned *in situ* or as loose gel in a sintered glass funnel. Suppliers usually offer specific guidelines for cleaning gels. Common general cleaning agents include nonionic detergents (e.g., 1% [v/v] Triton X-100) for lipids and 0.2–0.5 *M* NaOH for proteins and pyrogens (not recommended for silica-based matrices). In extreme circumstances, contaminating protein can be removed by use of enzymic digestion (pepsin for proteins and nucleases for RNA and DNA). The gel should be stored in a buffer with antimicrobial activity such as 20% (v/v) ethanol or 0.02–0.05% (w/v) sodium azide. NaOH is a good storage agent that combines good solubilizing activity with prevention of endotoxin formation. It may, however, lead to chemical breakdown of certain matrices.

4. Notes

1. If the protein elutes before the void volume ($V_e < V_0$), this suggests chaneling through the column because of improper packing or operation of the column. If the protein elutes after the total volume ($V_e > V_t$), then some interaction must have occured between the matrix and the protein of interest.

2. Size exclusion is particularly suited for the resolution of protein aggregates from monomers. Aggregates are often formed as a result of the purification procedures used. Size exclusion is often incorporated as a final polishing step to remove aggregates and act as a buffer-exchange mechanism into the final solution.

3. Desalting gels are used to rapidly remove low-molecular-weight material such as chemical reagents from proteins and for buffer exchange. As the molecules to be separated are generally very small, typically less than 1 kDa, gels are generally used with an exclusion limit of approx 2–5 kDa. The protein appears in the void volume (V_0) and the reagents and buffer salts are retained. Because of the distinct molecular-weight differences, the columns are shorter than other size-exclusion columns and operated at higher flow rates (e.g., 30 cm/h).

4. In some instances, the length of the column required to obtain a satisfactory separation exceeds that which can be packed into a commerically available column (>1 m). In these cases, columns can be packed in series. The tubing connecting the columns should be as narrow and as short as possible to avoid zone spreading.

5. The majority of protein separations performed using size exclusion are carried out in the presence of aqueous phase buffers. Size exclusion of proteins in organic phases (sometimes called gel permeation) is not normally undertaken but is sometimes used for membrane protein separations. The agarose and dextran-based matrices are not suitable for separations with organic solvents. Synthetic polymers and, to a certain extent, silica are suitable for separations in organic phases. For separation in acidified organic solvents, polyacrylamide matrices are particularly suitable. The separation of the proteins may be influenced by the denatured state of the protein in the organic phase. The equipment must be compatable with the solvent system (e.g., glass or Teflon).

6. Interactions with the matrix are commonly seen when charged proteins are being resolved. Protein–matrix interactions can be minimized by the use of ionic strength in excess of 0.1 *M*. If low ionic strength is necessary, then the risk of interaction can be reduced by manipulating the charge on the protein via the pH of the buffer. This is best achieved by keeping the mobile phase above or below the p*I* of the protein as appropriate. Protein–matrix interaction is a common cause of protein loss during size exclusion. This may be the result of complete retention on the column or retardation sufficient for the material to elute in an extremely broad dilute fraction, thus evading detection above the baseline of the buffer system. Another important consideration is when enzymes are detected off-line by activity assay. The active enzyme may resolve in to inactive subunits. In such cases, a review of the mobile phase is advisable.

7. In addition to the determination of the number of theoretical plates (*N*) described, the performance of the column can be assessed qualitatively by the shape of the eluted peaks (*see* **Fig. 5**). The theoretically ideal peak is sharp and triangular with an axis of symmetry around the apex. Deviations from this are seen in practice. Some peak shapes are diagnostic of particular problems that lead to broadening and poor resolution. If the downslope of the peak is significantly shallow, it is possible that the concentration of the load was too high or the material has disturbed the equilibrium between the mobile and stationary phases. If the downslope tends to symmetrical intially but then becomes shallow, it is common to assume that there is a poorly resolved component; however, it may be suspected that interaction with the matrix is taking place. A shallow upslope may represent insolubility of the loaded

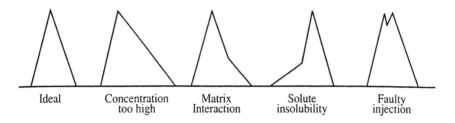

| Ideal | Concentration too high | Matrix Interaction | Solute insolubility | Faulty injection |

Fig. 5. Diagnosis of column performance by consideration of peak shape.

material. A valley betwen two closely eluting peaks may suggest poor resolution, but it can also be the result of a faulty sample injection.

References

1. Lathe, G. H. and Ruthven, C. R. (1956) Separation of substances and estimation of their relative molecular sizes by the use of columns of starch in water. *Biochem. J.* **62,** 665–674.
2. Porath, J. and Flodin, P. (1959) Gel filtration: a method for desalting and group separation *Nature* **183,** 1657–1659.
3. Moore, J. C. (1964) Gel permeation chromatography. I. New method for molecular-weight distribution of high polymers *J. Polym. Sci.* **A2,** 835–843.
4. Ettre, L. S. (1993) Nomenclature for chromatography. *Pure Appl. Chem.* **65,** 819–872.
5. Burnouf, T. (1995) Chromatography in plasma fractionation: benefits and future trends. *J. Chromatogr.* **664,** 3–15.
6. Katakam, M. and Banga, A. K. (1995) Aggregation of insulin and its prevention by carbohydrate excipients. *J. Pharm. Sci. Technol.* **49,** 160–165.
7. Sebille, B. (1990) The measurement of interactions involving proteins by size exclusion chromatography. *Chromatogr. Sci.* **51,** 585–621.
8. Beeckmans, S. (1999) Chromatographic methods to study protein–protein interactions. *Methods* **19,** 278–305.
9. Wen, J., Arakawa, T., and Philo, J. S. (1996) Size exclusion chromatography with on-line light scattering, absorbance and refractive index detectors for studying proteins and their interactions. *Anal. Biochem.* **240,** 155–166.
10. Uversky, V. N. (1993) Use of fast protein size-exclusion liquid chromatography to study the unfolding of proteins which denature through the molten globule. *Biochemistry* **32(48),** 13,288–13,298.
11. Uversky, V. N. (1994) Gel-permeation chromatography as a unique instrument for quantitative and qualitative analysis of protein denaturation and unfolding. *Int. J. Biochromatogr.* **1(2),** 103–114.
12. Ackland, C. E., Berendt, W. G., Freeza, J. E., Landgraf, B. E., Pritchard, K. W., and Ciardelli, T. C. (1991) Monitoring of protein conformation by high-performance size-exclusion liquid chromatography and scanning diode array second-derivative UV absorption spectroscopy. *J. Chromatogr.* **540,** 187–198.
13. Laurent, T. C. (1993) Chromatography classic: history of a theory. *J. Chromatogr.* **633,** 1–8.
14. Hagel, L. (1989) Gel filtration, in *Protein Purification* (Jansen, J.-C. and Ryden, L., eds.), VCH, New York, pp. 63–106.
15. Barth, H. G. (1998) Size exclusion chromatography. *Chromatogr. Sci.* **78,** 273–292.
16. Squire, P. G. (1985) Hydrodynamic characterization of random coil polymers by size exclusion chromatography. *Methods Enzymol.* **117,** 142–153.

17. Findlay, J. B. C. (1990) Purification of membrane proteins, in *Protein Purification Applications* (Harris, E. L. V. and Angal, S., eds.), IRL, Oxford, pp. 59–82.
18. Hagel, L. (1993) Size exclusion chromatography in an analytical perspective. *J. Chromatogr.* **648**, 19–25.
19. Stellwagen, E. (1990) Gel filtration. *Methods Enzymol.* **182**, 317–328.
20. Pohl, T. (1990) Concentration of proteins and removal of solutes. *Methods Enzymol.* **182**, 69–83.
21. Irvine, G. B. (1997) Size-exclusion high performance liquid chromatography of peptides—a review. *Anal. Chim. Acta.* **352**, 387–397.
22. Kopaciewicz, W. and Regnier, F. E. (1982) Non-ideal size exclusion chromatography of proteins: effects of pH at low ionic strength. *Anal. Biochem.* **126**, 8–16.
23. Dubin, P. L., Edwards, S. L., Mehta, M. S., and Tomalia, D. (1993) Quantitation of non-ideal behavior in protein size exclusion chromatography. *J. Chromatogr.* **635**, 51–60.
24. Lu, M. J. (1999) Preparation of beaded organic polymers and their applications in size exclusion chromatography, in *Column Handbook for Size Exclusion Chromatography,* Academic, New York, pp. 3–26.
25. Suzuki, H. and Mori, S. (1999) Shodex columns for size exclusion chromatography, in *Column Handbook for Size Exclusion Chromatography*, Academic, New York, pp. 171–217.
26. Alpert, A. J. (1999) Size exclusion high performance liquid chromatography of small solutes, in *Column Handbook for Size Exclusion Chromatography,* Academic, New York, pp. 249–266.
27. Meehan, E. (1999) Size exclusion chromatography columns from Polymer Laboratories, in *Column Handbook for Size Exclusion Chromatography,* Academic, New York, pp. 349–366.
28. Jansen, J. C. and Hedman, P. (1987) Large scale chromatography of proteins. *Adv. Biochem. Eng.* **25**, 43–97.

27

Fast Protein Liquid Chromatography

David Sheehan and Siobhan O'Sullivan

1. Introduction

High-resolution protein separation techniques critically depend on the availability of column packings of small average particle size. This gives a minimum of peak broadening on the column owing to the direct relationship between the theoretical plate height parameter, H, and particle size (lowest values of H give the highest resolution). High-performance liquid chromatography (HPLC) procedures exploit column packings with average diameters of as small as 5–40 μm. However, these are used in high-pressure systems (up to 400 bar) often with organic solvents and are generally limited to rather low sample loadings (*1*). To provide a more biocompatible high-resolution separation of biopolymers, including (although not exclusive to) proteins, Pharmacia (Uppsala, Sweden; now known as, Amersham Biosciences) developed fast protein liquid chromatography (FPLC) in 1982 (*2*). More recently, Regnier and his colleagues developed "perfusive" packings that contain large through-pores (*3*). These allow high-resolution chromatography with ion-exchange groups similar to those of Amersham Biosciences columns but at much faster flow rates, giving remarkably short separation times (e.g., 5–7 min). Perfusion chromatography columns are also compatible with the FPLC system.

Fast protein liquid chromatography provides a full range of chromatography modes, such as ion exchange, chromatofocusing (*4*), gel filtration, hydrophobic interaction (*5*), and reverse phase (*6*), based on particles with average diameter sizes in the same range as those used for HPLC separations. These columns can accommodate much higher protein loadings than conventional HPLC, however, and use a wide range of aqueous, biocompatible buffer systems. Although almost 2000 reports of the use of FPLC have appeared in the literature in the two decades since its introduction, two of the most popular FPLC modes are ion exchange and gel filtration. The objective of this chapter is to introduce the reader to these two FPLC chromatography modes as used in a typical protein chemistry laboratory. **References *4–6*** give examples of the use of the other main FPLC chromatography modes, which are not described in detail here.

2. Materials

1. The basic FPLC system consists of a program controller (LCC 500 Plus), two P-500 pumps (one each for buffers A and B), a mixer, prefilter, seven-port M-7 valve, assorted sample

From: *Methods in Molecular Biology, vol. 244: Protein Purification Protocols: Second Edition*
Edited by: P. Cutler © Humana Press Inc., Totowa, NJ

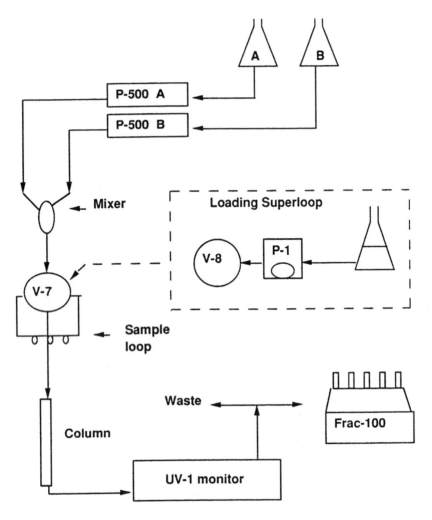

Fig. 1. Format used for FPLC chromatography. Information on other accessories and automation may be obtained from Pharmacia LKB. The superloop may be loaded using the arrangement shown within dashed lines

loops (0.025–10 mL), a column, a UV-1 ultraviolet (UV) monitor (fitted with an HR-10 flow cell and 280-nm filter), and a Frac-100 fraction collector. Associated equipment includes a superloop attached to an eight-port M-8 valve and a P-1 peristaltic pump for the introduction of samples up to 10 mL volume. All of this equipment is available from Amersham Biosciences. The most commonly used configuration for ion-exchange FPLC is shown in **Fig. 1**.

2. Mono Q HR5/5 anion-exchange, Mono S HR5/5 cation exchange, HR10/10 rapid desalting, and Superose 12 HR 10/30 gel filtration columns are obtained from Amersham Biosciences and are stored in 20% ethanol. Sephadex G-25 is also obtained from Amersham Biosciences. A glass column (45 × 3 cm in diameter) is required. Filters (0.22 μm) are from Millipore.

3. All reagents for buffer (*see* **Notes 1** and **2**) preparation are of analar grade. Buffers should be prepared in HPLC or milli-Q-grade water and require filtering and degassing immediately before use.

Buffer A: 10 mM Tris-HCl, pH 7.0.
Buffer B: 10 mM Tris-HCl, pH 7.0, 1 M NaCl.
Buffer C: 50 mM Tris-HCl, pH 7.0, 100 mM KCl.

3. Methods

3.1. Sample Preparation

Generally, FPLC chromatography is performed on material that has already been subjected to some preliminary chromatography steps (e.g., for ion exchange using cellulose matrices, *see* Chapter 14; for affinity chromatography, *see* Chapter 16). For ion-exchange chromatography, the sample must then be desalted into buffer A (*see* **Note 3**). This may be accomplished by passing it through a column of Sephadex G-25 (45 × 3 cm in diameter, total bed volume 318 mL).

1. New Sephadex G-25 resin is defined by aspiration before packing as a slurry in water into glass column with a no. 1 glass sinter.
2. The column is equilibrated in 3–4 column volumes of buffer A. Equilibration is complete when the pH and conductivity of the eluate are the same as buffer A at the same temperature. The sample (50 mL) is loaded under gravity flow. Buffer A is passed through the column. Protein is monitored by measuring A_{280} and collected as a single large peak.
3. Conductivity and pH of desalted protein should be the same as those of buffer A at the same temperature.

Sephadex G-25 gel filtration is performed at 4°C in a cold room or cabinet (*see* **Note 4**). The sample is centrifuged (bench centrifuge, 3–5 min) before application to the column.

3.2. Ion-Exchange FPLC

The FPLC system is operated usually at room temperature, taking care to hold samples on ice and to return eluted protein fractions to ice as soon as possible (*see* **Note 4**).

1. With no column in the system (this may be replaced with some tubing), prime P-500 pumps A and B with filtered (0.22-μm filter) and degassed buffers A and B, respectively.
2. Set pressure limits on both P-500 pumps well below the maximum for the column in use (*see* **Notes 5** and **6**).
3. Equilibrate Mono Q column (1-mL volume) with 5 vol buffer A and 10 vol buffer B followed by 5 vol buffer A.
4. Wash loading loop with buffer A.
5. Load 0.5–10 mL sample (approx 1 mg/mL) and wash with buffer A. Collect wash-through and assay for protein of interest.
6. If protein has not bound, immediately replace Mono Q with Mono S and repeat procedure described in **steps 3–5** (generally, at pH 7.0, proteins that do not bind to one resin will bind to the other). If the protein still does not bind, follow the procedure given in **Subheading 3.3**.
7. The mono Q (or Mono S) column is developed with a 0–100% gradient of buffer B. Elution of the protein is determined by A_{280} while the protein of interest is assayed.
8. Regenerate column by washing with 10 vol buffer B, followed by 5 vol buffer A. Column is now ready to receive another sample.

3.3. Scouting FPLC Methods

The aforementioned procedure describes a basic FPLC ion-exchange chromatography experiment. If the protein of interest has not previously been purified using FPLC, a new method of purification may be needed. Usually, 25- to 500-μL loadings are used to scout a new procedure. The general approach used is as follows.

1. By varying the % B at different time-points, create gradients of varying degrees of shallowness (down to and including isocratic). Knowing the % buffer at which the protein of interest elutes, separation from other proteins can be improved by varying the gradient program around this value *(7)*.
2. Having identified column (either Mono Q or Mono S) to which the protein of interest binds, carry out chromatography at different pH values using various buffer systems *(see* **Table 1**; *see* **Notes 1** and **2**). By carrying out rechromatography of a single sample at a number of different pH values on a Mono Q or Mono S column, this can often separate two proteins that may elute close to each other at a single pH.
3. Having identified a suitable buffer and gradient program for optimum separation of our protein of interest, the 500-μL sample loading loop may be replaced with the superloop. This allows loading of up to 10 mL (20 mg protein) of sample on the column. Fractions may be collected on the Frac-100 for further analysis.
4. Final adjustments to the gradient may be required to allow for the effect of scaling up the loading.

3.4. Gel Filtration FPLC

Gel filtration is often used as a final "polishing" step to remove minor contaminants from partially purified protein *(see* **Note 7**). It is also especially useful in determining native molecular weights for oligomeric proteins. Run times for samples may be as little as 30 min. Superose 12 has a fractionation range of 1–300 kDa. An identical procedure may be used for Superose 6 (fractionation range: 5–5000 kDa).

1. With no column in the system, prime both pumps A and B with buffer C.
2. Set pressure limits on both P-500 pumps to the same value (that for the column in use; *see* **Notes 5** and **6**).
3. Connect Superose 12 and operate system at a flow rate of 0.5 mL/min for 90 min to equilibrate column.
4. Wash sample loading loop with buffer C.
5. Load 200 μL sample (maximum) at approx 2 mg/mL.
6. Monitor A_{280}; collect peaks with Frac-100 fraction collector. Note elution volume of peak.

3.5. System Storage and Troubleshooting

When not in use the system should be stored in 20% ethanol to avoid microbial contamination *(see* **Note 8**). Performance and lifetime of system is aided by avoiding high-urea concentrations *(see* **Note 9**). Rechromatography usually improves purification greatly *(see* **Note 10**) while simultaneously loading pairs of standards speeds up calibration of gel filtration columns *(see* **Note 11**).

4. Notes

1. The purpose of using a fairly neutral pH to begin with is to allow assessment of both the anion- and cation-exchange columns at the same pH. It is much quicker to swap columns than to reprime the entire system with new buffers at different pH values. The procedure outlined in **Subheading 3.2.** takes less than 2 h to complete.
2. Although it is possible to carry out ion-exchange FPLC across a wide pH range, extremes of pH should be avoided if possible unless it is known that the protein of interest is stable at these pH values.

Table 1
Some Buffers Suitable for Ion-Exchange FPLC

pH Range	Buffer[a]	Column
3.8–4.3	50 mM Na formate-formic acid	Mono S
4.5–5.2	20 mM Na acetate-acetic acid	Mono S
5.5–6.0	20 mM Histidine-HCl	Mono Q
6.0–7.6	50 mM Na phosphate	Mono S/Mono Q
7.6–7.8	50 mM HEPES	Mono S
7.0–8.0	10 mM Tris-HCl	Mono Q
8.0–9.0	50 mM Tris-HCl	Mono Q
9.0–9.5	20 mM Ethanolamine-HCl	Mono Q

[a]These may be used as buffer A (*see* **Subheading 3.3.**).

3. If the protein of interest is stable, then desalting may also be achieved by dialysis against three changes of 1 L buffer A (minimum of 6 h/change). Centrifuge sample after dialysis to remove any precipitated material. If the protein is particularly unstable, then the rapid desalting HR 10/10 column may be included in the system before the ion-exchange column as shown in **Fig. 1**. This allows for desalting of sample in as little as 4 min.

4. Operation of the FPLC system in a cold room at 4°C may be essential for purification of some particularly labile proteins. However, permanent location in the cold may lead to deterioration of the system. A convenient approach to this problem is to mount the system on a trolley, with the components electrically connected to an extension lead on the trolley. This may be wheeled into the cold room, the extension lead connected to a power point, and the system operated once it has cooled to the temperature of the room. After use, the system may be easily wheeled out of the cold room again.

5. Flow rates of 1 mL/min are routinely achievable with newly purchased ion-exchange columns, but even with rigorous cleaning, back-pressure in the system soon rises appreciably after a number of uses. It is often better to accept a comparatively slow flow rate of 0.5 mL/min and modest back-pressures, because this seems to give better chromatography and to extend column lifetimes.

6. Decrease in flow rate usually improves the resolution of gel filtration chromatography.

7. Although gel filtration is often used toward the end of a purification procedure, in the particular case of immunoglobulin purification from serum, it may be used with advantage at an early stage of the purification *(8)*.

8. When FPLC is complete, wash the system with water. Fill pumps with 20% filtered, degassed ethanol, wash loading loop with 20% ethanol, and place arm of Frac-100 in the center of the carousel (to minimize risk of damage to optical sensor).

   Prime pumps with water to remove 20% ethanol in which the system is stored. This avoids any risk of NaCl precipitation on contact of buffer B with 20% ethanol.

9. It is sometimes desired to operate the system (gel filtration or ion exchange) in the presence of urea. It is not advisable to use high concentrations (e.g., 6–8 M) of urea though, because urea may easily precipitate, clogging valves and leading to excessive wear on seals and gaskets. Urea is rarely used at concentrations higher than 6 M, and special care has to be taken if the system is being operated at 4°C. It is essential when chromatography in the presence of urea is complete that the entire system be washed extensively with water before storing it in 20% ethanol. Otherwise, urea will precipitate on contact with the ethanol solution.

Also, it is important not to allow air (e.g., air bubbles) to come into contact with buffers containing urea, because this will lead to crystal formation. Once formed, crystals are very difficult to remove.

10. It is sometimes advisable to rechromatograph the peak of interest. This is achieved by rapidly desalting the peak into buffer A and reapplying it to the column. This often produces a considerable improvement in purification, because contaminants "move" to one side of the chromatogram away from the peak of interest.

11. Calibration of Superose columns should be performed whenever a different chromatography buffer is used. A quick way to achieve this is to load standard proteins with widely separated molecular masses (e.g., cytochrome-*c* and ovalbumin) in pairs (i.e., as a single sample).

References

1. Boyer, R. F. (1993) Separation and purification of biomolecules by chromatography, in *Modern Experimental Biochemistry,* Benjamin-Cummings, Redwood City, CA, pp. 90–102.
2. Richey, J. (1982) FPLC: a comprehensive separation technique for biopolymers. *Am. Lab.* **14,** 104–129.
3. Regnier, F. E. (1991) Perfusion chromatography. *Nature* **350,** 634–635.
4. Fagerstam, L. G., Lizana, J., Axio-Fredriksson, U.-B., and Wahlstrom, L. (1983) Fast chromatofocusing of human serum proteins with special reference to α_1-antitrypsin and G_c-globulin. *J. Chromatogr.* **266,** 523–532.
5. Fagerstam, L. G. and Pettersson, G. L. (1979) The cellulytic complex of *Trichoderma reesei* QM9414. An immunological approach. *FEBS Lett.* **98,** 363–367.
6. Jeppson, J.-O., Kallman, I., Lindgren, G., and Fagerstam, L. (1984) A new hemoglobin mutant characterised by reverse phase chromatography. *J. Chromatogr.* **297,** 31–36.
7. Fitzpatrick, P. J. and Sheehan, D. (1993) Separation of multiple forms of glutathione *S*-transferase from the blue mussel, *Mytilus edulis. Xenobiotica* **23,** 851–862.
8. Havarstein, L. S., Aasjord, P. M., Ness, S., and Endresen, G. (1988) Purification and partial characterisation of an IgM-like serum immunoglobulin from Atlantic salmon *(Salmo salar). Dev. Comp. Immunol.* **12,** 773–785.

28

Reversed-Phase Chromatography of Proteins

William A. Neville

1. Introduction

Modern liquid chromatography (LC), commonly referred to as high-performance liquid chromatography (HPLC), is over 30 yr old. Because of its age, many people might consider HPLC a mature technique and, as such, view the field as stagnant. However, nothing could be further from reality. The field of HPLC is changing rapidly although more subtly than it did in earlier years and it is interesting to note that today's trends are in essence driven by the same application benefits that were prevalent in the early years of LC (viz. improved quantitation, rapid determination of purity and structure, and preconcentration of samples). HPLC has gained wide acceptance in the modern analytical laboratory as well as in most other areas of science.

There are two broad areas in which HPLC is widely utilized: an application focus for applied uses and for investigative purposes. Forces with different objectives drive the trends in these two broad areas. In the applied area, scientists are generally looking to accomplish tasks more quickly and more productively. In addition to these goals, the investigative scientist wants the highest efficiencies available with an increased capability to accomplish the difficult task of analyzing increasingly small amounts of more complex and sometimes unknown samples. The investigative researcher will often trade off increased analysis time for increased efficiency.

Each of the application areas is influenced by a main objective that guides the work. In the applied area, HPLC methods are used to cost-effectively generate information to verify or confirm the expectations of the analyst. In the investigative area, HPLC activity is focused on providing definitive information. In the applied area, the desire is to confirm expectations as fast as possible and usually effort would be expended to develop an optimum validated method.

Obtaining definitive information usually takes precedence over the cost-effectiveness of the method because the specific work will be undertaken less frequently. For example, when a potential drug is discovered, the investigative work would focus on the isolation of the compound with the goal of identifying its structure. Replicate analyses would seldom be required. Once the compound is isolated, much of the applied work, the drug development process, confirms the presence of the compound based on less rig-

From: *Methods in Molecular Biology, vol. 244: Protein Purification Protocols: Second Edition*
Edited by: P. Cutler © Humana Press Inc., Totowa, NJ

orous information, such as retention time and ultraviolet (UV) spectra. This confirmation is based on a consistency of information, retention time match, spectral match and so forth. If the compound were present in a sample, it would thus elute at a specific retention time and have a specific UV-visible spectrum.

Reversed-phase HPLC (RP-HPLC) is now well established as a technique for isolation, analysis, and structural elucidation of peptides and proteins *(1,2)*. Its use in protein isolation and purification may have reached a peak as a result of recent developments in high-efficiency ion-exchange and hydrophobic interaction supports, which are now capable of equivalent levels of resolution to RP-HPLC without concomitant risk of denaturation and loss of biological activity. Nevertheless, there are many applications in which denaturation may be unimportant and high concentrations of organic modifier can be tolerated (e.g., purity analyses, structural studies, and micropreparative purification prior to microsequencing). It is a widely used tool in Biotechnology for process monitoring, purity studies, and stability determinations (*see* **Note 1**).

1.1. Principle of Separation

Application of a solvent gradient is generally superior to isocratic elution because separation is then achievable within a reasonable time-frame and peak broadening of later eluting peaks is minimized, thus increasing sensitivity. In gradient elution RP-HPLC, proteins are retained essentially according to their hydrophobic character. The retention mechanism can be considered either as adsorption of the solute at the hydrophobic stationary surface or as a partition between the mobile and stationary phases *(3)*.

In the first case, retention is related to total interfacial surface of the RP packing and is expressed by the adsorption coefficient K_A. Retention is based on a hydrophobic association between the solute and the hydrophobic ligands of the surface *(4)*. By increasing solvent strength of the mobile phase, attractive forces are weakened and the solute is eluted. The process can be regarded as being entropically driven and endothermic (i.e., both ΔS and ΔH are positive) *(5)*. Through the relation between the solute capacity factor k' and the mobile- and stationary-phase properties, solvophobic theory permits prediction of the effect of the organic modifier of the mobile phase, the ionic strength, the ion-pairing reagent, the type of ligand of the RP packing, and other variables in the chromatographic retention of proteins.

The second case assumes a partitioning of the solute between the mobile and stationary phases, the latter being regarded as a hydrophobic bulk phase. This system resembles an *n*-octanol/water two-phase system in which the selectivity is expressed by the partition coefficient P of the solute. Both retention principles have their pros and cons *(3)*. In essence, the effective process is largely dependent on the molecular organisation of the bonded *n*-alkyl chains of the RP packing in the solvated state and the size and conformation of the solute.

Several methods have been developed for correlating peptide structure with RP-HPLC retention. One method sums empirically derived retention coefficients, representing the hydrophobic contribution of each amino acid residue *(6)*. The assumptions behind this approach are that each amino acid residue contacts the adsorbent surface and that the total hydrophobicity of the solute determines RP-HPLC retention. This approach works well for peptides of 20 amino acids or less *(7)* but, generally, it is not accurate for predicting retention of larger proteins. This is caused by the peculiarities, especially of polar basic moieties, leading to either less retention or irreversible adsorption by residual matrix silanols.

Studies with peptides containing amphiphilic helices indicate that secondary structure plays a large role in determining the surface of a polypeptide that is exposed to the hydrophobic adsorbent *(8)*. Other studies have shown that proteins are less sensitive to changes in the HPLC support or bonded phase than are corresponding small molecules *(9)*. These results imply that protein retention in RP-HPLC is governed mainly by the protein surface and not by the support surface chemistry. Although some proteins show a loss of biological activity during RP-HPLC, others retain full biological activity. Thus, the hydrophobic forces necessary for RP binding compete with those required to maintain the protein's secondary and tertiary structures. Through steric, hydrophobic, and ionic constraints, these structural features define a chromatographic surface or "footprint" of the protein that, in turn, defines its RP binding. This concept of multipoint attachment of a protein to adsorbent surface is consistent with a relatively large chromatographic footprint for a protein compared with that of a small molecule.

1.2. RP-HPLC Columns

The reality of practical chromatography makes clear that no single column exists to separate all compounds. An appropriate choice of column matrix, mobile phases, and temperature of separation determines the optimization of chromatographic separation. RP-HPLC has rapidly become the most widely used tool for the separation, purification, and analysis of peptides. For proteins, RP-HPLC has suffered from problems associated with denaturation, including loss of activity, poor recoveries, wide and misshapen peaks, and ghost peaks *(10,11)*. Some of these problems may be column related *(12)* and insurmountable, although others can be overcome by selective optimization of extracolumn variables, such as sample pretreatment, mobile phase, and hardware considerations *(13)*.

Recent changes in Food and Drug Administration (FDA) rules simplify the regulatory treatment of "specified" biotechnology products. HPLC often plays a crucial role in characterizing biotechnology products as to purity, potency an identity. Keys to defining a "specified" product by HPLC are selectivity, stability, and reproducibility. Selectivity is the primary measure of column performance. To monitor biotechnology products for purity and identity, important impurities must be separated from the major product, and digest fragments with minor changes must be resolved from normal fragments. Resolving minor impurities such as deamidation products and oxidized methionine variants place the ultimate on HPLC column selectivity. Column stability is the second measure of column performance. Stable columns result in constant column selectivity and sample resolution over hundreds of sample injections, thus ensuring robust and reliable assays. HPLC column stability in polypeptide separations requires that columns retain their selectivity even under the somewhat harsh conditions of analysis (e.g., pH 2 or less). Batch-to-batch column reproducibility is the third measure of column performance. Column selectivity and sample resolution should remain the same when used columns are replaced with columns from a new batch.

As with small-molecule HPLC, there are a large number of RP-HPLC columns for use with peptides and proteins. In trying to select the best column for a new application or for broad applicability, one should consider such column variables as support, bonded

LIVERPOOL
JOHN MOORES UNIVERSITY
AVRIL ROBARTS LRC
TEL. 0151 231 4022

phase, pore size, particle size, and column dimensions. The choice of column packing is of key importance in liquid chromatography/mass spectrometry (LC/MS) method development. Factors to consider are the quality of the silica and the quality, functionality, and hydrophobicity of the bonded phase. High-purity silicas guarantee high performance with MS-compatible mobile phases and there are a wide range of functionalities available that provide plenty of alternatives for problematic separations. These factors are dealt with in the following headings.

1.2.1. Packing Support

Most commercial RP-HPLC columns for peptides and proteins are silica based, as a consequence of its use in the separation of small molecules over many years. Silica offers good mechanical stability and allows a wide range of selectivities by virtue of the bonding of various phases. Although it is well known that silica-based columns are not stable at basic pH, recent reports suggest that in the acidic mobile phases typically used, the bonded phase may also be slowly dissolved from the base silica. Degradation can thus affect not only the reproducibility, stability, and lifetime of a column but also recovery and, in addition, modify selectivity as the silica surface becomes uncovered. An added problem could be contamination of recovered products with silica and bonded-phase material. Classical drawbacks of RP-LC, such as asymmetrical peaks and irreproducible retention times, are caused by the "secondary" interaction of a basic nitrogen atom in an analyte with residual silanol groups on the silica surface of a packing material. Trace metals in the silica will increase this interaction as they increase the acidity of silanols. Optimizing the mobile phase to decrease this "secondary" interaction has limited capacity to solve this problem.

These problems have led to the development of very pure, low-trace-metal-content silicas (99.99% pure), which exhibit fewer acidic silanols. Nonporous silica (NPS) with particle sizes of 1.5 or 3.0 μm bonded with polymeric C_{18} is also being used to achieve high-efficiency separations. The highly uniform smooth/spherical particle and tight particle-size distribution provides consistent interstitial voids with no channeling, resulting in low-column dead volume, improved reproducibility, and higher stability. Short NPS C_{18} columns (4.6 mm inner diameter $\times$ 14 mm length) can be operated at quite high flow rate (e.g., 1–2 mL/min), without excessive back-pressures (<250 bar). However, columns packed with NPS material have the disadvantage of being easily overloaded with a sample at smaller injected amounts (mass) than totally porous particles because of the lower surface area of the NPS particle. Total carbon load of a NPS C_{18} column is approx 6%. These columns are compatible with conventional HPLC equipment, and with proper operation, they can be used successfully. In general, NPS columns provide better resolution than porous packings. It is advisable to operate with an integral 0.2-μm mobile-phase filter to avoid increasing back-pressure and eventual plugging of the column.

Because of the limitations associated with silica-based columns, polymer-based columns have gained increased popularity for peptide and protein separations by RP-HPLC. Synthetic polymer-based columns, such as divinylbenzene-crosslinked polystyrene are usually stable over the pH range 1.0–14.0, making them more widely applicable with the added advantage that they are easier to clean up after use with very acidic (e.g., 1 *M* sulfuric acid) or basic (e.g., 1 *M* sodium hydroxide) solutions. Although most polymer-based columns are still inferior to their silica-based counterparts in terms of mechanical strength, selectivity, and efficiency (*see* **Note 2**), newer materials such as PLRP-S (Polymer Laboratories, Church Stretton, Shropshire, UK) show superior performance in complex protein separations. Recent studies have shown that the use of polymer-based columns at high pH (>8.0) can offer unique selectivities and thus complement separations with classical acidic mobile phases. These advantages, coupled with superior stability, longer column life, and better reproducibility make polymer-based columns a good first choice for complex protein separations.

Finally, chromatographers are paying attention to restricted access media (RAM) for protein-bound drugs and metabolites. The interior of the pores is conventional C_{18}, but access to the pores is restricted so that proteins cannot enter. The protein–drug complex being in dynamic equilibrium and obeying the laws of mass action means that when the C_{18} binds the analytes, the drug complex dissociates. Ideally, the free protein elutes at V_0 (corresponding to interstitial volume) because it is excluded from the pores of the particles.

1.2.2. Particle Size

In the applied area of HPLC, a change in the practice will only occur if it offers a clear advantage over the existing method. To that end, the overriding recent trend is the need for increased productivity. Thus, the market drivers that effect change are those that will increase throughput, increase sensitivity, maintain reliability, and be cost-effective. Regulatory, economic, and environmental factors place demands on laboratories to produce large quantities of quality data and that projects be completed on time. Such pressures define the main goal and underline the need for multiple subgoals (e.g., reduced time, reduced analytical expense, faster analysis with no sacrifice of separation resolution, rugged and easily reproduced methods using existing equipment, and reduced solvent consumption).

With increased productivity as the main emphasis in a wide variety of industries, it is no surprise that the main trend in HPLC is focused on higher throughput using smaller particles and/or short columns using in-line (MS) for detection. This results in decreased retention times while maintaining efficiency.

As with small molecules, theory predicts that column performance should increase with decreasing particle size. This is generally the case for most commercially available RP-HPLC columns used for protein separations. Columns with particle sizes of 5–10 μm show little difference in performance. Columns with particle sizes of 2–3 μm are now available and can give much greater resolution. Theoretically, as particle size is decreased, column efficiency will increase as the square of the relative decrease in particle size. However, the operating pressure will increase by the square of the relative decrease in particle size (*see* **Note 3**). In addition, as the column is reduced in volume (shorter length), the band-broadening contribution of the instrument becomes important relative to the final volume of the eluted peak. The "best" column configuration is a trade-off between efficiency, pressure drop, and peak volume, compatible with current instrumentation. In effect, 3-μm particles in a column of length (approx 100 mm) is the best compromise for today's instrumentation. This conclusion is based on the fact that this particle size provides good efficiency and reasonable pressure drop (approx 130 bar) and is compatible with most instruments in use today that typically contribute a band-broadening volume of approx 70–100 μL. If the column can generate a peak volume of <70 μL, there would be no gain in overall performance because the instrument would broaden the peak more than the column. Smaller particles (e.g., 1 or 2 μm), contained in a well-packed short column and installed in a "conventional" HPLC system will not exceed the performance of the 3-μm column packing. Thus, the 3 μm packing results in the best match of column-to-instrument performance.

In 1990, <10% of HPLC separations were conducted with 3-μm particles, although it was well known even then that these small-particle/short-column configurations would produce faster separations without sacrificing resolution. The main reason that 3-μm columns were not more rapidly embraced by analysts was that these columns tended to plug faster and have shorter lifetimes than conventional 5-μm particle-size columns. In addition, column efficiencies were lower than expected and back-pressures tended to be higher than expected. In a practical sense, the movement to 3-μm columns in the applied area was slow because of a perceived lack of ruggedness. Using 3-μm particles required the use of 0.5-μm end frits to retain the packing in the column and these end frits were such good filters that they were prone to plug-in operation. However, recent interest in increased throughput using short columns has resulted in the trend toward using 3.5-μm particles packed in a column with standard 2-μm column end frits that are less prone to plugging. The practical use of these columns requires a much tighter particle-size distribution with an absence of fines <2 μm in a nominally 3.5-μm packing. Columns of 3.5-μm particles with no fines are operable at high flow rates with reasonable back-pressures, thus permitting faster analyses without loss of resolution and with the ruggedness of 5-μm columns.

1.2.3. Pore Size

High-performance liquid chromatography adsorbents are porous particles and the majority of the interactive surface is inside the pores. Consequently, proteins must enter a pore in order to be adsorbed and separated. The standard pore size for separation of small molecules is 80 Å, but for the separation of proteins, a pore size of 300 Å has become accepted as the norm. Proteins chromatograph poorly, in part because they are generally too large to enter pores of <100 Å. For proteins >100 kDa, pore sizes of >1000 Å may be even better *(12)*. For complex samples, if all other column variables were equal, it would be better to use a larger pore size to ensure that very large proteins do not encounter restricted diffusion or exclusion from the pores.

1.2.4. Bonded Phases

The majority of LC analyses are performed under reversed-phase (RP) conditions because of its applicability to a wide range of compounds. RP-HPLC adsorbents are formed by bonding a hydrophobic phase to the silica matrix by means of chlorosilanes, silicon-based molecules with chlorine as the reactive group and to which a hydrocarbon group is attached. A wide variety of reversed phases bonded to porous silica are available for the separation of proteins by RP-HPLC but the most common are the *n*-alkyl bonded phases, namely C_4, C_8, and C_{18}. In general, C_4 and C_8 phases are preferable for more hydrophobic samples and the C_{18} phase for hydrophilic samples. In general, significant differences may be observed in the separations achieved on nominally the same column but from different manufacturers. For proteins, a C_8 column is generally a good compromise. If the proteins of interest are too strongly retained, a C_4 column could be substituted. Subtle differences in RP surfaces sometimes results in diifferences in RP-HPLC selectivity for polypeptides that can be used to exploit and optimize specific separations. Additionally, in some situations, there is a need to use a column that exhibits very low "bleed." Under certain conditions, this concern is not easily met and it requires the use of unique bonded phases that do not hydrolyze during use.

In general, method development using LC/UV (ultraviolet) focuses on adjustment of separation parameters (e.g., mobile-phase composition, temperature, and pH) to optimize selectivity and specificity. In LC-API/MS the choice of mobile phase composition is restricted to volatile buffers and/or additives and their concentration minimized to ensure maximum analytical sensitivity. Reducing the buffer concentration necessitates the need for high-quality and highly reproducible stationary phases, particularly for the analysis of basic and polar compounds. In the analysis of polar compounds by LC/MS, reduced retention results from a lower percentage of organic modifier, which, in turn, has an adverse effect on sensitivity with nebulization techniques such as electrospray (ES). Thus, a stationary phase that provides increased retention may be necessary. Therefore, with limited scope to optimize the mobile phase, packings that provide generic applicability and exhibit symmetrical peak shape in a reproducible manner with weakly buffered mobile phases are desirable for LC/MS analysis. As previously stated, the high compound specificity and selectivity of MS detection reduces the requirement for highest efficiency if a packing with improved bonded-phase coverage, a modified end-capping, or otherwise modified surface (e.g., shielding from silanolic activity) is selected.

Retention is a function of the distribution coefficient between the stationary and mobile phases and their relative volume ratios. Hence, packings with a high surface area would typically show higher overall retention. Novel stationary phases, designed to exhibit increased retention of polar compounds, are typically capable of dispersive and dipole interactions. Strategies for conferring dipole character to a stationary phase include novel end-capping processes or substituting fluorine for hydrogen atoms to create a polar C–F bond.

In addition, a stationary phase with alternative functionality might be necessary to achieve required selectivity. Examples of alternative phases are cyano, hexyl, phenyl, hexyl/phenyl, perfluorinated, and so forth. Phenyl columns are slightly less hydrophobic than C_4 columns and may offer unique selectivity for some proteins. The hexyl/phenyl "mixed-mode" packing combines the separation mechanism of the C_6 packing with the added selectivity provided by interactions between the phenyl rings of the packing and polar groups in the analyte.

1.2.5. Column Dimensions

The two factors to consider when selecting a column size are the efficiency and sample-loading capacity. In the case of proteins, column length contributes little to efficiency. In addition, longer columns may have an adverse effect on protein recovery. When considering gradient rather than isocratic elution, resolution of a mixture of proteins depends directly on the particle diameter of the packing but less on the column length and flow rate for gradients of equal time. However, peak height, which will determine the detection limit, depends inversely on the particle diameter, flow rate, and, to a lesser extent, gradient time. An increase in column length would have little effect on resolution because by the time a protein band reaches the extra column length, it would be overtaken by the gradient and the organic modifier concentration would ensure that the protein spent little time in the adsorbed state. The effect of additional column length is thus largely passive and may even lead to increased band broadening. Long columns might be useful in isocratic protein separations, but if extra capacity were needed, a short fat column might be a better choice than a long thin column. With the advent of MS as a detection technique, there is now a general trend toward faster analyses using short columns packed with smaller particle adsorbents, including those <50 mm in length, which are used only to give minimal separation prior to introduction of effluent into the MS. As one would expect, these columns do not exhibit many plates; however the efficiency is adequate because the MS accomplishes compound selectivity and identification.

Having selected the column length, the choice of column diameter should be based on required sample capacity. Sample capacity is a function of column volume. For columns of equal diameter, longer columns maximize sample capacity. For analytical applications in the microgram to low milligram range, columns of 4.6 mm inner diameter (analytical) are best. Columns of 1.0 mm (microbore) to 2.1 mm (narrow bore) internal diameter are best suited to submicrogram levels of material, thus minimizing sample loss and also increasing sensitivity of detection.

Because of the goals of investigative work, trends in this area have tended toward increased capability and sophistication. These goals have resulted in an emphasis on generating more separating power through the use of higher-resolution columns with an increased number of theoretical plates, using conventional column configurations with more sophisticated detection for compound identification, or both. Because overall efficiency of the separation is dependent on both column and instrumentation contributions, column trends in this area require instrumentation improvements in order for the total benefit of the column to be appreciated.

Requirements in the investigative area have moved HPLC columns to narrow diameters (2.1 mm and lower), with lengths of 250 mm and longer, packed with small particles. Columns with small diameters (e.g., 0.5–2 mm) are preferred when high sensitivity is required or the amount of sample is limited. Columns of small diameter are often required when analyzing complex biological samples. Using these microbore/narrowbore columns increases the absolute sensitivity of an analyte. Because very high-efficiency columns spread the bands less, use of high-performance columns do require instruments with very low dispersion. As a result, many instrument manufacturers now offer low-dispersion instrument capability. These recently introduced systems are designed with smaller-diameter columns in mind. They find major application in biological research, in which small sample amounts are common and sensitivity requirements high. Traditional instrumentation would require extensive modification to handle the microvolume sample and solvent fluidics, to eliminate extra column dead volume, and to operate with small-diameter columns (*see* **Note 4**).

1.3. Mobile Phase

The mobile-phase composition is the most readily changed variable in a RP-HPLC separation. "Like dissolves like" is the basic concept for the selection of solvents in the eluent for LC. Controlling the solubility of analytes is the key to success. If the selected solvent or mixture of solvents does not interfere with detection, it is a good eluent. The selection of a suitable solvent for low-wavelength absorption detection and postcolumn derivatization detection is important to obtain high sensitivity. The selection of a volatile solvent is the key for preparative-scale LC and for MS detection.

In normal usage, the mobile phase consists of a mixture of water, a miscible organic solvent, and dissolved buffers and salts. A buffer, such as a low concentration of a strong acid or salt, is essential. If a protein is adsorbed in the absence of a salt or acid, no increase in the proportion of organic component will elute it. In addition, other components, either solvents or solutes, may be added in order to effect the separation. This may be to accelerate or delay elution, improve peak shape, or adjust the elution position of some components with respect to others, thus affecting selectivity. Other important variables are the pH of the mobile phase and the concentration and type of ions present. The ionic strength is a factor in limiting "mixed-mode" retention involving silica hydroxyls, but the choice of ions can also affect the solubility and stability of the protein in the mobile phase and, through ion pairing, the distribution of the protein between stationary and mobile phases. The net effect depends on the combination of column packing and mobile phase. Flow rate has little effect on protein separations. Protein desorption is the result of reaching a precise organic modifier concentration. Protein resolution, therefore, is relatively independent of mobile-phase flow rate. Higher flow rates may improve the solubility of hydrophobic proteins, although this would increase the amount of solvent to be removed if samples were collected.

Numerous studies have examined the effect of mobile phase on the retention and selectivity of proteins in RP-HPLC. Systematic studies by a number of researchers over the past 10 yr have resulted in a number of fairly standardized sets of elution conditions. Their behavior can be altered by the variation of a number of mobile-phase parameters. Altering the nature of the organic component of the mobile phase can alter column behavior but not usually in a predictable manner. The most successful way to alter column behavior is to manipulate the ionic component of the solvent system. By altering the nature and strength of the ion-pairing reagent and the pH of the mobile phase, it is possible to exploit both the acidic and basic character of proteins in a systematic and predictable way. Without prior knowledge of the primary structure of a protein, one can be certain of some hydrophobic character by virtue of the content of the lipophilic amino acids leucine, iso-leucine, methionine, phenylalanine, tyrosine, tryptophan, and valine. It will also have hydrophilic character because of the presence of aspartic acid, glutamic acid, arginine, and lysine. The ratio between hydrophobic and hydrophilic amino acids determines the initial behavior observed for any protein in the most common RP-HPLC solvent system in use today, namely aqueous acetonitrile containing 0.1% trifluoroacetic acid (TFA). To some extent, the hydrophilic character of solutes is suppressed at pH 2.0. All carboxylic acid groups are protonated under these conditions and their contribution to hydrophilic character is reduced. This, in turn, will improve peak shape. Basic amino acids (e.g., arginine, lysine, and histidine) will remain as cations that, in turn, will lead to a marked reduction in silanophilic solute–matrix interactions.

Selectivity can be optimized by adjusting mobile-phase composition, column packing, or temperature. The most successful option, particularly for ionizable compounds, is to optimize the mobile-phase composition in terms of solvent strength/type, buffer concentration/type, and pH. However, solvent choice is restricted with LC-API/MS, as adjustment of pH may affect sensitivity.

Electrospray ionization (ESI) transfers ions from the liquid phase into the gas phase; therefore, conditions that promote ionization in the liquid phase (i.e., correct pH) will generally provide best sensitivity. With these restrictions in mind, the preferred solution for LC/MS method development is to have a range of column-packing functionalities that can be evaluated against a generic mobile phase. These advantages are also observed in LC/MS because RP conditions are compatible with all interfaces and LC method development protocols are generally applicable to LC/MS. In LC-ESI/MS, signal intensity is sensitive to the concentration of organic modifier in the mobile phase. Organic solvents have lower surface tension and higher volatility than aqueous solvents. The higher the concentration of solvent, the greater the efficiency of desolvation. However, the combination of a high concentration of organic solvent and a weak buffer can lead to poor resolution unless a highly retentive stationary phase is used.

1.3.1. Organic Component of Mobile Phase

The most popular organic solvents are acetonitrile, methanol, 1-propanol, and 2-propanol. Acetonitrile is the most commonly used organic modifier because it is volatile and easily removed from collected fractions, has low viscosity minimizing column back-pressure, has little UV absorption at low wavelengths, is available at high purity, and has a long history of proven reliability in RP separations. Propanols are much more viscous than acetonitrile, giving higher column back-pressures (*see* **Note 5**), but these are not significant when low flow rates and short columns are used. The acetonitrile concentration needed to elute a protein is considerably higher than the equivalent 2-propanol concentration, perhaps half as high again, and thus proteins are likely to be seriously denatured. There is also evidence that acetonitrile is intrinsically a more powerful denaturant than the alcohols. Recoveries are often higher in propanol-containing mobile phases, especially for the more "difficult" model proteins such as ovalbumin. Adding 1–5% of 2-proanol to acetonitrile has been shown to increase protein recovery in some cases *(14)*. Acetonitrile and 2-propanol can be frozen in dry ice/alcohol mixtures and lyophilized. All solvents should be of the highest quality available (*see* **Note 6**). Methanol or other solvents offer little advantage over the more commonly used solvents and are not generally used for protein separations.

1.3.2. Minor Mobile-Phase Additives

The most popular minor components of the mobile phase are the perfluoroalkanoic acids (e.g., TFA and heptafluorobutyric acid). These appear to solubilize proteins in organic solvents. TFA is used in preference to other fully dissociated acids because it is completely volatile, will not corrode stainless steel, has little UV absorption at low wavelengths, and has a long history of proven reliability in RP-HPLC polypeptide separations. TFA at 0.1% (w/v) acts as a weak hydrophobic ion-pairing reagent. When chromatographing proteins, using TFA concentration at <0.1% may degrade peak shape, although new column developments allow the use of much lower TFA concentration. TFA forms such strong complexes with polypeptides that when using MS for detection, the electrospray signal and hence detection sensitivity is reduced (especially at concentrations typical for polypeptide separations). Recent developments in HPLC columns have resulted in columns with good polypeptide peak shapes using very low concentrations of TFA (*see* **Note 7**).

Most RP columns employed for RP-HPLC are silica based, and apart from the suppression of silanol ionization under acidic conditions (thereby suppressing undesirable ionic interactions with basic residues), silica-based columns are more stable at low pH. As mentioned previously (**Subheading 1.3.**), an acid pH is preferred although many polypeptides will lose their tertiary structure at pH 2.0–3.0 (*see* **Note 5**).

Other popular additives are pyridinium acetate buffers and buffers based on phosphoric acid, particularly triethylammonium phosphate (TEAP). TEAP is useful in circumstances where tailing is suspected because of nonspecific interactions with surface silanols. Proteins are eluted at a higher percentage of organic component with an equivalent concentration of phosphoric acid. Sulfate, phosphate, perchlorate, and chloride are all transparent in the far UV whereas acetate, formate, and fluoroalkanoic acids must be used at concentrations <20 mM if wavelengths below 220 nm are to be monitored. Pyridine-containing eluents cannot be monitored by UV. Ion-pairing reagents such as heptane and octane sulfonic acids, sodium dodecyl sulfate (SDS), or alkyl ammonium salts may be added to the mobile phase to selectively increase the retention of proteins carrying larger charges of opposite sign. Other additives might be needed depending on the nature of the protein being analyzed (*see* **Note 9**). It has recently been shown that the use of very low-metal-content silica (99.999% pure) and an inert bonded surface can reduce the need for expensive mobile-phase modifiers or ion-pairing reagents. Proprietary bonding/end-capping procedures are now such that physically and chemically stable bonded phases are routinely available.

1.3.3. Gradient Elution

The theory of gradient elution is now well established and a good understanding exists concerning the effects of gradient steepness on separation. Whereas isocratic elution is carried out with a mobile phase of fixed composition, in gradient elution the mobile phase is intentionally varied during a chromatographic run. Although this change in mobile-phase composition can take various forms, gradient elution almost always involves an increase in solvent strength from the beginning to the end of the separation, as the result of mixing a stronger solvent B with a weaker solvent A. Therefore, the capacity factor k (also known as the retention factor) for each sample component decreases with time after sample loading. The effective use of gradient elution requires an understanding of how separation varies with experimental conditions. Geng and Regnier *(15)* developed the stoichiometry factor, Z, which represents the number of solvent molecules displaced during the binding of a protein to the column packing. Thus, the value of Z is a measure of the size of the effective chromatographic footprint of that protein.

For RP-HPLC, Eq. (1) has been derived; it shows that Z is proportional to the term S:

$$Z = 2.3\varphi S \tag{1}$$

where φ is the volume fraction of the organic in the mobile phase (16). Under defined gradient conditions, every protein will have an S value described by

$$\log k' = \log k_w - S\varphi \tag{2}$$

where k' is the retention of the solute (capacity factor) and k_w is the value of k' when water is the mobile phase ($\varphi = 0$). Values of k_w and S are characteristic of each solute in a sample. Although empirically derived, Eq. (2) is a good approximation of RP-HPLC retention in the gradient mode. For a series of related proteins of nearly identical composition, the hydrophobic contribution to RP retention will be similar and, thus, observed changes in Z (and S) should reflect changes in the structure of the protein that contacts the adsorbent surface. Examination of Eq. (2) reveals why small changes in the organic composition of RP eluents result in large changes in protein retention as opposed to small-molecule separations. The separation of small molecules involves continuous partitioning of the molecules between the mobile phase and the hydrophobic stationary phase. Proteins, however, are too large to partition into the hydrophobic phase; they absorb to the hydrophobic surface of the packing after entering the column. Proteins may be thought of as "sitting" on the stationary phase with most of the molecule exposed to the mobile phase and only a part of the molecule, the "hydrophobic footprint," in contact with the RP surface.

Proteins will have a much larger chromatographic footprint (Z and S) than small molecules. RP-HPLC separates proteins based on subtle differences in their "hydrophobic footprints." Differences in "hydrophobic footprint" results from differences in amino acid sequences and differences in conformation. Geng and Regnier found that the Z number correlated with molecular weight for denatured proteins; however, proteins with intact tertiary structure eluted earlier than expected because of their smaller "hydrophobic footprints." Thus, changes in φ (% organic) will be amplified by large values of S, resulting in even larger changes in retention ($\log k'$). Many examples of the applicability of Eq. (2) for peptide and protein samples have been reported. Values of S and k_w for each solute can be obtained from two experimental gradient separations of the sample. Retention times in gradient elution can then be predicted as a function of gradient conditions. Two components that elute adjacent to one another in a chromatogram will often show significant changes in band spacing when isocratic solvent strength or gradient steepness is varied. The resolution of such band pairs can usually be accomplished when values of S for the two components differ by 5% or more. Gradient separation of proteins seems to follow different rules than are observed for the separation of small molecules; namely change in flow rate or column length has little effect on sample resolution, change in flow rate or column length has little effect on φ at elution (φ_e), and changes in column type or surface area have little effect on φ at elution (φ_e). For very large molecules (>10 kDa), isocratic elution with $k>0$ may be difficult to achieve; that is, for small changes in % B, retention seems to change rapidly from $k\sim0$ to no elution at all. This observation suggests that gradient elution differs fundamentally for large versus small molecules, and various alternative descriptions of large-molecule gradient elution have been suggested. The "on–off" model (*17*) assumes that each solute band is retained strongly at the column inlet until φ reaches a certain "critical" value φ_c; when $\varphi = \varphi_c$, the value of k for the band decreases from a very large value to 0 and the solute band then moves through the column with $k_a = 0$ and is eluted in a mobile-phase composition $\varphi_e = \varphi_c$. There is little further interaction with the reversed-phase packing surface. Components elute quickly once the critical organic concentration is reached. The sensitivity of the Z number to protein conformation (*16*) and the sudden desorption at the critical modifier concentration accounts for the sharp peaks, high resolution, and ability to resolve very closely related polypeptides achieved with this technique.

Because large polypeptides diffuse slowly, RP-HPLC results in broader peaks than obtained with small molecules. Peak widths of polypeptides eluted isocratically are a function of molecular weight with large proteins such as myoglobin having column efficiencies only 5–10% that obtained for a small molecule such as biphenyl. Isocratic elution is rarely used for polypeptide separations. Gradient elution of polypeptides, even with shallow gradients, is preferred because it results in much sharper peaks. Gradients of organic modifier concentration are normally used to obtain sharp peaks and optimum selectivity. Initial organic concentrations range from 1–5% for hydrophilic peptides to 10–50% for large or hydrophobic proteins. Whereas most peptides will elute in 50% organic modifier or less, large or hydrophobic proteins may require as high as 95% organic to elute. The gradient slope (percent change in organic modifier per unit volume or time) is normally around 1–2% per minute. However, very shallow gradients (e.g., 0.1–0.5% per minute) are often used to separate complex mixtures of very similar components.

1.4. Temperature

Protein samples will often contain many individual components. A complete separation of such samples poses a real challenge because statistical considerations suggest that one or more peak pairs will usually be poorly resolved. To overcome this dilemma, a systematic variation of separation selectivity would be required. This approach has been widely used for the separation of typical small-molecule samples. In the case of peptide and protein samples, however, the control of selectivity has received little attention. When a change in selectivity is desired, the usual approach is to change column or mobile phase. The use of elevated temperatures for RP-HPLC of protein samples has been advocated primarily as a means of increasing column efficiency, controlling retention reproducibility or shortening run time. For samples of this type, a few studies have shown that a change in column temperature can also affect separation selectivity. In RP-HPLC, retention time decreases as the temperature increases, and, if necessary, the exact relationship can be obtained by using a van't Hoff plot. Generally, temperature optimization can improve productivity for routine analyses.

Because of this trend, instruments with ovens are now available as standard. Today, an HPLC without a column oven would not effectively serve the applied user. A combination of low pH and higher temperature can often result in a very short lifetime for commonly used alkyl-silica columns that, in turn, limits the application of temperature optimization. This problem has recently been overcome by the development of "sterically protected" RP packings that are extremely stable at low pH (<2.0) and high temperature ($>80°C$). Di-isobutyl-octadecylsilanes (C_{18}) phases are less susceptible to loss of the bonded phase by acid hydrolysis than traditional dimethylsilanes, providing high column lifetime under these harsh conditions. Monolayer bonding with these bulky silanes results in a high degree of reproducibility from column lot to lot.

The combined use of temperature and gradient steepness would provide an efficient procedure for the control of peak spacing and optimization of separation. At the same time, this approach to selectivity control is more convenient than alternatives such as change of column or mobile phase because temperature and gradient steepness can be varied via the HPLC system controller.

1.5. Biological Activity and RP-HPLC

Reversed-phase HPLC may disrupt protein tertiary structure because of the hydrophobic solvents used for elution or because of the interaction with the hydrophobic surface of the material. Biological activity of proteins depends on tertiary structure and permanent disruption of tertiary structure eliminates biological activity. The amount of activity lost depends on the stability of the protein and on the elution conditions used (*see* **Note 1**). The loss of activity can be minimised by proper postchromatographic treatment. Small peptides and very stable proteins are less likely to lose activity than large enzymes. Denaturation of proteins on hydrophobic surfaces is kinetically slow. Reducing the residence time of the protein in the column generally reduces the loss of activity. Some solvents are less likely to cause a loss of activity than others. Isopropanol is the best solvent for retaining activity. Ethanol and methanol are slightly worse and acetonitrile causes the greatest loss of activity. Stabilizing factors, such as enzyme cofactors, added to the chromatographic eluent may stabilize proteins and reduce the loss of activity. The most important factor in maintaining or regaining activity is postcolumn sample treatment. Dissolution of collected protein in a stabilizing buffer often allows the protein to refold.

2. Materials

To illustrate the general points in **Subheading 1.**, the separation of a mixture of commercially available standard proteins is described.

2.1. Equipment

Analysis of proteins by RP-HPLC is suited to most commercially available equipment (*see* **Note 10**). Systems require a column packed with a suitable matrix giving adequate retention and resolution, a detector to monitor eluted proteins, a chart recorder or data system to view detector response, and a fraction collector for recovery of eluted proteins, if further analysis is intended. If high sensitivity is a requirement, then narrow-bore (2.1 mm) or microbore (1.0 mm) chromatography should be considered and specialized dedicated equipment may be necessary (*see* **Note 4**). Proteins are usually monitored by their absorbance in the UV either at 280 nm, which is suitable for proteins with aromatic amino acids or at 206 nm, a wavelength characteristic of the peptide bond (*see* **Note 11**). Microfraction collectors are also available with small-internal-diameter tubing to reduce the volume between detector and collector to a minimum. This allows a more accurate reflection of the real-time elution situation.

Today's general-purpose HPLC instrumentation is designed to deliver greater performance and improved operator interaction (user-friendliness) at a cost similar to or lower than predecessor instruments. This is in keeping with the drive toward higher productivity. All chromatographic systems now have computers to control operating parameters and to collect and store data. Many instruments can now not only be controlled by the operator but also by some sort of insertable PC disk, allowing a lesser skilled instrument user to automatically set up the desired method. In addition, modern instruments are also usually equipped with diagnostic software to help the operator identify the cause(s) of failure as well as alert the user when preventative maintenance should be undertaken. Because of increased regulatory oversight, gathering of as much information as possible is positively encouraged.

2.2. Column

A wide variety of suitable C_8 or C_{18} columns (analytical 4.6 mm $\times$ 150 mm) are commercially available for protein analysis. The price of modern columns is such that little advantage is gained by self-packing a column (*see* **Note 12**). The column must be able to withstand moderate pressures (350 bar) and be resistant to the mobile phase (pH typically 1.5–7.0 for silica columns). A guard column (containing the same or very similar packing material) should be used as a matter of course to prolong analytical column lifetime.

2.3. Injector

For most purposes, a conventional HPLC manual injection valve (e.g., Valco or Rheodyne) with a fixed volume loop (e.g., 10 µL) is adequate. For microbore analysis, an injector with a small internal loop is preferable. Most modern HPLC chromatographs have autosamplers fitted as standard, allowing injection volume to be varied between, for example, 0.1 and 500 µL. In addition, many autosamplers are capable of being cooled down to 4°C to minimize sample degradation before analysis.

2.4. Eluents

Use Analar-grade or Aristar-grade reagents and distilled/deionized water for all solutions (*see* **Note 13**). All eluents should be filtered through a 0.2–µm filter and degassed under vacuum before use.

1. Eluent A: 0.1% TFA (v/v). To 1 L of water, add 1 mL of TFA (*see* **Note 14**).
2. Eluent B: Acetonitrile:water (70:30) + 0.085% TFA (v/v). To 700 mL of acetonitrile, add 300 mL of water and 0.85 mL of TFA (*see* **Note 13**).

Eluents should be degassed by a slow stream of helium while on the chromatograph, if at all possible, to counteract the effect of dissolved air. Some chromatographic systems now incorporate vacuum degassing systems as standard. If a ternary system is available, acetonitrile can be programmed into a gradient to fully elute hydrophobic materials that might otherwise be retained by the column packing.

2.5. Detector

By far the most common detector used in chromatography generally is the variable wavelength UV detector, operated at 210–220 or 280 nm. Most proteins have an absorption maximum at about 280 nm because of the presence of aromatic amino acids, which falls to a minimum at 254 nm, a popular wavelength for fixed-wavelength HPLC detectors. Highest sensitivities are obtained by monitoring the strong absorption bands, which peak below 220 nm because of the "peptide bond" itself (*see* **Note 11**).

Proteins are usually eluted as fairly broad peaks (*see* **Fig. 1**, a protein test mixture of insulin, cytochrome-*c*, lactalbumin, carbonic anhydrase, and ovalbumin) and thus place fairly modest demands on the dimensions of a normal flow cell (typically 8 µL, with path length 1 mm).

An alternative to UV detection is to monitor any intrinsic fluorescence. The usual excitation wavelength (λ_{ex}) is 280 nm with emission (λ_{em}) monitored at either 320 nm (for tyrosine) or 340 nm (for tryptophan). The main advantage is a several-fold increase in sensitivity and an improvement in selectivity of detection, particularly in distinguishing protein peaks from "ghost" peaks originating from eluent impurities (*see* **Note 6**).

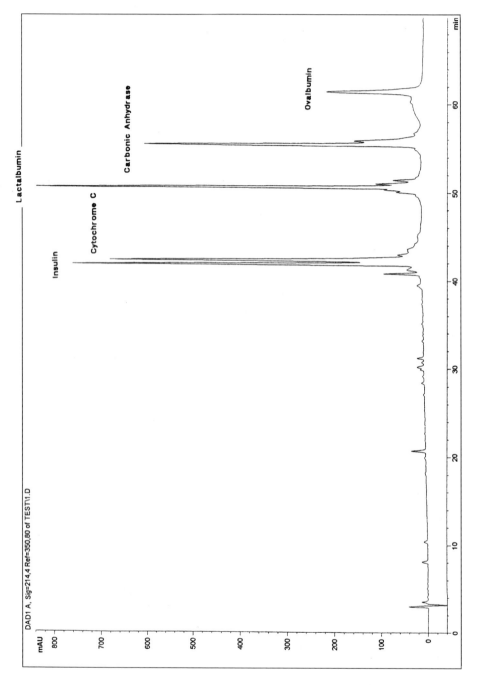

Fig. 1. The RP-HPLC of a protein test mixture. The conditions used for chromatography are described in **Subheading 3**.

When appropriate, the need for additional information to confirm a compound's presence at the desired level has led to the incorporation of photodiode array (PDA) detectors into many instruments. In its simplest mode, a PDA detector is easily utilized as a single-wavelength or multiwavelength detector. Protein and peptide separations can be monitored at 214 nm, which provides a general detection mode for all peptidic components, and at 280 nm, which selectively detects only components containing aromatic amino acids. For peptides, this dual-wavelength mode is useful in distinguishing peptides with Tyr, Trp, or Phe from those without, but proteins will always contain one or more aromatic amino acids. The benefit of simultaneous detection at 214 and 280 nm for proteins lies in the improved sensitivity at the lower wavelength, whereas the absorbance at 280 nm provides evidence that the UV spectrum of the detected species is consistent with that of a protein. Two-wavelength detection has also found application in quantitating protein solution concentration. Van Iersel et al. *(18)* compared several procedures developed for homogeneous cuvet-based assays to HPLC separations with PDA detection. The strategy involved one wavelength in the far UV (e.g., 205 nm) and a second wavelength in the aromatic region (e.g., 280 nm) to calculate a protein-extinction coefficient. Approximation of total protein concentration can be calculated from simple algebraic functions of the two absorbances. The advantage of an HPLC assay over solution techniques is the separation of the proteins of interest from other proteins and the removal of potential interference from buffer constituents, detergents, or stabilizers.

The authors reported that using Eq. (3) gave results superior to those from other methods:

$$A_{280/0.1\%} = 34.14 \left(\frac{A_{280}}{A_{205}} \right) - 0.02 \tag{3}$$

Simultaneous dual-wavelength detection is also easily used for calculating wavelength ratios. Most PDA detectors have software that makes it routine to calculate automatically the 280/214 nm absorbance ratio. This ratio will be characteristic for any single, homogeneous component and a resulting constant ratio across a peak is one of the simplest approaches to demonstrating peak purity.

The most aggressively pursued detection technique now being used is MS. Several years ago, HPLC-MS with an atmospheric-pressure ionization source (API) was beginning to be used in the metabolism laboratories of pharmaceutical companies. These workhorse instruments were large and expensive, but their cost was justified by the productivity increase in the drug discovery and development process. Today, the use of HPLC-MS as a tool for the metabolism investigator is universally accepted. Initially, it was not felt that HPLC was needed because samples could be introduced directly into the MS. However, it soon became apparent that some HPLC retention was needed to separate the substances of interest from other substances when present in complex matrices. With the advent of benchtop instruments, HPLC-MS is more convenient than with past instrumentation. This has quickened the trend toward the routine use of HPLC-MS in analysis of proteins.

2.6. Protein Sample

A mixture of test proteins, namely insulin (from bovine pancreas), cytochrome-*c* from horse heart), α-lactalbumin (from bovine milk), carbonic anhydrase (from bovine

Table 1
Column Characteristics

	Column diameter (mm)	Typical flow rate[a]	Sample capacity[b] (µg)	Max. practical sample load[c]
Capillary	0.075	0.25 µL/min	0.05	n/a[d]
	0.15	1 µL/min	0.2	n/a
	0.30	5 µL/min	1	n/a
	0.50	10 µL/min	2	n/a
Microbore	1.0	20–50 µL/min	0.05–10	0.5
Narrow bore	2.1	50–250 µL/min	0.2–50	2.5
Analytical	4.6	0.5–1.5 mL/min	1–200	10

[a]Actual flow rates can be a factor of 2 higher or lower depending on the method.

[b]Sample capacity is the quantity of protein that can be loaded onto the column without reducing resolution.

[c]Maximum practical sample load is approximately the maximum quantity of sample that can be purified with reasonable yield and purity on the column.

[d]n/a = not available.

erythrocytes), and ovalbumin (from hen egg white), can be used as a column test. Two of the proteins, insulin and cytochrome-*c*, are difficult to resolve. A partial to complete separation should be achievable when a column with a high enough number of theoretical plates is used. Of the five standard proteins, ovalbumin is not only the most hydrophobic but also the most difficult to recover. The utility of the column from a recovery perspective will be illustrated by the size of this peak.

3. Method

1. Set up appropriate pump, temperature, injection, detection, and integration methods, as required by the chromatographic system being used.
2. Install appropriate column (C_8- or C_{18}-bonded silica, 300 Å, 5 µm, 100–250 mm length, 2.1 or 4.6 mm inner diameter).
3. Equilibrate the column by pumping eluent A at 200 µL/min (narrow bore, 2.1 mm) or 1 mL/min (normal bore, 4.6 mm) for several column volumes. Monitor detector baseline (214 nm) until flat and then zero the detector. Inject 5 µL of HPLC-quality water and run a blank gradient to identify any system peaks (*see* **Note 3**) and to balance baseline ((TFA), if necessary (*see* **Note 11**).
4. Dissolve protein mixture in HPLC-quality water at a concentration of approx 2 µg/µL to produce a stock solution. The stock solution should be diluted further with eluent A (1:1) to produce a working solution containing 1 µg/µL of each protein.
5. Inject the sample (5 µL) at initial gradient conditions (*see* **Note 15**).
6. Start gradient to elute proteins.
7. At completion of analysis, flush column with 50:50 A:B before storage (*see* **Note 16**).

The elution profile obtained is shown in **Fig. 1**.

4. Notes

1. Perhaps the most compelling evidence that biological activity is not inevitably lost during RP-HPLC is the fact that several commercial biotherapeutics utilize RP-HPLC during purification of the marketed product.

2. Polymer columns unfortunately suffer from a relatively low resistance to pressure and, in general, pressure should be kept below 150 bar. Furthermore, the physico-chemical properties are governed by a solvent-dependent swelling and shrinking of the organic matrix that is associated with concomitant changes in pore diameters, leading to changes in mass transfer characteristics of the column. This effect may be encountered during gradient elution starting with an aqueous phase (by which little or no wetting of the matrix takes place) and termination with a pure organic solvent, (e.g., methanol, acetonitrile, 2-propanol). The pore diameter will continuously change and often may be the cause of poor chromatographic performance. In contrast, silica gel as the starting material for subsequent alkyl silylation offers the advantage that a great number of different alkyl substituents can be bound to the silica surface.

3. Column back-pressure is directly proportional to the column length. Any gain in column efficiency can be quickly outweighed by increased column back-pressure, susceptibility to plugging, and shorter column lifetime.

4. In order to obtain good performance from microcolumns, it is very important to use an HPLC system with very low dead volumes so that extracolumn peak broadening does not destroy the resolution achieved by the column. Consideration should be given to the minimization of dead volume in all parts of the chromatographic system (e.g., sample injector, detector flow cell, pump heads, and connection tubing). If there is a large dead volume between the point at which the gradient is mixed and the injector, then lag will be introduced into the gradient. Also, the gradient system must be capable of delivering the gradient accurately and reproducibly at low flow rates. A suitable mixing device, (volume <200 µL) may need to be added to the system to enable a reproducible gradient to be delivered at very low flow rates. A number of capillary chromatography (c-LC) systems are commercially available to allow chromatography to be routinely undertaken at very low flow rates.

5. The pressure of an acetonitrile-containing mobile phase decreases approximately linearly with increasing percentage of organic modifier, whereas aqueous mixtures of methanol, ethanol, 1-propanol, and 2-propanol, exhibit a more or less convex pressure dependence on the percentage of organic modifier.

6. Water needs to be free of UV-absorbing impurities, which can be adsorbed onto the column during equilibration and the early part of a gradient run and that may be eluted later as "ghost" peaks. Water should be freshly distilled and deionized and stored out of contact with plastic, thus avoiding dissolution of plasticizers.

7. In some cases TFA may be completely replaced with either formic or acetic acid while retaining good resolution. With recent column developments, it is possible to maintain good peak shapes using only 0.01%TFA (w/v). It should be noted that lower TFA concentration may affect selectivity.

8. This may not necessarily be fatal for resultant activity, but if the polypeptide consists of subunits stabilized by noncovalent forces, the individual chains will almost certainly be separated during chromatography.

9. Metal ions, such as calcium, may be included to enhance the stability of a protein and low concentrations of nonionic detergents may improve the behavior of hydrophobic membrane proteins. Chaotropic agents such as guanidinium hydrochloride or urea, at moderate concentrations, may be added to elute very hydrophobic proteins. Protein samples sometimes contain surfactants. Although surfactants usually degrade RP separations, they do not harm RP columns. Even trace amounts of SDS in a sample can significantly reduce separation efficiency. If surfactants are present in the sample to be analyzed, either the use of a C_4 column is recommended or removal of the surfactant prior to chromatography.

10. In many areas of HPLC, metal surfaces in contact with the sample stream have been reported to interfere with a wide variety of samples and mobile phases. The metallic surface

area exposed to the sample can be attributed to various components within the flow path (e.g., frit, column, in-line filter, guard column, and connecting tubing). The corrosion rate of metal components in an HPLC system determines the concentration of metal ions in the mobile phase. The concentration of metal ions may be accelerated by corrosive mobile phases such as high concentrations of acids and salts. Although titanium is considered inert to most mobile phases, only HPLC systems fabricated of polymer will be totally corrosion-free as well as biocompatible. The consequences of having metal present in an HPLC system are not always obvious. Tailing peaks, low recoveries, or even entire loss of sample can appear as a result of the interaction of the sample with metal components in the HPLC system if the sample irreversibly adsorbs to metal sites along the flow path. In less extreme cases, a hysteresis effect will be seen with recoveries increasing with sequential sample injections as the binding sites get used up. Changes in solvent composition will upset the equilibrium, producing "ghost" peaks and/or baseline drift.

Biological samples, in particular, are often sensitive to metal ions present in the flow path of conventional HPLC systems plumbed with metal columns, frits, tubing, and fittings. Biomolecules such as proteins can be "poisoned" or inactivated in the presence of these metal ions, posing a significant problem for a number of biotechnology applications. By using metal-free columns along with other nonreactive parts for your system (e.g., injectors, tubing, connectors, valves, solvent inlet, and in-lie filters), it is possible to plumb the entire HPLC system with inert materials for metal-sensitive chromatographic applications. Most consumable supply companies offer a comprehensive stock of biocompatible parts and accessories.

11. During a gradient, there is usually an associated change in background absorption. This may be caused directly by the increase in the concentration of organic component or by the effect of this concentration change on other mobile-phase components. In addition, refractive index changes may be detected in the flow cell. Usually, not much can be done about flow cell design, but careful choice of wavelength and pH can minimize the change in absorption spectrum of mobile-phase components. The change in dielectric constant as the solvent environment changes from aqueous to nonaqueous affects π–π electron interactions, which, in turn, affects the absorption spectrum in the 190- to 250-nm region, leading to a baseline shift during many RP separations. For TFA and acetonitrile, the optimum wavelength is 214 nm. Alternatively, the concentrations of other components in the mobile phase can be adjusted to compensate or a small amount of UV-absorbing solute can be added to one of the eluent reservoirs. For example, in a TFA/water/acetonitrile system monitored at 214 nm, the rise in background can be balanced by using 0.1% (v/v) aqueous TFA as eluent A and 70% aqueous acetonitrile, containing 0.085% (v/v) TFA as eluent B.

12. Packing columns is based on know-how. Packing being commercially important, know-how is not readily imparted. Packing LC columns is still not fully understood. Commercial columns are usually supplied with a test certificate and a guarantee to replace the column if performance does not match that specified. Although packed columns cannot be expected to last indefinitely, a usual lifetime of several months of continuous use and several hundred injections would be a reasonable objective.

13. The HPLC quality water is commercially available from a number of suppliers if access to high-quality distilled or distilled/deionized water is not possible. A definition of HPLC quality is when 40 mL of water is pumped onto the column which is then eluted with a linear water/acetonitrile gradient at 5% per minute, at a flow rate of 2 mL/min, and absorbance of the largest eluted peak, monitored at 254 nm, is <0.002 AU. It is important to use highest-purity TFA and to buy in small volumes. Poor quality or aged TFA may contain impurities that chromatograph in the RP system, causing spurious peaks to appear.

14. The RP solvents are by convention installed on the HPLC channels A and B. The A solvent by convention is the aqueous solvent (water) and the B solvent by convention is the organic

solvent (acetonitrile, 2-propanol, etc.). It is important to follow this convention because the terms A and B are commonly used to refer to the aqueous and organic solvents, respectively. The A solvent is generally HPLC-grade water with 0.1% TFA and the B solvent is generally an HPLC-grade organic solvent with 0.1% TFA.

15. Proper gradient conditions are critical in optimizing separations of proteins by RP-HPLC. The gradient time, range, and shape are all important and must be optimized for a given sample. A typical starting gradient would be a 2% per minute linear increase of acetonitrile from 5% eluent B to 95% eluent B over perhaps 45 min. It is advisable to load sample with a small percentage of acetonitrile in the eluent at initial conditions, to improve later gradient mixing. For the best reproducibility and equilibration, it is advisable to avoid extremes in organic modifier composition. It is recommended that gradients should not begin at less than 3–5% organic modifier concentration. Gradients beginning with less organic modifier may cause column equilibration to be long or irreproducible because of the difficulty in "wetting" the surface. It is recommended to end gradients at no more than 95% organic modifier, because high organic concentration may remove all traces of water from the organic phase, also making column equilibration more difficult. A washing step of 100% eluent B (70% acetonitrile) should then be included before returning to initial conditions, allowing sufficient equilibration before the next injection. Column re-equilibration time is an extremely important variable. This time must be sufficient to allow the initial mobile phase to equilibrate in the pores of the column packing (at least 3 column volumes) and should be exactly repeatable if precise retention times are required. A linear gradient of the above type (0.5–5% per minute increase in acetonitrile) should allow resolution of the proteins in the test mixture. For increased resolution of certain of the proteins, a number of isocratic steps may have to be included in the gradient profile.

16. Reversed-phase columns, if properly cared for, may give good performance for >1000 injections, depending on sample preparation and elution conditions. Performance may deteriorate for a number of reasons, including use of improper eluents, high pH, contamination of packing by strongly adsorbed sample constituents, insoluble materials from the solvent or sample, or simply age or extensive use. Column lifetime can be extended by filtering all solvents and samples and using an eluent filter and guard column. It is recommended that an eluent filter be used between the solvent delivery system and the injector to trap debris from the solvents, pumps, or mixing chamber. In addition, a guard column sited between injector and column should be considered if samples contain insoluble components or compounds that strongly adsorb to the column packing. Because of the nature of the RP surface, column performance (i.e., resolution and retention) may change slightly during the first few injections of a protein. A column can be conditioned by repeated injections of the protein until the column characteristics remain constant or by injection of a commonly available protein (e.g., bovine serum albumin) followed by running a quick gradient. If high back-pressure is encountered, disconnect the column from the injector and run the pumps to ensure that the back-pressure is not the result of the HPLC system. If column back-pressure is high, most HPLC columns can be reversed and rinsed to flush any contaminants retained by the inlet frit. Begin rinsing at a low flow rate (e.g., 10% of normal for 10 min) and then increase to normal flow rate. If the column is contaminated, wash with either 10–20 column volumes of a "strong" eluent or run 2–3 blank gradients without sample injection, to remove strongly adsorbed contaminants. If the loss in column performance appears to be the result of adsorbed protein, it is recommended to rinse the column with a mixture of 0.1 *M* nitric acid and 2-propanol (1:4, v/v). Rinse at a low flow rate (e.g., 20% of normal) overnight. If contamination is the result of lipids or very hydrophobic small molecules, washing with several column volumes of dichloromethane or chloroform may restore column performance. When changing from water to chloroform or dichloromethane

or back again, it is important to rinse the column with a mutually miscible, intermediate solvent such as 2-propanol or acetone between the two less miscible solvents. RP-HPLC columns are stable to all common organic solvents, including acetonitrile, methanol, ethanol, 2-propanol, and dichloromethane. Silica-based RP columns are stable up to pH 7.0 and are not harmed by common protein detergents such SDS.

The RP columns are generally stable to 60°C and up to 350 bar back-pressure. RP columns can be stored in organic solvent and water. For long-term storage, the ion-pairing reagent or buffer should be rinsed from the column and the organic content raised to at least 50% before the column is stoppered.

References

1. Hancock, W. S. and Harding, D. R. K. (1984) Review of separation conditions, in *CRC Handbook of HPLC for the Separation of Amino acids, Peptides and Proteins,* vol. 2 (Hancock, W.S., ed.), CRC, Boca Raton, FL, pp. 303–312.
2. Regnier, F. E. (1987) Peptide mapping. *LC/GC* **5(5),** 392–395.
3. Dill, K. A. (1980) The mechanism of solute retention in reversed-phase liquid chromatography. *J. Phys. Chem,* **91,** 1987–1992.
4. Melander, W. and Horvath, Cs. (1980) Reversed-phase chromatography, in *HPLC—Advances and Perspectives*, vol. 2 (Horvath, Cs., ed.), Academic, New York, p. 114.
5. Hearn, M. T. W. (1980) HPLC of peptides, in *HPLC—Advances and Perspectives,* vol. 3 (Horvath, Cs., ed.), Academic, New York, p. 99.
6. Guo, D., Mant, C. T., Parker, J. M. R., and Hodges, R. S. (1986) Prediction of peptide retention times in reversed-phase HPLC: I. Determination of retention coefficients of amino acid residues of model synthetic peptides. *J.Chromatogr.* **359,** 499–517.
7. Meek, J. L. and Rossetti, Z. L. (1981) Factors affecting retention and resolution of peptides in HPLC. *J.Chromatogr.* **211,** 15–28.
8. Heinitz, M. L., Flanigan, E., Orlowski, R. C., and Regnier, F. E. (1988) Correlation of calcitonin structure with chromatographic retention in HPLC. *J.Chromatogr.* **443,** 229–245.
9. O'Hare, M. J., Capp, M. W., Nice, E. C., Cooke, N. H., and Archer, B. G. (1982) Factors influencing chromatography of proteins in short alkylsilane-bonded large pore-size silicas. *Anal. Biochem.* **126,** 17–128.
10. Pearson, J. D., Lin, N. T., and Regnier, F. E., Separation of proteins, in *HPLC of Peptides and Proteins* (Hearn, M. T. W., Wehr, C. T., and Regnier, F. E., eds.), Academic, New York, p. 81.
11. Hearn, M. T. W. (1986) Reversed-phase chromatography, in *HPLC—Advances and Perspectives,* vol. 3 (Horvath, Cs., ed.), Academic, New York, pp. 87–91.
12. Burton, W. G., Nugent, K. D., Slattery, T. K., and Summers, B. R. (1988) Separation of proteins by reversed-phase HPLC: I. Optimising the column. *J. Chromatogr.* **443,** 363–379.
13. Nugent, K. D., Burton, W. G., Slattery, T. K., and Johnson, B. F. (1988) Separation of proteins by reversed-phase HPLC: II. Optimising sample pretreatment and mobile phase conditions. *J. Chromatogr.* **443,** 381–397.
14. Ford, J. C. and Smith, J. A. (1989) Synthetic peptide purification by application of linear solvent strength gradient theory. *J. Chromatogr.* **483,** 131–143.
15. Geng, X. and Regnier, F. E. (1984) Retention model for proteins in reversed-phase liquid chromatography. *J. Chromatogr.* **296,** 15–30.
16. Kunitani, M., Johnson, D., and Snyder, L. R. (1986) Model of protein conformation in the reversed-phase separation of interleukin-2 muteins. *J. Chromatogr.* **371,** 313–333.
17. Regnier, F. E. (1983) High performance liquid chromatography of biopolymers. *Science* **222,** 245–252.
18. Van Iersel, J., Frank, J., and Duine, J. A. (1985) Determination of absorption coefficients of

purified proteins by conventional UV spectrophotometry and chromatography combined with multiwavelength detection. *Anal. Biochem.* **151**, 196–204.

14. Glajch, J. L., Kirkland, J. J., and Kohler, J. (1987) Effect of column degradation on the reversed-phase HPLC separation of peptides and proteins. *J. Chromatogr.* **384**, 81–90.

15. Sagliano, J., Flold, T. R., Hartwick, R. A., Dibussolo, J. M., and Miller, N. T. (1988) Studies on the stabilisation of reverse-phases for liquid chromatography. *J. Chromatogr.* **443**, 155–172.

16. De Vos, F. L., Robertson, D. M., and Hearn, M. T. W. (1987) Effect of mass loadability, protein concentration and *n*-alkyl chain length on the reversed-phase behaviour of bovine serum albumin and bovine follicular fluid inhibin. *J. Chromatogr.* **392**, 17–32.

17. Guo, D., Mant, C. T., and Hodges, R. S. (1987) Effects of ion-pairing reagents on the prediction of peptide retention in reversed phase HPLC. *J. Chromatogr.* **386**, 205.

29

Extraction of Membrane Proteins

Kay Ohlendieck

1. Introduction

The landmark paper on biomembrane structure in 1972 by Singer and Nicolson *(1)* is still widely accepted as an excellent working hypothesis *(2)*, but a more rigid micro domain structure is now believed to exist in native membranes *(3–5)*. The original fluid mosaic model of biomembrane structure basically defined two classes of membrane proteins that are associated, to varying degrees, with the phospholipid bilayer *(1)*. In addition to peripheral and integral membrane proteins, a third class of membrane proteins is represented by lipid-anchored proteins *(6)*. In addition to these protein–membrane interactions, certain components of the membrane cytoskeleton and the extracellular matrix are also directly or indirectly associated via binding-proteins or receptor molecules with membranes *(7)*. Treatment of biological membranes with salt solutions or change in pH usually dissociates peripheral proteins because these extrinsic membrane proteins interact with the membrane surface mostly via electrostatic and hydrogen bonds. Integral membrane proteins that possess hydrophobic surfaces are more strongly associated with the bilayer and these intrinsic proteins extend across or are partially inserted into the lipid bilayer. Extraction of integral membrane proteins is commonly accomplished by solubilizing the protein-containing membrane fraction using a variety of detergents *(8)* (*see* **Fig. 1**).

1.1. Peripheral Membrane Proteins

Because peripheral membrane proteins are extracted by relatively mild treatments, care must be taken not to partially release these membrane proteins accidentally during subcellular fractionation procedures prior to purification. Isolation of a soluble form of an extrinsic membrane protein is usually achieved by a single technique or combinations of mild extraction procedures depending on the particular properties of the protein under investigation and the starting material (i.e., a particular subcellular fraction enriched in the peripheral protein) *(9–11)*. Most extraction procedures aim at the disordering of the structure of water, the disruption of weak electrostatic interactions and hydrogen bonds, and, occasionally, weak hydrophobic interactions in order to break the interactions between the extrinsic proteins and the membrane. Following extraction for 10–30 min at 4°C, the remaining membrane bilayer and its associated integral proteins

From: *Methods in Molecular Biology, vol. 244: Protein Purification Protocols: Second Edition*
Edited by: P. Cutler © Humana Press Inc., Totowa, NJ

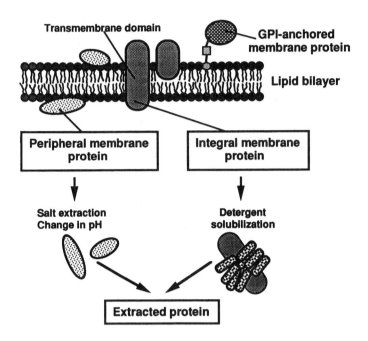

Fig. 1. Diagrammatic representation of the different kinds of membrane proteins and the most commonly employed methods to extract them from biological membranes.

are separated by centrifugation (30–60 min at 100,000*g*) and the released peripheral membrane proteins are recovered in the supernatant. The following list summarizes commonly used procedures for the extraction of peripheral membrane proteins:

1. Treatment with alkaline buffers (pH 8.0–12.0).
2. Treatment with acidic buffers (pH 3.0–5.0).
3. Use of metal chelators (10 m*M* EGTA or EDTA).
4. Treatment with high ionic strength (1 *M* NaCl or KCl).
5. Treatment with denaturing agents (i.e., urea).
6. Treatment with organic solvents (i.e., butanol).
7. Sonication of membrane fractions.

Usually, extraction procedures employing high ionic strength, alkaline, or acidic buffers and metal chelators result in a relatively distinct separation between solubilized peripheral proteins and membrane-associated integral membrane proteins. However, the nonspecific association of soluble proteins with membrane fractions and/or the entrapment of cytosolic proteins in membrane vesicles during subcellular fractionation, as well as the partial release of integral membrane proteins even under mild conditions of extraction and the existence of certain proteins in both integral and soluble isoforms makes the interpretation of these kinds of differential extraction procedures sometimes difficult *(9–11)*.

1.2. Integral Membrane Proteins

Extraction of integral membrane proteins is most conveniently achieved by the use of detergents. Detergents are amphipathic molecules that contain both hydrophobic and

Table 1
Properties of Commonly Used Solubilizing Detergents

Detergent	Relative molecular mass (monomer) (M_r)	Critical micelle concentration (cmc) (M)	Aggregation number
Triton X-100	625	3.0×10^{-4}	150
Tween 20	1320	0.9×10^{-5}	60
Brij 35	1200	9.0×10^{-5}	40
Lubrol PX	1000	4.0×10^{-6}	90
Octyl-β-D-glucoside	292	2.5×10^{-2}	90
Zwittergent 3-14	364	3.0×10^{-4}	83
CHAPS	615	1.4×10^{-3}	22
Cholate	430	1.4×10^{-3}	4
Sodium dodecyl sulfate	288	7.0×10^{-3}	62

Note: For heterogeneous compounds, average values are given. *See* **refs. 8** and *12–14* for a more elaborate listing of chemical properties of solubilizing detergents used in biological research.

hydrophilic moities, and the preferred form of detergent aggregation in water is the formation of micelles. Detergent micelles are characterized by a unique critical micelle concentration (cmc), and below the specific cmc value, individual detergent molecules predominate in solution. cmc values differ quite significantly between individual classes of detergents (i.e., the cmc for octylglucoside is 25 mM, whereas Triton X-100 exhibits a cmc of 0.3 mM) *(8,12–14)* (*see* **Table 1**). With respect to micelle size, certain detergents form relatively large micelles; that is, Triton X-100 forms micelles with approx 150 detergent molecules per micelle and a molecular mass of 90–95 kDa *(12)*. A detailed listing of detergents used in the biological sciences and a description of their individual properties can be found in a review on detergent structure by Neugebauer *(13)*. A recent article by LeMaire et al. *(8)* describes the molecular interaction between membrane proteins and lipids with solubilizing detergents. Ideally, the detergent of choice should not only be available in pure form and be relatively inexpensive for its usage in large-scale preparation, but it should also sufficiently solubilize the membrane protein under investigation without irreversibly denaturing it. Furthermore, easy removal of excess detergent from the solubilized protein fraction is an additional criterion for the choice of detergent, which will be discussed in Chapter 30. For comprehensive information on the availability, purity, unit prices, and sources of commonly used detergents for the solubilization of biological membranes, *see* the review article by Jones et al. *(14)*.

A good initial selection of suitable detergents for pilot experiments on the extraction of novel integral membrane proteins with little knowledge of their physical and chemical properties is summarized in **Fig. 2**. As extensively reviewed by Helenius and Simons *(12)*, detergents may be classified by their overall chemical structure as type A or type B detergents and further subdivided according to their electric charge as nonionic, ionic, or zwitterionic detergents. Whereas type A detergents exhibit flexible hydrophobic tails and hydrophilic head groups, type B detergents are more rigid and are cholesterol-based structures with amphiphilic properties (*see* **Fig. 2**). Commonly used type A detergents are the nonionic components Triton X-100, octylglucoside, and Lubrol PX, as well as the zwitterionic detergent sulfobetaine 14 (Zwittergent 3-14). The nonionic

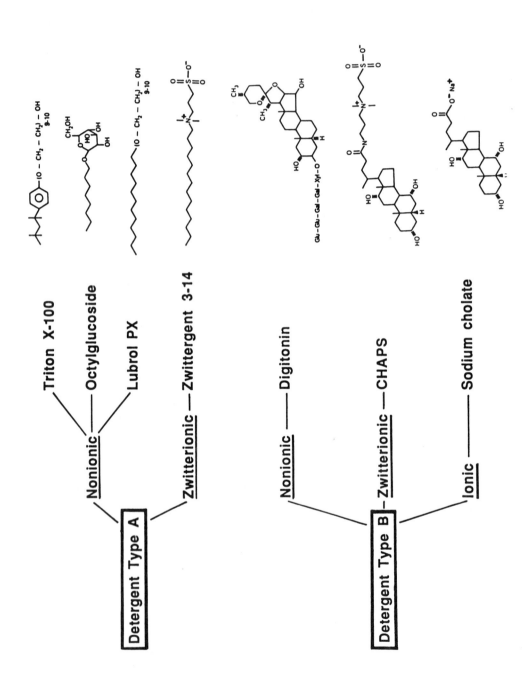

Fig. 2. Commonly used detergents employed in the solubilization of biological membranes.

detergent digitonin, the zwitterionic component CHAPS, and the ionic bile salt sodium cholate are typical type B detergents extensively used in biological research *(8,12–14)*. Many other detergents exist that are closely related to the described type A and type B detergents and are listed in recent review articles on membrane solubilization *(8,13,14)*. The ionic detergent sodium dodecyl sulfate (SDS), although widely used in analytical and preparative polyacrylamide gel electrophoresis techniques *(15)*, is not usually employed in the initial extraction of membrane proteins *(8)* and is, therefore, not further discussed in this chapter.

1.3. Experimental Design

Although the design of an extraction experiment may partially be based on the physical and chemical properties of the specific detergent used and the biological characteristics of the particular integral membrane protein and subcellular fraction under investigation, often a series of small-scale trial-and-error experiments is a good start to determine conditions for optimum solubilization. In pilot experiments, aliquots of the particular biological membrane are incubated with different concentrations of a variety of commonly used detergents. Initially, standard conditions are used with respect to buffer composition, salt solutions, temperature, and incubation time. In order to optimize the extraction process, these conditions may be varied. Most importantly, the denaturing properties of individual detergents, compatibility of detergents with divalent cations, the pH dependency of detergent solubility, spectral properties of detergents, as well as electrostatic and temperature effects on detergent behavior should be taken into account when choosing the appropriate buffer conditions and the ratio of detergent to protein *(8)*. Effective solubilization is usually achieved in a well-buffered, physiological pH environment and the addition of 0.15 M NaCl to the extraction mixture for optimum ionic strength. Membrane preparations are used at a protein concentration of 1–5 mg/mL and are solubilized by detergent concentrations of 0.1–5% (v/v). By convention, retention of a membrane protein in the supernatant following centrifugation for 60 min at 100,000g after solubilization defines this protein operationally as soluble *(8,13)*. However, in large-scale preparations shorter spins and lower g-forces might be sufficient to extract a high yield of a particular membrane protein. To determine whether the integral protein of interest was sufficiently solubilized by a particular concentration of a detergent, a relatively simple and fast assay procedure should be chosen. Enzyme assays or ligand-binding assays, as well as cell biological test systems are certainly very convenient for evaluating the efficiency of the extraction protocol and might also be useful in determining the potential loss of biological activity during the solubilization procedure. If a specific labeled probe or antibody is available to the integral membrane protein to be solubilized, overlay assays with immobilized protein fractions and immunoblot analysis using mini-gel systems can be performed reasonably quickly and with large numbers of samples.

To illustrate the use of commonly employed methods and reagents for the extraction of peripheral and integral membrane proteins, this chapter covers the initial isolation of a typical extrinsic protein, calsequestrin, and that of a typical intrinsic protein, Ca^{2+}-ATPase. Both proteins are major constituents of the sarcoplasmic reticulum from skeletal muscle. The Ca^{2+}-ATPase is localized to the longitudinal tubules and is responsible for the resequestration of calcium into the lumen of the sarcoplasmic reticulum following muscle contraction *(16–18)*. Calsequestrin represents a high-capacity, intermediate-affinity calcium-binding protein localized to the luminal site of the junctional sarcoplasmic reticulum and increases the calcium holding capacity of the terminal cisternae in muscle *(19–21)*. These properties make both proteins key components of calcium homeostasis and excitation–contraction coupling in muscle *(22)*.

2. Materials

2.1. Preparation of Sarcoplasmic Reticulum From Rabbit Skeletal Muscle

1. Dissection kit; good quality scissors for fine mincing of muscle tissue.
2. Ice bucket or tray, sturdy glass plate (size of the ice-containing bucket or tray).
3. Skeletal muscle from a New Zealand white rabbit (100 g).
4. Waring blender.
5. Buffer A (homogenization buffer): 10% (w/v) sucrose, 20 mM histidine, pH 7.0, 0.5 mM EDTA, 0.23 mM phenylmethylsufonyl fluoride (PMSF), 0.83 mM benzamidine (*see* **Note 1**).
6. Refrigerated high-speed centrifuge.
7. Cheesecloth.
8. Ultracentrifuge with fixed-angle rotor.
9. Small glass homogenizer (for resuspension of pellets).
10. Buffer B: 0.6 M KCl, 30 mM histidine, pH 7.0, 0.23 mM PMSF, 0.83 mM benzamidine.
11. Buffer C (storage buffer): 10% (w/v) Sucrose, 30 mM histidine, pH 7.4, 0.23 mM PMSF, 0.83 mM benzamidine.

2.2. Extraction of the Extrinsic Calcium-Binding Protein Calsequestrin From Sarcoplasmic Reticulum

1. Buffer D (extraction buffer): 100 mM Sodium carbonate, pH 11.4.
2. Ultracentrifuge and fixed-angle rotor.
3. Magnetic stirrer.
4. Concentrated HCl and a pH meter.
5. NaCl.
6. Morpholinopropanesulphonic acid (MOPS).
7. Dithiothreitol (DTT).

2.3. Assay of Calcium Binding to Calsequestrin

1. Standard dialysis tubing (with a molecular weight cutoff of approx 10,000).
2. Buffer E (to pretreat dialysis membrane): 50 mM EDTA, 100 mM sodium carbonate.
3. Heat plate and large glass beaker (to boil dialysis tubing).
4. Buffer F (calcium-binding buffer): 5 mM Tris-HCl, pH 7.5, 0.1 mM $^{45}CaCl_2$ (40,000 cpm/mL).
5. Scintillation fluid
6. Scintillation spectrometer

2.4. Extraction of the Intrinsic Enzyme Ca^{2+}-ATPase From Sarcoplasmic Reticulum

1. Buffer G: 10% (w/v) Sucrose, 50 mM sodium phosphate, pH 7.4, 1 M KCl, 0.23 mM PMSF, 0.83 mM benzamidine.
2. Ultracentrifuge and fixed-angle rotor.
3. Magnetic stirrer.
4. 10% (w/v) Sodium deoxycholate.

2.5. Assay of Ca²⁺-ATPase Activity

1. Stock solutions for assay buffer: 100 mM histidine, pH 7.4; 1 M KCl; 50 mM MgCl$_2$; 1 mM CaCl$_2$; 10 mM EGTA; 25 mM ATP.
2. Stock solutions for phosphate determination:
   a. Ammonium molybdate: 28.6 g Ammonium molybdate are dissolved in 500 mL of 6 M HCl.
   b. Polyvinyl alcohol: 11.6 g Polyvinyl alcohol are dissolved in 500 mL of boiling water and allowed to cool.
   c. Malachite green: 0.81 g Malachite green are dissolved in 1 L of distilled water.
   d. Mixed reagent: Two volumes of malachite green are mixed with 1 vol each of polyvinyl alcohol and ammonium molybdate and 2 vol of distilled water. Allow the reagent to stand at room temperature until it turns a golden-yellow color, which takes approx 30 min.
3. Potassium hydrogen phosphate standard (1.74–17.42 µg/mL distilled water).
4. Spectrophotometer.

3. Methods

Isolation of membrane vesicles derived from sarcoplasmic reticulum of skeletal muscle was based on the method of Eletr and Inesi *(23)*. Ca²⁺-ATPase activity and calcium binding was assayed according to Meissner et al. *(24)* and MacLennan *(25)*, respectively. Inorganic phosphate was determined by the method of Chan et al. *(26)* and measurement of protein concentration was performed according to Hartree *(27)* using bovine serum albumin as a standard. Extraction procedures for the extrinsic membrane protein calsequestrin and the intrinsic membrane protein Ca²⁺-ATPase were based on the protocols of Cala and Jones *(28)* and Warren et al. *(29)*, respectively. All preparative steps are performed at 4°C unless otherwise stated and all g-forces are given as $g_{av.}$

3.1. Preparation of Sarcoplasmic Reticulum From Rabbit Skeletal Muscle

1. Mince 100 g of trimmed white rabbit skeletal muscle (*see* **Note 2**) with fine scissors on a sturdy glass plate positioned on top of a tray of crushed ice. Disperse the tissue pieces in 300 mL of ice-cold buffer A.
2. Homogenize the muscle tissue in a Waring blender, at full speed, four times for 15 s.
3. Centrifuge the homogenate for 20 min at 15,000g and filter the supernatant through three layers of washed cheesecloth and recentrifuge at 40,000g for 90 min.
4. Resuspend the pellet in 50 mL of buffer B and incubate the suspension for 40 min on ice and then centrifuge at 15,000g for 20 min. Recentrifuge the supernatant at 40,000g for 90 min and resuspend the pellet in buffer C. At this stage, the crude sarcoplasmic reticulum preparation may be quick-frozen in liquid nitrogen and stored at −70°C until future usage (*see* **Note 3**).

3.2. Extraction of the Extrinsic Calcium-Binding Protein Calsequestrin From Sarcoplasmic Reticulum

1. Pellet freshly thawed sarcoplasmic reticulum, prepared as described in **Subheading 3.1.**, at 100,000g for 30 min and resuspend in ice-cold buffer D to a protein concentration of 1.5 mg/mL.

Table 2
Calcium Binding to Calsequestrin Extracted From Sarcoplasmic Reticulum

Step	Total protein (mg)	Specific Ca^{2+} binding (nmol Ca^{2+}/mg protein)
Sarcoplasmic reticulum	80	175
Carbonate solubilized supernatant	27	280
Purified calsequestrin	—	713

2. Incubate the suspension on ice for 30 min and then centrifuge at 100,000*g* for 30 min, yielding a pellet and the carbonate-extracted supernatant fraction containing calsequestrin.
3. Stir the supernatant magnetically and make the suspension to 50 m*M* MOPS, 0.5 *M* NaCl, and 1 m*M* DTT by the addition of solid reagents and adjust the pH to 7.0 by the dropwise addition of concentrated HCl (*see* **Note 4**).

3.3. Assay of Calcium Binding to Calsequestrin

1. Boil a sufficiently long piece of thin dialysis tubing for 10 min in buffer E, followed by boiling for 10 min in distilled water, and then rinse the tubing extensively in distilled water (*see* **Note 5**).
2. Place a 0.4-mg protein sample in a 1-mL volume and dialyze against 100 mL of buffer F for 24 h in the cold room at 4°C. Use all of the necessary precautions when handling the radioactive isotope (*see* **Note 6**).
3. Dissolve 0.1-mL samples from both inside and outside of the dialysis bag, each in 10 mL of scintillation fluid and count the specific radioactivity in a scintillation spectrometer (*see* **Note 7**).
4. Assuming that the dialysis reached an equilibrium, calculate Ca^{2+} binding per milligram of protein from the increased radioactivity within the dialysis bag (*see* **Table 2**).

3.4. Extraction of the Intrinsic Enzyme Ca^{2+}-ATPase From Sarcoplasmic Reticulum

1. Centrifuge freshly thawed sarcoplasmic reticulum, prepared as described in **Subheading 3.1.**, at 100,000*g* for 30 min and resuspend the pellet in ice-cold buffer G.
2. Add sodium deoxycholate from a 10% (w/v) detergent stock solution slowly under magnetic stirring and on ice to the membrane suspension at a ratio of 0.4 mg detergent/mg protein.
3. Centrifuge the resulting mixture at 100,000*g* for 60 min and collect the clear supernatant containing the solubilized Ca^{2+}-ATPase (*see* **Note 8**).

3.5. Assay of Ca^{2+}-ATPase Activity

1. In a microcuvet, mix 10 µL of protein sample with 390 µL of distilled water and with 100 µL each of the assay stock solutions, including histidine, KCl, $MgCl_2$, as well as calcium (to measure total ATPase activity) or EGTA (to measure basal ATPase activity).
2. Initiate the enzyme reaction by the addition of 200 µL of ATP stock solution.
3. After an incubation time of 5 min, terminate the reaction by the addition of 2 mL of mixed reagent (malachite green–molybdate–polyvinyl alcohol reagent) (*see* **Note 9**).
4. Immediately after the addition of the mixed reagent, measure color development at 630 nm against a mixed reagent blank.
5. Compare your measurements with a potassium dihydrogen phosphate standard graph.

Table 3
Enzyme Activity of Ca²⁺-ATPase Extracted From Sarcoplasmic Reticulum

Sample	Protein (mg)	Ca^{2+}-ATPase activity (μmol P_i/mg protein/min)
Sarcoplasmic reticulum	52	2.1
Detergent solubilized supernatant	29	3.9
Purified Ca^{2+}-ATPase	—	8.1

6. Calculate the Ca^{2+}-ATPase activity by subtracting the basal ATPase activity (in the presence of EGTA) from the total ATPase activity (in the presence of calcium) (*see* **Note 10** and **Table 3**).

4. Notes

1. No general composition of a protease inhibitor cocktail suitable for all homogenization or extraction procedures can be given. To prevent proteolytic degradation of calsequestrin and the Ca^{2+}-ATPase during isolation from rabbit skeletal muscle, the addition of EDTA, PMSF, and benzamidine to the isolation buffer appears to be sufficient. Other widely used protease inhibitors include antipain, aprotinin, bestatin, chymostatin, E-64, leupeptin, pefabloc, pepstatin, and phosporamidon. Some companies (i.e., Boehringer-Mannheim) sell protein inhibitor sets for small-scale pilot experiments in order to determine the minimum concentration and composition of a protease inhibitor cocktail sufficient for preventing proteolytic degradadtion in a novel purification protocol.

2. Skeletal muscle tissue should be quickly trimmed of fat tissue. Avoid contaminating the isolation buffer and blender with rabbit hair. Fine mincing of muscle tissue should preferentially be performed in a cold room at 4°C with the muscle positioned on top of an ice-cold glass plate.

3. To further purify the sarcoplasmic reticulum, the crude preparation of **Subheading 3.1.** may be centrifuged through a 26–40% sucrose gradient for 150 min at 100,000g. The upper half of the sucrose density gradient contains the purified sarcoplasmic reticulum vesicles with considerably lower amounts of contaminating mitochondrial membranes and vesicles derived from the transverse tubular membrane system.

4. Solubilized calsequestrin, extracted by alkaline treatment as described in **Subheading 3.2.**, can be very simply purified to homogeneity using phenyl–sepharose chromatography *(28)*. Bound calsequestrin is eluted from the column by 10 mM CaCl$_2$ and this results in a homogeneous protein preparation.

5. Alternatively, gentler washing of dialysis tubing may be performed at 60°C by incubation for 2 h in buffer E followed by extensive washing in distilled water. To avoid time-consuming washing of dialysis tubing in order to remove impurities and heavy metals, more expensive brands of tubing can be purchased that have been prewashed and sterilized (Spectrum, Los Angeles, CA). However, irrespective of the brand of tubing, handle dialysis tubing at all times with gloves to avoid contamination of protein samples with proteases and other impurities from your hands.

6. When using ^{45}CaCl$_2$ solutions, handle the radioactive isotope with extreme caution. Work exclusively in an area of the cold room designated for radioactive work and confine radioactive experiments to a minimum space. Wear double plastic gloves and protective clothing at all times when handling radioactive solutions and clearly mark all radioactive glassware and equipment. When removing labeled sample from the inside of the tubing fol-

lowing equilibrium dialysis, ask someone to help you in order to avoid spillage of radioactive solution when transfering it to the scintillation cocktail.

7. If only small quantities of membrane vesicles are available for analysis, the equilibrium dialysis system can be miniaturized. Place 0.1 mg protein into a Slide-A-Lyzer Mini Dialysis MWCO-10000 cassette system (Pierce & Warriner, Chester, Cheshire, UK) with a maximum volume of 0.25 mL and dialyze against 100 mL of 5 mM Tris-Cl, pH 7.5, 0.1 mM $^{45}CaCl_2$ (20,000 cpm/mL) for 24 h at 4°C. The subsequent calculation of ion binding is then performed in the same way as described in **Subheading 3.3.**

8. Solubilized Ca^{2+}-ATPase may be purified to homogeneity by centrifuging the solubilized preparation (**Subheading 3.4.**) through a linear 20–60% sucrose density gradient at 100,000g for 24 h *(29)*. The peak fractions with the highest specific Ca^{2+}-ATPase activity are usually found in the lower part of the sucrose gradient and are greatly depleted of excess detergent.

9. An alternative way to determine Ca^{2+}-ATPase activity is a coupled enzyme assay. Instead of measuring the production of inorganic phosphate, the increase of ADP is measured by coupling the ATPase reaction to the widely employed pyruvate kinase/lactate dehydrogenase system. The production of NAD can conveniently be measured at 340 nm.

10. Contaminating mitochondrial ATPase activity, possibly present in crude membrane preparations, can be inhibited by the addition of 5 mM NaN_3. The presence of inorganic phosphate in the protein sample prior to initiation of the enzyme reaction or nonenzymatic hydrolysis of ATP can be accounted for by control assays using the sample but no ATP or just ATP with no sample.

Acknowledgments

Research in the author's laboratory has been supported by project grants from the European Commission, Enterprise Ireland, and the Irish Health Research Board.

References

1. Singer, S. J. and Nicholson, G. L. (1972) The fluid mosaic model of the structure of cell membranes. *Science* **175**, 720–731.
2. Jacobson, K., Sheet, E. D., and Simson, R. (1995) Revisiting the fluid mosaic model of membranes. *Science* **268**, 1441–1442.
3. Kinnunen, P. K. (1991) On the principles of functional ordering in biological membranes. *Chem. Phys. Lipids* **57**, 375–399.
4. Tocanne, J. F., Cezanne, L., Lopez A., et al. (1994) Lipid domains and lipid/protein interactions in biological membranes. *Chem. Phys. Lipids* **73**, 139–158.
5. Somerharju, P., Virtanen, J. A., and Cheng, K. H. (1999) Lateral organisation of membrane lipids. The superlattice view. *Biochim. Biophys. Acta* **1440**, 32–48.
6. Varma, R. and Mayor, S. (1998) GPI-anchored proteins are organized in submicron domains at the cell surface. *Nature* **394**, 798–801.
7. Ohlendieck, K. (1996) Towards an understanding of the dystrophin–glycoprotein complex: linkage between the extracellular matrix and the membrane cytoskeleton in muscle fibres. *Eur. J. Cell Biol.* **69**, 1–10.
8. LeMaire, M., Champeil, P., and Moller, J. V. (2000) Interaction of membrane proteins and lipids with solubilizing detergents. *Biochim. Biophys. Acta* **1508**, 86–111.
9. Penefsky, H. S. and Tzagoloff, A. (1971) Extraction of water-soluble enzymes and proteins from membranes. *Methods Enzymol.* **22**, 204–219.
10. Thomas, T. C. and McNamee, M. G. (1990) Purification of membrane proteins. *Methods Enzymol.* **182,** 499–520.

11. Scopes, R. K. (1994) *Protein Purification: Principles and Practice*, 3rd ed., Springer-Verlag, New York.

12. Helenius, A. and Simons, K. (1975) Solubilization of membranes by detergents. *Biochim. Biophys. Acta* **415**, 29–79.

13. Neugebauer, J. M. (1990) Detergents: an overview. *Methods Enymol.* **182**, 239–252.

14. Jones, O. T., Earnest, J. P., and McNamee, M. G. (1987) Solubilization and reconstitution of membrane proteins, in *Biological Membranes, a Pactical Approach* (Findlay, J. B. C. and Evans, W. H., eds.), IRL, Oxford, pp. 139–172.

15. LaemmLi, U. K. (1970) Cleavage of structural proteins during the assembly of the head of bacteriophage T4. *Nature (London)* **227**, 680–685.

16. MacLennan, D. H. (1970) Purification and characterization of an adenosine triphosphate from sarcoplamic reticulum *J. Biol. Chem.* **245**, 4508–4518.

17. MacLennan, D. H., Brandl, C. J., Korczak, G., and Green, N. M. (1985) Amino-acid sequence of a $Ca^{2+}+Mg^{2+}$-dependent ATPase from rabbit sarcoplasmic reticulum, deduced from its complementary DNA sequence. *Nature* **316**, 696–701.

18. MacLennan, D. H., Rice, W. J., and Green, N. M. (1997) The mechanism of Ca^{2+} transport by sarco(endo)plasmic reticulum Ca^{2+}-ATPases. *J. Biol. Chem.* **272**, 28,815–28,818.

19. MacLennan, D. H. and Wong, P. T. S. (1971) Isolation of a calcium-sequestrating protein from sarcoplasmic reticulum. *Proc. Natl. Acad. Sci. USA* **68**,1231–1235.

20. Fliegel, L., Ohnishi, M., Carpenter, M. R., Khanna, V. K., Reithmeier, R. A. F., and MacLennan, D. H. (1987) Amino acid sequence of rabbit fast-twitch skeletal muscle calsequestrin deduced from cDNA and peptide sequencing. *Proc. Natl. Acad. Sci. USA* **84**, 1167–1171.

21. Yano, K. and Zarain-Herzberg, A. (1994) Sarcoplasmic reticulum calsequestrins: structural and functional properties. *Mol. Cell Biochem.* **135**, 61–70.

22. Murray, B., Froemming, G. R., Maguire, P. B., and Ohlendieck, K. (1998) Excitation–contraction–relaxation cycle: role of Ca^{2+}-regulatory membrane proteins in normal, stimulated and pathological skeletal muscle fibres. [review]. *Int. J. Mol. Med.* **1**, 677–697.

23. Eletr, S. and Inesi, G. (1972) Phospholipid orientation in sarcoplasmic membranes: spin-label ESR and proton NMR studies. *Biochim. Biophys. Acta* **282**, 174–179.

24. Meissner, G., Conner, G. E., and Fleischer, S. (1973) Isolation of sarcoplasmic reticulum by zonal centrifugation and purification of Ca^{2+}-pump and Ca^{2+}-binding proteins. *Biochim. Biophys. Acta* **298**, 246–269.

25. McLennan, D. H. (1975) Isolation of proteins of the sarcoplasmic reticulum. *Methods Enzymol.* **32**, 291–302.

26. Chan, K. M., Delfert, D., and Junger, K. D. (1986) A direct colorimetric assay for Ca^{2+}-stimulated ATPase activity. *Anal. Biochem.* **157**, 375–380.

27. Hartree, E. F. (1972) Determination of protein: a modification of the Lowry method that gives a linear photometric response. *Anal. Biochem.* **48**, 422–427.

28. Cala, S. E. and Jones, L. R. (1983) Rapid purification of calsequestrin from cardiac and skeletal muscle sarcoplasmic reticulum vesicles by Ca^{2+}-dependent elution from Phenyl–Sepharose. *J. Biol. Chem.* **258**, 11,932–11,936.

29. Warren, G. B., Toon, P. A., Birdsall, N. J. M., Lee, A. G., and Metcalfe, J. C. (1974) Reconstitution of a calcium pump using defined membranecomponents. *Proc. Natl. Acad. Sci. USA* **71**, 622 626.

30

Removal of Detergent From Protein Fractions

Kay Ohlendieck

1. Introduction

The solubilization of biological membranes by detergents plays an important role in the identification, characterization, and extraction of integral membrane proteins. The solubilization process involves a number of intermediate states starting with the destabilization of membrane lipids and followed by the creation of membrane fragments, which finally results in the formation of protomers *(1)*. The most commonly used detergents in the biological sciences are described in Chapter 29 and several review articles exist on solubilization procedures *(2–4)*. Because the initial extraction of intrinisic membrane proteins usually involves high detergent concentrations, excess detergent has to be removed or exchanged for another type of detergent at later stages of preparative or analytical procedures involving the solubilized protein fraction. The efficiency of certain chromatographic procedures is often significantly improved by lowering the overall detergent concentration or by exchanging one type of detergent for another. Lectin chromatography, a powerful technique to affinity-purify subsets of glycoproteins (*see* Chapter 18), appears to be especially sensitive to high concentrations of a variety of detergents *(5)*. Furthermore, because high concentrations and/or certain types of detergent interfere with many physical and chemical analyses and detergents also exhibit undesirable side effects on highly sensitive cell biological assays, detergent removal is, in many cases, of central importance for retaining the biological activity of isolated membrane proteins. Another important area of detergent removal is reconstitution studies with hydrophobic membrane proteins that exhibit vectorial transport *(4)*.

This chapter summarizes some of the techniques employed in removing or exchanging detergents used in the solubilization of biological membranes. For a more in-depth discussion, see recent reviews on procedures of detergent removal *(6–9)*. Technical bulletins from companies selling commercially available resins for detergent removal (i.e., Biobeads SM-2 [Bio-Rad], Extracti-Gel D [Pierce)], or SDS-Out precipitation kit [Pierce]) usually contain a list of references relevant to conditions recommended for detergent removal. It is well worth studying these general guidelines and recommendations prior to attempting to remove or exchange detergent from a scarce, novel protein sample. Detergents are amphipathic molecules and their preferred form of aggregation

From: *Methods in Molecular Biology, vol. 244: Protein Purification Protocols: Second Edition*
Edited by: P. Cutler © Humana Press Inc., Totowa, NJ

Table 1
Techniques Used for the Removal or Exchange
of Detergents From Protein Fractions

1. Equilibrium dialysis
2. Batch or column chromatography
   - Hydrophobic adsorbtion
   - Ion-exchange chromatography
   - Gel filtration
   - Affinity chromatography
   - Lectin chromatography
3. Sucrose density gradient centrifugation
4. Protein precipitation procedures
5. Phase partioning
6. Electroelution techniques

Note: Suitability of individual techniques for detergent removal depends on the critical micelle concentration, aggregation number, and hydrophobic-lipophile balance of the individual detergent (*see* text).

in water is the formation of micelles whose size and molecular weight may vary considerably between different types of detergent (*1–4*). Thus, the most important properties of a detergent with respect to removal are its unique critical micelle concentration (cmc), its hydrophile–lipophile balance (HLB), as well as its micellar molecular weight (mMW) determined by the aggregation number of detergent molecules (*8*).

The most commonly employed techniques for detergent removal from protein fractions are summarized in **Table 1** and are based on physical and chemical differences between protein–detergent complexes and detergent micelles. The suitability of individual techniques depends on the unique properties of the detergent used and the choice of procedure is, furthermore, strongly dependent on the concentration range of the protein fraction to be depleted of detergent. Probably the simplest and least labor-intensive method of removing detergent from protein fractions is dialysis (*see* **Fig. 1**). Using large external volumes and frequent changes of dialysis medium, ionic detergents with a relatively high cmc can quite successfully be removed. However, it can take considerable time to reach an equilibrium and this can lead to undesirable side effects, such as protein degradation. Excellent examples of detergent removal, routinely performed by dialysis, are reconstitution experiments (*4*). Alternatively, incorporation of hydrophobic membrane proteins into lipid vesicles can be achieved by gel filtration or dilution procedures.

A variety of chromatographic techniques are available to remove or exchange detergents. This work is more detailed than dialysis, but takes less overall time and thereby tends to keep protein degradation during detergent removal to a minimum. Hydrophobic adsorption chromatography is certainly a very convenient way of exchanging different classes of detergent (*see* **Fig. 1**). The exchange of alkyl detergents (i.e., octyl glucoside and dodecyl sulfate) for Triton X-100-type detergents using Phenyl–Sepharose (Pharmacia) was reported by Robinson et al. (*10*). Reasonably inexpensive matrixes for the specific binding of detergents are commercially available from Bio-Rad (Bio-Beads SM-2) and Pierce (Extracti-Gel D). Both adsorbents, however, require a high enough protein concentration to avoid losses in recovering protein samples during detergent re-

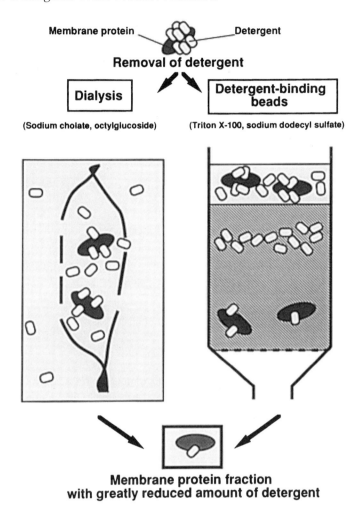

Fig. 1. Commonly used methods to remove detergent from protein fractions.

moval. The preincubation of columns with bulk carrier proteins in order to saturate non-specific protein-binding sites might at least partially solve this problem. Alternatively, protein–detergent complexes could be bound to an affinity matrix and eluted following washing and/or exchange with a different detergent. Affinity matrixes with immobilized ligands for receptor binding or a variety of lectin columns highly specific for binding to subsets of glycosylated membrane proteins are suitable for these kinds of procedure. However, high detergent concentrations often adversely affect ligand–protein interactions, and dilution prior to application to the affinity column is beneficial. Another way to bind charged membrane proteins and remove excess detergent by extensive washing is ion-exchange chromatography. Elution of the bound protein fraction can be achieved by increasing the ionic strength or addition of an ionic detergent *(6,8)*. Furthermore, if the difference in size between detergent micelles and protein–detergent complexes is large enough, gel filtration chromatography can be employed to exchange detergents in a protein fraction *(6,8)*. The use of sucrose gradient centrifugation in the removal of excess detergent was demonstrated by Warren et al. *(11)*, who could successfully separate excess deoxycholate from an integral membrane protein by this method. Precipitation of solubilzed membrane proteins from aqueous solutions can be achieved by treatment with polyethylene glycol, and phase partioning can also be exploited to precipitate hydrophobic integral proteins *(6,8)*. Finally, electroelution is a widely used method to recover protein samples separated by SDS-polyacrylamide gel electrophoresis (SDS-PAGE) (*see* Chapter 34) and many reasonably priced electroelution units are now commercially available.

To illustrate the practical aspects involved in the removal of detergent from protein fractions, equilibrium dialysis and detergent adsorbtion chromatography are described in more detail in **Subheading 3**. For a typical protocol of detergent-exchange chromatography, see the article by Robinson et al. *(10)*. Because the unique properties of individual classes of integral membrane proteins and their interaction with a variety of ionic, nonionic, and zwitterionic detergents cannot be predicted adequately, the procedures described are only general outlines and do not describe the removal of detergent from a specific solubilized membrane protein. *See* quoted research papers and review articles for details on specific requirements with respect to individual membrane proteins.

2. Materials

2.1. Dialysis

1. Dialysis tubing or dialysis casset systems (with a molecular mass cutoff of approx 10,000).
2. Wash buffer: 100 mM NaCO$_3$, 50 mM EDTA.
3. Hot plate and large glass beaker.
4. Reliable, leakproof plastic clamps for closing dialysis tubing.
5. Dialysis buffer: 20 mM Tris-HCl, pH 7.4, 0.15 M NaCl.
6. Large beaker (4–6 L).
7. Small plastic funnel.
8. Magnetic stirrer and suitable large stir bar.

2.2. Detergent Adsorption Chromatography

1. Small columns (1- to 5-mL bed volume).
2. Detergent adsorbtion matrix (i.e., macroporous Bio-Beads SM-2 [BioRad] or Extracti-Gel D [Pierce]) or any other commercially available detergent adsorption matrix.
3. Blocking buffer: 0.1% (w/v) Bovine serum albumin in 50 mM Tris-HCl, pH 7.4, 0.15 M NaCl.
4. Washing buffer: 50 mM Tris-HCl, pH 7.4, 0.15 M NaCl.
5. Small peristaltic pump and suitable tubing.

3. Methods

All procedures are performed in a cold room at 4°C unless otherwise stated. Because highly concentrated stock solutions of detergents are potential skin irritants, proper protective clothing and gloves should be worn during handling of these chemicals. Face masks should be used when weighing out powered forms of potential lung irritants such as sodium dodecyl sulfate. Furthermore, the toxic nature of substances such as digitonin should be taken into account when working with high concentrations of this cardiac glycoside.

3.1. Dialysis

1. Take a sufficiently long piece of standard dialysis tubing and boil it for 10 min in washing buffer, followed by boiling for 10 min in distilled water, and extensive washing in distilled water. Alternatively, use prewashed dialysis tubing or a Slide-A-Lyzer Mini Dialysis cassette systems from Pierce. For example, the miniaturized MWCO-10000 cassette allows for a maximum volume of 0.25 mL and can thus be used for the dialysis of relatively small biological samples.
2. Transfer the solubilized membrane protein fraction with the aid of a small funnel into the dialysis tubing, which is securely closed at the lower end (*see* **Note 1**) or with a clean syringe into the dialysis cassette system.
3. Close the dialysis tubing after removal of any air bubbles possibly introduced during transfer of the detergent-containing suspension and allow for a small increase in volume during equilibrium dialysis. Generally, dialysis tubing is very sturdy and leakage is not a problem as long as the tubing is tightly closed.
4. Dialysis should be performed with large external volumes (4–6 L) and adequate stirring, as well as frequent exchanges of the external solution (*see* **Note 2**).
5. At the end of the dialysis, wash the outside of the tubing or dialysis cassette and carefully remove the dialysed protein fraction (*see* **Note 3**).

3.2. Detergent Adsorption Chromatography

1. The protein fraction to be treated with respect to detergent removal or detergent exchange should have a relatively high protein concentration (*see* **Note 4**) and be of large enough relative molecular mass to avoid entrapment in the pores of the affinity matrix (*see* **Note 5**).
2. Wash the detergent-removing column matrix first with distilled water, then equilibrate it thoroughly with blocking buffer (*see* **Note 6**), and wash with 2 column volumes of washing buffer.
3. Apply the protein sample, preferentially dissolved in the washing buffer or another suitable buffer for optimum detergent binding, to the equilibrated column.
4. Collect 0.5- to 1-mL fractions and combine the protein peak fractions. The protein concentration can conveniently be measured by microprotein assays to avoid substantial losses resulting from assaying. Peak fractions may then be concentrated by ultrafiltration prior to subsequent analytical or preparative procedures.

4. Notes

1. Dialysis tubing should be carefully closed by tight knotting and preferentially secured by special leakproof plastic clamps (Spectrum, Los Angeles, CA).
2. Because reaching equilibrium in dialysis procedures can be quite time-consuming, the placement of a suitable matrix to bind detergent outside of the dialysis tubing might accelerate the process. Whereas nonionic detergent might be trapped by a hydrophobic adsorbtion matrix, ionic detergents might be removed by the use of an ion-exchange matrix.
3. Following equilibrium dialysis, avoid losing dialyzed samples when removing them from the tubing. Dialysis bags can be expanded following extensive dialysis and should be carefully opened surrounded by a larger, clean glass beaker to avoid any accidental spillage.
4. Generally, a very important requirement for avoiding substantial losses of protein during procedures of detergent removal is a high enough protein concentration of the starting material. Especially chromatographic techniques and precipitation procedures might result in a severe loss of proteins from very dilute solutions. Ideally, protein concentrations above 1 mg/mL should be used to avoid these problems. Thus, concentrating a protein fraction prior

to detergent removal by a suitable type of ultrafiltration is recommended, although these techniques are usually also not without danger of losing protein samples resulting from nonspecific binding to membrane filters. These kinds of problems have to be worked out with every new class of solubilized membrane protein and no general strategy with respect to optimizing overall protein recovery can be given.

5. Using hydrophobic adsorbtion chromatography to remove or exchange detergent, the molecular weight of the protein to be recovered should be large enough to avoid entrapment of smaller peptides within the pores of the support matrix. The chromatographic matrix of commercially available columns exhibits exclusion limits between 2 and 10 kDa.

6. If dilute starting material cannot be concentrated, the detergent-exchange columns should be pretreated with solutions of bulk protein to avoid substantial protein losses. Bovine serum albumin is usually very useful in blocking nonspecific binding sites for protein on affinity matrixes. This precaution should significantly lower the loss of dilute protein samples during detergent removal or exchange.

Acknowledgments

Research in the author's laboratory has been supported by project grants from the European Commission, Enterprise Ireland, and the Irish Health Research Board.

References

1. LeMaire, M., Champeil, P., and Moller, J. V. (2000) Interaction of membrane proteins and lipids with solubilizing detergents. *Biochim. Biophys. Acta* **1508**, 86–111.
2. Helenius, A. and Simons, K. (1975) Solubilization of membranes by detergents. *Biochim. Biophys. Acta* **415**, 29–79.
3. Neugebauer, J. M. (1990) Detergents: an overview. *Methods Enzymol.* **182**, 239–253.
4. Jones, O. T., Earnest, J. P., and McNamee, M. (1987) Solubilization and reconstitution of membrane proteins, in *Biological Membranes, a Practical Approach* (Findlay, J. B. C. and Evans, W. H., eds.), IRL, Oxford, pp. 139–177.
5. Lotan, R., Beattie, G., Hubbel, W., and Nicolson, G. L. (1977) Activities of lectins and their immobilized derivatives in detergent solutions. Implications on the use of lectin affinity chromatography for the purification of membrane glycoproteins. *Biochemistry* **16,** 1787–1794.
6. Hjelmeland, L. M. (1990) Removal of detergents from membrane proteins. *Methods Enzymol.* **182**, 277–282.
7. Furth, A. J., Bolton, H., Potter, J. and Priddle, J. D. (1985) Separating detergents from proteins. *Methods Enzymol.* **104,** 318–328.
8. Furth, A. J. (1980) Removing unbound detergent from hydrophobic proteins. *Anal. Biochem.* **109**, 207–215.
9. Moriyama, H., Nakashima, H., Makino, S., and Koga, S. (1984) A study on the separation of reconstituted proteoliposomes and unincorporated membrane proteins by use of hydrophobic affinity gels, with special reference to band 3 from bovine erythrocyte membranes. *Anal. Biochem.* **139**, 292–297.
10. Robinson, N. C., Wiginton, D. and Talbert, L. (1984) Phenyl–Sepharose-mediated detergent-exchange chromatography: its application to exchange of detergent bound to membrane proteins. *Biochemistry* **23**, 6121–6126.
11. Warren, G. B., Toon, P. A., Birdsall, N. J. M., Lee, A. G., and Metcalfe, J. C. (1974) Reconstitution of a calcium pump using defined membrane components. *Proc. Natl. Acad. Sci. USA* **71**, 622–626.

31

Purification of Membrane Proteins

Kay Ohlendieck

1. Introduction

In contrast to soluble proteins, the isolation of a peripheral or integral membrane protein requires extraction of the biological membrane containing the protein of interest prior to purification. The most commonly used extraction procedures to isolate membrane proteins are described in Chapter 29. Following initial solubilization in high concentrations of detergent *(1)*, intrinsic proteins can then be purified to homogeneity using a variety of biochemical techniques in the presence of relatively low concentrations of detergent. Solubilized membrane proteins from diverse sources and with different properties have successfully been purified by a combination of standard techniques such as density gradient centrifugation, ion-exchange chromatography, gel filtration, lectin chromatography, and different forms of the affinity chromatography method. Usually, separation techniques based on biological differences such as affinity chromatography using specific ligands, probes, or antibodies highly specific for a membrane protein result in a higher yield and purity of the isolated protein than chromatographical methods based on physical differences, such as size and charge of the membrane protein *(2,3)*. An especially powerful separation technique for the initial purification of membrane-associated glycoproteins is lectin affinity chromotography. As described in more detail in Chapter 18, this method explores the highly specific interaction between *N*- and *O*-linked oligosaccharide chains on glycosylated proteins with immobilized lectins, which usually results in a remarkable enrichment of integral glycoproteins *(4)*.

Although the ultimate experiments to be carried out with an isolated membrane protein may influence the overall strategy of purification, it is desirable that the molecules of interest be obtained in their native, biologically active form in high yield and purity. However, biochemical, biophysical, or physiological experiments with homogeneous membrane proteins might require a gentler purification scheme than, for example, the preparation of only partially purified antigen samples for the production of monoclonal antibodies. Once a suitable source, such as a subcellular membrane fraction enriched in the membrane protein of interest, is found (*see* Chapter 29), small-scale pilot experiments should be performed to test the effects of various detergents on the biological activity of the membrane protein to be purified. With respect to the choice of detergent, a

From: *Methods in Molecular Biology, vol. 244: Protein Purification Protocols: Second Edition*
Edited by: P. Cutler © Humana Press Inc., Totowa, NJ

reagent with a high critical micelle concentration (cmc) (i.e., cholates) should be tried first, because these detergents can be more easily removed following purification than detergents with a lower cmc (*see* Chapter 30).

To illustrate the basic strategy for purifying an integral membrane protein, for monitoring its purification, and analyzing the purity of the final product, this chapter will address the purification of the sea urchin egg receptor for sperm. This highly glycosylated, integral surface protein represents a novel class of cell recognition molecules (*5*) and exists in its native configuration as a disulfide-bonded homotetrameric complex (*6*). Each sea urchin egg contains approx 1.25×10^6 receptor molecules (*7*) whose subunits exhibit an apparent molecular weight of 350 kDa (*8*). Based on the previous biochemical characterization of proteolytic receptor fragments and information on the primary structure deduced from its cDNA sequence (*9*), a strategy for purifying the sperm receptor to homogeneity was determined. The purification protocol, typical for the isolation of integral membrane proteins, includes cell homogenization, subcellular fractionation, solubilization of membrane fractions, lectin affinity chromotography, ion-exchange chromotography, dialysis, ultrafiltration, and lyophilization (*7*). Because this membrane protein does not exhibit enzyme activity, purification was monitored using immunoblot analysis with an antibody raised to a recombinant protein representing the extracellular sperm-binding domain of the receptor molecules. The purity of the final product was analyzed by gradient sodium dodecyl sulfate-polyacrylamide gel electrophoresis (SDS-PAGE) in combination with sensitive silver staining (*6,7*).

Thus, the techniques described in **Subheading 3.** and elaborated on in the Notes give a good overview of how to approach the purification of a novel protein based on limited information about its physical and biological properties. The flowchart in **Fig. 1** summarizes the different steps in the purification of the sea urchin sperm receptor typical for the isolation of an integral membrane protein. Other excellent examples of detailed descriptions of the methods employed in the purification of membrane proteins (i.e., rat liver 5' nucleotidase and rabbit skeletal muscle Ca^{2+}-ATPase) appeared previously in this series (*10,11*).

2. Materials

2.1. Isolation of Crude Egg Surface Membrane Complex

1. Sea urchin eggs (50 mL of settled, dejellied, and washed eggs from *Strongylocentrotus purpuratus*) (*see* **Note 1**).
2. Small plastic beakers (for collection of gametes).
3. 6°C Water bath for storage of sturdy 2-L glass beakers.
4. 120 μm Nitex nylon membrane (Tekton, Inc., Elmsford, NY).
5. 0.5 *M* KCl.
6. Buffer A (artificial seawater): 0.48 *M* NaCl, 10 m*M* KCl, 27 m*M* MgCl$_2$, 29 m*M* MgSO$_4$, 11 m*M* CaCl$_2$, 2 m*M* NaHCO$_3$, pH 7.5 (*see* **Note 2**).
7. Buffer B (homogenization buffer): 0.5 *M* NaCl, 10 m*M* KCl, 25 m*M* NaHCO$_3$, 63 m*M* NaOH, 25 m*M* ethylene glycol-bis(β-aminoethyl ether) *N,N,N',N'*-tetracetic acid (EGTA), 63 m*M* NaOH, pH 8; supplemented with 1 m*M* of each of aprotinin, soybean trypsin inhibitor, antipain, leupeptin, benzamidine, and phenylmethanesulfonyl fluoride (PMSF) (Sigma Chemical Co.) (*see* **Note 3**).
8. Buffer C: 20 m*M* Tris-HCl, 0.15 *M* NaCl, pH 7.4, supplemented with the same protease inhibitor cocktail as described for buffer B.

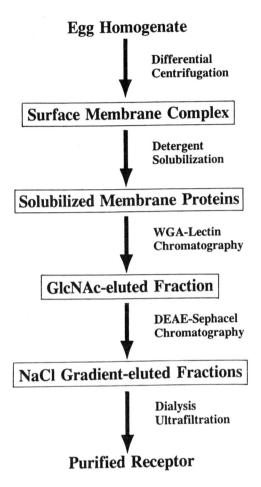

Fig. 1. Flowchart of the receptor purification protocol. The sea urchin egg receptor for sperm is purified to homogeneity from *Strongylocentrotus purpuratus* egg homogenenates. Following isolation of a crude surface membrane complex by subcellular fractionation, membranes are solubilized and a subset of glycoproteins purified by lectin affinity chromatography. The eluted glycoprotein fraction is then bound at low ionic strength to an ion-exchange matrix and eluted with a linear salt gradient. The peak fractions containing a homogeneous receptor are dialyzed, concentrated, and lyophilized, resulting in a preparation of purified sperm receptor.

9. Hand-operated Potter–Elvehjem homogenizer (25-mL volume).
10. Low-speed bench centrifuge with 4 × 200 mL rotor.
11. Standard bright-field microscope.
12. Water aspirator.

2.2. Solubilization of Sperm Receptor Complex

1. 4% (w/v) Octylgluoside (Boehringer-Mannheim) in buffer C.
2. Ultracentrifuge with a 30-mL fixed-angle rotor and appropriate tubes.

2.3. Lectin Affinity Chromatography

1. Wheat germ agglutinin–agarose (EY Labs, San Mateo, CA) matrix with a bed volume of approx 10 mL (*see* **Note 4**).
2. Wash buffer: 0.1% (w/v) Octylglucoside in buffer C.

3. Elution buffer:0.5 *M* N-Acetylglucosamine, 0.1% (w/v) octylglucoside in buffer C.
4. End-over-end mixer.

2.4. Ion-Exchange Chromatography

1. DEAE–Sephacel column (Pharmacia) with a bed volume of approx 10 mL.
2. Wash buffer: 0.1% (w/v) Octylglucoside in buffer C.
3. Elution buffer A: 0.5 *M* NaCl, 0.1% (w/v) octylglucoside, 20 m*M* Tris-HCl, pH 7.4, supplemented with the above-described protease inhibitor cocktail.
4. Elution buffer B: 4 *M* NaCl, 0.1% (w/v) octylglucoside, 20 m*M* Tris-HCl, pH 7.4, supplemented with the protease inhibitor cocktail described in **Subheading 2.1.**
5. 30-mL Linear gradient maker, small magnetic pellet, and magnetic stirrer.
6. Small peristaltic pump and suitable elastic tubing.
7. Fraction collector (optional).

2.5. Dialysis, Ultrafiltration, and Lyophilization

1. Standard dialysis tubing (with a molecular weight cutoff of approx 10,000).
2. Ultrafiltration filter (i.e., pre-equilibrated Amicon PM-30 membrane filter).
3. Ultrafiltration apparatus (i.e., 50-mL volume Amicon concentrator).
4. Lyophilizer.

3. Methods

Perform all manipulations of eggs at 6°C and subsequent homogenization and chromatographical purification steps in a cold room at 0–4°C. All *g*-forces are given as g_{av}.

3.1. Isolation of Crude Egg Surface Membrane Complex

1. To release mature gametes, treat 10–20 adult sea urchins with intracoelomic 0.5 *M* KCl injection and collect eggs in small plastic beakers of artificial seawater. After microscopical examination of individual batches of eggs, combine batches of morphological integrity and wash them three times by settling in 2 L of artificial seawater.
2. Dejelly washed eggs by 10 slow passages through a 120-μm Nitex membrane and then wash twice in filtered artificial seawater and once in 2 L of buffer B (*see* **Note 5**).
3. Resuspend the dejellied, washed, and settled eggs in 100 mL of ice-cold buffer B, supplemented with the protease inhibitor cocktail (*see* **Subheading 2.1.**), and homogenize by approx 10 gentle strokes using a hand-operated glass homogenizer and Teflon pestle. When breakage of eggs is adequate, the majority of eggs can be seen to release cytoplasmic organelles such as yolk platelets when viewed under a light microscope (*see* **Note 6**).
4. After 10-fold dilution, centrifuge the ghost membranes for 2 min at 1000*g*, carefully remove the yellowish supernatant with the help of a water aspirator and combine the white membrane pellets.
5. Wash the pellet twice in buffer B and, finally, resuspend it in 10 mL of buffer C (*see* **Note 7**).
6. If the egg surface complex cannot be processed immediately, it may be quick-frozen in liquid nitrogen and stored at −80°C until usage.

3.2. Solubilization of Sperm Receptor Complex

1. Solubilize the egg surface complex membranes at a final protein concentration of 1 mg/mL by adding an equal volume of 4% (w/v) octylglucoside in buffer C and incubate this mixture for 60 min on ice.
2. Remove insoluble material by centrifugation at 105,000*g* for 60 min and decant the supernatant containing the solubilized membrane protein fraction.
3. Dilute the supernatant 1:4 (v/v) with buffer C.

3.3. Lectin Affinity Chromatography (see Note 8)

1. Equilibrate the wheat germ agglutinin matrix by settling five times in 20 vol of buffer C.
2. Add the diluted, solubilized egg surface membrane complex suspension to the settled lectin matrix and incubate gently for 3 h or overnight in the cold room on an end-over-end mixer.
3. Allow the lectin matrix to settle on ice and carefully remove the supernatant. Wash the matrix three times with 10 vol of wash buffer.
4. Add 1.5 vol of elution buffer to the lectin matrix and incubate the mixture for 20 min in the cold room on a mixer.
5. Allow the matrix to settle on ice, carefully recover the supernatant containing the eluted glycoprotein fraction, and proceed immediately to **Subheading 3.4**. For regenerating the wheat germ agglutinin matrix, *see* **Note 9**.

3.4. Ion-Exchange Chromatography

1. Apply the glycoprotein fraction to 10 mL of DEAE–Sephacel ion-exchange resin that had been pre-equilibrated with wash buffer. Gently shake this mixture for 2 h in the cold room.
2. Allow the ion-exchanger matrix to settle, remove the supernatant, and wash the resin three times with 10 vol of wash buffer.
3. Pour the resin into a clean glass column and elute bound protein with a linear NaCl gradient of increasing ionic strength. The gradient is formed in a small gradient maker under adequate stirring using elution buffer A and B.
4. Collect 3-mL fractions and store them on ice prior to analysis.
5. Analyze the fractions by immunoblot and/or lectin blot analysis. The sea urchin egg receptor for sperm exhibits a characteristic brownish color in silver-stained SDS polacrylamide gels, indicative of highly glycosylated proteins. In addition, the receptor protein band of approx 350 kDa is strongly labeled by peroxidase-conjugated wheat germ agglutinin lectin (*see* **Fig. 2**).

3.5. Dialysis, Ultrafiltration, and Lyophilization

1. Combine the peak fractions containing homogeneous sperm receptor and dialyze them overnight against distilled water. For the preparation of dialysis tubing, *see* Chapter 29.
2. Concentrate the dialyzed receptor preparation using ultrafiltration with a PM-30 membrane. Following removal of the concentrated protein solution from the ultrafiltration chamber, rinse the filter twice with 1 mL of buffer to elute some of the protein possibly sticking to the membrane.
3. Quick-freeze the concentrated receptor preparation in liquid nitrogen and lyophilize the sample. Resuspend the freeze-dried receptor preparation in a small volume of buffer suitable for the subsequent analytical procedures (i.e., filtered artificial seawater for fertilization inhibition assays or sample buffer for SDS-PAGE [sodium dodecyl sulfate-polyacrylamide gel electrophoresis]).
4. **Figure 2** shows a silver-stained SDS polyacrylamide gel and corresponding immunoblot and lectin blot of the different preparative steps involved in the isolation of the integral receptor glycoprotein (*see* **Note 10**).

4. Notes

1. Sea urchins are common marine organisms that can be held in captivity under simple conditions and mature gametes can be obtained in large quantities. In mature animals, intracoelomic KCl injection may release 10^7 eggs during a single spawning. The most commonly used sea urchin is *Strongylocentrotus purpuratus* which can be obtained from Marinus, Inc., Long Beach, CA. This species produces mature gametes from late autumn to early summer.

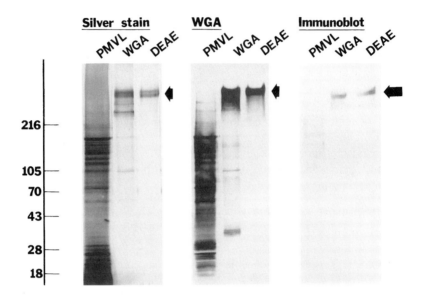

Fig. 2. Analysis of sperm receptor purification. Shown is the SDS-PAGE analysis of the different steps in the purification of the receptor using subcellular fractionation of the crude egg surface (PMVL; plasma membrane/vitelline layer), followed by lectin affinity chromatography (WGA, wheat germ agglutinin), and, finally, ion-exchange chromatography (DEAE–Sephacel). The silver-stained and the WGA lectin-blotted 350-kDa protein band corresponds to the receptor band visualized by immunoblotting. The double protein band in silver staining probably reflects two receptor isoforms in the vitelline envelope–egg surface complex with slightly different relative molecular mass. Molecular mass standards ($M_r \times 10^{-3}$) are indicated on the left

2. Artificial seawater should be made from high-quality water and filtered through 0.22-µm filters (Millipore) prior to use on isolated gametes. Ready-made mixtures of artificial seawater (i.e., "Instant Ocean" from Aquarium Systems, Mentor, OH) may be purchased from a pet shop.

3. The composition of the protease inhibitor cocktail used during the purification of the sea urchin egg receptor for sperm may not be suitable for other purification protocols. Because certain protease inhibitors are relatively expensive, small-scale pilot experiments should be performed to determine the lowest concentration of protease inhibitors necessary to prevent protein degradation during preparative steps.

4. Immobilized wheat germ agglutinin (WGA) is an expensive affinity matrix and can alternatively be prepared quite simply from crude wheat germ purchased in a health store. Following extraction, WGA can be purified to homogeneity by affinity chromatography using *N*-acetylglucosamine agarose (Sigma) *(12)*. Purification of the lectin can be monitored by SDS-PAGE analysis combined with immunoblot analysis using a polyclonal antibody to WGA commercially available from Sigma. Purified WGA can then be immobilized by standard procedures using cyanogen bromide-activated agarose (Pharmacia).

5. If suitable nylon membranes are not available, dejellying of sea urchin eggs can also be achieved by acid treatment. A 10% solution of eggs is treated for 2 min at pH 5.5 with the addition of 0.1 *M* HCl under gentle stirring with a plastic spatular. The solution is then neutralized by the addition of 2 *M* Tris-HCl, pH 8.0 and washed three times in filtered artificial seawater.

6. It is important to monitor visually the generation of ghost membranes following homogenization. Figure 2 in **ref. 7** illustrates the typical appearance of ghost membranes. Depend-

ing on the type of hand-operated homogenizer used, the breakage of the majority of eggs (>95%) can vary from batch to batch. On the one hand, the breakage of sea urchin eggs should be adequate to release the cytoplasmic content, but, on the other hand, the surface membrane should not be greatly fragmented because this leads to loss in yield of suitable ghost membranes.

7. If a small yellow pellet is seen at the bottom of the final membrane complex pellet, this is caused by eggs that were inadequately broken. Avoid resuspending this part of the otherwise whitish egg surface membrane complex.

8. Alternative to the described batch method, binding and elution of the glycoprotein fraction may be performed by column chromatography. However, elution of the bound glycoprotein fraction using a linear gradient of *N*-acetylglucosamine does not result in a major separation of WGA-binding proteins. Because the batch method is technically simpler, we use this procedure for routine purification of the sperm receptor. However, the disadvantage of batch methods is a possible gradual loss of matrix when removing the eluted supernatant fraction. This can be a problem with expensive affinity resins, and in that case, column chromatography would be advantageous.

9. Regeneration of a lectin matrix is usually achieved by salt washes. Treat the wheat germ agglutinin matrix, following elution with *N*-acetylglucosamine, with three 1.5-*M* NaCl washes. Then, re-equilibrate the WGA matrix with incubation buffer or phosphate-buffered saline complemented with 5 m*M* NaN_3 for storage at 4°C.

10. For a description of the standard techniques used in the analysis of the sperm receptor, *see* **ref. 7**. Gel electrophoretic procedures combined with analytical blotting techniques are ideal for monitoring the purification of a membrane protein because they are not only comparatively inexpensive and simple methods but also relatively fast and reliable. Silver staining of proteins separated by gel electrophoresis is a very useful method for estimating the purity of the final preparation because it is highly sensitive in the detection of contaminating proteins.

Acknowledgments

The author thanks Dr. W.J. Lennarz, State University of New York, in whose laboratory experiments purifying the sea urchin sperm receptor were conducted. This research was funded by the National Institutes of Health (HD 18590 to W.J.L.).

References

1. Helenius, A. and Simons, K. (1975) Solubilizationn of membranes by detergents. *Biochim. Biophys. Acta* **415**, 29–79.
2. Scopes, R. K. (1994) *Protein purification: Principles and Practice,* 3rd ed., Springer-Verlag, New York.
3. Thomas, T. C. and McNamee, M. G. (1990) Purification of membrane proteins. *Methods Enzymol.* **182**, 499–519.
4. Lis, H. and Sharon, N. (1986) Lectins as molecules and as tools. *Annu. Rev. Biochem.* **55**, 35–67.
5. Ohlendieck, K. and Lennarz, W.J. (1996) Molecular mechanisms of gamete recognition in sea urchin fertilization. *Curr. Topics Dev. Biol.* **32**, 39–58.
6. Ohlendieck, K., Dhume, S. T., Partin, J. S., and Lennarz, W. J. (1993) The sea urchin egg receptor for sperm: isolation and characterization of the intact, biologically active receptor. *J. Cell Biol.* **122**, 887–895.
7. Ohlendieck, K., Partrin, J. S., and Lennarz, W. J. (1994) The biologically active form of the sea urchin egg receptor for sperm is a disulfide-bonded homo multimer. *J. Cell Biol.* **125**, 817–824.

8. Kitazume-Kawaguchi, S., Inoue, S., Inoue, Y., and Lennarz, W. J. (1997) Identification of sulfated oligosialic acid units in the O-linked glycan of the sea urchin egg receptor for sperm. *Proc. Natl. Acad. Sci. USA* **94**, 3650–3655.

9. Ohlendieck, K. and Lennarz, W. J. (1995) Role of the sea urchin egg receptor for sperm in gamete interactions. *Trends Biochem. Sci.* **20**, 29–33.

10. Luzio, J. P. and Bailyes, E. M. (1993) Isolation of a membrane-bound enzyme, 5′nucleotidase, in *Biomembrane Protocols: I. Isolation and Analysis* (Graham, J. M. and Higgins, J. A., eds.), Humana, Totowa, NJ, pp. 229–242.

11. East, J. M. (1994) Purification of a membrane protein (Ca^{2+}/Mg^{2+}-ATPase) and its reconstitution into lipid vesicles, in *Biomembrane Protocols: II. Architecture and Function* (Graham, J. M. and Higgins, J. A., eds.), Humana, Totowa, NJ, pp. 87–92.

12. Vretblad, P. (1976) Purification of lectins by biospecific affinity chromatography. *Biochim. Biophys. Acta* **434,** 169–176.

32

Lyophilization of Proteins

Ciarán Ó'Fágáin

1. Introduction

Lyophilization, or freeze-drying, is a method for the preservation of labile materials in a dehydrated form. It can be particularly suitable for high-value biomolecules such as proteins. The process involves the removal of bulk water from a frozen protein solution by sublimation under vacuum with gentle heating (primary drying). This is followed by controlled heating to more elevated temperatures for removal of the remaining "bound" water from the protein preparation (secondary drying). Residual moisture levels are often lower than 1% *(1)*. If the freeze-drying operation is carried out correctly (*see* **Subheading 3.**), the protein will preserve all or most of its initial biological activity in the dry state. This dry state offers many advantages for long-term storage of the protein in question.

An isolated protein in an aqueous system can suffer various adverse reactions over time because of physical, chemical, and biological factors. Typical physical phenomena are aggregation *(2,3)* and precipitation. There are many deleterious chemical reactions involving the side chains of amino acid residues, notably asparagine, aspartic acid, and cysteine/cystine *(2–4)* or the glycation of lysine residues with reducing sugars (the Maillard reaction) *(5,6)*. Biological deterioration can result from loss of an essential cofactor or from the action of proteolytic enzymes, either endogenous or arising from microbial contamination. Water participates directly in many of the chemical reactions and in proteolysis. In any case, it provides a medium for molecular movement and interactions. For these reasons, removal of water effectively prevents deterioration of the protein. The freeze-dried preparation will be much less bulky than the original solution and can conveniently be stored in a laboratory freezer or refrigerator (or perhaps even at room temperature). When one wishes to use the protein preparation, one can rehydrate it simply by the addition of an appropriate volume of pure water or suitable buffer solution.

At very low temperatures, a liquid may behave in one of two ways. *Eutectic* solutions undergo a sharp liquid–solid freezing transition over a very narrow temperature range, whereas *amorphous* liquids are characterized by a *glass transition* in which viscosity increases dramatically with cooling and the solution takes on the macroscopic properties

From: *Methods in Molecular Biology, vol. 244: Protein Purification Protocols: Second Edition*
Edited by: P. Cutler © Humana Press Inc., Totowa, NJ

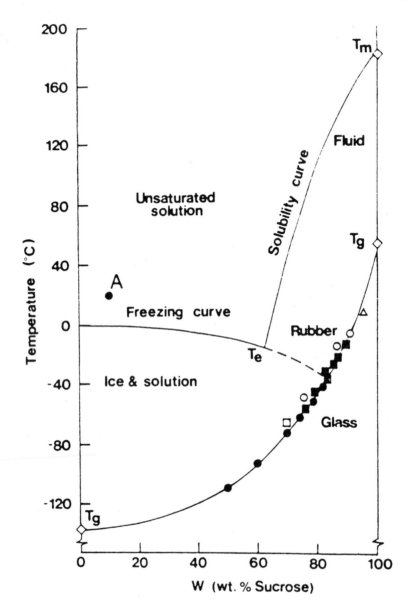

Fig. 1. Solid–liquid-state diagram for the sucrose–water system, showing the glass transition T_g/concentration profile, the equilibrium solubility curve, running between the melting point and the eutectic temperature, T_e, and the region of metastability (supersaturation) beyond T_e (broken line). Symbols represent experimental T_g data taken from different sources. Point A represents a typical dilute solution that is to be dried to a stable solid product. (From **ref. 7.**)

of a solid, even though it has not crystallized. Below this glass transition temperature (T_g') virtually no adverse chemical or biological reaction can take place. Above T_g', however, the viscous, rubbery material is very prone to deterioration *(7)* (*see* **Fig. 1**). Understanding these low-temperature features of liquids is important for effective freeze-drying. Eutectic and glass transitions will greatly influence the freeze-drying protocol (the way the freeze-dryer is run) and the choice of substances used as preservatives or excipients in the product formulation subjected to lyophilization. Understanding of freeze-drying as a process has grown in recent years *(7–9)* and knowledge of the effects of additives has also grown *(10,11)*. Together, these advances have allowed lyophilization to be undertaken on a more rational and less strictly empirical basis. These matters are discussed in more detail in **Subheading 3**. Reference *12* contains a useful treatment that touches on the underlying theory. Reference *8* outlines principles for process design, and Oetjen has published a detailed monograph *(13)*.

Effective freeze-drying is time-consuming and cannot be accomplished in a hurry. Process times of 72 h or greater are not unusual, depending on the nature of the product formulation and the properties of its constituents. The complex lyophilizer apparatus will have high capital and running costs (the latter because of its power consumption and the time required). Therefore, freeze-drying is usually reserved for high-value proteins or is used in cases where alternative product presentations (such as ammonium sulfate precipitates) are unsuitable or give insufficient shelf lives.

A typical freeze-dryer comprises a *set of shelves* that can be cooled to about −40°C or heated above ambient temperature, a *condenser* usually cooled to −60°C, and a high-performance *vacuum pump*. The shelves allow rapid freezing of the protein solution and, later, gentle programmed heating for primary and secondary drying. The vacuum pump reduces the pressure within the system to allow sublimation of the bulk water (i.e., primary drying) and the condenser acts as a lower-temperature trap for the sublimed water that collects in the form of ice (i.e., it provides a temperature gradient within the apparatus). This explains why the condenser temperature must be considerably less than that of the shelves.

A complex interplay of chemical and physical phenomena takes place during freeze-drying. The product yield (i.e., the percentage recovery of the initial active protein) depends on the formulation in which the protein is placed prior to lyophilization *(7,11)* and its ease of rehydration and its stability on long-term storage (or "shelf life") are influenced by the processing regime *(7)*. **Subheading 3.1.** describes, in broad terms, the operation of a typical freeze-dryer apparatus. Some critical factors concerning the operations of freezing and primary and secondary drying are outlined in **Subheadings 3.2.–3.4**. **Note 1** outlines some reports of the freeze-drying-induced activation of enzymes suspended in nonaqueous media.

2. Materials

Freeze-dryers differ greatly in their specifications depending on the model or manufacturer. Regardless of the type, one must use it at all times in accordance with the manufacturer's instructions. The method discussed in **Subheading 3.1.** assumes that the reader studies this chapter together with the user's handbook.

The most basic lyophilizer equipment will consist of a condenser/vacuum pump unit to which one may attach a centrifugal test tube holder or a manifold for the drying of multiple round-bottomed flasks. Freezing is accomplished in a separate cooling bath, usually filled with an alcohol. Such equipment can be used successfully for small-volume samples, but fine and reproducible control of the overall process may not be possible. Neither can one attain continuous recording of sample temperature. Higher-grade equipment, with temperature-programmable shelves and a number of temperature probes, is preferred. Shelf-equipped freeze-dryers are especially suitable for use with rubber-capped pharmaceutical vials or with microtiter plates. Often, an externally operated screw press will be present, allowing one to seal vials (partially closed with rubber stoppers so as not to restrict gaseous movement) under vacuum before releasing air into the chamber.

Obviously, one should thoroughly familiarize oneself with the machine to be used. The user should consult the machine's handbook and should know the locations of the main components and indicators, together with appropriate settings for the meters, controls, and valves.

The characteristics of the vials (usually glass) in which product is lyophilized will influence the process. The vial diameter, glass type, and bottom shape and thickness will all affect the rate of heat transfer from the shelves to product. The vials should withstand freezing and pressure changes and should be uniform with respect to internal diameter and bottom thickness. The vial bottoms should be completely flat to make good contact with the shelves. The same model of vial should be used consistently, as a change of vial type will introduce a variable into an optimized process *(7)*. Stoppers also should be chosen with care, as they can influence the residual moisture content of the freeze-dried product *(13)*.

Ensure that any other vessels used (e.g., round bottomed flasks) are of sufficient quality to withstand the temperatures and pressures associated with freeze-drying.

3. Method

3.1. Operation of Freeze-Dryer

These directions for freeze-drying are of necessity very general. Exact details will depend on the material and apparatus in question and on the user's requirements. Good Manufacturing Practice, ISO 9000, or regulatory authorities will likely impose many procedural and formulation disciplines on pilot or manufacturing operations. Such regulatory concerns are beyond the scope of this chapter.

3.1.1. Startup

Ensure that the valve connecting the vacuum pump to the drying chamber is closed. Start the vacuum pump and allow it to evacuate. Observe the decrease in pressure on the vacuum indicator. It is important to start the pump first so as to reach a steady-state high vacuum long before evacuation of the main chamber. It is also important that the pump warms up thoroughly before the condenser or shelves are cooled: water in any form reaching the pump can cause damage. A warm-up time of 30 min will usually suffice. Close the condenser drain outlet, which should always be left open when the freeze-dryer is not in use. (If not, open it, allow any water to drain completely, and then close tightly once again.) Switch on the condenser and allow to cool to −60°C. At this point, it may be convenient to cool the shelves slightly: this will aid the overall chilling process later. However, do not cool the shelves more than a few degrees below ambient temperature, to prevent atmospheric condensation. Condensation will add to the water load that must be removed during the lyophilization process.

3.1.2. Filling and Loading

Now, loading of the shelves with the preparation to be lyophilized can begin (*see* **Note 2**). Fill only minimal amounts of material into each container to ensure a high ratio of surface area to volume: This will aid effective freeze-drying. Fill depths should not exceed 20 mm *(8)*. The product matrix will tend to inhibit sublimation of water vapor from the surface of the ice crystal. This resistance depends on the depth of liquid and on the solids content of the product *(9)*. A solid content of about 10% (w/w) is usually best *(8)*. Also, a good head space in the vial or ampoule will allow easier and better gaseous movement. One can conveniently load filled vials into metal trays at the bench and then place the trays into the drying chamber. The trays should have level bottoms to make good contact with the freeze-dryer shelves. Many vials have narrow necks that one closes with special rubber stoppers. The design of these stoppers allows one to cap the vials partially while maintaining contact between their contents and the atmosphere. (Sealing of the vials takes place later under vacuum.) If one wishes to use such closures, one should partially stopper the vials before loading them into the freeze-dryer. If using microtiter plates, place them within special metal frames that then come into contact with the shelves.

The freeze-dryer will likely have a number of flexible temperature probes; carefully clean each of these. Place some of the probes so as to monitor the temperatures of the shelves. Dip others directly into the material to be lyophilized. A temperature record of the actual solution (as distinct from the shelf underneath it) is well worth the loss of a small number of product vials. Arrange the temperature probes in different locations throughout the chamber: in the centers of shelves and also at the edges. Conditions will not be homogeneous across all vials and monitoring should be as complete as equipment will allow. Insert the temperature probes right to the bottom of vials. Ensure that the probes remain in contact with the intended sample or shelf and do not become displaced during product loading or manipulation, or during lyophilization.

3.1.3. The Freeze-Drying Operation

Once loading is complete, close and secure the door of the drying chamber. Now, the freezing process may begin. Bring the material well below its freezing temperature (or glass transition temperature) as quickly as possible. This is particularly important for a number of reasons (*see* **Subheading 3.2.**) Allow the shelf/product temperature to fall further to a steady −40°C before drawing the vacuum. Check that the drying chamber door is properly closed and sealed, then evacuate the chamber by opening the vacuum valve. The vacuum gauge reading will indicate atmospheric pressure as the valve is opened and may take a few minutes to show a vacuum once again as the air within the drying chamber evacuates. Soon, however, the pressure within the drying chamber will decrease and a steady-state high vacuum will result.

There now exists a high vacuum within the drying chamber together with a temperature gradient from the shelves/product (−40°C) to the condenser (−60°C). These conditions permit sublimation of the bulk water in the product over a period of hours as the shelves are gently heated. The sublimed water will collect as ice on the condenser. Sublimation, or primary drying, removes only the bulk water in the system; it is insufficient to remove the "bound" water closely associated with the protein molecules. Removal of this bound water, or secondary drying, requires heating to elevated temperatures. This is usually applied through the shelves on which the product rests (and has previously been frozen). One can program the shelves to heat to a particular temperature at a defined rate appropriate for the product undergoing lyophilization. One must select the heating regime with particular care (*see* **Subheadings 3.3.** and **3.4.**)

3.1.4. Termination of Run and Removal of Product

One may terminate the freeze-drying process when the cake has a good appearance and the product has reached a sufficiently low steady-state percentage moisture (determined in a separate series of experiments). Seal vials, partially capped with rubber stoppers before insertion into the chamber, by operating the special screw press. This operation will insert the stoppers fully before air is admitted into the drying chamber. Sealing of vials while still under vacuum has an important advantage over the use of an inert, moisture-free gas such as nitrogen. Air will rush in when a vial, sealed under vacuum, is uncapped. The sound of the in-rushing air will be absent from a defective vial that has failed to seal or where the seal has broken down. In this way, the user will immediately know of the defect and be aware that the vial contents may have deteriorated *(1)*.

Close the valve to the vacuum pump firmly. Slowly release the vacuum within the drying chamber by a minimal opening of the air inlet valve: Air admission should be as gentle as possible to prevent undue sudden stresses on sealed vials and also to prevent disturbance or upset of chamber contents by a strong jet of incoming air. The vacuum/pressure gauge will show a rise in drying chamber pressure; this will eventually equalize to atmospheric pressure and one can then open the chamber door. Vials of freeze-dried product can be removed for storage (preferably at refrigerated temperatures). Store microtiter plates immediately in a desiccator or (better) vacuum-sealed within foil packets containing desiccant sachets.

3.1.5. Shutting Down

If the freeze-dryer is to be reused immediately, one must be certain that the condenser's ice capacity will withstand the accumulation of ice from two runs (*see* **Note 2**). If there is no more material for lyophilization, shut down the apparatus carefully. Remove any material spilled in the drying chamber and clean the shelves carefully according to the manufacturer's instructions. Leave the chamber door slightly ajar to allow circulation of air and to prevent sticking and compression of the door seals. Switch off the condenser and open its drain outlet. (The drain should remain open until the freeze-dryer is next used.) Over a period of hours, the ice on the condenser surfaces will melt and drain away. Once the condenser temperature has returned to ambient and all of the melted ice has drained away, allow the vacuum pump to run for a further 3 h before shutting it down. This is to prevent any damage to the pump resulting from occurrence or accumulation of condensation. To maintain good pump performance, change the oil often according to the handbook's directions.

3.1.6. Using Simpler Freeze-Dryers

Use of simpler apparatus with manifold or centrifuge accessories is carried out in much the same stepwise fashion as earlier. In these cases, aliquots of the product are frozen in individual open-necked flasks or tubes. Switch on the vacuum pump and condenser and it allow it to run as noted earlier. Freeze flask contents by immersion in an alcohol cooling bath or in liquid nitrogen (follow the normal precautions). Swirl or rotate the flask during the freezing step to effect even distribution of the product over the widest possible surface area. This will minimize the depth of material through which water loss must occur (*see* **Subheading 3.1.2.** and **ref. 9**. Connect the frozen material directly to the manifold assembly (or load into the centrifugal tube dryer) and draw a

FREEZING

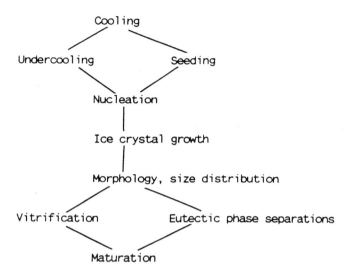

DRYING

Ice sublimation (Product Temperature, Chamber pressure)

Desorption (Temperature profile, Residual moisture)

STORAGE   Residual moisture/temperature control

Fig. 2. Physical principles of freeze-drying. (From **ref. 7.**)

vacuum in the chamber as quickly as possible so as to minimize back-melting of ice while the sample is still under atmospheric pressure. The rapid reduction of pressure by the vacuum pump will aid sublimation and help minimize melting to water. Whereas control of temperature and heating rates are more problematic with such accessories, visual inspection of the material and of the dried cake is much easier than in an enclosed chamber.

3.2. Freezing

In freezing, liquid water crystallizes to yield solid ice. Water, in common with other substances, including many solutes, is a *eutectic* material with a sharp transition between the liquid and solid states. Noncrystalline materials undergo a *glass transition* and become rigid at a certain low temperature (T_g'). This is the end result of a notable increase in viscosity with decreasing temperature. The material becomes increasingly rubbery until rigidification occurs at T_g' and liquid movement effectively ceases (*see* **Note 3** and **Figs. 1** and **2**). Even for eutectics, ice formation usually does not occur at the thermodynamic freezing point: The actual temperature of freezing is generally 10–15°C lower than this, so that supercooling (or undercooling) is required *(7,9)*. It is generally satisfactory to cool the product to just below 0°C initially, but not so low as to induce crystallization. When all of the product has cooled to this set temperature, reduce the shelf temperature sharply to crystallize bulk water to ice. Typical temperatures lie in the range −20 to −30°C. Allow time for ice to form in all containers before proceeding to the next stage, which involves further shelf cooling below the lowest eutectic temperature (for crystalline materials) or below the glass transition temperature ($T_g{}'$, for amorphous substances) *(see* **Note 4**). A usual final temperature is −40°C. The product is reliably solidified only below these temperatures and only now can primary drying begin. It is vital to maintain the freezing temperature below $T_g{}'$ before and during primary drying. If the temperature rises above this value, the material ceases to be a solid and becomes a viscous rubber very prone to deleterious reactions, with resulting losses of activity. The rubber will also undergo a mechanical *collapse*, take on a different appearance, and be difficult to rehydrate *(7)*. The collapsed product is always unacceptable even if reasonable biological activity remains. Thus, accurate measurement of $T_g{}'$ is of great importance; in most cases (where the protein of interest is in dilute solution), it depends on the types and proportions of the excipients and salts in the product formulation *(8)*.

Other notable changes occur in the product during freezing, many of which can lead to significant protein denaturation. Secondary structural alterations have been detected in lyophilized proteins by Fourier transform infrared spectroscopy *(3,14)*. The protein of interest may undergo inactivation, as indicated by poor recovery of prelyophilization activity. As the bulk water freezes to ice, the amount of liquid water remaining naturally decreases. This leads to great increases in the concentrations of solutes such as salts. Such freeze-concentration can have far-reaching effects. Concentration will increase the rates of unwanted chemical reactions such as oxidations. Buffer components may crystallize differentially, leading to pronounced pH shifts; also, pK_a values are temperature dependent. For phosphate buffers, one should use potassium or mixed salts in preference to sodium salts *(7,15)*. For all of these reasons, one should accomplish the freezing steps (crystallization of bulk water and cooling below the eutectic or $T_g{}'$ temperatures) in any lyophilization process as quickly as possible.

Suitable excipients or protective substances can lessen or overcome some of these damaging freezing effects: They can particularly influence melting or collapse temperatures *(16)*. Excipients may be classed as bulking agents, tonicity modifiers, buffers, and cryoprotectants/lyoprotectants *(16)*. One should use bulking agents when the product's solids content is low. They help to ensure the development of a plug of dried material that will be of good appearance. Bulking agents can also prevent blowout: Where the material has a very low solids content of the order of 1%, disruption of cake structure by the issuing water vapor can result in loss of the dried product from the container along with the vapor *(11)*. Typical bulking agents include polyols and certain sugars, mostly nonreducing.

Buffer substances must be chosen with great care. One must consider possible variations with temperature in pK_a and solubility as well as compatibility with the protein(s) in question and other constituents. Useful information can be found in **ref.** *17* and in monographs and articles dealing with the handling of proteins.

Many bulking agents will double as tonicity modifiers in preparations for administration in vivo to ensure that the preparation is isotonic with body fluids. Any agent used for this purpose must, of course, be pharmaceutically acceptable.

Cryoprotectants and lyoprotectants are substances that stabilize a protein against the deleterious effects of freezing and of lyophilization/storage, respectively. These protecting additives appear to be preferentially excluded from the protein surface. This results in a strengthening of the water "shell" surrounding the protein *(18)*. Sugars, certain salts, and polyols can each be beneficial (although polyethylene glycol [PEG]–dextran mixtures may give undesirable and damaging phase separations at low temperatures; *see* **ref. 14***)*. In general, nonreducing sugars are preferable because, unlike reducing sugars, they cannot participate in Maillard reactions. Xylitol's T_g' is −46.5°C in a freeze concentrate of 42.9 wt % water, whereas T_g' values for sorbitol, sucrose, and trehalose are −43.5°C, −32°C, and −29.5°C at water weight % values of 18.7, 35.9, and 16.7, respectively *(7)*. Trehalose proved effective in the lyophilization of a membrane protein *(19)*. Note that mannitol and lactose may separate as crystals from a frozen solution under certain conditions *(8)*. This is undesirable and so these compounds should be used with caution. Glycerol has a notably low T_g' value of −65°C at 46 wt % water *(7)*. This means that glycerol-containing formulations must reach very low temperatures to become glassy. Glycerol, therefore, is not an ideal lyophilization excipient, as such temperatures may be difficult to achieve. Volatile excipient compounds, such as ammonium bicarbonate, will be removed with the subliming ice and, therefore, will not occur within the final product *(8)*. Hydroxy-β-cyclodextrin (10 m*M*) stabilized lactate dehydrogenase during freeze-thawing and freeze-drying *(20)*. Other cyclodextrin derivatives proved useful with α-chymotrypsin *(21)*. Useful discussions and examples of cryoprotectants occur in **refs. 10**, **18**, and **22***)*; *see also* **Note 3** of this chapter and **Subheading 3.3.** of Chapter 33.

3.3. Primary Drying

Primary drying is the sublimation under vacuum of bulk ice from the product to the much colder condenser typically held at −60°C. Sublimation will be faster at higher shelf temperatures, so heat the shelves to a few degrees below T_g' to quicken the process (*see* **Note 3**). NEVER allow the shelf temperature to equal or exceed T_g', however: if it does, collapse and product deterioration may occur. A usual safety margin is 2–5°C below the eutectic or collapse temperature *(9)*. (The term "collapse temperature" is usually equivalent to T_g' for an amorphous substance.) The rate of sublimation depends on the vacuum and on the condenser temperature and thus is largely determined by the characteristics of the lyophilizer apparatus *(7)*. Drying often becomes easier as the temperature approaches T_g' or the eutectic temperature; the sublimation rate can increase by about 13% for each 1°C increase in temperature *(9)*. Resistance to drying also decreases with decreasing product thickness (i.e., filling height) and with increasing vial diameter (which influences the area of the drying surface) *(9)*. It is important that the temperature of the actual product remains constant (and therefore is monitored) throughout primary drying. The sublimation rate will decrease as primary drying proceeds and therefore the degree of product cooling resulting from sublimation will decrease also. The purpose of shelf warming is to counter this sublimation cooling. Accordingly, be sure to adjust the heat input to the shelves as the product dries to prevent a net rise in product temperature. An unchecked increase in product temperature could lead to collapse. One can use chamber pressure to control the length of the primary drying cycle

LIVERPOOL JOHN MOORES UNIVERSITY
LEARNING SERVICES

(9). Even under uniform conditions, primary drying times may vary by up to 10% for a given process. Make sure to include a delay period (ascertained empirically and occupying a period of hours) at the end of the primary drying cycle to ensure that all ice has sublimed satisfactorily and to avoid collapse *(9)*. **Note 5** refers to a model describing the primary and secondary drying stages of pharmaceutical product lyophilization in vials.

3.4. Secondary Drying

Secondary drying removes the remaining unfrozen, bound water from the lyophilizing material, yielding a final product, low in residual moisture, that will be stable for an extended period without deterioration. The unfrozen water is removed by heating the shelves on which the primarily dried product rests. The rule of thumb for initiation of secondary drying is the equivalence of product and shelf temperatures. For safety, however, one should include a delay period at the end of primary drying in any freeze-drying protocol to prevent collapse of any vials that have not quite finished sublimation *(9)*: *see also* **Subheading 3.3.** The partial pressure of water within the drying chamber drops at the end of primary drying as the last of the ice sublimes. If it can be monitored, this drop in water partial pressure can be a good indicator of the completion of primary drying. Even during secondary drying, with much of the original water gone, the preparation's temperature should never increase above T_g', something that is seldom appreciated *(7)*. T_g', however, increases as the residual water content drops *(9)*. One can, therefore, increase the temperature (within limits) during secondary drying. (Overheating during primary and/or secondary drying will likely be deleterious, however.) The vacuum need not be exhaustive during secondary drying; indeed, one should use a pressure in the region of 0.2 torr for secondary drying *(9)*.

3.5. Quality Indices

Most freeze-dryers will have a built-in chart recorder to provide a profile of shelf temperatures (and perhaps other parameters such as vacuum) during the lyophilization run. Study this carefully and retain with other batch information. Note the cake shape and texture and any variations between vials, especially those located at different positions within the chamber. Choose a representative number of vials from different chamber locations for scrupulous moisture determination (*see* **Note 6**). Measure the yield or recovery (%) of the initial biological activity by appropriate assay following rehydration of a representative number of vials. While waiting to assay, note the time required for complete product rehydration. Also note whether any turbidity remains on rehydration, or after what interval turbidity appears in a clear sample *(7)*. Persistence of the rehydrated biological activity can be measured at suitable or convenient time intervals. Accelerated degradation methods can predict shelf lives of the lyophilized preparation at temperatures of interest for long-term storage. These methods involve extrapolation of an Arrhenius plot. Accelerated degradation protocols can be of value *provided* the activity decay is first order at each of the temperatures tested and all data are scrupulously accurate and precise. **Reference 23** gives guidelines for the proper use of accelerated methods; *see also* **Subheading 3.6.** of Chapter 33.

Note that the formulation used greatly influences yield, and the process parameters affect ease of rehydration and shelf life *(7)*.

4. Notes

1. Although enzymes can remain active in nonaqueous media (often showing increased thermostability and an ability to catalyze novel reactions), their levels of activity may be considerably less than in water (*25* and references therein). Lyophilization is commonly used to remove an enzyme from aqueous systems, permitting resuspension of the enzyme in the nonaqueous medium. The reduced activity levels observed in organic media have been ascribed to reversible denaturation occurring on lyophilization. A reversibly denatured lyophilized enzyme can regain its optimal conformation (and hence full activity) upon rehydration in water or buffer; an enzyme suspended in a nonaqueous medium (where water is limiting) cannot. For certain oxidases, two distinct types of excipient were found to minimize this activity loss (i.e., to enhance activity in organic media). One group comprised substrate: like compounds, which presumably bind to the active site, whereas the second comprised general lyoprotectants such as polyethylene glycol. The effects were additive when both excipient types were used together *(25)*. Hydrolases show increased activity in organic solvents following lyophilization from salt- (especially KCl) containing mixtures and an optimum lyophilization time has been demonstrated *(26)*. In general, salts classified as kosmotropic seem to be the most beneficial *(27)*.

2. The volumes of bulk liquid subjected to freeze-drying must never exceed the manufacturer's recommendations. If the condenser's ice capacity is reached or exceeded, the degree of product drying will be insufficient and many problems can result.

3. The glass transition temperature, T_g', and the product water content are critical parameters in the freeze-drying process. It is essential to prevent or minimize damage to the protein of interest during freezing. Inclusion of excipients (additives) with high T_g' values in the protein formulation to be freeze-dried can be very useful. The mixture will form a glass at relatively low temperatures, minimizing freezing damage. A high T_g' will also allow the use of higher temperatures during primary drying with less danger of product collapse. Any constituent that will lower the unbound water content of the freeze concentrate will help shorten the secondary drying operation *(7)*, but uncrystallized salts will decrease T_g', because any salt will bring about a depression of freezing point. Thus, the salts content of the product formulation should be as low as is practicable *(7)*. The optimum temperature for freezing and primary drying depends on the ratio protein: protectant additive: salt in the freeze concentrate rather than in the initial solution *(7,11)*. The ratio protein: other solids in the freeze concentrate influences T_g' *(7)*.

   There have been reports of protein damage as a result of mechanical stresses at the interfaces of separated liquid phases arising from freeze concentration effects (*14* and references therein). Sucrose and trehalose exerted little protection against this phenomenon, despite being good glass formers. Rapid cooling below the glass transition temperature appeared to minimize damage from this cause, because the protein spends less time in a freeze-concentrated solution before attainment of the glassy state *(14)*.

4. The glass transition temperature can be determined by differential scanning calorimetry, although a sensitive instrument is required *(7)*. Electrical resistance measurements can sometimes be useful but are not suitable for nonelectrolyte mixtures *(8)*. However, microscopic observation of freeze-drying over a range of temperatures is arguably the most direct, sensitive, and unambiguous method for determination of the collapse temperature *(9)*.

5. **Reference *24*** presents and solves a non-steady-state, spatially multidimensional model describing the dynamic behavior of the primary and secondary stages of pharmaceutical product freeze-drying in vials. The model does not claim to be an optimal process control program, however, and each user must decide if it is appropriate for his/her situation.

6. The residual moisture content of, and its distribution throughout, the lyophilized preparation will dictate its long-term stability. Uneven moisture distribution between vials often leads to biphasic activity–loss profiles on extended storage *(7)*. Each 1% of moisture can depress T_g' by more than 10°C *(12)*. Significant aggregation of lyophilized recombinant human serum albumin occurred within hours upon incubation at 37°C and 96% relative humidity *(3)*, indicating just how critical the residual moisture content can be.

Acknowledgment

Dr. F. Franks permitted the reproduction of figures from his work and indicated some useful references.

References

1. Thuma, R. S., Giegel, J. L., and Posner, A. H. (1987) Manufacture of quality control materials, in *Laboratory Quality Assurance* (Howanitz, P. J. and Howanitz, J. H., eds.), McGraw-Hill, New York, pp. 101–123.

2. Liu, W. R., Langer, R., and Klibanov, A. M. (1991) Moisture-induced aggregation of lyophilized proteins in the solid state. *Biotech. Bioeng.* **37,** 177–184.

3. Costantino, H. R., Langer, R., and Klibanov, A. M. (1995) Aggregation of a lyophilized pharmaceutical protein, recombinant human albumin. *Bio/Technology* **13**, 493–496.

4. Volkin, D. B. and Middaugh, C. R. (1992) The effect of temperature on protein structure, in *Stability of Protein Pharmaceuticals, Part A: Chemical and Physical Pathways of Protein Degradation* (Ahern, T. J. and Manning, M. C., eds.), Plenum, New York, pp. 215–247.

5. Hageman, M. J. (1992) Water sorption and solid-state stability of proteins, in *Stability of Protein Pharmaceuticals, Part A: Chemical and Physical Pathways of Protein Degradation* (Ahern, T. J. and Manning, M. C., eds.), Plenum, New York, pp. 273–309.

6. Quax, W. J. (1993) Thermostable glucose isomerases. *Trends Food Sci. Tech.nol.* **4**, 31–434.

7. Franks, F. (1990) Freeze drying: from empiricism to predictability. *Cryo-Letters* **11,** 93–110.

8. Franks, F. (1998) Freeze drying of bioproducts: putting principles into practice. *Eur. J. Pharm. Biopharm.* **45,** 221–229.

9. Pikal, M. J. (1990) Freeze-drying of proteins. Part 1: process design. *BioPharm.* **3,** 18–27.

10. Carpenter, J. F. and Crowe, J. H. (1988) The mechanism of cryoprotection of proteins by solutes. *Cryobiology* **25,** 244–255.

11. Pikal, M. J. (1990) Freeze-drying of proteins. Part 2: formulation selection. *BioPharm* **3,** 26–30.

12. Franks, F., Hatley, R. H. M., and Mathias, S. F. (1991) Materials Science and the production of shelf-stable biologicals. *Pharm. Technol. Int.* **3,** 24–34.

13. Oetjen, G.-W. (1999) *Freeze Drying*, Wiley–VCH, Weinheim, p. 115.

14. Heller, M. C., Carpenter, J. C., and Randolph, T. W. (1999) Protein formulation and lyophilization cycle design. *Biotech. Bioeng.* **63,** 166–174.

15. Franks, F. and Murase, N. (1992) Nucleation and crystallisation in aqueous systems during drying: theory and practice. *Pure Appl. Chem.* **64,** 1667–1672.

16. Hanson, M. A. and Rouan, S. K. R. (1992) Formulation of protein pharmaceuticals, in *Stability of Protein Pharmaceuticals, Part B: In Vivo Pathways of Degradation and Strategies for Protein Stabilization* (Ahern, T. J. and Manning, M. C., eds.), Plenum, New York, pp. 209–233.

17. Blanchard, J. S. (1984) Buffers for enzymes. *Methods Enzymol.* **104,** 404–415.
18. Timasheff, S. N. (1992) Stabilization of Protein Structure by Solvent Additives, in *Stability of Protein Pharmaceuticals, Part B: In Vivo Pathways of Degradation and Strategies for Protein Stabilization* (Ahern, T. J. and Manning, M. C., eds.), Plenum, New York, pp. 265–285.
19. Sode, K. and Yasutake, N. (1997) Preparation of lyophilised pyrroloquinoline quinone glucose dehydrogenase using trehalose as an additive. *Biotechnol. Techn.* **11,** 577–580.
20. Izutsu, K., Yoshioko, S., and Terao, T. (1994) Stabilising effect of amphiphilic excipients on the freeze-thawing and freeze-drying of lactate dehydrogenase. *Biotechnol. Bioeng.* **43,** 1102–1107.
21. Ooe, Y., Yamamoto, S., Kobayashi, K., and Kise, H. (1999) Increase of catalytic activity of α-chymotrypsin in organic solvent by co-lyophilisation with cyclodextrins. *Biotechnol. Lett.* **21,** 385–389.
22. Schein, C. H. (1990) Solubility as a function of protein structure and solvent components, *Bio/Technology* **8,** 308–317.
23. Kirkwood, T. B. L. (1984) Design and analysis of accelerated degradation tests for the stability of biological standards. III. Principles of design. *J. Biol. Stand.* **12,** 215–224.
24. Sheehan, P. and Liapis, A. I. (1998) Modeling of the primary and secondary drying stages of the freeze drying of pharmaceutical products in vials. *Biotechnol. Bioeng.* **60,** 712–728.
25. Dai, L. and Klibanov, A. M. (1999) Striking activation of oxidative enzymes suspended in non-aqueous media. *Proc. Natl. Acad. Sci. USA* **96,** 9475–9478.
26. Ru, M. T., Dordick, J. S., Reimer, J. A., and Clark, D. S. (1999) Optimising the salt-induced activation of enzymes in organic solvents. *Biotechnol. Bioeng.* **63,** 233–241.
27. Ru, M. T., Hirokane, S. Y., Lo, A. S., Dordick, J. S., Reimer, J. A., and Clark, D. S. (2000) On the salt-induced activation of lyophilised enzymes in organic solvents. *J. Am. Chem. Soc.* **122,** 1565–1571.

33

Storage of Pure Proteins

Ciarán Ó'Fágáin

1. Introduction

There is often a need to store an isolated or purified protein for varying periods of time. If the protein in question is to be studied, it will take some time to characterize the properties of interest. If the protein is an end product or is for use as a tool in some procedure, it will likely be used in small quantities over a period of time. It is vital, therefore, that the protein retains as much as possible of its original, post-purification, biological (or functional) activity over an extended period of storage. This storage period or "shelf life" may vary from a few days to more than 1 yr. Shelf life can depend on the nature of the protein and on the storage conditions. This chapter explores the means by which activity losses occur on storage and discusses a range of measures to prevent or lessen the inactivating events. The chapter also describes the use of accelerated storage (accelerated degradation) testing for the prediction of shelf lives at particular temperatures.

Apart from extremes of temperature and pH (which will, naturally, be avoided as conditions for routine or long-term storage), a variety of factors can lead to loss or deterioration of a protein's biological activity. These include proteolysis, aggregation, and certain chemical reactions *(1)*. Proteolysis may arise from microbial contamination or be an intrinsic hazard in the case of proteolytic enzymes. A protein may lose an essential cofactor or the subunits of oligomeric proteins may dissociate from each other, resulting in each case in loss of activity *(1)*. Adsorption to surfaces may also lead to inactivation *(1,2)*.

The purely chemical reactions are few and well defined. Deamidation of glutamine and asparagine can occur at neutral to alkaline pH values, whereas peptide bonds involving aspartic acid undergo cleavage under acidic conditions. Cysteine is prone to oxidation, as are tryptophan and methionine. Alkaline conditions lead to reduction of disulfide bonds and this is often followed by β-elimination or thiol-disulfide exchange reactions *(1)*. Where reducing sugars are present with free protein amino groups (N-termini or lysine residues), there may be destructive glycation of the amino functions by the reactive aldehyde or keto groups of the sugar (the Maillard reaction) *(3)*. Elevated temperatures favor all of these reactions, but it is important to note that aggregation and deleteri-

From: *Methods in Molecular Biology, vol. 244: Protein Purification Protocols: Second Edition*
Edited by: P. Cutler © Humana Press Inc., Totowa, NJ

ous chemical reactions can occur at moderate temperatures also. Virtually complete aggregation of lyophilized bovine serum albumin occurred over 24 h at 37°C, following addition of just 3 μL of physiological saline. The degree of aggregation was less, but still significant, at lower temperatures. The underlying cause was the formation of an intermolecular disulfide bond by a thiol-disulfide exchange. Ovalbumin, glucose oxidase, and β-lactoglobulin underwent similar aggregation under the same conditions but by a somewhat different mechanism *(4)*; *see* **Note 1**. These findings underline how important it is to ascertain correct storage conditions for the protein of interest. Simple reliance on the laboratory refrigerator to minimize activity losses may not suffice over an extended period.

Other factors may work against the refrigerator. Certain proteins are more stable at room temperature than in the refrigerator and are said to be *cold labile*. This cold denaturation has been well characterized for myoglobin and a few other proteins *(5,6)*. It is a property of the protein itself and is distinct from freezing inactivation (*see* **Subheading 3.2.** of Chapter 32). This phenomenon arises from the fact that it is thermodynamically possible for a protein to unfold at low as well as at high temperatures (*see* **ref. 7** for a summary of the notable features of cold denaturation). Although cold denaturation is reversible *(7)*, it is inconvenient and can waste valuable time as the protein refolds (as well as causing laboratory scientists some unpleasant shocks!).

Storage concerns a protein's *long-term* or *kinetic* protein stability. Kinetic stability is distinct from (and need not correspond with) *thermodynamic* stability. Thermodynamic stability refers to a protein's conformational stability in terms of the change in heat capacity or the Gibbs (free) energy on reversible unfolding. (Unfolding can be reversible in vitro, but other events occurring subsequent to unfolding can lead to irreversible inactivation.) Kinetic stability measures the persistence of activity with time (or, to put it another way, the progressive loss of function). It can be represented by the scheme

$$N \xrightarrow{k_{in}} I \qquad (1)$$

where N is the native, functional protein, I is an irreversibly inactivated form and k_{in} is the rate constant for the inactivation process. The equation $V_{in} = -d[N]/dt = k_{in}[N]$ describes the process mathematically, where V_{in} is the experimentally observed rate of disappearance of the native form *(1)*. Often, the activity loss will be first order although more complex inactivation patterns are well documented *(8)*. Note that an apparently unimolecular first-order time-course of inactivation may mask a more complex set of inactivating molecular events *(8)*.

This article does not consider the special requirements of pharmaceutical regulatory authorities, good manufacturing practice (GMP), or ISO 9000 specifications. Readers needing to meet these directives and standards should consult materials produced by the appropriate authorities in order to ensure compliance. Finally, mention of suppliers' names does not imply endorsement of any particular product(s).

2. Materials

All containers used for storage of pure proteins should be of good quality and should tolerate temperatures as low as −20°C or even −80°C if freezer storage is desired or necessary. A number of manufacturers (such as Sarstedt, Germany, or Nunc, Denmark; there are many others) supply presterilized screw-cap plasticware with good mechanical and low-temperature properties. Clean glassware exhaustively and sterilize it by dry heat. Screw caps or rubber stoppers that will not withstand dry heat can be autoclaved (*see* **Note 2**).

Membrane or cartridge filters, of pore size 0.22 μm for sterile filtration, are available from companies such as Gelman or Sartorius, with or without a Luer-lock attachment for extra secure attachment to a hand-held syringe.

Use an ordinary domestic refrigerator for storage at temperatures of a few degrees Celsius. Modern machines are often combined with a freezer unit that can easily maintain temperatures as low as −20°C. Storage at −70°C or below will require a specialized low-temperature freezer.

A number of standard constant-temperature laboratory incubators will be required if accelerated storage testing is to be performed.

Highly purified forms of the following chemicals will be useful:

Antimicrobials: Sodium azide, thiomersal.
Protease inhibitors: Phenylmethylsulfonyl fluoride (PMSF).
Chelators: Ethylenediamine tetraacetic acid (EDTA).
Salts: Ammonium sulfate; other ammonium, citrate, sulfate, acetate, and phosphate salts.
Osmolytes: Sucrose, glucose, trehalose, xylitol, glycerol, polyethylene glycol (PEG).
Reducing agents: Dithiothreitol (DTT), 2-mercaptoethanol.

3. Method

3.1. Prevention of Bacterial Contamination

3.1.1. Antimicrobials

Microbial contamination can lead to significant losses of a pure protein by proteolysis. Contamination can be detected by standard microbiological plating techniques, but the aim must always be to avoid it in the first place. Even if one can achieve successful elimination or removal of contaminating micro-organisms, the protein of interest may already have lost at least some activity or may have deteriorated in ways difficult to detect. The addition of antimicrobial compounds such as sodium azide or thiomersal (sodium merthiolate, a mercury-containing compound) can prevent microbial growth. (Both of these compounds are poisonous; handle them with care.) Add sodium azide to a final concentration of 0.1% (w/v) or thiomersal to a final concentration of 0.01% (w/v). Note that azide will inactivate heme-containing proteins such as peroxidase (*see also* **Note 3**).

3.1.2. Filtration

Where one wishes to ensure sterility or to avoid use of the antimicrobials discussed in **Subheading 3.1.1.**, filtration offers a useful alternative. (One can use both strategies together, of course.) A filter of pore size 0.22 μm will exclude all bacteria; indeed, this method is used in industry to sterilize labile materials that cannot be autoclaved or irradiated. Disposable filter cartridges are widely available in a variety of configurations; most have very low protein-binding capacities. Typically, one draws the solution to be sterilized into a syringe and then removes the needle or tube. Next, connect the filter to the syringe nozzle, ensuring that it is firmly mounted. Uncap a suitable sterile storage container directly beneath the filter outlet (using standard aseptic manipulations to avoid contamination of container or cap) and depress the syringe plunger to force the protein solution through the sterilizing filter into the container. Recap immediately. It will not be possible to "flame" plastic containers in a Bunsen burner as part of the aseptic technique: It is much better practice to perform filtration operations of this sort in a Class 2 laminar flow microbiological safety cabinet, the design of which prevents contamination of sample. Following the manufacturer's instructions closely, turn on the cabinet's fans and allow to run for at least 10 min to allow adequate filtration of cabinet air. Open and remove the front door. Swab the cabinet's internal surfaces and the outer surfaces of storage containers brought inside the cabinet with 70% alcohol and allow to evaporate. Carry out the filtration maneuvers, remove the storage containers, and dispose of waste materials appropriately. Swab the internal surfaces of the cabinet with alcohol once again, replace the front cover, and allow to run for 10 min (or according to the user's handbook) before shutting down. It is not always possible to use a filter as fine as 0.22 μm directly; *see* **Note 4**.

3.2. Avoidance of Proteolysis

It can be difficult to remove proteases completely during purification of a target protein. Unless the object protein is completely pure (homogeneous), even very small amounts of contaminating proteolytic enzymes can cause serious losses of activity during extended storage periods. The molecular diversity of proteases complicates the situation: There are exopeptidases, which remove amino acid residues from the N- or C-termini, and endopeptidases, which cleave internal peptide bonds within protein chains. In addition, there are four types of protease classified by their molecular reaction mechanisms: the serine, cysteine (or thiol), acid, and metalloproteases *(9)*. Use EDTA in the concentration range 2–5 mM to complex the divalent metal ions essential for metalloprotease action. Pepstatin A is a potent but reversible inhibitor of acid proteases. It is used at concentrations of around 0.1 μM, as are similar protease inhibitors. The compound PMSF reacts irreversibly with the essential serine in the active site of serine proteases, inactivating them. It can also act on some thiol proteases. Use it at a final concentration of 0.5–1 mM, having dissolved it first in a solvent such as acetone (it is poorly soluble in water) *(10)*. One must, of course, ensure before addition that none of these compounds will adversely affect the protein of interest (*see also* **Note 5**).

If the protein of interest is itself a proteolytic enzyme, use of protease inhibitors is not feasible. One may need to store such a protein in dried form (*see* **Subheading 3.5.**) or as a freeze-dried preparation (Chapter 32). In some cases, it may be possible simply to place the enzyme in a solution having a pH value far removed from the protease's optimum. Trypsin, for example, is most active at mildly alkaline pH values. Stock solutions of trypsin for daily use are often prepared in 1 mM HCl, where the very acid pH value renders the enzyme effectively incapable of catalysis. This helps prevent autolysis during the course of the experiment. The trypsin molecule does not inactivate under these conditions and is fully active on dilution into a suitable assay solution *(11)*. Of course, one should ascertain by experiment whether the protease of interest will tolerate such storage conditions.

3.3. Use of Stabilizing Additives

3.3.1. Background

It has long been known that inclusion of low-molecular-weight substances such as glycerol or sucrose in protein solutions can greatly stabilize the critical protein's biological activity. However, it took some time for the exact mechanism of this stabilization to be ascertained. Timasheff and colleagues have shown that these types of substance are preferentially excluded from the vicinity of the protein molecules, because their binding to the protein is thermodynamically unfavorable *(12,13)*. ("Preferential exclusion" means that there is less of the solute [additive] immediately surrounding the protein than there is in the bulk solution; it does not necessarily mean that *no* solute molecules can penetrate the protein molecule's hydration shell *[14]*). This preferential hydration of the protein molecule arises from a polyol-induced increase in the surface tension of the solvent water *(15)*. Interaction between the protecting additive and the peptide backbone is unfavorable *(16)*. Loss of the protein's compact, properly folded structure (denaturation) will increase the protein–solvent interface. This, in turn, will tend to increase the degree of thermodynamically unfavorable interaction between the additive and the protein molecule. The result is that the protein molecule is stabilized by the additive. A recent study indicates that naturally occurring stabilizing additives increase a protein's T_m (the temperature at which 50% of the protein molecules are unfolded) but do not affect the protein's denaturation Gibbs energy ($\Delta G_D°$) *(17)*. This means that the intrinsic conformational stability of the protein molecule itself is not increased, but its unfolding is greatly disfavored by virtue of the medium's containing the stabilizing additives.

It is important to note that the additives discussed next are *generally* applicable as stabilizing agents for proteins, but a given substance may not be effective for a particular protein. Both sucrose and PEG, for instance, are good stabilizers of invertase but have denaturing effects on lysozyme; the same additive has contrary effects on the two enzymes *(18)*.

3.3.2. Addition of Salts

Certain salts can significantly stabilize proteins in solution. The effect varies with the constituent ions' positions in the Hofmeister lyotropic series, which relates to ionic effects on protein solubility *(19,20)*. This series ranks both cations and anions in order of their stabilizing effects. In the following, the most stabilizing ions are on the left and those on the right are actually destabilizing:

$$(CH_3)_4 N^+ > NH_4^+ > K^+, Na^+ > Mg^{2+} > Ca^{2+} > Ba^{2+},$$
$$SO_4^{2-} > Cl^- > Br^- > NO_3^- > ClO_4^- > SCN^- \tag{2}$$

The "stabilizing" ions force protein molecules to adopt a tightly packed, compact structure by "salting out" hydrophobic residues. This helps prevent the unfolding that is the initial event in any protein deterioration process (*see* **Subheading 1.**). Most stabilizing ions seem to act via a surface tension effect *(12,13)*. Ions can also stabilize proteins by shielding surface charges and they can act as osmolytes by affecting the bulk properties of water *(21)*. Note that ammonium sulfate, which is widely used as a stabilizing additive and as a noninactivating precipitant, comprises two of the most stabilizing ions from the above list, the NH_4^+ cation and the SO_4^- anion. To stabilize proteins in solution while avoiding precipitation, add ammonium sulfate to a final concentration in the range 20–400 mM *(21)*. One can do this by adding a minimal volume of a stock solution of ammonium sulfate of known molarity or by the careful addition of solid ammonium sulfate. Sprinkle the solid salt, a few grains at a time, into the protein solution. Ensure that each portion of ammonium sulfate added dissolves fully before the addition of the next lot. This procedure will prevent accumulation of high local salt concentrations, which are undesirable. In addition to ammonium sulfate, salts containing citrate, sulfate, acetate, phosphate, and quaternary ammonium ions are generally useful *(21)*. Note, however, that the nature of the counterion will influence the overall effect of such compounds on protein stability *(12,13)*.

Charged compounds other than salts may benefit protein stability. Polyethyleneimine is a cationic polymer with numerous uses, including protein stabilization *(22)*. Both high- and low-molecular-weight fractions of polyethyleneimine, when included at 0.01–1% (w/v) concentrations, greatly increased the shelf lives of dehydrogenases and hydrolases stored at 36°C. The effect seems to be kinetic rather than thermodynamic, as the denaturation temperature of lactate dehydrogenase was unaffected by the presence of polyethyleneimine *(22)* (*see* **Note 6**). The cationic surfactant benzalkonium chloride (0.01% or 0.1%) maintained the activity of bovine lactoperoxidase stored at 37°C, pH 7.0 for much longer than that of a control sample; however, this stabilizing effect was not observed at pH 6 *(23)*.

This discussion assumes that ions (or other substances) used or added do not act as substrates, activators, or inhibitors of the enzyme or protein in question. It also assumes that added ions do not interfere with, or precipitate, possibly essential ions already in solution. The ammonium ion, for instance, is actually a substrate for the enzyme glutamate dehydrogenase.

3.3.3. Use of Osmolytes

Osmolytes are a diverse group of substances comprising such compounds as polyols, monosaccharides and polysaccharides, neutral polymers (such as PEG), amino acids (and their derivatives), and methylamines (such as sarcosine and trimethylamine N-oxide, TMAO) *(16)*. They are not strongly charged and have little effect on enzyme activity below 1 *M* concentration. In general, they affect water's bulk solution properties and do not interact directly with the protein *(21)*.

Use polyols and sugars at high final concentrations; typical figures range from 10% to 40% (w/v) *(13,21)*. Sugars are reckoned to be the best stabilizers, but reducing sugars can react with protein amino groups, leading to inactivation *(3,13)*. This problem can be avoided by using nonreducing sugars or the corresponding sugar alcohols. Glycerol is a very widely used low-molecular-weight polyol. Its advantages include its ease of removal by dialysis and its noninterference with ion-exchange chromatography *(21)*. However, glycerol suffers from two significant disadvantages: It is a good bacterial substrate *(21)* and it greatly lowers the glass transition temperature (T_g') of materials to be preserved by lyophilization (*see* **Subheading 3.2.** of Chapter 32) or drying (*see* **Subheading 3.5.**). The five-carbon sugar alcohol, xylitol, can often replace glycerol; it can be recycled from buffers and is not a convenient food source of bacteria *(21)*.

Polymers such as PEG are generally added to a final concentration of 1–15% (w/v). They increase the viscosity of the single-phase solvent system and thus help prevent aggregation. Note, however, that higher polymer concentrations will promote the development of a two-phase system. The protein of interest will concentrate in one of these phases and this may actually lead to aggregation *(21)*.

Amino acids with no net charge, notably glycine and alanine, can act as stabilizers if used in the range 20–500 mM *(21)*. Amino acids and derivatives occur as osmolytes in nature *(13)*. Some related compounds, such as γ-aminobutyric acid (GABA) and TMAO, can be good stabilizers in the 20–500 mM range used for amino acids *(13,21)*.

3.3.4. Substrates and Specific Ligands

The addition of specific substrates, cofactors, or competitive (reversible) inhibitors to purified proteins can often exert great stabilizing effects. (Indeed, their inclusion may be necessary where an essential metal ion or coenzyme is only loosely bound to the apoprotein.) Occupation of the target protein's binding or active site(s) by these substances leads to minor but significant conformational changes in the polypeptide backbone. The protein adopts a more tightly folded conformation, reducing any tendency to unfold *(24)* and (sometimes) rendering it less prone to proteolytic degradation. Occlusion of the protein's active site(s) by a bound substrate molecule or reversible inhibitor will protect those amino acid side chains critical for function. A starch-degrading amyloglucosidase enzyme (from an *Aspergillus* species) stored in the presence of 14% (w/v) partial starch hydrolysate was 80% more stable over a 24-wk period at ambient temperature than the corresponding enzyme preparation stored in the hydrolysate's absence *(25)*.

Note that dialysis (or some other procedure for the removal of low-molecular-mass substances) may be necessary to avoid carryover effects of the substrate or inhibitor when the protein is removed from storage for use in a particular situation where maximal activity is desired.

3.3.5. Use of Reducing Agents and Prevention of Oxidation Reactions

The thiol group of cysteine is prone to destructive oxidative reactions. One can prevent or minimize these by using reducing agents such as 2-mercaptoethanol (formerly β-mercaptoethanol) (a liquid with an unpleasant smell) or dithiothreitol ("Cleland's reagent", or DTT, a solid with little odor). Add 2-mercaptoethanol to reach a final concentration of 5–20 mM; then, keep the solution under anaerobic conditions. To achieve these anaerobic conditions, gently bubble an inert gas such as nitrogen through the solution before adding the reducing agent. Fill the solution to the brim of a screw-cap container to minimize head space and the chances of gaseous exchange. Dithiothreitol is effective at lower concentrations; usually, 0.5–1 mM will suffice *(10)*. Indeed, Schein has advised that the DTT concentration should not exceed 1 mM; it can act as a denaturant at higher temperatures and is not very soluble in a high-salt solution *(21)*. Note that these reducing agents are themselves prone to oxidation. (This is why solutions containing them must be stored under "anaerobic conditions.") Dithiothreitol oxidizes to form an internal disulfide that is no longer effective but which will not interfere with protein molecules *(10)*. 2-Mercaptoethanol, on the other hand, participates in intermolecular reactions and can form disulfides with protein thiol groups *(10)*. Such thiol-disulfide exchanges are highly undesirable and may actually lead to inactivation or aggregation. It is probably best to add reducing agents only in situations where they are known (or can be demonstrated) to be beneficial *(19)*.

Much of the oxidation of thiol groups is mediated by divalent metal ions that can activate molecular oxygen. Complexation of free metal ions (where they are not themselves essential for activity) can prevent destructive oxidation of thiol groups. Polyethyleneimine at 1% (w/v) concentration protected the sulfhydryl groups of lactate dehy-drogenase against oxidation and prevented the consequential aggregation of the protein, even in the presence of Cu^{2+} ions; the protecting effect was ascribed to metal chelation by polyethyleneimine *(22)* (*see* **Subheading 3.2.** regarding the use of EDTA to complex metal ions).

3.3.6. Extremely Dilute Solutions

Very dilute protein solutions are highly prone to inactivation. This is especially true of oligomeric proteins, where dissociation of subunits can occur at low concentration. The individual polypeptide chains comprising the oligomer may lack activity alone and/or may denature with consequent loss of activity. Protein solutions of concentration less than 1–2 mg/mL should be concentrated as rapidly as possible *(10)* by ultrafiltra-tion or sucrose concentration (*see* **Note 7**).

Where rapid concentration is not possible, inactivation may be prevented by the addition of an exogenous protein such as bovine serum albumin (BSA), typically to a final concentration of 1 mg/mL. Alkaline phosphatase from *Escherichia coli* is unstable at room temperature at concentrations less than 10 µg/mL but can be stabilized by the addition of BSA *(26)*. Scopes has discussed possible reasons for the undoubted benefits of BSA addition *(10)*. It may seem foolish to deliberately add an exogenous, contami-nating protein such as BSA to a pure protein preparation, but occasionally this may be the price to be paid in order to avoid inactivation.

3.4. Low-Temperature Storage

Refrigeration at 4–6°C is often sufficient for the preservation of a protein's biologi-cal activity provided the hints in **Subheadings 3.1.–3.3.** are followed judiciously. Many proteins are supplied commercially in 50% glycerol or as slurries in approx 3 *M* am-monium sulfate. Freezing of such preparations is not necessary and should be avoided. Occasionally, one may observe the phenomenon of cold denaturation; this has been dis-cussed briefly in **Subheading 1.**

Some proteins can deteriorate at "refrigerator" temperatures and require storage at temperatures lower than 0°C. Usually, temperatures between −18°C and −20°C, at-tainable by a domestic freezer, will allow for stable storage (*see* **Note 8**). Sometimes, however, it may be necessary to use temperatures below −20°C. In these cases, it is nor-mal to use a low-temperature laboratory freezer designed to maintain temperatures in the range −70°C to −80°C (*see* **Note 9**).

Most protein solutions will undergo freezing to a solid at temperatures below 0°C. (Mixtures containing high concentrations of glycerol will remain liquid at −20°C: *see* **Subheading 3.3.3.** of this chapter and **Subheading 3.2.** of Chapter 32.) The events oc-curring on freezing of a protein-containing mixture or biological system are much more complex than the simple macroscopic phase change would suggest. Differential freez-ing of particular components of the mixture can lead to enormous concentration effects and to dramatic changes of pH at low temperatures. These processes can lead, in turn, to a notable degree of protein inactivation. The subject of freezing damage and its avoid-ance is discussed in Chapter 32 (**Subheading 3.2.**). Note that the problem can often be minimized by judicious choice of stabilizing additives as discussed in **Subheading 3.3.**; *see also* **ref. 14.**

Prevention of freezing will, of course, avoid freezing damage. It is possible to undercool liquids without freezing by preventing the nucleation of ice crystals. This means that proteins can be stored well below 0°C in the liquid phase. The preparation of protein-containing aqueous–organic emulsions that can maintain complete biological activity in the liquid state over extended periods at −20°C has been described *(27)*. The method is very useful for small volumes of valuable proteins, avoids the need to use additives, and is more economical than freeze-drying. The same process is used for many different proteins and one can remove portions of a sample without any effect on the activity of the remainder. The actual storage temperature matters little, provided the upper temperature is less than 4°C and the lower temperature remains above −40°C, the nucleation temperature for ice crystal formation *(28)*.

3.5 Drying for Stable Storage

The advantages of water removal as a protein storage/stabilization strategy have been set out in the **Subheading 1.** of Chapter 32. Lyophilization can remove more than 95% of water from a protein preparation, but there is the risk of freezing damage (*see* **Subheading 3.2.** of Chapter 32). It is possible to design protein-compatible formulations with glass transition temperatures ($T_g{'}$) typically as high as 37°C *(29)*. With these high $T_g{'}$ values, controlled evaporative drying can be used in place of lyophilization to stabilize proteins in the solid state. Worthwhile evaporation rates will occur below these high $T_g{'}$ values at reduced pressure. Evaporation is faster, less costly, and more easily controlled than freeze-drying *(28,29)*. The high $T_g{'}$ values also mean that one can sometimes store the resulting dried product at ambient temperature; as long as room temperature does not exceed the glass transition temperature, the protein formulation will not undergo a glass/rubber transition during storage at room temperature. The glass-forming compounds are typically carbohydrates; maltose and maltohexose are particularly valuable, and sucrose can be useful if the moisture content can be reduced to 2% or less *(29)*. Reconstitution of the solid protein preparation is accomplished simply by rehydration with added water or buffer. The method has been patented and is described in **ref. 30**. An alternative formulation for vacuum-drying of proteins, involving the use of a cationic, soluble polymer (e.g., diethylaminoethyl dextran) and the sugar alcohol lactitol has also been patented *(31)* and published *(32)*.

3.6. Stability Analysis and Accelerated Degradation Testing

3.6.1. Background

The distinction between conformational and kinetic stability has already been drawn in **Subheading 1.** [Very recently a three-parameter model permitting description of the decay of enzyme activity irrespective of the underlying mechanism has been described *(33)*.] Kinetic stability is usually measured at elevated temperatures *(1)* but the inactivating event(s) at high temperatures may not mirror that/those at the much lower temperatures used for storage. It is not feasible, however, to monitor stability in real time at the actual storage temperature; the experiment would take too long. Inaccuracy may result over shorter intervals, because only minimal losses, virtually indistinguishable from the starting activity, would be apparent.

Fortunately, there is a methodology that can, in many cases, overcome these difficulties, namely accelerated degradation (or accelerated storage) testing. This involves the periodic assay of samples incubated at different temperatures and the use of the Arrhenius equation to predict "shelf lives" at temperatures of interest. The Arrhenius equation is conveniently expressed in logarithmic form as $\ln k = -E_a/RT + \ln A$, where k is the first order rate constant of activity decay, E_a is the activation energy, R is the gas constant and T is the temperature in Kelvin. This log form of the Arrhenius equation yields a straight line plot of $\ln k$ against $1/T$ with slope $-E_a/R$ *(34)*. Extrapolation of this plot can give the rate constant (and hence the useful life) at a particular temperature; *see* **Note 10**. Accelerated storage testing has been used as a practical means of quality assurance for biological standards *(35)* and has been employed in some scientific investigations *(36,37)*.

3.6.2. Setting Up a Test

An accelerated storage test must use the minimum amount of the (often precious) test protein that is compatible with achievement of precise and accurate results. Because each experiment will occupy a period of weeks, it is important that every test yield a meaningful outcome. Accordingly, take great care in setting up an accelerated degradation test run.

It is essential to prevent microbial contamination of test samples (*see* **Subheading 3.1.**), especially because one assumes that no (or very little) sample degradation takes place during the period of the experiment. Place test samples at a series of elevated temperatures (e.g., 48°C, 45°C, 42°C, 37°C, 33°C, 30°C) and at a suitable, lower reference temperature—say, 4°C (*see* **Note 8**). Ensure that the reference temperature will not cause freezing of liquid samples. (The liquid–solid phase change will introduce a further variable and may also lead to freezing damage, as discussed in **Subheading 3.2.** of Chapter 32.) *See also* **Note 11**. Remove samples at intervals from the various incubation conditions, bring them to the same temperature and assay for activity under standardized, optimal conditions. Ensure that stock solutions used in assays are carefully prepared and standardized; it is likely that different batches of assay solutions will be needed over the period of the accelerated storage test, so variations must be minimized. Be equally scrupulous with regard to procedural details of the assay and the performance of the instruments used.

Kirkwood has made some practical recommendations for successful accelerated storage testing *(38)*, which are summarized here. Use at least three elevated temperatures plus a low reference temperature. Set up 10 or more samples at each of these temperatures. The following schedule avoids waste of material resulting from testing at inappropriate intervals. It also allows checking of the order of reaction. An acceptable result is very likely if enough material is placed on test at the beginning.

1. At intervals, test samples stored at the highest temperatures against the low-temperature reference samples. Ignore intermediate temperatures until a loss of 25% or greater (against the low-temperature reference sample) has occurred at the highest temperature. Now, test a second sample to confirm the first result. If the results disagree, continue this **step 1** testing. If they agree, move on to **step 2**.
2. Test the next two highest temperatures against a reference sample. Fit all available data to the Arrhenius equation. This procedure will not give a final precise result, but it will facilitate progress to **step 3**.

3. Use the approximate result from **step 2** to ascertain further periods of storage at all temperatures such that measurable activity losses will have occurred at the intermediate temperatures. Assay samples from all available temperatures and check all available data for their ability to fit the Arrhenius equation. This should result in a reasonable estimate of the low-temperature degradation rate, especially if multiple assays have been performed. If too little degradation has occurred to yield a precise result, one can repeat **step 3** for a longer time *(38)*.

3.6.3. Analysis of Results

It is clear that "raw" experimental data must undergo a number of transformations for use in the Arrhenius plot, notably conversion to natural log or reciprocal forms. Error relationships are often significantly affected by these sorts of transformations. Use a computer for all statistical fitting to minimize such errors. Use of good quality replicate results is very important.

The activity decay must be first order at all of the temperatures used in the Arrhenius plot. Verify this by fitting the time-course of activity loss at each test temperature to a first-order exponential decay. Assess goodness of fit by inspection of the graphic fit and of parameters such as standard errors or chi-square values. First-order exponentials yield straight-line plots when transformed into semilogarithmic form, unlike higher-order functions (i.e., a plot of ln[Activity] against time is linear for a first-order decay). Deviations from a first-order function are more likely to occur at higher temperatures *(35)*.

Using the k values determined at different temperatures, plot the Arrhenius graph (ln k is the ordinate, $1/T$ the abscissa). A linear plot of negative slope results. Extrapolate the line to a temperature of interest (e.g., 0°C, 4°C, or 25°C; these are respectively 273, 277, and 298 K) and estimate the value of k at this temperature. It is easy to estimate a true half-life at this temperature by using the equation $t_{1/2} = 0.693/k$. Specially designed programs for the analysis of accelerated storage data are available (*see* **Note 12**).

4. Notes

1. Loss or decrease of the protein's biological or functional activity will be the main and most important index of deterioration. Often, however, the degree or time-course of activity loss will not give any indication of the underlying molecular cause (although aggregation may be readily visible). **Reference *1*** provides a useful table of methods to identify the molecular changes leading to inactivation of the protein.

2. The type of stopper used can influence the residual moisture contents of freeze-dried materials. Such effects can arise from the nature of the stopper material itself and also from its prehistory *(39)*. Such considerations will likely apply also to dry preparations for storage, so stoppers should be chosen with care.

3. Thiomersal and azide are totally unacceptable in any product for administration. Do not discard azide compounds or azide-containing solutions down laboratory sinks. Not only is azide toxic, but it can accumulate in old lead piping, leading to the formation of potentially explosive compounds.

4. Some biological matrices, particularly sera, will not filter effectively through a 0.22-μm filter alone. One may need to prefilter the material initially through a coarser 0.45-μm filter to which the desired 0.22-μm filter is connected in series. Alternatively, one can accomplish the finer filtration as a separate operation. One can best filter larger volumes (hundreds of milliliters, or liters) using a stack of filters clamped in a special filtration unit. A filter as coarse as 1 μm may be used directly in contact with the solution of interest, the stack com-

prising progressively finer filters until the sterilizing 0.22-μm filter is encountered at the bottom of the stack. Technical representatives of filtration manufacturers can give specialist's advice for individual cases.

5. Many suppliers offer specific inhibitors of proteases or classes of protease. These inhibitors are often peptides or proteins (e.g., aprotinin, soybean trypsin inhibitor). A cocktail of protease inhibitors is available in tablet form from Roche Molecular Biochemicals under the trade name "Complete." The product is stated to give effective inhibition of serine, cysteine, and metalloproteases during protein extractions from a variety of tissues and sources.

6. Curiously, however, lactate dehydrogenase activity levels at pH 5 decreased with increasing concentrations of polyethyleneimine. In contrast, the polymer stimulated activity at pH 7.2 and 9.0 *(22)*.

7. Many different vessels and membranes for laboratory-scale ultrafiltration, with a range of defined molecular-weight cutoffs, are commercially available. These may comprise permanent stirred pressure cells with replaceable membranes (for volumes in the range 10–500 mL) or disposable centrifugal concentrators (for volumes up to 10 mL) *(21)*. The Millipore Corporation is a prominent supplier of such apparatus. Schein gives some useful observations on ultrafiltration and suggests some other means of achieving protein concentration *(21)*.

    Sucrose concentration is an effective and rapid means of concentrating a dilute protein solution. Place the solution of interest into a suitably treated, softened dialysis tube and secure the ends tightly. Tear off a piece of aluminum foil such that the dialysis tube will rest on the foil with roughly 5–6 cm to spare all round. Shake some solid sucrose onto the foil, then rest the dialysis tube on top of the sucrose. Shake more sucrose on top of the dialysis tube, wrap the foil around the sucrose and dialysis tube to form a parcel and place in the refrigerator. Water from the dilute protein solution will move by osmosis through the pores of the dialysis tube to the surrounding solid sucrose, leading to concentration of the protein. Examine the dialysis tubing every 15–20 min. The sucrose surrounding the dialysis tubing will gradually form a viscous liquid that can be removed periodically and replaced with fresh solid material. Volume reduction can take place quite quickly. The method has the drawback that sucrose will enter the dialysis tube in amounts not readily calculable. (The sucrose will likely help to stabilize the protein, of course.) If the presence of sucrose is undesirable, gently pull the dialysis tube between finger and thumb to force its contents into one end. Knot or clamp the dialysis tube tightly as close as possible to the concentrated solution and then dialyze the shortened dialysis tube against a suitable buffer to remove the sucrose. Note that the dialysis tube will swell in dilute buffer as water moves by osmosis into the protein solution, which will have a high sucrose concentration. The dialysis tube must be clamped very tightly and as short as possible to prevent undue "redilution" of the sucrose-concentrated protein solution.

8. It is a good idea to place a maximum/minimum thermometer inside the refrigerator, freezer, or incubator(s) close to the containers of interest in order to record any significant variations of temperature that may occur over an extended period. (Ensure first that the thermometer will withstand the low or high temperatures!)

9. Low-temperature freezers typically function between $-70°C$ and $-80°C$. These temperatures are extremely cold and can inflict a "cold burn" on exposed skin. Always wear insulating or autoclave gloves when handling low-temperature items: Latex laboratory gloves are not sufficient.

10. The Arrhenius equation is empirical, but a similar equation results from the applications of Eyring's transition state theory; *see* **ref. *34*** for a fuller discussion of this point.

11. Amorphous solid preparations will follow Arrhenius kinetics provided they remain in the glassy state. However, if any of the elevated temperatures used exceeds the glass transition

temperature (T_g'), the product will become rubbery and will no longer obey the Arrhenius equation. (In fact, it will deteriorate much faster than predicted by the Arrhenius equation.) *See* **ref.** *29* for a description of this phenomenon and of an alternative kinetic scheme. Other situations in which deviations from Arrhenius kinetics may occur are outlined in **ref.** *40*; *see also* Note 6 of Chapter 32.

12. There follows a brief description, for information only, of two program for analysis of accelerated degradation data. Their underlying statistical methodology is beyond the scope of this chapter and the present author makes no endorsement or judgment of their merits or claims. Each user must determine the suitability of any package for his/her purposes.

    The DEGTEST program carries out a three-stage analysis of data. First, the % activities at each temperature (relative to the low-temperature reference sample) are used to give estimates of relative degradation rates, assuming first-order decay. Second, these rates are fitted to the Arrhenius equation and the goodness of fit is ascertained. Third, the decay rate at the proposed storage temperature is predicted from the Arrhenius fit and the statistical precision of this value is calculated. The program can fit to the Arrhenius equation or to the more rigorous Eyring equation *(41)*. In practice, however, little difference is observed between the two equations *(42)*.

    The POTENCYLOSS program is based on a linear model for stability prediction that uses a double logarithmic primary plot together with an altered secondary plot corresponding to the Arrhenius plot. The program can distinguish between zero- and first-order data and gives loss rate and shelf-life parameters at a desired designated temperature. In addition, one can obtain estimates of activity values at any temperature or future time period *(43)*.

References

1. Mozhaev, V. V. (1993) Mechanism-based strategies for protein thermostabilisation. *Trends Biotech.* **11,** 88–95.
2. Sluzky, V., Klibanov, A. M., and Langer, R. (1992) Mechanism of insulin aggregation and stabilisation in agitated aqueous solutions. *Biotech. Bioeng.* **40,** 895–903.
3. Quax, W. J. (1993) Thermostable glucose isomerases. *Trends Food Sci. Technol.* **4,** 31–34.
4. Liu, W. R., Langer, R., and Klibanov, A. M. (1991) Moisture-induced aggregation of lyophilised proteins in the solid state. *Biotech. Bioeng.* **37,** 177–184.
5. Privalov, P. L. (1990) Cold denaturation of proteins. *Crit. Rev. Biochem. Mol. Biol.* **25,** 281–305.
6. Franks, F. and Hatley, R. H. M. (1991) Stability of proteins at subzero temperatures: thermodynamics and some ecological consequences. *Pure Appl. Chem.* **63,** 1367–1380.
7. Franks, F. (1993) Conformational stability of proteins, in *Protein Biotechnology: Isolation, Characterization and Stabilization* (Franks, F., ed.), Humana, Totowa, pp. 395–436.
8. Sadana, A. (1988) Enzyme deactivation. *Biotech. Adv.* **6,** 349–446.
9. Beynon, R. J. and Bond, J. S. (1987) *Proteolytic Enzymes: A Practical Approach*, Oxford University Press, Oxford, (2nd ed. 2001).
10. Scopes, R. K. (1994) *Protein Purification: Principles and Practice*, 2nd ed., Springer-Verlag, Berlin, pp. 317–324.
11. Erlanger, B. F., Kokowsky, N., and Cohen, W. (1961) The preparation and properties of two new chromogenic substrates of trypsin. *Arch. Biochem. Biophys.* **95,** 271–278.
12. Timasheff, S. N. and Arakawa, T. (1997) Stabilization of protein structure by solvents, in *Protein Structure: A Practical Approach*, 2nd ed. (Creighton, T. E., ed.), IRL, Oxford, pp. 349–364 (1st ed., 1989, pp. 331–345).
13. Timasheff, S. N. (1992) Stabilization of protein structure by solvent additives, in *Stability*

of Protein Pharmaceuticals, Part B (Ahern, T. J. and Manning, M. C., eds.), Plenum, New York, pp. 265–285.

14. Carpenter, J. F. and Crowe, J. H. (1988) The mechanism of cryoprotection of proteins by solutes. *Cryobiology* **25,** 244–255.

15. Kaushik, J. K. and Bhat, R. (1998) Thermal stability of proteins in aqueous polyol solutions. *J. Phys. Chem.* B **102,** 7058–7066.

16. Qu, Y., Bolen, C. L., and Bolen, D. W. (1998) Osmolyte-driven contraction of a random coil protein. *Proc. Natl. Acad. Sci. USA* **95,** 9268–9273.

17. Anjum, F., Rishi, V., and Ahmad, F. (2000) Compatibility of osmolytes with Gibbs energy of stabilisation of proteins. *Biochim. Biophys. Acta* **1476,** 75–84.

18. Combes, D., Yoovidhya, T., Girbal, E., Willemot, R.-M., and Monsan, P. (1987) Mechanism of enzyme stabilization. *Ann. NY Acad. Sci.* **501,** 59–62.

19. Tombs, M.P. (1985) Stability of enzymes. *J. Appl. Biochem.* **7,** 3–24.

20. Klibanov, A. M. (1983) Stabilisation of enzymes against thermal inactivation. *Adv. Appl. Microsc.* **29,** 1–25.

21. Schein, C. H. (1990) Solubility as a function of protein structure and solvent components. *Biotechnology* **8,** 308–317.

22. Andersson, M. M. and Hatti-Kaul, R. (1999) Protein stablising effect of polyethyleneimine. *J. Biotech.* **72,** 21–31.

23. Marcozzi, G., Di Domenico, C., and Spreti, N. (1998) Effects of surfactants on the stabilisation of the bovine lactoperoxidase activity. *Biotech. Prog.* **14,** 653–656.

24. Volkin, D. B. and Klibanov, A. M. (1989) Minimizing protein inactivation, in *Protein Function: A Practical Approach* (Creighton, T. E., ed.), pp. 1–24 IRL Press, Oxford.

25. Shah, N. K., Shah, D. N., Upadhyay, C. M., Nehete, P. N., Kothari, R. M. and Hegde, M. V. (1989) An economical, upgraded, stabilized and efficient preparation of amyloglucosidase. *J. Biotech.* **10,** 267–276.

26. Reid, T. W. and Wilson, I. W. (1971) *E. coli* alkaline phosphatase, in *The Enzymes* (Boyer, P. D., ed.), Academic, London, vol. 4, pp. 373–415.

27. Hatley, R. H. M., Franks, F.and Mathias, S. F. (1987) The stabilisation of labile biomolecules by undercooling. *Process. Biochem.* **22,** 169–172.

28. Franks, F. (1993) Storage stabilization of proteins, in *Protein Biotechnology: Isolation, Characterization and Stabilization* (Frank, F., ed.), Humana, Totowa, NJ, pp. 489–531.

29. Franks, F., Hatley, R. H. M., and Mathias, S. F. (1991) Materials science and the production of shelf-stable biologicals. *Pharm. Tech. Int.* **3,** 24–34.

30. Franks, F. and Hatley, R. H. M. (1992) Storage of Materials. US patent 5,098,893.

31. Gibson, T. D. and Woodward, J. R. (1991) Enzyme stabilisation. PCT/GB91/00443, Publ. No. Wo91/14773.

32. Gibson, T. D., Higgins, I. J., and Woodward, J. R. (1992) Stabilisation of analytical enzymes using a novel polymer-carbohydrate system and the production of a stabilised, single reagent for alcohol analysis. *Analyst* **117,** 1293–1297.

33. Aymard, A. and Belarbi, A. (2000) Kinetics of thermal inactivation of enzymes. *Enzyme Microb. Tech.* **27,** 612–618.

34. Eisenberg, D. S., Crothers, I., and Donald, M. (1979) *Physical Chemistry.* Benjamin-Cummings, Menlo Park, CA, pp. 239–245.

35. Jerne, N. K. and Perry, W. L. M. (1956) The stability of biological standards. *Bull. World Health Organ.* **14,** 167–182.

36. Malcolm, B. A., Wilson, K. P., Matthews, B. W., Kirsch, J. F., and Wilson, A. C. (1990) Ancestral lysozymes reconstructed, neutrality tested, and thermostability linked to hydrocarbon packing. *Nature* **345,** 86–89.

37. Ó Fágáin, C., O'Kennedy, R., and Kilty, C. (1991) Stability of alanine aminotransferase is enhanced by chemical modification. *Enzyme Microb. Tech.* **13,** 234–239.
38. Kirkwood, T. B. L. (1984) Design and analysis of accelerated degradation tests for the stability of biological standards III. Principles of design. *J. Biol. Stand.* **12,** 215–224.
39. Oetjen, G.-W. (1999) *Freeze Drying*, Wiley–VCH, Weinheim, pp. 115.
40. Franks, F. (1994) Accelerated stability testing of bioproducts: attractions and pitfalls. *Trends Biotech.* **12,** 114–117.
41. Kirkwood, T. B. L. and Tydeman, M. S. (1984) Design and analysis of accelerated degradation tests for the stability of biological standards II. A flexible computer programme for data analysis. *J. Biol. Stand.* **12,** 207–214.
42. Tydeman, M. S. and Kirkwood, T. B. L. (1984) Design and analysis of accelerated degradation tests for the stability of biological standards I. Properties of maximum likelihood estimators. *J. Biol. Stand.* **12,** 195–206.
43. Nash, R.A. (1987) A new linear model for stability prediction. *Drug Dev. Ind. Pharm.* **13,** 487–499.

34

Electroelution of Proteins From Polyacrylamide Gels

Michael J. Dunn

1. Introduction

The ability of polyacrylamide gel electrophoresis (PAGE) techniques to resolve the individual components present in complex mixtures of several thousand proteins has resulted in this group of methods being one of the most widely used laboratory techniques in biochemistry. Although gel electrophoresis procedures can provide characterization of proteins in terms of their charge (pI), size (M_r), relative hydrophobicity, and abundance, they give no direct clues as to their identities or functions. A major obstacle to successful structural, functional, or immunochemical chemical characterization is efficient recovery of the separated proteins from the polyacrylamide gel, as most procedures are not compatible with the presence of the gel matrix. The most popular approach used to overcome this problem for the recovery of intact proteins is Western electroblotting (*see* Chapter 35). In this technique, the proteins separated by one-dimensional (1-D) or two-dimensional (2-D) PAGE are transferred ("blotted") onto the surface of an appropriate inert membrane support, on which the total protein pattern can be visualized using an appropriate total protein stain (*see* Chapter 35). For subsequent detection and characterization of proteins using polyclonal or monoclonal antibodies (immunoblotting), the blot membrane is destained and then probed with the relevant antibody. For subsequent microchemical characterization, the protein band or spot of interest is excised. The protein, while still on the surface of the inert membrane support, can then be subjected to the appropriate microchemical characterization technique, such as microsequencing by automated Edman degradation *(1)*, amino acid compositional analysis *(2)*, or identification and characterization by mass spectrometry *(3)*.

An alternative approach that has been extensively used is electroelution. In this technique, protein zones are localized after electrophoresis by staining with Coomassie brilliant blue R-250. Protein-containing gel pieces are then excised and placed in an electroelution chamber where the proteins are transferred in an electric field from the gel into solution and concentrated over a dialysis membrane with an appropriate molecular-weight cutoff. Although this method can have a high efficiency (>90%) of protein recovery, it suffers from a number of disadvantages, including (1) relatively slow isolation of a small number of samples, (2) contamination of the eluted protein with sodium

From: *Methods in Molecular Biology, vol. 244: Protein Purification Protocols: Second Edition*
Edited by: P. Cutler © Humana Press Inc., Totowa, NJ

dodecyl sulfate (SDS), salts, and other impurities, (3) peptide chain cleavage during staining or elution, and (4) chemical modification during staining or elution, which can lead to N-terminal blockage *(4)*. For these reasons, Western blotting has almost universally replaced electroelution as the method of choice for isolation of proteins for microchemical analysis. However, electroelution is still often used for the purification of gel-separated proteins for functional studies such as activity assays, binding studies, and solubility studies.

In the simplest system for electroelution, gel slices containing the protein of interest are mixed with a small volume of polyacrylamide or agarose gel and cast into a glass tube (such as that used for rod gel electrophoresis). A small sack made from dialysis tubing and containing electrophoretic buffer is then fastened with elastic bands over the end of the tube. The tube is then inserted into a rod gel electrophoresis apparatus and the protein eluted into the dialysis sack, from which it can subsequently be recovered. Electroelution is generally more efficient if 0.1% (w/v) SDS is added to the elution buffer, as it helps to maintain proteins in solution and provides them with a high negative charge density, thereby increasing their electrophoretic mobility.

More sophisticated variants of this approach have been described *(4)* and commercial equipment is now available, which allows several different gel segments or even whole gels without prior cutting into sections *(5)* to be eluted into a series of elution chambers. In this chapter, the use of one such commercial device, the Bio-Rad Model 422 Electro-Eluter, to recover proteins separated by SDS-PAGE is described.

2. Materials

Prepare all solutions freshly from analytical-grade reagents and dissolve in deionized water.

1. Fixing and destaining solution: 450 mL Methanol, 100 mL acetic acid, made up to 1 L in water.
2. Staining solution: 0.2 g Coomassie brilliant blue R-250 made up to 100 mL in water.
3. Electroelution apparatus: Bio-Rad Model 422 Electro-Eluter apparatus for the simultaneous elution of up to six samples (*see* **Fig. 1**).
4. Membrane caps (for collection of electroeluted proteins): Two types are available from Bio-Rad; clear (exclusion limit 12,000–15,000 kDa) or green (exclusion limit 3500 kDa).
5. Elution buffer: A buffer system generally suitable for electroelution of proteins is 25 mM Tris base, 192 mM glycine, pH 8.8, containing 0.1% (w/v) SDS (*see* **Note 1**). For subsequent concentration of eluted proteins, it is preferable to use a volatile, basic buffer such as 50 mM NH_4HCO_3 or 50 mM N-ethylmorpholine acetate, pH 8.5, containing 0.1% (w/v) SDS (*see* **Note 1**).

3. Method

3.1. Gel Staining

After the protein sample has been separated by SDS-PAGE, the protein bands of interest must be localized for excision and elution. This is usually achieved by staining the gel with Coomassie brilliant blue R-250 (*see* **Note 2**).

1. Following separation of the proteins by SDS-PAGE, place the gel in fixing solution for at least 1 h at room temperature with gentle agitation.

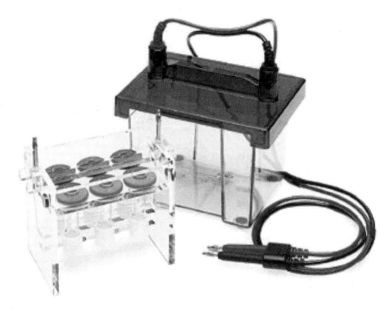

Fig. 1. The Bio-Rad Model 422 Electro-Eluter.

2. Place the gel in staining solution for at 1 h at room temperature with gentle agitation.
3. Transfer the gel to destaining solution and agitate gently at room temperature. After about 24 h and with several changes of destaining solution, the gel background will become colorless, leaving the separated proteins as blue colored bands.

3.2. Electroelution

1. Soak the membrane caps in protein elution buffer for at least 1 h at 60°C (*see* **Note 3**).
2. Take the number of glass tubes required for the gel slices to be eluted and place one frit in the bottom of each tube (*see* **Fig. 2**). Insert the glass tube and frit into the Electro-Eluter module and fill any empty grommet holes with stoppers.
3. Place one prewetted membrane cap in the bottom of each silicone adaptor (*see* **Fig. 2**). Fill the adaptor with elution buffer and pipet the buffer up and down to remove any air bubbles around the dialysis membrane.
4. Slide the silicone adaptor with the membrane cap onto the bottom of the glass tube with frit. Pull the silicone adaptor partially on and off a few times to ensure that all the air bubbles are expelled (*see* **Fig. 2**).
5. Fill each tube with elution buffer and place the gel slice into the tube (*see* **Note 4**).
6. Place the entire module into the buffer chamber. Fill the lower buffer chamber with 600 mL elution buffer (*see* **Note 5**). Fill the upper buffer chamber with 100 mL of elution buffer.
7. Place a magnetic stirring bar in the bottom buffer tank (*see* **Note 6**).
8. Attach the lid with cables, orienting the red to red (anode) and black to black (cathode).
9. Perform electroelution at 8–10 mA/glass tube for 3–5 h (*see* **Note 7**).
10. After the elution is completed, remove the Electro-Elutor module from the buffer tank.
11. Remove one stopper (if used) or one glass tube assembly from the upper buffer chamber and allow the upper buffer to drain.
12. Using a plastic pipet, remove the buffer left in the tube to the level of the frit, and discard the liquid (*see* **Note 8**).

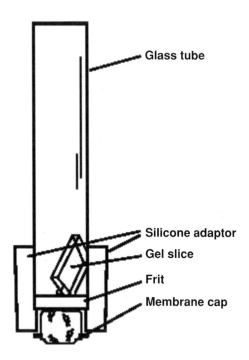

Fig. 2. Side view of Model 422 Electro-Eluter assembly. The vertical glass tube is filled with elution buffer. The negative electrode is at the top. The positive electrode is below the membrane cap. Proteins are carried by the electrical current out of the gel slice, through the frit, and into the membrane cap. The proteins are retained by a dialysis membrane that is molded into the cap.

13. Remove the silicone adaptor together with the membrane cap from the bottom of the glass tube. The liquid level should be slightly above the membrane cap.
14. Using a new plastic pipet, remove the remaining liquid in the membrane cap into a microfuge tube. The volume should be around 400 µL. Rinse the membrane cap with 200 µL fresh elution buffer and add to the sample in the microfuge tube. This solution contains the eluted protein.
15. Repeat **steps 12–14** for each glass tube.

3.3. Protein Precipitation

After electroelution, the protein enriched solution also contains SDS, Coomassie brilliant blue R-250 stain, buffer salts, and other contaminants. These can be readily removed by precipitation of the protein before being redissolved for subsequent analysis.

1. Lyophilize the electroeluted protein sample or dry using a rotary vacuum evaporator.
2. Add 50 µL of water.
3. Add 450 µL of ice-cold acetone acidified to a final concentration of 1 mM HCl and incubate for 3 h at −20°C.
4. Pellet the precipitated protein using a microcentrifuge.
5. Wash the pellet three times with 100 µL ice-cold acetone.
6. Air-dry the pellet, which can then be redissolved in the appropriate solvent for subsequent analysis.

4. Notes

1. Electroelution is generally more efficient in the presence of 0.1% SDS, as it helps to maintain proteins in solution and provides them with a high negative charge density, thereby increasing their electrophoretic mobility. It can be reduced or omitted in the case of some highly soluble and highly charged proteins.
2. The yield of fixed and stained protein is typically around 70–90% and some protein may remain in the gel slice. In the case of poor recovery following electroelution of fixed and stained proteins, it may be better to visualize the separated proteins using a copper-chloride negative stain that does not require fixation (*6*).
3. Membrane caps may be soaked for longer than 1 h. Gloves must be worn when handling the membrane caps to prevent the dialysis membranes from becoming contaminated. Membrane caps can be reused for at least five complete runs without a decrease in protein yield. In this case, they must be stored in elution buffer containing 0.05% sodium azide at 4°C.
4. To increase sample recovery, several bands may be excised from the gel, pooled together, and minced, so long as the height of the gel within the glass tube is approx 1 cm or less. If the glass tube is filled higher than 1 cm, then the time required for elution will increase.
5. The level of the lower buffer must be above the top of the silicone adaptors or bubbles may form on the bottom of the dialysis membrane.
6. Vigorous stirring during the run will prevent air bubbles from sticking to the bottom of the dialysis membrane.
7. The time taken for electroelution is dependent on the molecular weight of the protein to be eluted and the gel percentage used for separation.
8. Take care not to dislodge the silicone adaptor or the membrane cap. Make sure that the liquid below the frit is not disturbed or shaken up during this process.

References

1. Patterson, S. D. (1994) From electrophoretically separated protein to identification: Strategies for sequence and mass analysis. *Anal. Biochem.* **221**, 1–15.
2. Yan, J. X., Wilkins M. R., Ou, K., et al. (1996) Large-scale amino-acid analysis for proteome studies. *J. Chromatogr. A* **736**, 291–302.
3. Patterson, S. D., Aebersold, R., and Goodlett, D. R. (2001) Mass spectrometry-based methods for protein identification and phosphorylation site analysis, in *Proteomics, From Protein Sequence to Function* (Pennington, S.R. and Dunn, M.J., eds.), BIOS Scientific, Oxford, pp. 87–130.
4. Aebersold, R. (1991) High sensitivity sequence analysis of proteins separated by polyacrylamide gel electrophoresis, in *Advances in Electrophoresis*, vol. 4 (Chrambach, A., Dunn, M. J., and Radola, B. J., eds.), CH, Weinheim, pp. 81–168.
5. Radko, S. P., Chen, H. T., Zakharov, S. F., et al. (2002) Electroelution without gel sectioning of proteins from sodium dodecyl sulfate–polyacrylamide gel electrophoresis: fluorescent detection, recovery, isoelectric focusing and matrix assisted laser desorption/ionisation–time of flight of the electroeluate. *Electrophoresis* **23**, 985–992.
6. Lee, C., Levin, A., and Branton, D. (1987) Copper staining: a five-minute protein stain for sodium dodecyl sulfate-polyacrylamide gels. *Anal. Biochem.* **166**, 308–312.

35

Electroblotting of Proteins From Polyacrylamide Gels

Michael J. Dunn

1. Introduction

Since the first complete genome sequence, that of the bacterium *Haemophilus influenzae*, was published in 1995 *(1)*, a flurry of activity has seen the completion of the genomic sequences for more than 149 organisms (16 archael, 114 bacterial, and 19 eukaryotic). An up-to-date list of completed genomes is maintained on the GOLD website (Genomes OnLine Database, http://igweb.integratedgenomics.com/GOLD). Early in 2001, a major milestone was reached with the completion of the human genome sequence *(2,3)*. A major challenge in the postgenome era will be to elucidate the biological function of the large number of novel gene products that have been revealed by the genome sequencing initiatives, to understand their role in health and disease, and to exploit this information to develop new diagnostic and therapeutic agents. The assignment of protein function will require detailed and direct analysis of the patterns of expression, interaction, localization, and structure of the proteins encoded by genomes; the area now known as "proteomics" *(4)*.

The ability of polyacrylamide gel electrophoresis (PAGE) techniques to resolve individual proteins of complex mixtures has resulted in this group of methods being indispensable to the protein biochemist. In particular, two-dimensional gel electrophoresis (2-DE) can routinely separate up to 2000 proteins from whole-cell and tissue homogenates, and the use of large format gels separations of up to 10,000 proteins have been described *(5,6)*. For this reason, 2-DE remains the core technology of choice for protein separation in the majority of proteomics projects. Combined with the currently available panel of sensitive detection methods *(7)* and computer analysis tools *(8)*, this methodology provides a powerful approach to the investigation of differential protein expression.

Although gel electrophoresis procedures can provide characterization of proteins in terms of their charge (pI), size (M_r), relative hydrophobicity, and abundance, they give no direct clues as to their identities or functions. Fortunately, over the last years, a variety of sensitive methods have become available for the identification and charcaterization of proteins separated by gel electrophoresis. Conventional methods include reactivity with specific monoclonal and polyclonal antibodies, microsequencing by automated Edman degradation *(9)*, and amino acid compositional analysis *(10)*. More recently, techniques

From: *Methods in Molecular Biology, vol. 244: Protein Purification Protocols: Second Edition*
Edited by: P. Cutler © Humana Press Inc., Totowa, NJ

based on the use of mass spectrometry for mass peptide profiling and partial amino acid sequencing have made this group of technologies the primary toolkit for protein identification and characterization in proteomics projects *(11)*.

A major obstacle to successful chemical characterization is efficient recovery of the separated proteins from the polyacrylamide gel, as most procedures are not compatible with the presence of the gel matrix. The two major approaches used to overcome this problem for the recovery of intact proteins are electroelution (*see* Chapter 34) and Western electroblotting. In Western blotting, proteins separated by one-dimensional (1-D) or 2-D PAGE are blotted onto an appropriate membrane support, on which the total protein pattern can be visualized using an appropriate total protein stain. For subsequent detection and characterization of proteins using polyclonal or monoclonal antibodies (immunoblotting), the blot membrane is destained and then probed with the relevant antibody. For subsequent microchemical characterization, the protein band or spot of interest is excised. The protein, while still on the surface of the inert membrane support, can then be subjected to the appropriate microchemical characterization technique.

The most popular method for the transfer of electrophoretically separated proteins to membranes is the application of an electric field perpendicular to the plane of the gel. This technique of electrophoretic transfer, first described by Towbin et al. *(12)*, is known as Western blotting. Two types of apparatus are in routine use for electroblotting. In the first approach (known as "tank" blotting), the sandwich assembly of gel and blotting membrane is placed vertically between two platinum wire electrode arrays contained in a tank filled with blotting buffer. The disadvantages of this technique are as follows:

1. A large volume of blotting buffer must be used.
2. Efficient cooling must be provided if high current settings are employed to facilitate rapid transfer.
3. The field strength applied (V/cm) is limited by the relatively large interelectrode distance.

In the second type of procedure (known as "semidry" blotting), the gel-blotting membrane assembly is sandwiched between two horizontal plate electrodes, typically made of graphite. The advantages of this method are as follows:

1. Relatively small volumes of transfer buffer are used.
2. Special cooling is not usually required although the apparatus can be run in a cold room if necessary.
3. A relatively high field strength (V/cm) is applied because of the short interelectrode distance resulting in faster transfer times.

In the following sections, both tank and semidry electroblotting methods for recovering proteins separated by 1-D or 2-D PAGE for subsequent immunological or chemical characterization will be described. In addition, alternative total-protein-staining procedures compatible with either immunodetection or chemical characterization techniques are given. Electroblotting is ideal for use with immunodetection protocols (immunoblotting), probing with a variety of other ligands (e.g., lectins), and the recovery of gel-separated proteins for automated Edman sequencing and amino acid compositional analysis. It has also often been used (usually with trypsin) for subsequent peptide mass profiling by matrix-assisted laser desorption/ionization mass spectrometry (MALDI-MS). However, on-membrane digestion has now largely been superseded by methods of in-gel digestion as the latter process gives better overall sensitivity *(11)*.

2. Materials

2.1. Electroblotting

Prepare all buffers from analytical-grade reagents and dissolve in deionized water. The solutions should be stored at 4°C and are stable for up to 3 mo.

1. Blotting buffers are selected empirically to give the best transfer of the protein(s) under investigation. The following compositions are commonly used:
   a. For immunoprobing of proteins with p*I*'s between pH 4 and 7: Dissolve 2.42 g Tris base and 11.26 g glycine and make up to 1 L (*see* **Note 1**). The pH of the buffer is pH 8.3 and should not require adjustment.
   b. For chemical characterization of proteins with p*I*'s between pH 4.0 and 7.0 (*see* **Note 2**): Dissolve 6.06 g Tris base and 3.09 g boric acid and make up to 1 L (*see* **Note 1**). Adjust the solution to pH 8.5 with 10 *N* sodium hydroxide *(13)*.
   c. For chemical characterization of proteins with p*I*'s between pH 6.0 and 10.0 Dissolve 2.21 g of 3-[cyclohexyl-amino]-1-propanesulphonic acid (CAPS) and make up to 1 L (*see* **Note 1**). Adjust the solution to pH 11.0 with 10 *N* sodium hydroxide *(14)*.
2. Filter paper: Whatman 3MM filter paper cut to the size of the gel to be blotted.
3. Transfer membrane:
   a. For immunoprobing: Hybond-C Super (Amersham Biosciences) cut to the size of the gel to be blotted (*see* **Note 3**).
   b. For chemical characterization: FluoroTrans (Pall) cut to the size of the gel to be blotted (*see* **Note 4**).
4. Electroblotting equipment: A number of commercial companies produce electroblotting apparatus and associated power supplies. For tank electroblotting, we use the TE 42 Transphor Unit (Amersham Biosciences), whereas for semidry electroblotting, we use the NovaBlot apparatus (Amersham Biosciences).
5. Rocking platform.
6. Plastic boxes for gel incubations.

2.2. Protein Staining

2.2.1. For Immunoprobing

1. Instaview Nitrocellulose staining kit (BDH) (*see* **Note 5**). The kit comprises three components:
   a. 50× Stain concentrate.
   b. 100× Enhancer concentrate.
   c. 10× Destain concentrate.

2.2.2. For Chemical Characterization

1. Destain: 450 mL Methanol, 100 mL acetic acid made up to 1 L in deionized water.
2. Stain: 0.2 g Coomassie brilliant blue R-250 made up to 100 mL in destain.

3. Method

3.1. Electroblotting

3.1.1. Semidry Blotting

1. Following separation of the proteins by gel electrophoresis, place the gel in equilibration buffer and gently agitate for 30 min at room temperature (*see* **Note 6**).
2. Wet the lower (anode) plate of the electroblotting apparatus with deionized water.

3. Stack six sheets of filter paper wetted with blotting buffer on the anode plate and roll with a glass tube to remove any air bubbles.

4. Place the prewetted transfer membrane (*see* **Note 7**) on top of the filter papers and remove any air bubbles with the glass tube.

5. Place the equilibrated gel on top of the blotting membrane and ensure that no air bubbles are trapped.

6. Apply a further six sheets of wetted filter paper on top of the gel and roll with the glass tube.

7. Wet the upper (cathode) plate with deionized water and place on top of the blotting sandwich.

8. Connect the blotter to the power supply and transfer at 0.8 mA/cm^2 of gel area (see **Note 8**) for 1 h at room temperature (*see* **Note 9**).

3.1.2. Tank Blotting

1. Following separation of the proteins by gel electrophoresis, place the gel in equilibration buffer and gently agitate for 30 min at room temperature (*see* **Note 6**).

2. Place the anode side of the blotting cassette in a dish of blotting buffer.

3. Submerge a sponge pad, taking care to displace any trapped air, and place on top of the anodic side of the blotting cassette.

4. Place two pieces of filter paper onto the sponge pad and roll with a glass tube to ensure air bubbles are removed.

5. Place the prewetted transfer membrane (*see* **Note 7**) on top of the filter papers and remove any air bubbles with the glass tube.

6. Place the equilibrated gel on top of the blotting membrane and ensure that no air bubbles are trapped.

7. Place a sponge pad into the blotting buffer, taking care to remove any trapped air bubbles, and then place on top of the gel.

8. Place the cathodic side of the blotting cassette on top of the sponge and clip to the anode side of the cassette.

9. Remove the assembled cassette from the dish and place into the blotting tank filled with transfer buffer.

10. Connect to the power supply and transfer for 6 h (1.5 mm thick gels) at 500 mA at 10°C (*see* **Note 9**).

3.2. Protein Staining

3.2.1. For Immunoprobing

1. Prepare a working staining solution by the addition of 1 mL of 100× enhancer concentrate and 2 mL of 50× stain concentrate to 97 mL deionized water. Mix well.

2. Place the membrane to be stained in the staining solution and agitate for 2–5 min.

3. While the membrane is staining, prepare a working solution of the enhancing solution by the addition of 1 mL of 100× enhancer concentrate to 99 mL deionized water and mix well.

4. Pour off the staining solution from the membrane and replace with the working enhancing solution. Agitate until a clear background is obtained (usually less than 5 min) (*see* **Note 10**).

5. At this stage, the membrane may be dried and stored, or destained (*see* **step 6** and **Note 11**).

6. Prepare a working destaining solution by addition of 10 mL of 10× destain concentrate to 90 mL deionized water.

7. Place the membrane in the destain solution and agitate until all the color has been removed (typically, 5–10 min).

8. An example of a membrane stained by this method is shown in **Fig. 1**.

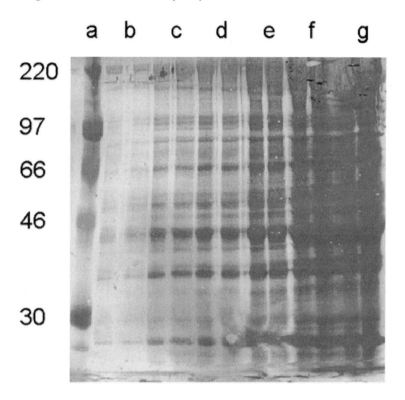

Fig. 1. The SDS-PAGE separation of human heart proteins, electroblotted onto nitrocellulose and visualized by staining with Instaview Nitrocellulose. *Lane a* contains the M_r marker proteins and the scale at the left indicates $M_r \times 10^{-3}$. The amount of sample protein applied to each lane was *b* 1 µg, *c* 5 µg, *d* 10 µg, *e* 25 µg, *f* 50 µg, and *g* 100 µg.

3.2.2. For Chemical Characterization

1. Remove the blotting membrane from the sandwich assembly.
2. Place the membrane into a dish containing the Coomassie blue staining solution for 2 min and agitate gently on the rocking platform.
3. Place the membrane into destaining solution and agitate for 10–15 min (or until the background is pale).
4. Wash the membrane with deionized water and place on filter paper and allow to air-dry.
5. Place the membrane into a clean plastic bag and seal until required for further analysis. The membrane can be stored in this state at room temperature for extended periods without any apparent adverse effects on subsequent chemical characterization.
6. An example of a membrane stained by this method is shown in **Fig. 2**.

4. Notes

1. Methanol (10–20%, v/v) is often added to transfer buffers because it removes SDS from protein–SDS complexes and increases the affinity of binding of proteins to blotting membranes. However, methanol acts as a fixative and reduces the efficiency of protein elution, so that extended transfer times must be used. This effect is worse for high-molecular-weight proteins, so that methanol is best avoided if proteins greater than 100 kDa are to be transferred.

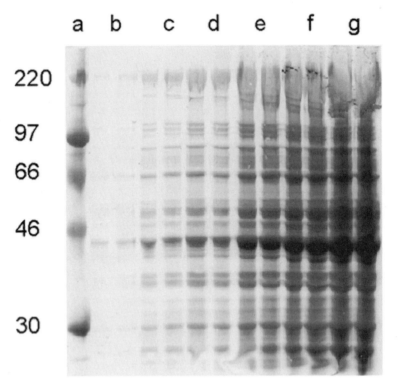

Fig. 2. The SDS-PAGE separation of human heart proteins, electroblotted onto polyvinyl difluoride (PVDF) and visualized by staining with Coomassie brilliant blue R-250. *Lane a* contains the M_r marker proteins and the scale at the left indicates $M_r \times 10^{-3}$. The amount of sample protein applied to each lane was *b* 1 µg, *c* 5 µg, *d* 10 µg, *e* 25 µg, *f* 50 µg, and *g* 100 µg.

2. The use of transfer buffers containing glycine or other amino acids must be avoided for proteins to be subjected to microchemical characterization.

3. Nitrocellulose is the most popular support for electroblotting, as it is compatible with most general protein stains, it is relatively inexpensive, and it has a high protein-binding capacity (249 µg/cm² *[15]*). Hybond-C Super is a 0.45-µm-pore-size supported nitrocellulose membrane, which is more robust than an unsupported matrix. Nitrocellulose membranes of smaller pore size are available (0.1 and 0.2 µm) and can give better retention of small proteins (1500 kDa). Polyvinylidene difluoride (PVDF) membranes, which have a high mechanical strength and a binding capacity similar to that of nitrocellulose (172 µg/cm² *[15]*), are compatible with most immunoblotting protocols.

4. Nitrocellulose membranes are not compatible with the reagents and organic solvents used in automated protein sequencing. A variety of alternative (e.g., glass fiber-based and polypropylene based) membranes have been used for chemical characterization *(16)*, but PVDF-based membranes (FluoroTrans, Pall; ProBlott, Applied Biosystems; Immobilon-P and Immobilon-CD, Millipore; Westran, Schleicher and Schuell; Trans-Blot, Bio-Rad), are generally considered to be the best choice for this application *(14)*. Nitrocellulose can be used as a support in applications such as internal amino acid sequence analysis and peptide profiling, where the protein band or spot is subjected to proteolytic digestion prior to characterization of the released peptides.

5. A variety of methods providing different levels of sensitivity have been described for the detection of total protein patterns on nitrocellulose and PVDF membranes following electroblotting *(17)*. However, most of these methods (including Coomassie brilliant blue R-250, Amido black 10B, India ink, colloidal gold) are incompatible with subsequent immunoprobing. Instaview Nitrocellulose (Merck) is a rapid and sensitive (10 ng/band) method for visualization of protein patterns on nitrocellulose or PVDF membranes and the membrane can be rapidly destained without loss of immunoreactivity. This allows a second probing of the membrane after initial protein detection, either with specific activity stains or antibody-detection methods.

6. Gels are equilibrated in blotting buffer to remove excess SDS and other reagents that might interfere with subsequent analysis (e.g., glycine). This step also minimizes swelling effects during protein transfer. Equilibration may result in diffusion of zones and reduced transfer efficiencies of high-molecular-weight proteins. It is important to optimize the equilibration time for the protein(s) of interest.

7. Nitrocellulose membranes can be wetted with blotting buffer, but PVDF-based membranes must first be wetted with methanol prior to wetting with the buffer.

8. The maximum mA/cm^2 of gel area quoted applies to the apparatus we have used. This should be established from the manual for the particular equipment available.

9. Blotting times need to be optimized for the particular proteins of interest and according to gel thickness. Larger protein usually need a longer transfer time, whereas smaller proteins require less time. Proteins will also take longer to be transferred efficiently from thicker gels. The transfer time cannot be extended indefinitely ($>$ 3 h) using the semidry technique, as the small amount of buffer used will evaporate. If tank blotting is used, the transfer time can be extended almost indefinitely ($>$ 24 h), providing that the temperature is controlled.

10. With PVDF membranes, a blue background staining may still be observed.

11. The destaining step completely destains the protein bands and the background of the blot, allowing the subsequent use of specific methods of probing the membrane using normal protocols.

References

1. Fleischmann, R. D., Adams, M. D., White, O., et al. (1995) Whole-genome random sequencing and assembly of *Haemophilus influenzae* Rd. *Science* **269**, 496–512.
2. Venter, J. C., et al. (2001) The sequence of the human genome. *Science* **291**, 1304–1351.
3. International Human Genome Sequencing Consortium (2001) Initial sequencing and analysis of the human genome. *Nature* **409**, 860–922.
4. Banks, R., Dunn, M. J, Hochstrasser, D. F., et al. (2000) Proteomics: new perspectives, new biomedical opportunities. *Lancet* **356**, 1749–1756.
5. Görg, A., Obermaier, C., Boguth, G., et al. (2000) The current state of two-dimensional electrophoresis with immobilized pH gradients. *Electrophoresis* **21**, 1037–1053.
6. Dunn, M. J. and Görg, A. (2001) Two-dimensional polyacrylamide gel electrophoresis for proteome analysis, in *Proteomics, From Protein Sequence to Function* (Pennington, S. R. and Dunn, M. J., eds.), BIOS Scientific, Oxford, pp. 43–63.
7. Patton, W. F. (2001) Detecting proteins in polyacrylamide gels and on electroblot membranes, in *Proteomics, From Protein Sequence to Function* (Pennington, S. R. and Dunn, M. J., eds.), BIOS Scientific, Oxford, pp. 65–86.
8. Pleissner, K. P., Oswald, H., and Wegner, S. (2001) Image analysis of two-dimensional gels, in *Proteomics, From Protein Sequence to Function* (Pennington, S. R. and Dunn, M. J., eds.), BIOS Scientific, Oxford, pp. 131–149.

9. Patterson, S. D. (1994) From electrophoretically separated protein to identification: strategies for sequence and mass analysis. *Anal. Biochem.* **221**, 1–15.

10. Yan, J. X., Wilkins M. R., Ou, K., et al. (1996) Large-scale amino-acid analysis for proteome studies. *J. Chromatogr. A* **736**, 291–302.

11. Patterson, S. D., Aebersold, R., and Goodlett, D. R. (2001) Mass spectrometry-based methods for protein identification and phosphorylation site analysis, in *Proteomics, From Protein Sequence to Function* (Pennington, S. R. and Dunn, M. J., eds.), BIOS Scientific, Oxford, pp. 87–130.

12. Towbin, H., Staehelin, T., and Gordon, G. (1979) Electrophoretic transfer of proteins from polyacrylamide gels to nitrocellulose sheets: procedure and some applications. *Proc. Natl. Acad. Sci. USA* **76**, 4350–4354.

13. Baker, C. S, Dunn, M. J., and Yacoub, M. H. (1991) Evaluation of membranes used for electroblotting of proteins for direct automated microsequencing. *Electrophoresis* **12**, 342–348.

14. Matsudaira, P. (1987) Sequence from picomole quantities of proteins electroblotted onto polyvinylidene difluoride membranes. *J. Biol. Chem.* **262**, 10,035–10,038.

15. Pluskal, M. G., Przekop, M. B., Kavonian, M. R., Vecoli, C., and Hicks, D. A. (1986) Immobilon PVDF transfer membrane: a new membrane substrate for Western blotting of proteins. *BioTechniques* **4**, 272–283.

16. Eckerskorn, C. (1994) Blotting membranes as the interface between electrophoresis and protein chemistry, in *Microcharacterization of Proteins* (Kellner, R., Lottspeich, F., and Meyer, H. E., eds.), VCH, Weinheim, pp. 75–89.

17. Dunn, M. J. (1993) *Gel Electrophoresis: Proteins*. Bios Scientific, Oxford.

36

Two-Dimensional Polyacrylamide Gel Electrophoresis for Proteome Analyses

Neil A. Jones

1. Introduction

The term "proteome" can be defined, in an analogous manner to the term "genome," as the complement of the proteins in a biological system. Undertaking a proteomic analysis enables the identification and quantitation of proteins on a proteomewide scale, thereby allowing comparisons between different proteomes. The current methodology of choice for proteome analysis is based on protein separation by two-dimensional polyacrylamide gel electrophoresis (2-D PAGE). 2-D PAGE technology, by virtue of differences in protein charge and molecular mass, allows the resolution of many thousands of proteins at the same time (*see* **Fig. 1**).

2-D PAGE has been utilized as a micropreparative technique to allow high-resolution separation and therefore purification of a physicochemically defined form of the same protein. For example, posttranslationally modified forms of a protein can be separated from each other. In this way, antisera can be produced directly from protein spots that have been excised from a gel (*1*). Recently, the use of mass spectrometry to identify and characterize proteins separated by and excised from polyacrylamide gel (*2*) has facilitated the rapid cloning and expression of proteins of interest. This, in turn, has allowed for the rapid production of either antisera to whole proteins, or of defined regions of a protein by using short peptides as immunogens.

Sample preparation is the important first step in the 2-D PAGE process. In ideal circumstances, all proteins in the biological sample under investigation should be solubilized, fully denatured, and reduced in the lysis buffer. A universally applicable sample preparation buffer has yet to be developed to achieve this aim, and, in particular, hydrophobic membrane proteins are often under represented on 2D gels.

Denatured and reduced proteins are separated in the first dimension within a pH gradient according to their isoelectric point (isoelectric focusing, IEF). In the original methodology, the pH gradient was created by mobile carrier ampholytes (*3*). Extensive optimization was required to obtain linear and reproducible pH gradients; this meant that batch-to-batch and lab-to-lab comparison of 2D gels was often problematic. The development of immobilized pH gradient (IPG) strips (*4*) followed by their commercializa-

From: *Methods in Molecular Biology, vol. 244: Protein Purification Protocols: Second Edition*
Edited by: P. Cutler © Humana Press Inc., Totowa, NJ

pH 4   Isoelectric focusing   pH 7

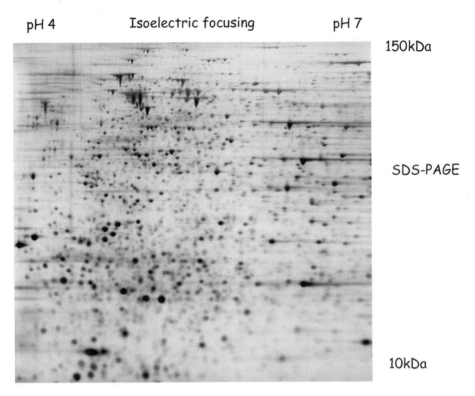

150kDa

SDS-PAGE

10kDa

Fig. 1. Typical two-dimensional sodium dodecyl sulfate-polyacrylamide gel. A total cellular protein extract from *Saccharomyces cerevisiae* was separated using pH 4–7 immobilized pH gradient (IPG) strips in the first dimension and 12% SDS-PAGE in the second dimension. Proteins were visualized by silver staining.

tion, initially by Amersham Biosciences and later by Bio-Rad, has greatly reduced these problems, making isoelectric focusing technically less demanding and more robust and allowing larger experiments to be undertaken with more reproducibility in a shorter timeframe. In addition, a greater variety of pH separations in, for example, wide (pH 3.0–10.0), medium (pH 4.0–7.0), and narrow (pH 4.0–5.0) ranges are now possible.

Protein separations in the second dimension, by sodium dodecyl sulfate–polyacrylamide electrophoresis (SDS-PAGE), are achieved using any of a number of commercially available apparatus (Amersham Biosciences, Bio-Rad, Genomic Solutions). These allow up to 14 large format gels to be poured and run at the same time, thereby ensuring greater reproducibility of protein spot patterns within an experimental run. After staining of the resultant gels for protein and image acquisition, computer-aided spot pattern analysis allows identification of quantitative changes of protein levels in different samples. The identification of the protein from the gel spots is the achieved using mass spectrometry techniques.

The aim of this chapter is to provide working protocols to obtain high-quality 2D gels. The protocols are based on the use of commercially available IPG strips and gel running apparatus and the use of a standard sample preparation procedure.

2. Materials

Best-quality reagents and very high-quality water (dH$_2$O; $>$18 MΩ) should be used for all buffers; at least AnalR quality is recommended.

Second dimension (SDS-PAGE) gel apparatus may be purchased from Amersham Biosciences, Bio-Rad, or Genomic Solutions.

2.1. First Dimension: Isoelectric Focusing

1. Sample preparation buffer/IPG strip rehydration buffer (100 mL): 7 *M* urea, 2 *M* thiourea, 4% CHAPS. Dissolve 42.04 g urea and 15.22 g thiourea in distilled water. Adjust volume to approx 90 mL. Add 4 g CHAPS, make the volume up to 100 mL, aliquot (5-mL or 10-mL volumes), and store at −80°C.
2. Dithiothreitol (DTT) (100 m*M*) and Pharmalytes (1% v/v) should be added after a protein estimation of the solubilized sample has been undertaken.

2.2. Second Dimension: SDS-PAGE

1. 1.5 *M* Tris-HCl, pH 8.8: Dissolve 171.21 g Tris-HCl in approx 800 mL dH$_2$O, pH to 8.8 using concentrated hydrochloric acid and make up to 1 L with dH$_2$O.
2. 10× SDS-PAGE running buffer (5 L): Consists of 1.87 *M* glycine, 0.25 *M* Tris-base, 1% SDS. Dissolve 700 g glycine, 152 g Tris-base, and 50 g SDS in dH$_2$O and make to 5 L with dH$_2$O.
3. Agarose overlay (100 mL): Mix 1 g agarose in 100 mL running buffer. Add Bromophenol blue to 0.03% (30 mg/100 mL).
4. 10% (w/v) SDS (100 mL): Dissolve 10 g SDS in 90 mL dH$_2$O and make up to 100 mL.
5. 10% (w/v) Ammonium persulfate (APS): Dissolve 1 g APS in 10 mL dH$_2$O, make up the volume required, use fresh, and discard excess.
6. IPG strip equilibration buffer 1 (500 mL): 50 m*M* Tris-HCl (pH 6.8), 6 *M* urea, 30% (v/v) glycerol, 1% (w/v) SDS, 1% (w/v) DTT. Mix 50 mL of 0.5 *M* Tris-HCl, pH 6.8, 180 g urea, 300 mL of 50% (v/v) glycerol, 5 g SDS and make up to 500 mL with dH$_2$O. Add 5 g DTT just before use. The solution (with omission of DTT) may be made and stored at 4°C for up to 1 wk.
7. IPG strip equilibration buffer 2 (500 mL): 50 m*M* Tris-HCl, pH 6.8, 6 *M* urea, 30% (v/v) glycerol, 1% (w/v) SDS, 4.8% (w/v) iodoacetamide. Mix 50 mL of 0.5 *M* Tris-HCl, pH 6.8, 180 g urea, 300 mL of 50% (v/v) glycerol, 5 g SDS, and make up to 500 mL with dH$_2$O. Add 24 g iodoacetamide just before use and keep shielded from bright light. The solution (with omission of iodoacetamide) may be made and stored at 4°C for up to 1 wk.

3. Methods

3.1. Sample Preparation

It is important to work as quickly as possible during all stages of any sample preparation procedure. The sample of material from which protein to be extracted should be kept on ice during the sample preparation procedure to minimize any proteolytic activity (*see* **Notes 1–4**). Protein from 1 × 10^6 to 1 × 10^7 cells can be very effectively prepared in 50–100 μL of sample buffer. Solubilization of protein from a tissue can be achieved by using 2 vol of buffer to 1 vol of tissue. It may be necessary to sonicate for 3× 10 s to fragment cellular DNA, which may hinder efficient protein separation during subsequent isoelectric focusing. If sonication is required (a good indication will be that the solubilized protein sample is viscous), it is important to chill the sample between each sonication to avoid excessive heat generation. The sample should then be agitated for a minimum of 2 h to allow maximal solubilization of cellular proteins. Following a 10-min spin at 13,000g in a microfuge to pellet cell debris, the protein concentration in the soluble sample should then be estimated. Commercial kits from Bio-Rad and Pierce work very well (*see* **Note 5**). At this point, the samples may be aliquoted (in general, loading 100 µg of protein will allow optimal separation) and stored at −80°C for future use.

3.2. First Dimension: Isoelectric Focusing

The use of isoelectric focusing apparatus should be undertaken by following the manufacturer's instructions. Such apparatus to run IPG strips are available from Amersham Biosciences, Bio-Rad and Genomic Solutions. In the first instance, the in-gel rehydration method of sample loading is recommended. This entails soaking the dehydrated IPG strip in a defined volume of sample buffer containing the protein sample to be separated (*see* **Notes 6** and **7**). The volume of buffer used and amount of protein loaded is dependent on the size of the IPG strip used (*see* **Note 8**). Separation of proteins by isoelectric focusing should be undertaken for a fixed quantity of kVh, as recommended in *the manufacturer's recommendatio*ns.

1. Place the coffin onto the IPGPhor apparatus. Pipet the sample along the whole length of the coffin. Care should be taken not to introduce any air bubbles.
2. Remove the protective backing film from the IPG strip and lay the strip, gel side down, onto the sample fluid in the coffin, to enable it to soak up the sample. At this point, care should be taken not to displace the fluid over the top of the strip and that the full length of the strip is in contact with the sample fluid.
3. Overlay the strip and sample fluid with drystrip cover fluid to prevent dehydration. Run the strips following the manufacturer's recommendations for the type of IPG strip used.
4. After isoelectric focusing is completed, the IPG strips can be run by SDS-PAGE immediately. Alternatively, IPG strips may be stored at −80°C for extended periods with little loss of resolution. Prior to freezing, excess immersion oil should be removed from the strips and the strips placed in a drystrip reswelling chamber (Amersham Biosciences).

3.3. Second Dimension: SDS-PAGE

Several manufacturers produce excellent gel running apparatus; these include the DALT and ETTAN DALT from Amersham Biosciences, the Protean II from Bio-Rad, and the Investigator system from Genomic Solutions (*see* **Note 9**). Assembly of the glass plates and gel pouring is fully explained in the manufacturer's instructions for each apparatus. Glass plates should be washed, rinsed with dH$_2$O, and then dried before use. These apparatus allow the pouring of up to 14 gels at the same time. It is highly recommended that when pouring such large numbers of gels, any buffer that contains acrylamide, a known neurotoxin, should be handled using suitable personal protection equipment (lab coats, gloves, and face mask). Before transfer of IPG strips to the second-dimension gel, efficient solubilization of proteins within the strip is required. The strips should, therefore be incubated in a solution containing SDS and DTT and then containing SDS and iodoacetamide to maintain protein reduction and to allow coating of the proteins with negatively charged SDS. This allows efficient transfer of proteins out of the IPG strip and into the second-dimension gel.

The following is a recipe that will allow pouring of 14 slab gels (of 1 mm thickness) using the Amersham Biosciences gel pouring apparatus and following that manufacturer's instructions. The same volume of either acrylamide (30:0.8) or Duracryl (30:0.65) may be used. The gel solution consists of 12.5% acrylamide, 375 mM Tris-HCl, and 0.1% SDS.

1. Gently mix 416 mL acrylamide, 250 mL of 1.5 M Tris-HCl, pH 8.8, 10 mL of 10% SDS, and 470 mL dH$_2$O. APS (4 mL of 10% solution) and TEMED (400 µL), which act as catalysts for the polymerization process, should be added just before pouring of the gels. Avoid vigorous mixing and introduction of air into the solution, which can affect the polymerization process and may affect the reproducibility of protein separation.

2. After pouring of the acrylamide solution between the glass plates, water-saturated butan-2-ol is overlaid; this ensures that a flat top gel surface is achieved when the gel polymerizes. Cover the gel casting chamber with cling film or parafilm to minimize butanol evaporation. The gels should be left without disturbance for at least 3 h at room temperature to allow polymerization to take place. This ensures that no unpolymerized acrylamide remains within the gel matrix, which could result in acrylamide adduction on proteins during separation.

3. After the gels have been allowed to polymerize and the casting chamber dissembled, excess polymerized acrylamide, which may protrude from the bottom of the glass plates, should be removed using a scalpel blade. The top of the gels should be thoroughly washed with dH$_2$O to remove all traces of butanol and nonpolymerized acrylamide. The gel cassettes cab then be stacked vertically in a rack or in the gel tank and overlaid with gel running buffer to prevent drying out.

4. Prepare the agarose solution by melting in a microwave and keep the solution at around 50°C in a water bath. Add a small amount of bromophenol blue to the dissolved hot agarose solution to obtain a light blue colour.

5. Place the strips in the drystrip reswelling tray on a rocking shaker. Strips should be first equilibrated for up to 20 min in equilibration buffer 1 and then this buffer replaced with equilibration buffer 2 and incubated for a further 20 min.

6. After this 30-min equilibration time, the equilibrated IPG strips may be transferred from the equilibration chamber onto the top edge of the SDS gel between the glass plates, with the anode end of the IPG strip toward the left-hand side. Ensure that excess equilibration solution has been removed from the strip. It is easiest if the plastic backing is placed along the glass plate closest to the operator.

7. Gently push the strip down on to the top of the SDS gel with a thin spatula, making sure that there is full contact between the IPG strip and the entire length of the SDS gel and that no air bubbles are trapped between the two gels.

8. Overlay the two gels with 2 mL of liquid agarose, again making sure that no air bubbles are introduced. Allow the agarose to set and repeat with the remaining IPG strips.

9. The gel cassettes can be transferred into the running chamber containing an appropriate volume of running buffer; again, refer to the manufacturer's recommendations.

10. After addition of an appropriate volume of running buffer into the top tank, the apparatus can be closed and set to run at suitable constant current or voltage to allow the gels to run overnight.

11. When the bromophenol blue front reaches the bottom of the gels, the run is complete. The gels can then be removed from between the glass plate sandwich and then be fixed in an

LIVERPOOL
JOHN MOORES UNIVERSITY
AVRIL ROBARTS LRC
TEL. 0151 231 4022

appropriate way dependent on the method of protein staining (*see* **Note 10**). The gel running tank should be rinsed after use in dH$_2$O to remove any residual running buffer.

4. Notes

1. Urea is sensitive to heat; therefore, warming of the buffer or cell lysate above 30°C must be avoided to prevent possible modifications of proteins by carbamylation with subsequent shifts of their isoelectric focus points.
2. Protease inhibitors should not normally be needed because of the strong denaturing activity of urea and thiourea; in addition, some protease inhibitors, because of their mode of action, can alter the pI of the proteins.
3. The exact methodology used to prepare a protein sample for separation by 2-D PAGE will always depend on the cell or tissue source and should be determined empirically. The use of ionic buffers should be avoided if possible, and where their use is required, any excess buffer should be removed from a cell pellet before solubilization in sample buffer. For example, cells obtained from an in vitro tissue culture experiment may be solubilized simply adding the sample buffer to a cell pellet.
4. The recipe for the lysis/rehydration buffer described in this chapter represents an improvement on a sample buffer widely used for general proteome analysis of whole cells that contains 9 *M* urea as a denaturant. The buffer described here, in contrast, contains both the denaturants urea (7 *M*) and thiourea (2 *M*). The lysis buffer also contains the zwitterionic detergent CHAPS and DTT as a reducing agent. Other nonionic or zwitterionic detergents (e.g., Triton X-100 or NP-40) can also be used as a substitute for CHAPS. The reducing agent DTT is charged, and prolonged focusing may result in loss of DTT out of the pH gradient, which may reduce solubility of some proteins. Tributyl phosphine has been shown to reduce this problem *(5)*.
5. A detergent-compatible protein assay protocol should be used, a number of which are now commercially available (Amersham Biosciences, Genotech, and Bio-Rad).
6. I have described a general procedure for the efficient running pH 4.0–7.0 IPG strips. In experiments where separations at basic pHs are needed, modifications to the protocol may be required for optimal separation. For example, for separation in the pH 6.0–11.0 range, a cup-loading method seems to result in better separation than the in-gel strip rehydration method described.
7. At the start of a project, when little information is available about the proteome of the sample to be analyzed, a broad pH range IPG strip (e.g., pH 3.0–10.0) should be initially chosen. This will give an overall impression of the range of pI of the proteins present within the sample. In this respect, the use of 7-cm strips can allow a rapid assessment as to whether the sample preparation procedure has allowed efficient extraction of protein from the sample as well as information of the pI profile of the proteins. For analytical gels, upon which subsequent image analysis will be performed, 18-cm or 24-cm strips are recommended, as these provide maximal resolution of separation. Based on initial range-finding experiments, narrower pH range IPG strips may be chosen for use, in order to improve the resolution of the separation in the pH ranges of interest.
8. The amount of protein loaded onto an IPG strip depends on the type of detection system and what downstream processing is required. In general, for optimal separation resolution, up to 100 µg total protein lysate may be loaded onto a pH 4.0–7.0 IPG strip. This amount of protein will be easily visualized using, for example, Sypro Ruby or silver staining for protein detection. Protocols have been published recently that describe methods for preparative 2-D PAGE that allow up to 10 mg protein to be loaded onto IPG strips *(6)*.

9. To enable a broad separation of proteins, 12.5% acrylamide gels are recommended for preliminary proteomics experiments. Using this format, a second-dimension gel will allow separation of proteins in the mass range 10–200 kDa.

10. The choice of which stain to use will be dependent on the sample and what sort of information is required from the gels run. Standard protocols for silver, Coomassie blue, and fluorescent staining of proteins in 2D gels are widely available.

References

1. Amero, S. A., James, T. C., and Elgin, S. C. R. (2002) Production of antibodies using proteins in gel bands, in *The Protein Protocols Handbook* (Walker, J. M., ed.), Humana, Totawa, NJ.

2. Shevchenko, A., Chernushevich, I., Wilm, M., and Mann, M. (2000) De novo peptide sequencing by nanoelectrospray tandem mass spectrometry using triple quadrupole and quadrupole/time-of-flight instruments. *Methods Mol. Biol.* **146**, 1–16.

3. O'Farrell, P. H. (1975) High resolution two-dimensional electrophoresis of proteins. *J. Biol. Chem.* **250**, 4007–4021.

4. Bjellqvist, B., Ek, K., Righetti, P. G., et al. (1982) Isoelectric focusing in immobilized pH gradients: principle, methodology and some applications. *J. Biochem. Biophys. Methods* **6**, 317–339.

5. Herbert, B. R., Molloy, M. P., Gooley, A. A., Walsh, B. J., Bryson, W. G., and Williams, K. L. (1998) Improved protein solubility in two-dimensional electrophoresis using tributyl phosphine as reducing agent. *Electrophoresis* **19**, 845–851.

6. Sabounchi-Schutt, F., Astrom, J., Olsson, I., Eklund, A., Grunewald, J., and Bjellqvist, B. (2000) An immobiline DryStrip application method enabling high-capacity two-dimensional gel electrophoresis. *Electrophoresis* **21**, 3649–3656.

37

Microscale Solution Isoelectrofocusing

A Sample Prefractionation Method for Comprehensive Proteome Analysis

Xun Zuo and David W. Speicher

1. Introduction

Proteomics is a relatively new research discipline that deals with systematic, large-scale analyses of proteins in biological systems *(1,2)*. Two-dimensional polyacrylamide gel electrophoresis (2-D PAGE) is currently the most commonly used method for quantitatively comparing changes of protein profiles (proteomes), despite a number of limitations, as described below. The basic method utilizes isoelectrofocusing (IEF) in a polyacrylamide gel containing either soluble ampholytes *(3–5)* or immobilines *(6)* under denaturing conditions, followed by a second dimension on an orthogonal sodium dodecyl sulfate (SDS) gel. Unfortunately, existing 2-D PAGE methods have inadequate resolution and insufficient dynamic range when used for separating complex proteomes *(7,8)*. A typical "full-size" 2-D gel (approx 18 × 20 cm) can resolve only approx 1500 protein spots from a cell extract using high-sensitivity stains, whereas about 10,000 genes are typically expressed at one time in a single mammalian cell *(9)*. However, the total number of protein components in higher eukaryotic cells greatly exceeds the number of expressed genes as a result of mRNA alternative splicing and posttranslational modifications *(10)*. Hence, it is highly likely that at least 20,000–50,000 unique protein components comprise typical proteomes from individual mammalian cell types, and the total unique protein species represented in a single tissue probably exceed 100,000. In addition, the quantitative dynamic range (i.e., the difference between the least and the most abundant proteins) in mammalian cells is about 5 to 6 orders of magnitude, whereas the dynamic range in serum is at least 10 orders of magnitude *(11)*. In contrast, current chromogenic gel stains have only dynamic ranges of approximately two orders of magnitude and fluorescent stains have dynamic ranges of approximately three orders of magnitude.

One interesting approach for enhancing the capacities of 2-D gels involves parallel separation of replicate aliquots from unfractionated samples on a series of narrow pH range immobilized pH gradient (IPG) gels *(12–14)*. The main benefit of narrow pH range IPG gradients is that the total number of protein spots per pH unit that can be separated should increase because of higher spatial resolution. However, in practice, this

From: *Methods in Molecular Biology, vol. 244: Protein Purification Protocols: Second Edition*
Edited by: P. Cutler © Humana Press Inc., Totowa, NJ

approach results in only moderate increases in proteins detected compared to a single broad pH range gel. These narrow pH range IPG gels are limited to low protein loads of unfractionated cell extracts if high resolution and reproducible separations are to be maintained. Narrow pH range IPG gels readily produce variable, unreliable separations when high protein loads of unfractionated complex samples are analyzed, because proteins with p*I*s outside the IPG strip usually causes massive precipitation and aggregation on the gel *(15–18)*.

The great complexity of eukaryotic proteomes together with the limited resolution and detection dynamic range of current 2-D gel methods clearly indicates that improved separation strategies are required. One strategy for enhancing the separation capacity of 2-D gels is to prefractionate complex samples to reduce complexity and facilitate detection of low-abundance proteins *(8)*. Unfortunately, most prefractionation methods such as gel filtration, ion exchange, subcellular fractionation, or selective solublization are low-resolution techniques that result in substantial cross-contamination of many protein components among two or more fractions. In addition, these methods typically produce large sample volumes necessitating sample concentration prior to 2-D gel analysis and overall yields are often low and variable.

The limitations of chromatographic and other alternative methods for proteome prefractionation suggest that high-resolution preparative IEF separations that are closely analogous to the actual analytical IPG separation might be advantageous when combined with the use of narrow pH range IPG strips. Although such a prefractionation strategy does not provide a true three-dimensional separation, it does simplify samples and allow high-resolution separations at higher protein loads on slightly overlapping narrow pH range gels *(15–17)*. However, most existing preparative IEF methods are not well suited for high-sensitivity protein profiling analyses because they often require large sample volumes, produce dilute large volume fractions, exhibit poorer resolution than the subsequent IPG gels, and sometimes require expensive, complex instrumentation. For example, one commonly used preparative electrophoresis device is the Rotofor (Bio-Rad), which utilizes soluble ampholytes to separate proteins in solution, as originally described by Bier et al. *(19)*. The apparatus consists of a rotating chamber divided into 20 compartments by open grids designed to minimize convection currents. The use of soluble ampholytes limits fractionation to the middle pH range, there is substantial cross-contamination of many proteins between any two adjacent fractions, and fraction volumes are relatively large. Another interesting prefractionation method uses the IsoPrime device (Amersham Biosciences), which has thin acrylamide partitions containing immobilines at different pH's that segregate proteins by p*I* as originally described by Righetti et al. *(20)*. This "multicompartment electrolyzer" was developed primarily for final purification of individual proteins under native conditions starting with partially purified proteins. Its complex plumbing system for recirculating samples through the separation chambers and the large fraction volumes make it difficult to use with denaturing conditions or with limiting amounts of samples. Recently, Herbert and Righetti briefly described a related solution IEF device, the multicompartment eletrolyzer (MCE), and used it to prefractionate *Escherichia coli* and human serum samples prior to 2-DE gel analysis *(21)*. This device appears simpler than the Isoprime unit, but it still contains large sample chambers (approx 100 mL), which are not compatible with samples available in small quantities.

The inadequacy of existing preparative IEF methods led to our development of an efficient high-resolution sample prefractionation that could be seamLessly interfaced with subsequent narrow pH range 2-D gel analysis for comprehensive protein profile comparisons. This microscale solution isoelectrofocusing (μsol-IEF) method divides complex proteomes into well-separated, concentrated fractions that can be directly applied to IPG gels without concentration. The μsol-IEF device utilizes the basic separation principle originally described by Righetti et al. *(20),* but unlike the IsoPrime IEF instrument, it is miniaturized, the number of chambers can be easily varied, fraction volumes are small (500 μL), cross-flow has been eliminated, the acrylamide partition membranes have larger pores and stronger membrane supports, and other design differences. The μsol-IEF method and device *(15)* and applications of this novel approach for complex proteomes *(16,17)* were recently published. Complex proteomes such as mammalian cell extracts or serum were separated into well-resolved pools according to their p*I*s, and total recoveries in the fractionated samples were typically greater than 80%. The μsol-IEF fractions could be directly applied to and separated on narrow pH range IPG strips using protein loads that were more than 10 times higher than for unfractionated samples. The combination of μsol-IEF prefractionation, high protein loads on narrow range IPG gels, and use of multiple protein stains can dramatically increase the detection dynamic range and the total number of proteins detected (*see* **Note 1**). In addition, this approach greatly conserves proteome samples compared with direct analyses of unfractionated samples on a series of narrow pH 2-D gels.

In this chapter, we describe the μsol-IEF method and its use for comprehensive protein profile analysis of complex proteomes. The well-separated μsol-IEF fractions can be directly applied to narrow pH range 2-D gels. Alternatively, μsol-IEF fractions can be directly analyzed by multidimensional liquid chromatography/mass spectrometry/ mass spectrometry (LC-MS/MS) methods or by 1-D SDS-PAGE followed by LC-MS/MS instead of by 2-D PAGE.

2. Materials

The μsol-IEF method is used to separate complex protein extracts such as whole-cell lysates or biological fluids using tandem small-volume liquid-filled chambers separated by thin porous acrylamide gel disks or membrane partitions that contain immobilines at specific pHs. After isoelectrofocusing to equilibrium, each chamber contains all proteins with p*I*s between the pHs defined by the two boundary membrane partitions. A photograph of a simple prototype device constructed from commercially available components is presented in **Fig. 1**, and a schematic illustration for the components of the separation device is presented in **Fig. 2**.

1. Electrophoresis chambers and tank: μsol-IEF device can be constructed using a variable number of tandem Teflon dialysis chambers and an electrophoresis tank (Harvard Bioscience, Holliston, MA, USA) (*see* **Fig. 1**). Volumes of separation chambers can be varied by using different-volume Teflon chambers or by connecting two chambers in tandem without an intervening membrane partition. For most applications, between five and eight separation chambers with 500 μL per chamber are appropriate (*see* **Fig. 2**). Two terminal Teflon chambers of the same size as the separation chambers contain IEF electrode buffers (*see* **Note 2**).
2. Membrane partitions and dialysis membranes: In general, adjacent separation chambers are bounded by 3%T acrylamide gels and terminal separation chambers by 10%T gels, all containing immobilines at specific pHs (*see* **Notes 3** and **4**). The terminal separation chambers

A

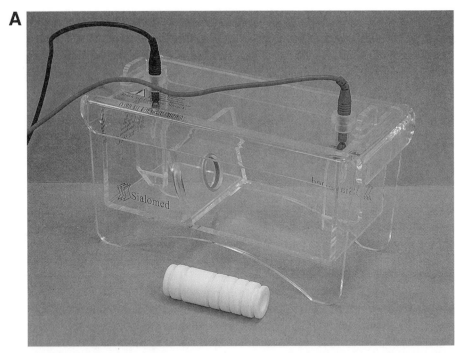

B

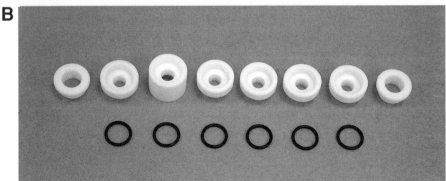

Fig. 1. Photograph of a prototype μsol-IEF device. Teflon chambers and an electrophoresis tank originally designed for electroelution of proteins or nucleic acids from gels (Harvard Bioscience) were adapted for μsol-IEF. (**A**) An electrophoresis tank and an assembled six-chamber separation unit in front of the tank. (**B**) Component parts for a six-chamber separation unit including (left to right, top row): a terminal Teflon cap with hole, a Teflon link chamber, a Teflon coupling chamber, four link chambers, and a terminal Teflon cap with hole. Bottom row: six O-rings (12 mm inner diameter). To assemble the unit, a 12-mm-diameter immobiline gel membrane is inserted into each O-ring and the membrane/O-ring assembly is placed between two adjacent 500-μL Teflon chambers (*see* **Fig. 4**).

    are protected from the bulk electrode buffers by electrode chambers containing 5 kDa cutoff membranes (MWCO 5K Dialysis Membranes, Harvard Bioscience) (*see* **Fig. 2** for schematic and **Note 2**).

3. Membrane supports: Two alterative types of hydrophilic material can be used to provide mechanical strength to the acrylamide/immobiline membrane partitions (i.e., GF/D glass

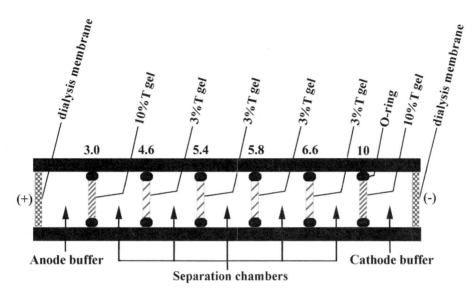

Fig. 2. Schematic illustration of a μsol-IEF device. The device consists of seven chambers separated by six gel membranes containing immobilines at different pHs. The pH values of membranes are shown above the partitions. The positions of the O-rings and the concentrations of acrylamide gels for the membranes are as indicated. The terminal electrode chambers are sealed with 5-kDa cutoff dialysis membranes. Using this device, a proteome sample can be fractionated into five pH pools between pH 3.0 and 10.0 in the five central separation chambers, and very acidic (pI <3.0) and very basic proteins (pI >10.0) can be recovered from the anode and cathode buffer chambers, respectively.

fiber filters [Whatman] or hydrophilic porous polyethylene [medium pores, 45–90 μ dry blend surfactant (DBS); POREX, Lecester, MA, USA]).

4. O-rings: Buna O-Rings (12 mm inner diameter, 2 mm thickness) (Scientific Instrument Services, Ringoes, NJ, USA) are used between separation chambers to form a seal between the chambers and the acrylamide/immobiline membrane disks (*see* **Fig. 2**).

5. Immobilines: Also called acrylamido buffers, they are provided by Amersham Biosciences as 0.2 M solutions in water or isopropanol. Currently, the six immobilines (pKs 3.6, 4.6, 6.2, 7.0, 8.5, and 9.3) available from Amersham Biosciences allow preparation of μsol-IEF partition membranes at any specific pH between 3.0 and 10.0.

6. Sample buffer: 7 M Urea, 2 M thiourea, 2% NP-40, 5 mM TBP (tributyl phosphine) or 50 mM dithiothreitol (DTT), 10% sorbitol, and 1% IPG-buffer (*see* **Note 5**). Urea and IPG buffer are from Amersham Biosciences; thiourea, NP-40 (IGEPAL CA-630), and sorbitol are from Sigma (St. Louis, MO, USA); TBP and DTT are from Bio-Rad Laboratories (Hercules, CA, USA). The sample buffer is used to solubilize protein samples and to fill any μsol-IEF separation chambers that do not receive sample. The buffer can be stored at −80°C for three months without losing effectiveness.

7. Electrode buffers are premade IEF anode buffer (7 mM phosphoric acid) and cathode buffer (20 mM lysine/20 mM arginine) from Bio-Rad. These buffers are used to fill the terminal μsol-IEF chambers and the bulk chambers of the electrophoresis tank. Similar to the sample buffer, 10% sorbitol is typically included in the electrode buffer. The buffers are stored at 4°C.

8. Gel reagents: Acrylamide, bisacrylamide, ammonium persulfate, and TEMED are from Bio-Rad Laboratories. Glycerol (87% solution) is from Amersham Biosciences.

9. Stock solutions for casting acrylamide/immobiline partition membranes:

    30%T/8%C Acrylamide/bisacrylamide: dissolve 55.2 g acrylamide and 4.8 g bisacry-lamide in a final volume of 200 mL Milli-Q water and filter using a 0.22-μm membrane. Store solution at 4°C for up to 3 mo.

    50% Sorbitol: Dissolve 50 g sorbitol in a final volume of 100 mL Milli-Q water and fil-ter using a 0.22-μm membrane. Store solution at 4°C for up to 2 wk.

    40% Ammonium persulfate: Dissolve 0.4 g of ammonium persulfate in a final volume of 1 mL double-distilled water. Make fresh immediately before each use.

    87% Glycerol: Purchased in this form. Store at room temperature.

3. Methods

3.1. Alternative Experimental Setups for μsol-IEF Prefractionation

The total number of separation chambers, the pHs of acrylamide/immobiline parti-tion membranes, and separation chamber volumes can be readily altered to fit require-ments of specific proteome studies. For example, albumin in serum constitutes more than 50% of total protein and normally severely restricts the amount of serum that can be applied to a 2-D gel (*16,17*). Therefore, a useful experimental design is to isolate albumin in a chamber with a final very narrow pH range to enhance detection of low abundant proteins in other fractions; that is, mouse serum albumin with heterogeneous p*I*s from approx pH 5.4 to 5.8 can be sequestered in a single narrow range pool using a total of five separation chambers as illustrated in **Fig. 2**. Although 500-μL separation chamber volumes have been used for most experiments, volumes can be readily adjusted to fit different experimental designs. To avoid the need to concentrate samples, frac-tionated sample volumes should be equal to or less than the IPG gel rehydration vol-umes for the desired number of replicate 2-D gel and the maximum sample loads de-sired. For example, a 500-μL fraction volume with a 200-μL rinse of the separation chamber from μsol-IEF prefractionation of 2.0 mg of a cell lysate will allow duplicate narrow pH range gels are to be run at high protein loads proportional to 1.0 mg of ini-tial sample when 18-cm IPG strips are used (2 × 350 μL per strip).

3.2. Preparation of Acrylamide/Immobiline Partition Membranes at Desired pHs

Partition membranes can be cast using different acrylamide concentrations and thick-nesses using either glass fiber filters or porous polyethylene for mechanical strength. In general, acrylamide concentrations should remain low so that pore sizes are maximized (*see* **Note 4**) and porous polyethylene produces more structurally robust membranes. The procedure for membrane preparation includes two steps as described next.

3.2.1. Preparation of Immobiline Mixtures at Desired pHs

A computer program, "Doctor pH," developed by Giaffreda et al. (*22*) is available from Amersham Biosciences and can be used to produce immobiline recipes to make μsol-IEF partition membranes at specific pHs (*see* **Notes 3** and **6**). Precalculated recipes using the Doctor pH software have also been published (*23*). For example, to make a membrane at pH 5.5, we choose the pH 5.5 recipe from the pH 4.0–7.0 gradient formula table (*23*). The following amounts of 0.2 *M* stock immobilines are pipetted into a 10-mL graduated cylinder: 268 μL of p*K* 3.6, 305 μL of p*K* 4.6, 199 μL of p*K* 6.2, 95 μL

Table 1
Recipes for μsol-IEF Partition Membrane Gel Solutions

	3%T/8%C gel	10%T/8%C gel
Immobiline mixture (from **Subheading 3.2.1.**)	7.5 mL	7.5 mL
Acrylamide/bis (30%T/8%C)	2.5 mL	8.33 mL
Sorbitol (50%)	5.0 mL	5.0 mL
NP-40	0.5 mL	0.5 mL
Glycerin (87%)	3.45 mL	3.45 mL
Ammonium persulfate (40%)	30 μL	20 μL
TEMED	15 μL	10 μL
Milli-Q water	5.95 mL	0.36 mL
Total volume	25 mL	25 mL

of pK 7.0, and 280 μL of pK 9.3. Milli-Q water is then added to a total volume of 7.5 mL, the sample is degassed, and the pH at room temperature is measured. If the solution pH is not pH 5.50, a small amount of the immobiline solution at the appropriate pH extreme (either 3.6 or 9.3) is added until pH 5.50 is reached. The pH is then adjusted to approx 6.5 by adding 1 M Tris-base to facilitate gel polymerization. The Tris-base does not affect the final membrane pH because it is washed out of the gel partition after polymerization.

3.2.2. Preparation of μsol-IEF Gel Membranes

Large-pore acrylamide/immobiline partition membranes are produced using low total acrylamide (%T) and high crosslinker (%C) concentrations to facilitate effective separation of large proteins and minimize protein precipitation on the partition membranes. However, the membrane must be sufficiently strong so that the membrane does not rupture during electrophoresis. Typically, 3%T/8%C gels are used between separation chambers and 10%T/8%C gels are used between the terminal separation chambers and the small-volume electrode buffer chambers (*see* **Note 4**). Recipes for acrylamide/immobiline partition membranes are shown in **Table 1** (*see* **Note 7**). The indicated 25 mL of gel solution is sufficient for casing approx 24 disk membranes at a given pH. After the gel solution is degassed, it is applied onto the membrane supporters (either glass fiber or porous polyethylene). As noted earlier, in general porous polyethylene is preferred because its superior mechanical strength. The method for casting gels on polyethylene supports is described next.

3.2.3. Preparing Polyethylene Disks and Casing Acrylamide/Immobiline Partition Membranes

The hydrophilic porous polyethylene sheet is cut into 12-mm-diameter disks using a stainless-steel core borer or a die punch. A plastic cover from a 1-mL Rainin peptide tips box (approx 10.4 × 12.6 cm) is a convenient holder for casing acrylamide/immobiline partition membranes. The gels are cast between two layers of gel support membrane (Bio-Rad) that has been coated with Repel-Silane (Amersham Biosciences) using the following procedure.

The polyethylene disks are placed on silanized support film and surrounded by 2-mm slab gel spacers (Bio-Rad) inside the pipet tip lid (*see* **Fig. 3**). The gel solution is then

Fig. 3. Casting μsol-IEF acrylamide/immobiline partition membranes using porous polyethylene supports. The container (approx 10.4 × 12.6 cm) is a plastic cover from a box of 1.0-mL Rainin pipet tips. One silanized gel support film is laid on the bottom of the container and four 2-mm spacers are placed around the perimeter, with the polyethylene disks distributed on top of the film and inside the spacers.

pipeted onto the polyethylene disks until the solution is completely absorbed inside the polyethylene pores. Additional gel solution is added to cover the remaining areas of the film until a 2-mm thickness (height of spacers) is achieved. The solution is then covered by another silanized support film. After the gel has polymerized (approx 1 h at 23°C), the gel disks are cut from the surrounding polymerized gel and removed. Membrane disks with continuous and homogeneous gel covering both sides are selected and excess gel on the surfaces of the polyethylene disk is removed using a scalpel or razor blade. Cleaned disks are immediately transferred into individual wells in a 24-well tissue culture plate (Fisher Scientific). The membrane disks are washed three times in 2 mL of 12% glycerol, 10% sorbitol, 2% NP-40 for 30 min with shaking at 23°C to remove Tris buffer and other polymerization byproducts. Washed membranes are stored in the same solution containing 2 m*M* sodium azide at 4°C for up to 1 mo.

3.3. Loading Samples and Assembling μsol-IEF Chambers

The day before μsol-IEF fractionation, partition membranes at the desired pHs are selected, storage solution is removed, and the membranes are soaked in 2 mL of sample buffer overnight at 4°C.

The next morning, the protein sample is prepared in sample buffer in a final volume equal to the number of chambers that will be loaded with sample (*see* **Notes 8** and **9**). For example, if a seven-chamber separation with 500-μL-volume chambers will be used

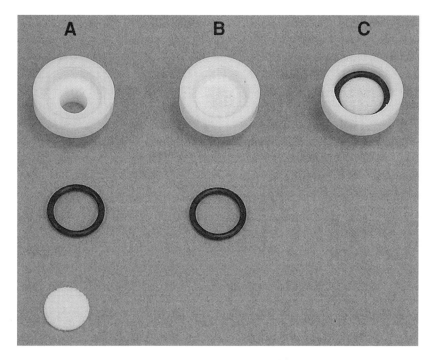

Fig. 4. Assembly of a partition membrane in a separation chamber. (**A**) The individual components before assembly, including a Teflon link chamber, an O-ring, and an acrylamide/immobiline partition membrane using a porous polyethylene support. (**B**) The first step in assembly, at which the partition membrane is placed inside the chamber. (**C**) The assembled unit where the O-ring has been inserted between the membrane and the outer edge of the separation chamber. At this stage, another chamber would be screwed into the top of the illustrated assembly, which will compress the O-ring to form seals between the two Telfon chambers and the porous polyethylene disk.

and the sample will be loaded in the five central chambers, the sample should be diluted to 2.5 mL. Because the device shown in **Fig. 1** does not have sample fill ports, each chamber must be filled sequentially during assembly. Typically, the unit is assembled starting with the cathode electrode chamber by inserting the dialysis membrane in the end of the chamber using a cap with a hole, the chamber is filled with cathode electrode buffer, and the lowest pH acrylamide/immobiline partition membrane is inserted into the top of the chamber (*see* **Fig. 4**). The next chamber is screwed into the unit, sample or sample buffer only is added to fill the chamber, the acrylamide/immobiline partition membranes with the next lowest pH is inserted as shown in **Fig. 4**, and the process is repeated until assembly is completed with the anode buffer chamber, dialysis membrane, and cap. The assembled unit is inserted into the hole in the compartment separation plate (*see* **Fig. 1Λ**) and the two compartments are filled with anode and cathode electrode buffers accordingly.

3.4. Electrophoretic Conditions

The µsol-IEF prefractionation requires a power supply with a capacity of at least 500 V such as the PS500X power supply (Amersham Biosciences). Typically, samples are focused using a four-step constant-voltage procedure (initial and final observed current reading are indicated): (1) 50 V for 30 min (initial, approx 1–2 mA; final, approx 1 mA), (2) 100 V for 1 h (initial, approx 2–3 mA; final, approx 1 mA), (3) 200 V for 1 h (initial, approx 2–3 mA; final, approx 1 mA), (4) 500 V (initial, approx 2–3 mA), focusing is terminated when current falls below approx 0.5 mA. The total focusing time is about 3–3.5 h, depending on the ionic strength of the sample and the amount loaded. The current should be maintained below 3 mA to prevent excessive heating and to avoid rupture of partition membranes. This method works well for relatively simple samples such as serum or prokaryote lysates.

An alternative focusing protocol utilizing a higher-voltage power supply with a capacity for at least 2000 V that can be programmed for constant current is recommended for more complex proteomes such as mammalian cell extracts. In this case, a convenient focusing protocol is to use 1 mA consistent current with a maximum voltage of 2000 V. Typically, when 2 mg of a human cell extract is fractionated using seven separation chambers, the prefractionation is completed when 1200 V and 1 W are achieved for a total focusing time of approx 4–4.5 h.

3.5. Collection of Fractionated Samples After μsol-IEF Prefractionation

When units without fill ports are used, retrieval of separated fractions involves reversal of the sample loading/assembly procedure. To minimize cross-contamination and loss of fractionated samples during removal, before opening each chamber, about half of the sample is withdrawn by using a 1-mL insulin syringe with attached 1/2-in. 28-gage needle to pierce the terminal dialysis membrane or acrylamide/immobiline partition membrane; the pierced membrane is then removed and the remainder of the liquid in the chamber is transferred to a microfuge tube: Finally, the walls and membrane surfaces of each chamber are rinsed with 200 μL sample buffer and the rinse is combined with the fractionated sample. This removal process is then repeated for each successive chamber. A few proteins, primarily those with pIs equal to the pHs of partition membranes, are retained in the partition membranes after isoelectrofocusing. To extract these proteins, the partition membranes are extracted two times with 350 μL sample buffer for 30 min with shaking at 23°C and these extracts are pooled. Samples can either be used immediately for subsequent analysis (*see* **Notes 1** and **10**) or stored in aliquots at −80°C until required.

4. Notes

1. After initial screening of μsol-IEF fractions using 7 cm long 10% SDS minigels, as described here, all fractions, including the terminal electrode chamber solutions containing very acidic and very basic proteins, are analyzed by full-sized (20-cm-long), large-pore 1-D SDS gels. When different experimental samples are run in parallel, these gels allow quantitative comparisons of proteins >100 kDa by comparing densitometric scans using 1-D gel analysis software. This is an important complement to the more laborious narrow pH range 2-D gel analyses because large proteins are not reliably recovered on 2-D gels, but reliable quantitative differences of large proteins can be identified on such 1-D gels *(16)*.

   To effectively compare <100-kDa proteins, high resolution (18 × 20 cm or larger) narrow pH range 2-D gels are run in duplicate. To maximize separation and spot resolution, custom IPG strips must be prepared for most μsol-IEF fractions because, ideally, the pH range of the IPG strip should be slightly greater (typically, ± 0.1 pH units) than the pH range of the μsol-IEF. The slightly wider IPG strip range prevents losses of proteins at the

IPG strip ends while maximizing separation distance as much as possible. Typically 10- to 30-fold higher protein loads can be applied to these narrow pH ranges when samples are µsol-IEF prefractionated compared to optimal loads without prefractionation *(16,17)*. The ability to load much larger proportional amounts of samples enables detection of much lower abundance spots. In addition, duplicate gels can be stained with different sensitivity stains such as Colloidal Coomassie, Sypro Ruby, and silver stains *(17)*. This multistain approach increases the dynamic range of detection by allowing quantitative comparisons of major proteins using the lower-sensitivity stains, and detection and comparison of minor proteins with the high-sensitivity stain. If µsol-IEF pH ranges are chosen so that the complexity of each resulting narrow pH range 2-D gel is similar, it should be practical to detect 1500–2500 protein spots per 2-D gel. When eukaryotic proteomes are divided into 7 or more fractions, it should be feasible to detect and quantitatively compare at least 10,000–15,000 proteins. Although µsol-IEF prefractionation increases the number of types of 2-D gels needed per sample compared with use of a single broad-range 2-D gel, the scope of 2-D analysis is no greater than that currently being used by investigators running multiple broad- and narrow-range gels without prefractionation *(12–14)*.

2. Dialysis membranes (5 kDa cutoff) and terminal small-volume electrode solutions were found to protect terminal separation chambers from pH extremes and mechanical stress from electro-osmosis. These terminal chambers with terminal dialysis membranes also prevent proteins with pIs beyond the pH range of the separation chambers from migrating into the electrophoresis tank.

3. The most critical parameters in µsol-IEF are the properties of the acrylamide/immobiline partition membranes because they are responsible for separating the proteins by defining discrete pHs within the membranes. It is therefore important to cast membranes with reproducible pHs and buffering capacities to ensure the producible fractionations. Although only one or two immobilines are required to define any given pH, typically more than two immobilines are mixed to ensure more uniform buffering capacity *(22)*. Ideally, the final concentration of combined immobilines in a gel membrane is approx 10–20 mM and should not exceed 40 mM, because higher concentrations cause gels to swell as a result of the high ionic strength and osmosis. In addition, as immobiline concentrations increase, their efficiency of incorporation can decrease and cause errors for the membrane pH *(22)*.

4. Partition membrane pores should be as large as possible to facilitate rapid transfer of all proteins, especially at high protein loads. In addition, large pores should help ensure that large proteins are effectively separated rather than precipitating on the membrane surfaces. In initial experiments of µsol-IEF, we used 5% gels for separation membranes and 10% gels for electrode membranes, similar to the gels used for separations with the IsoPrime device *(20,27,28)*. However, we found that 5% gels caused many proteins, especially higher-molecular-weight proteins with pIs not equal to membrane pHs, to precipitate on or in the partition membrane. When 3% gels were used as partition membranes, yields and separation of fractions were substantially improved, and only proteins with pIs equal to membrane pHs were retained in the 3% gel matrix *(15)*. Presumably, even lower acrylamide concentrations and thinner gel membranes would be advantageous, but more porous gels or thinner membranes are currently too fragile for reliable use. However, alternative formulations and supports with larger pores and improved mechanical strength are being investigated.

5. The µsol-IEF prefractionation method is typically conducted under denaturing conditions similar to that used for subsequent IPG-IEF (e.g., 7 M urea, 2 M thiourea, 2% NP-40 [or 2–4% CHAPS], IPG-buffer [0.5–2.0%, typically 1%], 5 mM TBP [or 100 mM DTT]). Using similar conditions for µsol-IEF and IEF on subsequent IPG strips has two advantages. First, µsol-IEF fractions can be applied directly to IPG gels and the fractionation by pI in solution should closely match subsequent focusing in IPG strips. Second, reagents that

promote sample solubility and effective focusing in one IEF system should produce similar beneficial effects in the other system. Specifically, the 2 M thiourea and 7 M urea combination used here has been shown to be superior to 9 M urea alone for solublizing sample proteins and IEF in IPG strips *(24)*. Similarly, TBP was thought to be superior to DTT because it had less electrical charge and it was a more powerful reducing agent that is compatible with thiourea *(25)*. In contrast, DTT has a mild negative charge and might migrate out of high pH separation chambers during µsol-IEF. However, TBP has poor stability and is rapidly degraded during isoelectric focusing. The presence of soluble ampholytes (wide ph range IPG buffer) is necessary in both systems, otherwise the current during electrophoresis would be too low to efficiently separate all the protein components into fractions. The IPG sample buffer can also minimize precipitation/aggregation of proteins during IEF. Typically, 1% IPG buffer (pH 3–10, Amersham Biosciences) is used for µsol-IEF. Because ampholytes are fractionated along with the proteins during µsol-IEF, subsequent IEF in IPG strips is improved if fresh ampholytes are added to the sample. The minimum amount of fresh ampholytes needed for effective IPG strip IEF has not been determined and probably varies with type of sample, protein load, and pH range of the fraction. In practice, some fresh ampholytes are routinely added via the 200 µL of fresh sample buffer used to rinse the µsol-IEF chamber. Similarly, if less than the maximum fractionated sample volume is loaded onto subsequent IPG strips, there is a further dilution of the sample with fresh sample buffer containing unfractionated ampholytes. Addition of sorbitol to the sample buffer and the electrode buffer as well as to the gel membranes appears to reduce electro-osmosis during µsol-IEF similar to its previously reported role in gel-based preparative IEF *(26)*.

6. In some experiments, we observed apparent major deviations for the actual pH ranges of µsol-IEF fractionated samples compared with the designed pH ranges when effectiveness of fractionation was evaluated on pH 4–7 IPG gels (Amersham Biosciences). However, consultation of the manufacturer's website showed that pH 4–7 IPG gels have an actual pH range of about 4.05 to about 6.5. When the actual pH gradient of the pH 4–7 gels or 1.0 pH unit wide narrow range gels were used to evaluate µsol-IEF fractions, the deviations between expected pHs and observed pHs were much less. Nonetheless, most µsol-IEF membrane pH's showed fairly consistent shifts of approx 0.1–0.2 pH units below the expected pHs. There are several potential reasons for this minor deviation in pH. First, the µsol-IEF partition membrane pH was determined in the absence of urea, whereas the IPG strip pHs are in the presence of urea/thiourea. Second, different immobilines may be incorporated into the gel matrix with varying efficiencies, which would skew the actual pH. The additives present in the partition membrane gel solution may further skew immobiline incorporation and resulting membrane pH. In practice, the minor pH differences observed are not a major concern because membrane pHs appear to be reproducible between batches when the same procedure is followed, and reproducibility is more important than absolute accuracy of the pH value. In fact, the pH values on polymerized immobiline gels cannot be measured directly even with an advanced surface electrode *(23)*. Therefore, when deviations are observed between different immobiline gels such as µsol-IEF membranes and IPG gels, it is not straightforward or simple to determine which apparent pH is more accurate.

7. Surprisingly, we found that the addition of 2% NP-40 to gel solutions prior to polymerization of partition membranes decreased precipitation of proteins on membranes and improved sample fractionation. This result was unexpected because the partition disks are equilibrated overnight with sample buffer prior to use and it was expected that even detergent micelles should have adequate time to equilibrate into the gel pores under these conditions. The positive effect of adding the NP-40 directly to the gel solution may either more effectively include detergent within the gel matrix or the detergent may affect the polymerization reaction

and contribute to formation of a more porous structure. Addition of sorbitol was found to be advantageous when glass fiber filters were used as the mechanical support because it reduced electro-osmosis and strengthened the memranes. It has been retained as an additive with the porous polyethylene support. These additives in the membrane acrylamide solution have been incorporated into the sample buffer at the same concentrations.

8. Samples can be loaded into any single separation chamber, into several separation chambers or into all separation chambers. In general, loading a sample into all separation chambers allows it to be diluted to the largest possible volume. However, it requires that a portion of the most acidic and most basic proteins must travel a larger distance than if the entire sample is placed in one or more central chambers. In addition, there may be sample-specific benefits to selected sample loading. For example, it might be possible to focus serum proteins faster or more effectively if the entire sample is placed in the chamber with the pH range matching albumin's pI.

9. The maximum sample loading capacity for μsol-IEF prefractionation depends on the sample type, loading position, and the number of separation chambers used. Typically, at least 3 mg of fairly simple samples such as bacterial extracts or serum can be well fractionated when three to five separation chambers of 500 μL each are used and the sample is loaded into all separation chambers. For a more complex sample such as mammalian cell extracts, sample loads of about 2 mg are suitable for a device having seven 500-μL separation chambers. Higher sample loads tend to result in protein deposition on the partition membrane surfaces and result in incomplete separation. The pIs of some proteins will exactly match the pH of any partition membrane, which may be a critical factor in limiting sample load, either by blocking the pores with subsequent deposition of proteins on the membrane surfaces or by overwhelming the buffering capacity of the immobilines in the membrane. Increasing the partition gel pore size, cross-sectional area, or buffering capacity of the membranes might further increase protein load capacity. However, the present load capacity matches well with the maximum feasible loads for subsequent narrow pH range 2-D gels *(16,17)*.

10. To evaluate effectiveness of μsol-IEF prefractionations, proportional amounts of all fractionated samples and partition membrane elutions are initially analyzed by 1-D SDS minigels. This is a rapid method to evaluate overall protein recovery and distribution among fractions using densitometry, and it provides a quick rough evaluation of the effectiveness and reproducibility of the separation. In some experiments, fractionated samples and extracts from partition membranes are then evaluated on broad or mid pH range IPG-based 2-D minigels to verify that good separation was obtained and to determine whether the sample boundaries occur near the expected pHs. For example, when we analyzed *E. coli* lysates, fractionated samples and proteins eluted from the separation membranes were analyzed using pH 4.0–7.0 IPG gels followed by 10% SDS gels. The results showed that the proteome was separated into three well-resolved pools with very minimal overlapping spots between fractions, and the proteins retained in the membranes were primarily those having pIs equal to the membrane pHs *(15)*.

References

1. Wilkins, M. R., Sanchez, J. C., Gooley, A. A., et al. (1996) Progress with proteome projects: why all protcins cxprcsscd by a genome should be identified and how to do it. *Biotechnol. & Genet. Eng. Rev.* **13**, 19–50.
2. Fields, S. (2001) PROTEMICS: proteomics in genomeland. *Science* **291**, 1221–1224.
3. Klose, J. (1975) Protein mapping by combined isoelectric focusing and electrophoresis in mouse tissue: a novel approach to testing for induced point mutations in mammals. *Humangenetik* **26**, 231–243.

4. O'Farrell, P. H. (1975) High resolution two-dimensional electrophoresis of proteins. *J. Biol. Chem.* **250,** 4007–4021.

5. Scheele, G. A. (1975) Two-dimensional gel analysis of soluble proteins: characterization of guinea pig exocrine pancreatic proteins. *J. Biol. Chem.* **250,** 5375–5385.

6. Bjellqvist, B., Ek, J., Righetti, P. G., et al. (1982) Isoelectric focusing in immobilised pH gradients: principle, methodology and some applications. *J. Biochem. Biophys. Methods* **6,** 317–339.

7. Quadroni, M. and James, P. (1999) Proteomics and automation. *Electrophoresis* **20,** 664–677.

8. Williams, K. L. (1999) Genomes and proteomes: towards a multidimensional view of biology. *Electrophoresis* **20,** 678–688.

9. Miklos, G. L. G. and Rubin, G. M. (1996) The role of the genome project in determining gene function: insights from model organisms. *Cell* **86,** 521–529.

10. Godley, A. A. and Packer, N. H. (1997) The importance of protein co- and post-translational modifications in proteome projects, in *Proteome Research: New Frontiers in Functional Genomics* (Wilkins, M. R., Williams, K. L., Appel, R. D., and Hochestrasser, D. F., eds.), Springer-Verlag, Berlin, pp. 65–91.

11. Herbert, B. R., Sanchez, J. C., and Bini, L. (1997) Two-dimensional electrophoresis: the state of the art and future directions, in *Proteome Research: New Frontiers in Functional Genomics* (Wilkins, M. R., Williams, K. L., Appel, R. D., and Hochestrasser, D. F., eds.), Springer-Verlag, Berlin, pp. 13–33.

12. Wasinger, V. C., Bjellqvist, B., and Humphery-Smith, I. (1997) Proteomic "contigs" of *Ochrobactrum anthropi*, application of extensive pH gradients. *Electrophoresis* **18,** 1373–1383.

13. Wildgruber, R., Harder, A., Obermaier, C., et al. (2000) Towards higher resolution: two-dimensional electrophoresis of *Saccharomyces Cerevisiae* proteins using overlapping narrow immobilized pH gradients. *Electrophoresis* **21,** 2610–2616.

14. Corthals, G. L., Wasinger, V. C., Hochstrasser, D. F., and Sanchez, J. (2000) The dynamic range of protein expression: a challenge for proteome research. *Electrophoresis* **21,** 1104–1115.

15. Zuo, X. and Speicher, D. W. (2000) A method for global analysis of complex proteomes using sample prefractionation by solution isofocusing prior to two-dimensional electrophoresis. *Anal. Biochem.* **284,** 266–278.

16. Zuo, X., Echan, L., Hembach, P., et al. (2001). Towards global analysis of mammalian proteomes using sample prefractionation prior to narrow pH range two-dimensional gels and using one-dimensional gels for insoluble and large proteins. *Electrophoresis* **22,** 1603–1615.

17. Zuo, X. and Speicher, D. W. (2002). Comprehensive analysis of complex proteomes using microscale solution isoelectrofocusing prior to narrow pH range two-dimensional electrophoresis. *Proteomics* **2,** 58–68.

18. Righetti, P. G., Castagna, A., and Herbert, B. (2001) Prefractionation techniques in proteome analysis: a new approach identifies more low-abundance proteins. *Anal. Chem.* **73,** 320–326.

19. Bier, M., Egen, N. B., Allgyer, T. T, Twitty, G. E., and Mosher, R. A. (1979) New developments in isoelectric focusing, in *Peptides: Structure and Biological Functions* (Gross, E. and Meienhofer J., eds.), Pierce Chemical Co., Rockford, IL, pp. 79–89.

20. Righetti, P. G., Wenisch, E., and Faupel, M. (1989) Preparative protein purification in a multicompartment electrolyser with immobiline membranes. *J. Chromatogr.* **475,** 293–309.

21. Herbert, B. and Righetti, P. G. (2000) A turning point in proteome analysis: sample prefractionation via multicompartment electrolyzers with isoelectric membranes. *Electrophoresis* **21,** 3639–3648.

22. Giaffreda, E., Tonani, C., and Righetti, P. G. (1993) A pH gradient simulator for electrophoretic techniques in a windows environment. *J. Chromatogr.* **630,** 313–327.

23. Pharmacia Biotech (1997) *Isoelectric Membrane Formulas for IsoPrime Purification of Proteins*, IsoPrime Protocol Guide #1, Pharmacia Biotech, Uppsala, Sweden.
24. Rabilloud, T., Adessi, C., Giraudel, A., and Lunardi, J. (1997) Improvement of the solubilization of proteins in two-dimensional electrophoresis with immolilized pH gradients. *Electrophoresis* **18,** 307–316.
25. Herbert, B. R., Molloy, M. P., Walsh, B. J., et al. (1998) Improved protein solubility in 2-D electrophoresis using tributyl phosphine. *Electrophoresis* **19,** 845–851.
26. Oh-Ishi, M., Satoh, M., and Maeda, T. (2000) Preparative two-dimensional gel electrophoresis with agarose gels in the first dimension for high molecular mass proteins. *Electrophoresis* **21,** 1653–1669.
27. Righetti, P. G., Wenisch, E., Jungbauer, A., Katinger, H., and Faupel, M. (1990) Preparative purification of human monoclonal antibody isoforms in a multi-compartment electrolyser with immobiline membranes. *J. Chromatogr.* **500,** 681–696.
28. Wenisch, E., Righetti, P. G., and Weber, W. (1992) Purification to single isoforms of a secreted epidermal growth factor receptor in a multicompartment electrolyzer with isoelectric membranes. *Electrophoresis* **13,** 668–673.

38

Practical Column Chromatography

Shawn Doonan

1. Introduction

Many of the most powerful methods of protein fractionation involve the use of column chromatography. Other chapters in this volume describe specific chromatographic techniques, but the focus is mainly on the particular application rather than on the generalities of how to do column chromatography in practice. This chapter is intended to fill in some of the practical gaps in terms of what materials and equipment are needed and how they should be used.

There is no doubt that for convenience and, to some extent, for quality of the results obtained, it is ideal to set up a laboratory with a completely automated set of fractionation equipment and to use commercial chromatography columns and ancillary equipment. A basic set of equipment and selection of columns can be purchased for US$13,000–$15,000, but a fully automated system can cost twice this amount. This is not always realistic. For example, if the intention is to do a one-off purification for a particular purpose, then the expense involved in setting up a dedicated laboratory will not be justified. It is, therefore, important to realize that much can be achieved with a minimum of equipment and that lack of more sophisticated facilities should not deter one from attempting protein purification by chromatographic methods; in principle, all that is needed is a column and some plastic tubing! These low-cost options are dealt with in what follows.

The main suppliers of fractionation equipment are Amersham Biosciences and Bio-Rad; Watson-Marlow specialize in peristaltic pumps. Their product ranges can be viewed on-line at their websites and their catalogs should be consulted for prices. Detailed descriptions of how to set up a particular piece of equipment are always provided by the supplier and are not given here.

No reference is made to specialist equipment for fast protein liquid chromatography (FPLC) or for high-performance liquid chromatography (HPLC), because these topics are covered in Chapters 27 and 28, respectively. Similarly, Fig. 2 in Chapter 26 gives a diagram of the basic arrangement for column chromatography, and this is not repeated here.

From: *Methods in Molecular Biology, vol. 244: Protein Purification Protocols: Second Edition*
Edited by: P. Cutler © Humana Press Inc., Totowa, NJ

2. Equipment and Materials

2.1. Columns

For most applications, columns should have a length-to-diameter ratio in the range 5/1 to 20/1 (*see* **Note 1**). Commercial columns are usually supplied with a variable-flow adaptor that allows complete enclosure of the matrix bed, so that sample and buffer application is directly onto the bed with no dead space. Flow adaptors can be used at both ends of the column, giving greater flexibility in bed height, although normally a fixed-end piece is used at the bottom of the column. Some columns come with thermostatic jackets for temperature control by circulation of coolant from an external source. If they do not, then the column must be run in a cold room if temperature control is required. Packing reservoirs, which effectively extend the length of the column for ease of packing, are also available.

An inexpensive alternative is provided by standard laboratory chromatography columns. These are essentially glass tubes drawn out to a taper at one end and with a sintered glass disk (usually of porosity P160; pore size, 100–160 µm) sealed in above the taper. Columns 30–50 cm long and with diameters of 10, 20, and 30 mm are widely available at a cost of about US$20–30 each. Larger sizes can be made by purchasing Buchner filter funnels with diameters in the range 30–95 mm (porosity P160), and then getting a professional glass blower to join a piece of glass tube of the same diameter and of the desired length onto the funnel. A 70-mm × 50-cm column (bed volume, 2 L) made in this way is very useful for large-scale work and easier to handle than commercial columns of a similar size. The major problem with them is the dead space under the sinter in which substantial mixing can occur, but for initial relatively crude separations, this is not important. At the other end of the range for very small-scale work, a Pasteur pipet or small syringe plugged with glass wool makes a perfectly adequate column (*see* **Note 2**).

2.2. Pumps

Columns can be operated under gravity flow with control exercised by the height of the solvent reservoir above the column and/or by a screw clip on the tubing at the column outlet. Ideally, however, a peristaltic pump should be used. This should give a continuous range of flow rates from 0.5 to 500 mL/h, which is adequate for most purposes (*see* **Note 3**). Pumps can usually be controlled from the fraction collector or from a chromatography controller for automatic shutdown. More expensive piston pumps that give greater flow rates and can exert moderately high pressures are available, but they are expensive and there are few situations in which their use is essential.

2.3. Fraction Collectors

Although it is obviously possible, it is very tedious to collect fractions from a column by hand, so a simple fraction collector is an important item of equipment (*see* **Note 4**). These usually carry between 100 and 200 standard test tubes. Collection can be on the basis of time or drop count, and there is usually a facility for marking tube changes on an attached recorder. Shutdown can be programmed and linked to switching off the peristaltic pump, so that flow through the column is terminated. More sophisticated fraction collectors have additional capabilities for control of ancillary equipment (pumps, monitors, switch valves) and for analysis of chromatograms. Some have advanced control functions, including a peak slope detection system, which allows for accurate collection of poorly resolved peaks. These features might be in demand in a laboratory that is dedicated to protein purification, but they are luxuries that can be dispensed with, and for most applications, a basic instrument is perfectly adequate. Certainly for an active laboratory, two less expensive fraction collectors would be a much better use of resources than one top-of-the-range model.

2.4. Monitors

Fractions from a column can be analyzed for protein content by manual absorption measurements at an appropriate wavelength (generally 280 nm; *see* **Note 5**), but it is more convenient to use a monitor attached to a chart recorder for continuous display of the elution profile. Ultraviolet (UV) monitors generally use a mercury discharge lamp with filters to select wavelengths of 254, 280, or 405 nm. It is important that a monitor be able to be used in a cold room. More sophisticated models provide a greater range of wavelengths (including 206 and 226 nm, which are useful for detecting proteins at very low levels) and have the capability of connection to a peak integrator, but the cost is higher. A chart recorder is, of course, required for display of the output signal. Single-channel and dual-channel monitors are available. The latter are to be preferred if it is intended to monitor more than one property of the column effluent. For example, flow-through conductivity and pH monitors are available, which are useful for monitoring gradients (particularly the former, because salt gradients are much more commonly used than are pH gradients), but neither can be considered essential pieces of equipment.

2.5. Gradient Makers

Many chromatographic procedures require elution of adsorbed protein with a gradient usually of increasing salt concentration. Commercial gradient mixers can be purchased. They consist essentially of two reservoirs of equal diameter connected at the base, such that as solvent is withdrawn from the reservoir originally containing the start buffer, the limit buffer flows in to maintain equal levels in the reservoirs (*see* **Note 6**). Mixing is provided by a motor-driven paddle or magnetic stirrer.

Such devices can also be easily made by a competent glass blower. All that is required is two flat-bottomed glass cylinders, one (for the limit buffer) with a single outlet near the base and the other (for the start buffer) with two such outlets. The two reservoirs are connected by a short piece of flexible tubing that can be closed off with a screw clip. Buffer is drawn off from the reservoir initially containing the start buffer through plastic tubing connected to its second outlet. Mixing in this reservoir is provided by a magnetic stirrer and pellet. It is useful to have a range of reservoirs available, with volumes in the range 100 mL to 2 L.

It is even easier to construct a system with two measuring cylinders connected by a syphon of narrow-bore plastic tube, buffer being withdrawn from the cylinder originally containing the start buffer through a second plastic tube. With this arrangement, it is essential to ensure that the plastic tubings do not slip out during gradient operation. This can be done by passing the plastic tubings through lengths of glass tube and fixing the latter in place with tape to the tops of the measuring cylinders. Beakers can be used instead of measuring cylinders, but they should be tall and narrow. Otherwise, small differences in the level of the beakers will cause a large movement of buffer from one to the other.

Yet another possibility is to use a peristaltic pump to deliver the limit buffer into a mixing vessel from which solvent is withdrawn onto the column by a second pump. If the flow rate of the pump delivering the limit buffer is one-half of the rate at which solvent is removed onto the column, then the gradient will be linear, as with the above devices. The problem with this method, of course, is that it ties up a second peristaltic pump.

2.6. Valves

It is convenient, but by no means essential, to include three- or four-way valves in the solvent stream before and after a column. These can be used for sample application and for diversion of the effluent stream to waste, respectively. Manual valves are relatively inexpensive. If the chromatography system is to be automatically controlled, then the valves must be solenoid driven and controlled by the fraction collector.

2.7. Tubing

It is important to have available a supply of flexible plastic tubings for transporting solvents from one part of the chromatographic system to another. These are marketed by most suppliers of laboratory materials. Standard polyethylene tubing with an internal diameter of 1 mm and external diameter of 1.8 mm is generally useful. Polyvinyl chloride tubings with larger internal diameters (1.6–4.0 mm) are useful when faster flow rates are required. A supply of flexible Microperpex (silicone rubber) tubings with internal diameters of 1.3, 2.7, and 4.0 mm are very useful for making joins between lengths of capillary tubings or joining these to glass tubing. For example, a 1-cm length of the 1.3-mm Microperpex tubing makes a good airtight connection between lengths of the standard 1.8-mm polyethylene tubing.

3. Methods

3.1. Packing Columns

1. Suspend the appropriate amount of matrix (*see* **Note 7**) in equilibration buffer. Check the pH, and if it is far removed from the target value, then adjust it by careful addition of the acidic or basic component of the buffer.
2. Resuspend the matrix by gentle stirring and allow to settle until a firm bed has formed. If there is fine material in suspension, then remove this by aspiration with a water vacuum pump or by syphoning. Repeat this step until no more fines remain (*see* **Note 8**).
3. Choose a column of such a size that the matrix bed will fill it sufficiently for the adaptor to reach the top of the bed (if using a commercial column with adaptor) or to about 80% of its height. Clamp the column vertically in a position where it is protected from drafts or sources of radiant heat (convection currents because of these will cause uneven column packing). Ideally, the column should be packed at the temperature at which it is to be run.
4. Seal off the outlet tube of the column and pour in some equilibration buffer. Open the outlet and allow buffer to flow through to remove air from the bottom net or sinter and from the outlet tubing. Clamp off the outlet again.
5. Suspend the settled matrix in about an equal volume of buffer and pour the slurry into the chromatography column, being careful not to trap any air bubbles (pouring it down a glass rod touching the side of the column can be helpful). If using a column-packing extension or reservoir, the remainder of the slurry can also be poured.

6. Open the bottom tube of the column and allow buffer to flow through at a rate about 20% greater than that at which it is intended to run the column (*see* **Note 9**); the flow rate can be controlled either with a peristaltic pump or by adjustment to the pressure on a screw clip attached to the outlet tubing.

7. As packing proceeds, a discontinuity will be seen between the packed bed and the suspension of matrix above it. If using a packing extension, then allow packing to continue until this discontinuity has reached the desired height and remove any remaining suspension from the column. If not using a packing extension, then as clear liquid forms above the suspension, remove it by aspiration and replace it with fresh suspension until the bed has reached the desired height (*see* **Note 10**). Stop flow through the column.

8. If using a column with a flow adaptor, then layer buffer on top of the matrix bed, being careful not to disturb it, and fill the column to the top. Slide the upper adaptor across the top of the column so that no air is trapped between the net and the liquid meniscus, push the adaptor just into the column, and tighten the compression nut so that the sealing O-ring makes contact with the sides of the column. Push the adaptor slowly down the column until the net is in contact with the matrix bed; during this process, excess buffer will exit though the tubing of the flow adaptor, the end of which should be submerged in the buffer reservoir. When the adaptor is in place, then tighten the compression nut so that the O-ring makes a firm contact with the sides of the column. If using a column without an adaptor, then carefully layer buffer on the top of the matrix (the depth depending on the size of the column) and place a tightly fitting rubber bung with a glass tube passing through it in the top; the latter should have a length of capillary tubing passing through, such that the tubing inside the column ends somewhat above the matrix bed (and ideally, touches the wall of the column) and the other end is immersed in a buffer reservoir. The whole arrangement must be airtight (*see* **Note 11**).

9. Flow equilibration buffer through the column either under the control of a peristaltic pump (*see* **Note 12**) or under gravity; in the latter case, flow rate can be controlled by a combination of adjusting the height of the buffer reservoir above the top of the column and adjustment of a screw clip on the outflow tubing. Generally, 2 or 3 column volumes of buffer are sufficient for equilibration, but this should be confirmed by measuring the pH of the effluent and, if possible, the conductivity.

3.2. Sample Application

1. The sample should have been equilibrated in application buffer either by dialysis or by gel filtration (*see* Chapter 11). Except when using gel filtration (*see* Chapter 26), there is usually no limit on the volume of sample that can be applied (*see* **Note 13**); protein loading will, of course, depend on the size of the column (*see* **Note 7**).

2. If using a flow adaptor and peristaltic pump, stop flow through the column by switching off the pump (*see* **Note 14**). Transfer the inlet tubing to the sample container and restart the flow when sample will be sucked onto the column. Be careful not to introduce air bubbles; if there is a bubble at the end of the inlet tubing when it is transferred to the sample container, then briefly reversing the direction of the peristaltic pump with the tubing immersed in the sample will get rid of it. Stop the pump, transfer the inlet tube to the buffer reservoir, and restart the flow.

3. If not using a flow adaptor, then remove the rubber bung from the top of the column and continue buffer flow until the meniscus is flush with the top of the bed. Stop the flow. Layer sample carefully on top of the bed using a pipet or a syringe, taking great care not to disturb the surface; allowing the sample to flow down the wall of the tube helps. Restart the flow. If the volume of sample is large, then replace the bung in the column and place the

inlet tube in the sample container; the sample will be sucked onto the column by vacuum. Run the sample completely on to the matrix bed (*see* **Note 15**), stop the flow through the column, and wash the sample in with a few milliliters of buffer. Layer buffer over the matrix bed, replace the bung, and restart the flow with the inlet tube in the buffer reservoir.

3.3. Column Development

1. Wash the column with starting buffer until no more protein is eluted (*see* **Note 16**). The flowthrough peak can often be collected as a single fraction (*see* **Note 17**).
2. Elute remaining protein by application of one or more step changes in eluent or by application of a gradient (*see* **Note 18**). If using a column with a flow adaptor, then be careful not to introduce air bubbles into the line when switching from the equilibration buffer reservoir to the gradient maker (*see* **Subheading 3.2., step 2**). If using a standard laboratory column, then the gradient can be applied by simply transferring the column inlet tube to the outlet of the gradient maker; some mixing between the gradient and the liquid layer over the matrix bed is unavoidable. Make sure that the mixing paddle or the magnetic stirrer is switched on.
3. Collect appropriate size fractions (*see* **Note 19**).
4. Combine fractions using appropriate criteria (*see* **Note 20**).
5. Regenerate and store the column as necessary (*see* **Note 21**).

4. Notes

1. This is the optimum range for applications such as ion-exchange and other types of adsorption chromatography. Short, fat columns give better flow rates and are less prone to development of high back-pressures. They are, however, less easy to pack evenly. Poor packing can lead to distortion of bands of material passing through the column, which, in turn, can give poor resolution if two bands are eluting close together. Long, thin columns (up to 100/1 length/diameter) are appropriate for true partition chromatography applications, such as gel filtration.
2. The great advantage of commercial columns is in the design of the end pieces (i.e., the parts in contact with the matrix bed). These have very low liquid volumes, so the possibility of mixing (and of dilution) of eluted materials is minimized. This is important in the most demanding applications (e.g., in gel filtration or in adsorption chromatography where poorly resolved peaks are eluted by application of a gradient). With standard laboratory columns, it is impossible to avoid some dead volume and consequent mixing in the space under the sintered support. In addition, there is always a liquid layer over the matrix bed, which interferes with application of gradients. That being said, there are few occasions where the refinements offered by commercial columns are essential (gel filtration being one such), and it is perfectly feasible to carry out protein purification procedures using unsophisticated homemade columns. (Note that if money is no object, several companies market prepacked chromatography columns. These offer consistency of behavior as well as convenience, but, of course, at a price. Catalogs should be consulted for a full list of columns available.)
3. Very large columns (e.g., 1-L bed volume) of ion-exchange materials can be run at flow rates of up to 1–2 L/h. It is perfectly feasible to run these under gravity.
4. When using large columns at high flow rates, particularly at the beginning of a purification procedure where the separation required may be relatively crude, the fractions can be quite large and may conveniently be taken by hand. For example, with a 7 × 50-cm column of CM cellulose (volume-2 L) developed with a 4-L salt gradient at a flow rate of 2 L/h, it might be adequate to collect 20 fractions of 200 mL each over the 2-h period of the chromatographic run; this is more easily done manually than by adapting a laboratory-scale fraction collector to cope with the large-volume fractions.

5. Nearly all proteins absorb light at 280 nm because of their content of the aromatic amino acids tryptophan and tyrosine. A very useful rule of thumb is that in most cases, an absorption of 1 in a 1-cm cell corresponds roughly to a protein concentration of 1 mg/mL (or $A^{1\%}_{280nm} = 10$), although for proteins devoid of tryptophan, this may be wrong by a factor of 4. However, a plot of A_{280nm} against volume provides a useful elution profile for a column. The absorbance at 254 nm is generally about one-half of that at 280 nm and, hence, gives lower sensitivity; it is used only because there is a strong line in the mercury emission spectrum at this wavelength. For very low protein concentrations, absorbance at 226 or 206 nm can be used, because these wavelengths are in the region of absorption by the peptide bond and the absorbance is very high. The problem is that many buffers absorb at these wavelengths (particularly at 206 nm), thus limiting the usefulness of the detectors. Even with detection at 280 nm, the possibility of buffer absorption should be considered if the buffering species is aromatic or contains conjugated double bonds. Low-level and constant buffer absorption can usually be blanked out, but problems will arise if a gradient of increasing buffer concentration is used.

6. The shape of the gradient obtained depends of the cross-sectional areas of the two reservoirs. It can be shown *(1)* that

$$C = C_l - (C_l - C_i) (1 - v)^{A_1/A_2} \tag{1}$$

where C is the concentration after a fraction v of the total gradient has been withdrawn, C_l and C_i are the limit and initial concentrations, respectively, and A_1 and A_2 are the cross-sectional areas of the mixing vessel and of the reservoir for the higher concentration solution, respectively. In the simplest case, where $C_i = 0$ and the cross-sectional areas are equal, this reduces to

$$C = C_l v \tag{2}$$

(i.e., a linear gradient). Concave or convex gradients can be obtained by appropriate choice of the ratio A_1/A_2, but this is rarely done. For example, if the protein of interest elutes early in a particular gradient, then better resolution could, in principle, be obtained either by using a concave gradient or, more simply, a linear gradient with a lower final concentration. If the protein of interest elutes late in the gradient, then improved results could be obtained with a convex gradient or, more simply, by using an initial concentration [C_i in Eq. (1)] >0. It should be noted that the method of making gradients with two interconnected vessels works by maintaining the heights of the solutions in the vessels equal; hence, with vessels of the same cross-sectional areas, this means starting with equal weights of the two solutions. To a first approximation, this can be taken as equality of volumes, but if the densities are markedly different (e.g., 0 and 1 *M* NaCl solutions), then some of the strong solution will move into the mixing vessel as soon as the connection is made; this is only likely to be important if the protein of interest elutes very early in the gradient.

7. The amount required will depend on the type of chromatography to be done, on the capacity of the matrix (see previous chapters on specific techniques), and on the state of purity of the sample. To some extent, amounts have to be determined by trial, but there are some principles and practices that can be useful in deciding the approach to take.

   In ion-exchange chromatography, for example, the protein-binding capacities of the matrices vary in the range 10–30 mg/mL of column bed; more precise values are given in the suppliers' literature. If the chosen protocol is for the protein of interest to be retained on the column and then eluted by a single step change in the elution buffer, then the whole of the binding capacity of the column may be used because no further separation can be achieved once the protein is displaced from the matrix. If the intention is to do gradient elution, however, then only a fraction (approx 20–30%) of the binding capacity of the column should be used so that the remaining matrix is available for ion exchange as the displaced material passes down the column. Similarly, if conditions are chosen so that the protein of interest passes through the column without retention, then the whole capacity of the matrix can be used to bind impurities.

An estimate of the amount of matrix required can be obtained in a trial using batch adsorption. For example, take 10 aliquots of the equilibrated protein solution and add increasing quantities of equilibrated matrix to each. Mix gently for 10 min and then sediment the matrix by centrifugation. Determine the amount of the protein of interest remaining in the supernatant in each case. This will allow an estimate of the minimum amount of matrix required to bind all of the protein of interest; hence, the amount to used in practice can be chosen on the basis of the outlined considerations. If the objective is to bind impurities, but not the target protein, then a similar experiment can be used to determine the amount of matrix that gives maximum removal of protein from solution without loss of target protein. It is important to bear in mind that the ability of a matrix to bind a particular protein will depend on the state of purity of the sample. Hence, if conditions for adsorption and the binding capacity of a matrix for a particular protein have been determined using a partially purified sample, then the capacity may be considerably less if a cruder preparation is used. In these circumstances, it may well happen that more strongly binding impurities saturate the column and displace the protein of interest, so that it emerges without adsorption to the column contrary to expectations. Conversely, if a protein is partially purified using a particular ion-exchange material followed by rechromatography on the same material under the same conditions, then a smaller amount of matrix will be required for the second step, because much of the impurities will already have been removed.

There is a temptation to "play it safe" and use much larger columns than are actually required; this should be resisted. The consequences will be reduced yield resulting from nonspecific adsorption of protein on the matrix and unnecessary dilution of the active fraction. There will also be a cost penalty, which can be severe with the more expensive matrices.

8. Care should always be taken to remove fines. If this is not done, then the fine particles will pack into the interstices between matrix particles and block flow of solvent through the column. If this happens, then the only recourse is to repack the column.

9. If the flow rate of the column is too great, then equilibration of protein between the matrix and the solvent may not be achieved; in addition, compaction of the matrix bed may occur. On the other hand, low flow rates will lead to peak spreading by diffusion with consequent dilution and loss of resolution and possibly to loss of activity because of the extended time of the procedure. Flow rates are generally in the range 0.1–0.5 cm/min (i.e., 0.1–0.5 cm^3/cm^2 of cross-sectional area/min).

10. It is important not to allow all of the matrix to settle and then replace the supernatant liquid with more matrix suspension, and so on. Because the coarser particles settle faster, this would result in a series of discontinuities between bands of fine and coarse material in the packed bed with the consequence of inferior chromatographic performance and the possibility of poor flow rates.

11. With this arrangement, as liquid is removed from above the matrix bed, the partial vacuum produced will cause buffer to be sucked out of the reservoir and into the column; there needs to be sufficient buffer above the matrix bed before flow is started to ensure that the syphoning starts before the bed goes dry. The tube carrying the buffer should extend to close to the top of the bed, so that incoming buffer does not "bomb" the bed and disrupt the surface. The arrangement must, of course, be airtight. This can be arranged by pushing one end of a piece of flexible plastic tubing onto the glass tube and then inserting one or more short lengths of tightly fitting plastic tubings into the other end until the bore of the innermost

insert is somewhat less than the external diameter of the capillary tubing to be used to connect to the reservoir. The capillary can then be pushed through this last insert by a sufficient distance to end just above the matrix bed. If very small columns are being used with rubber bungs that are too small to bore easily, then a hypodermic needle can be substituted for the glass tube.

12. If a flow adaptor is being used, then the peristaltic pump can be placed before or after the column (except during packing, of course, when it must be after). One disadvantage of placing it after the column is that some mixing of the eluent will occur because of the volume of the pump tubing and the pumping action; this can cause loss of resolution. Another problem arises if it is attempted to pump liquid through the column at a rate greater than its natural flow rate when air bubbles will form in the outlet tubing. Nevertheless, it is common practice to leave the peristaltic pump positioned after the column after packing is complete and when the column is run.

13. Large volumes of dilute protein solution do not usually pose a problem, particularly if the protein of interest binds to the column. Indeed, adsorption of the protein onto a column from a large volume of solution followed by stepwise or gradient elution provides concentration as well as purification.

14. Having a three-way valve in the buffer line (before the peristaltic pump), with the tubing from the third port dipped into the sample reservoir, is useful here, because simply turning the valve will direct sample rather than buffer onto the column. If the sample tubing contains air, then reversing the direction of flow of the pump temporarily will displace the air by buffer from the column; reversing it again will then load the sample onto the column.

15. Some problems can arise at this stage. Particularly if the protein solution is concentrated and the column diameter is large, imperfections in packing sometimes result in the protein solution channeling or slipping between the matrix bed and the walls of the column; this is easy to see if the solution is colored. The only remedy is to stop flow through the column, stir up the top section of the bed (to below the point of visible channeling), and allow it to settle again before applying the rest of the sample. Another problem can arise with crude protein samples that either have not been properly clarified before application to the column or from which protein precipitates (because of the particular pH, and so forth) during application. The particulate matter will form a layer on top of the matrix bed; this will slow down or stop buffer flow or may lead to channeling. The situation may be recoverable by stirring the top of the bed to disperse the layer of particulate matter, but subsequent running of the column is unlikely to be ideal. This emphasizes the desirability of using properly clarified protein solutions for chromatography. When flow adaptors are being used, particulate matter can block the applicator net; the only solution if this occurs is to dismantle the adaptor and clean or replace the net.

16. This can be judged by the trace on a recorder returning to baseline or by manual measurements of $A_{280\,nm}$. This should occur with a volume somewhat larger than the liquid volume of the column, unless material is partially adsorbed—in which case, the volume may be considerably greater; conditions should be chosen, if possible, to avoid the latter situation.

17. This is obviously the case if the protein of interest is retained on the matrix (although it is always advisable to retain the unadsorbed fraction until it is confirmed that the target protein has indeed bound; failure to equilibrate the matrix or the sample properly or overloading the column can lead to unexpected behavior, and it is easier to rechromatograph the fraction than to start again from the beginning). Note that diversion of the flowthrough fraction to a collecting reservoir is one of the functions that can be programmed into more sophisticated fraction collectors equipped with solenoid valves.

If conditions have been chosen such that the protein of interest is not adsorbed onto the matrix, then the breakthrough material can usually still be collected in a single fraction, because it should have a uniform composition. One circumstance where this is not so is if either the target protein or some of the impurities are retarded, but not retained by the column. The target protein will then be enriched in either the back part or the front part of the breakthrough peak, and individual fractions will need to be taken for optimal purification. Similarly, if the column is overloaded so that the protein of interest, which was intended to be retained, is displaced by more tightly binding contaminants (*see* **Note 7**), then the target protein will be concentrated in the back part of the breakthrough peak and potential purification will be lost if individual fractions are not collected. This situation should be suspected if the breakthrough peak is markedly asymmetric with increasing protein content in the back half.

18. The total volume of the gradient will generally be in the range of four to eight times the volume of the matrix bed. The most commonly used salt for gradients is NaCl. In the absence of previous information on the salt concentration required to elute a particular protein, a gradient of 0–1.0 *M* should be tried. From this trial, the salt concentration at the maximum of the eluted protein peak can be calculated from the equation in **Note 6**. This value plus about 20% can then be used as the limit of the gradient for subsequent experiments. If the target protein elutes at relatively high salt concentration, then the gradient can be started at a value >0. For example, if the protein elutes at 0.6 *M* NaCl, then a gradient from 0.4 to 0.75 *M* would probably give good results. It is important not to have the protein eluting at the limit of the gradient because tailing will then occur.

    To set up the gradient, pour equal volumes (*see* **Note 6**) of the start buffer and the start buffer containing salt into the mixing chamber and into the second reservoir, respectively, making sure that the connection between the vessels is closed off and that the vessels are at the same height. If it is intended to use a nonlinear gradient with vessels of unequal cross-section (*see* **Note 6**), then the vessels must be filled to the same level. At the time when the gradient maker is connected to the column, open the connection between the vessels and switch on the stirring motor. If the gradient device consists of two cylinders or beakers that need to be connected by a syphon tube, then fill them to the desired level, insert the tubing into one of the vessels, suck liquid up it with a syringe, squeeze the tube so that liquid cannot escape, and place the end below the surface of the liquid in the second vessel.

19. It is not usually worthwhile collecting a large number of small fractions; this simply increases the amount of analysis to be done. Fraction volumes should be between 1/50th and 1/20th of the total elution volume of the column, depending on the degree of resolution expected. Take larger rather than smaller volumes, unless there is a good reason to do otherwise.

20. Chromatography at early stages of a purification is unlikely to result in the protein of interest being obtained in a symmetric, well-resolved protein peak. The combination of fractions is therefore likely to be a payoff between yield and purification. The first requirement is to construct a protein profile (if not using an automatic monitor) by plotting $A_{280\ nm}$ against elution volume or fraction number. Next, the fractions containing the target protein must be established using a quantitative assay, and the activity profile superimposed on the protein profile. From this, it will be possible to see how to combine the fractions in such a way as to optimize both purification and yield. For example, it is usually worth rejecting fractions at the beginning and/or end of the peak of active material if their contents of contaminating protein are very high. It is also very important to determine the total yield of the target protein, because a particular chromatographic procedure may result in substantial purification, but may still be unacceptable if it results in a low overall yield. The importance of having a quantitative assay for the protein of interest is difficult to overestimate, particularly for the analysis of fractions from chromatographic procedures. In addition, the assay should be as rapid as possible, because large numbers of fractions may have to be analyzed.

Time devoted to developing a rapid quantitative assay will be well spent if a substantial amount of purification work is to be undertaken. A good example of what can be done with a little ingenuity is given in **ref. 2**.

21. With most matrices, regeneration can be done by washing with 1 or 2 column volumes of a solution of high salt concentration (approx 1 M), followed by a similar volume of equilibration buffer; manufacturers' instructions should be followed if different from the above. The column should then be stored in equilibration buffer, containing an antibacterial agent, such as sodium azide (0.2% [w/v]). Matrices used to fractionate crude protein mixtures may retain some protein and pigmented material after such a treatment. This can usually be removed by washing with 0.1 M NaOH (if the matrix is stable under these conditions) before re-equilibration. As a last resort, the discolored matrix can be removed from the top of the column and replaced. Some matrices (particularly those based on Sephadex) shrink considerably when subjected to solutions of high ionic strength, and the bed may not return to its original dimensions on re-equilibration. If this occurs, then the column should be repacked.

References

1. Bock, R. M. and Ling, N.-S. (1954) Devices for gradient elution in chromatography. *Anal. Chem.* **26,** 1543–1546.
2. MacGregor, S. E. and Walker, J. M. (1994) A microtiter plate assay for the detection of inhibitors of the Na$^+$, K$^+$-ATPase. *Appl. Biochem. Biotechnol.* **49,** 135–141.

39

Detection Methods

Jacek Mozdzanowski and Sudhir Burman

1. Introduction

Chromatography as a term was introduced by the Russian scientist Tswett, who separated plant pigments on a column containing calcium carbonate. It was based on the Greek word for color—"chromatos" and "graphein"—to write, together meaning "writing with color" because the separated pigments were seen on the column as zones of different colors. With time, the term "chromatography" became a description of separation of mixtures of different compounds. The current definition might be that "**Chromatography** is a separation method in which a mixture is applied as a narrow initial zone to a stationary, porous sorbent and the components are caused to undergo differential migration by the flow of the mobile phase, a liquid or a gas." Most of the compounds separated in chromatographic processes in the early days of chromatography were colored compounds, which could be detected visually. This quickly became a limiting factor for the technique. Rapid development of chromatography created the need for more accurate detection techniques for a variety of analytes. Initially, this was achieved by chemical reactions of selected reagents with colorless analytes. The colored products of such reactions were detected and quantified spectrophotometrically using visible light, filters, and detecting devices connected to a recorder. A well-known example of such an approach is amino acid analysis first presented by Moore and Stein (1). The separation and quantification was achieved by the use of ion-exchange chromatography and the detection was possible after the amino acids reacted with ninhydrin, yielding colored products. In the late 1960s and early 1970s, introduction of detection methods utilizing ultraviolet (UV) absorbance of many organic compounds permitted for a great expansion of analytical applications of chromatography. Expansion in knowledge of spectroscopic properties of organic compounds and technological progress created sensitive UV detectors for chromatographic systems and permitted detection of compounds with different light absorption characteristics. The need for detection of non-UV absorptive compounds led to the introduction of additional detection methods based on fluorescence, electrochemistry, refractive index, light scattering and other physico-chemical properties of compounds (2,3)

One of the fields where chromatography helped advance scientific knowledge is protein chemistry. Isolation, analysis, and characterization of proteins utilizes a wide vari-

From: *Methods in Molecular Biology, vol. 244: Protein Purification Protocols: Second Edition*
Edited by: P. Cutler © Humana Press Inc., Totowa, NJ

ety of chromatographic techniques such as size exclusion, hydrophobic interaction, ion exchange, reversed phase, and others. These techniques are applied for both the analytical- and industrial-scale uses, and the detection systems are critical for the proper determination of analytes. Most proteins are colorless and this limits the use of detection systems based on visible light, although it should be noted that detection in the visible light range may be very useful for some protein specific modifications such as the Maillard reaction *(4)*. The predominant detection system for proteins is based on the fact that all proteins and peptides absorb UV light at the 215-nm wavelength. At this wavelength, proteins can be detected "as is," without any chemical modification and with little interference from many of the solvents/buffers used for chromatographic separations. Proteins containing tryptophan, tyrosine, or phenylalanine residues show additional UV absorption at 280 nm, which permits specific identification of peptides containing these amino acid residues.

Current analytical procedures for characterization and analysis of proteins require not only their detection but also substantial analytical work to detect posttranslational modifications, confirm some of the structural properties such as disulfide bonds, and, in the case of biopharmaceutical products, detect process residuals. The compounds that have to be detected/analyzed include non-UV absorbing carbohydrates, polyethylene glycols, and a number of nonprotein process impurities, which frequently absorb very little or no UV light. In some cases, the non-UV absorbing compounds may also be chemically unreactive, which complicates the possible chemical tagging with the UV absorbing or fluorescent tag. In such cases, detection techniques based on different principles may be used for sensitive detection of proteins and related compounds. Examples of such techniques are fluorescence-, electrochemical-, refractive index-, and light-scattering-based detection systems.

2. Methods

2.1. Detection in Visible Light

The vast majority of proteins are colorless and cannot be detected by absorbance of visible light. Despite this fact, detection in the visible region may be useful for the evaluation of protein purity and the detection of possible chemical modifications, which may occur in formulated proteins. Myoglobin, hemoglobin, or cytochrome(s) could be detected in the visible light. The same is true of Maillard reaction products, where the reducing sugar(s) used for protein formulation may react with amino groups of lysines yielding strongly colored products. It is true that detection of proteins using visible light only yields very limited information, but if the detection is performed with the diode array detector (*see* **Subheading 2.2.**), it may be a good idea to collect spectra including the visible light. Such spectra may, in some cases, help to identify colored proteins and/or impurities. Most of the modern detectors are designed to cover both the visible and UV range, and for this reason, these detectors will be described in much greater detail in the following subsection dedicated to UV detection.

2.2 Detection in UV Range

2.2.1. UV Detectors

Early UV detectors were based on the use of filters and the detection was performed at 254 nm. Although this wavelength permitted the detection of many organic compounds, it remained inefficient for many others. Technological development and better understanding of UV spectral properties of organic compounds led to 215 nm as a wavelength that was absorbed by a much larger group of compounds and provided much higher sensitivity. It should be noted that relatively few organic solvents are UV transparent at 215 nm and the choice of appropriate solvents used for chromatographic elution is essential for proper detection of analytes.

Ultraviolet-visible (UV-Vis) detectors currently available on the market include single-wavelength, variable-wavelength, multiple-wavelength, and diode array detectors. The single-wavelength detectors are designed to measure response at a fixed wavelength and are the least expensive and easiest to operate. They require limited manipulation by the operator and usually offer an option of exchangeable flow cells. They are well suited for a manufacturing environment in which the process of protein purification is well optimized and troubleshooting is minimal. Variable-wavelength detectors have the advantage of selecting a single wavelength from approx 200 nm to 800 nm and they also offer exchangeable flow cells. Variable-wavelength detectors are well suited for the protein purification development lab and manufacturing environment in which the primary objective of detection is to determine yields based on absorbance. Multiple-wavelength detectors permit simultaneous measurements of absorbance at several different wavelengths, although do not allow acquisition of UV spectra. In protein detection/analysis, these detectors are very beneficial when used for detection at both 215 and 280 nm. Multiple-wavelength detectors utilize the diode array technology, and for this reason, most vendors discontinued their manufacture and offer the diode array detector instead. Diode array detectors (DAD) offer the detection at multiple wavelengths and usually offer a multiple choice of settings/options (different flow cells, bandwidth, reference wavelength, slit width). They also have the ability to acquire spectra in both UV and visible regions, providing the most comprehensive information for the analyst. In DAD, the UV and visible light beam passing through the sample in the flow cell are split into the electromagnetic spectrum (on the same principle as sunlight may be split into a rainbow by a prism). The spectrum is analyzed by hundreds of single diodes, each of which analyzes a very narrow band of split light. Some of the adjustments required for optimal performance of a DAD are unique to this type of detectors, but some of them apply to more simplified instruments, such as variable-wavelength or multiple-wavelength detectors. We will discuss the issues related to the optimal performance of detectors using DAD as an example. We will try to point out what adjustments apply to simpler UV-Vis detectors.

All proteins and peptides strongly absorb UV light around the 215-nm wavelength and may be detected with relatively high sensitivity. The potential problem is related to the fact that molar absorptivity of proteins and peptides at 215 nm is highly variable and cannot be reliably used for quantification based on the standard(s), which are different from the analyte. Absorption at 280 nm a provides a much better basis for quantification and can be reliably predicted (calculated) based on the amino acid composition of a given protein or peptide *(5)*. The UV absorption of proteins/peptides at 280 nm is directly related to the presence of aromatic amino acids in the sequence (tryptophan, tyrosine, and phenylalanine) and the vast majority of proteins contain at least some of these amino acids. Unfortunately, many of short peptides do not and cannot be detected and quantified at 280 nm absorbance. Contribution of disulfide bonds to 280 nm absorbance is usually too small to permit the detection of disulfide-bonded proteins/peptides in the absence of aromatic amino acids.

Most proteins can be solubilized in water-based buffers, sometimes with the addition of detergents. At the same time, most water solutions of salts, acids, and bases used for preparation of buffers do not absorb UV light at 215 nm. This permits the use of 215-nm UV detection of proteins with a large variety of chromatographic techniques, such as reversed phase, ion exchange, size exclusion, hydrophilic interaction chromatography (HILIC) and others, and makes UV detection of proteins at 215 nm the first choice for every protein chemist/chromatographer. One of the important factors for every chromatographic method is the peak resolution observed on the chromatogram. Although the choice of chromatographic procedure is the most critical factor, the choice of a detector and the settings for detection may greatly influence the results of separation of analytes.

2.2.2. Flow Rate

Ultraviolet absorption of the analyte is directly proportional to its concentration. Decreasing the internal diameter of chromatographic column or varying the flow rate of the mobile phase may increase concentration of an analyte in the eluate. These are not detector-specific parameters and will not be discussed here, but one should keep them in mind when optimizing chromatographic procedures for optimal sensitivity of detection.

2.2.3. Flow Cell

There are two factors related to a flow cell that will influence chromatographic outcome of a separation. These are the cell volume and the path length. Both parameters are given for a given flow cell and cannot be adjusted. This means that a flow cell should be carefully chosen for the specific chromatographic system. Most of the new generation UV-Vis detectors have an option of exchanging flow cells with different path length and/or cell volume depending on the application. The switching of the flow cells is straightforward and usually no tools are needed. The cell volume may influence the peak resolution, and the path length will influence the sensitivity of detection. Most of the flow cells have a path length of 10 mm, which provides for sensitive detection. A shorter path length may be used to minimize the cell volume or if high sensitivity is not desired (i.e., if the analyte is highly concentrated). The longer path length increases the light lost because of the dispersion inside the cell and as a result produces higher baseline noise. For this reason flow cells with a path length above 10 mm are not available for common high-performance liquid chromatography (HPLC) detectors.

Flow cell volume may have a significant impact on the peak resolution. The choice of flow cell volume depends on the flow rate used for chromatographic separation and should be at least an order of magnitude smaller than the flow rate expressed as volume per minute (see **Tables 1** and **2**). It is worth noting that the cell volume should also be based on a width of the narrowest peak expected during the separation. For example, if the flow rate is 200 μL/min and the narrowest peak width is 6 s (which corresponds to a 20-μL volume of eluate containing the peak), a 20-μL flow cell will broaden the peak to at least 12 s. If one imagines the analyte as a "plug" of higher light absorbance moving through the cell, the absorbance will be registered by a detector from the moment the front end of the "plug" enters the cell to the moment the back-end leaves it. In reality, additional mixing inside the cell would broaden the peak even more. An additional factor related to the cell volume is the back-pressure that will be created by the cell be-

Table 1
Example Chart Illustrating the Choice of Flow Cells Appropriate for Specific Flow Rates of Eluent

Typical column length	Typical peak width	Recommended flow cell				
$T \leq 5$ cm	0.025 min	500 nL flow cell				High-pressure flow cell for pressures above 100 bar
10 cm	0.05 min		Semi-micro flow cell			
20 cm	0.1 min			Standard flow cell		
≥ 40 cm	0.2 min					
	Typical flow rate	0.01 ... 0.2 mL/min	0.2 ... 0.4 mL/min	0.4 ... 0.4 mL/min	1 ... 2 mL/min	0.01 ... 5 mL/min
	Internal column diameter	0.5 ... 1 mm	2.1 mm	3.0 mm	4.6 mm	

Note: This chart refers to flow cells available from Agilent as defined in **Table 2**, but may be used as a guide for a large variety of flow cells and chromatographic conditions. It should be noted that this example does not include flow cells required in preparative chromatography. Preparative chromatography employs high flow rates and requires specialized flow cells designed for such conditions.
Source: From Agilent, 1100 Series Diode Array and Multiple Wavelength Detector Reference Manual, part G1315-90004, with permission.

Table 2
Examples of Flow Cells Available From Agilent

Flow cell type	Cell volume	Part number	Path length (nominal)	Path length (actual)	Correction factor
Standard flow cell	13 μL	G1315-60012	10 nm	9.80 ± 0.07 mm	10/9.8
Semimicro flow cell	5 μL	G1315-60011	6 nm	5.80 ± 0.07 mm	6/5.8
500 nL Flow cell list	0.5 μL	G1315-68714	10 nm	10.00 ± 0.02 mm	10/10
High-pressure flow cell	1.7 μL	G1315-60015	6 nm	5.75 ± 0.07 mm	6/5.75

Source: From Agilent, 1100 Series Diode Array and Multiple Wavelength Detector Reference Manual, part G1315-90004, with permission.

cause of its limited volume and orifice diameter. A cell with a low cell volume used for separation requiring high flow rates may create a back-pressure exceeding the limits of the flow cell. As a result, the flow cell may be destroyed. Many of the modern analytical detectors have flow cells with volumes within the 5- to 20-μL range. Such flow cells are usually sufficient for 0.2- to 1.0-mL/min flow rates, typical for many analytical chromatographic separations. Separations requiring higher flow rates, such as preparative separations, will require flow cells with larger volumes, and separations using low flow rates, such as microbore and capillary HPLC, will require special flow cells with ultralow volumes. Flow cells used with relatively low flow rates may require additional back-pressure in a cell to avoid any potential air bubbles that may be produced during a rapid decompression of a mobile phase in the cell. An increased back-pressure in the cell is usually achieved by installing a piece of a small-inner-diameter tubing on the outlet side of the cell.

The chromatographic separation of proteins and peptides may require flow rates from a few microliters per minute in the case of capillary HPLC to hundreds of milliliters per minute in case of preparative chromatography, and these different flow rates require the proper selection of a flow cell.

2.2.4. Response Time (Data Rate)

Response time indicates how fast the detector responds to a change in absorbance in a flow cell. The detector averages the incoming signal over a specified period of time and this average becomes the signal sent from the detector to a recording device. Proper setting of the response time may significantly decrease the baseline noise because the random noise is averaged and thus greatly reduced (*see* **Fig. 1**). At the same time, too long a response time will negatively impact the true representation of peak shape and peak area, and, because of that, the peak resolution. Currently, most of analytical data originating from a UV-Vis detector is stored digitally, and instead of a response factor, the setting is expressed as the data rate (frequency) with which the averaged signal is collected (as a single point/value). The peak is created by a way of connecting single points representing absorption averaged over a specified time period. With too few points collected, the peak shape will not be represented accurately. It should be noted that the response time and the data rate are not the same. Usually, the maximum response time is defined as one-third of the narrowest peak width, and, at the same time, a peak should be represented by at least 10–15 data points from its beginning to the end. The easiest approach to determine the data rate is by finding the narrowest peak on a chromatogram and measuring this peak's width in seconds. Dividing it by 15 will give a good guideline for the data frequency (data rate) expressed in points per second (Hz). For example, if the narrowest peak on a chromatogram elutes in 10 s, the data collection rate should be set at 1.5 Hz (1.5 points per second), or higher. It should be noted that the data rate set much higher than necessary will not only increase the baseline noise but will also increase digital storage space needed to save the data. It should also be noted that with some HPLC systems, the data rate is selected on the recording device, not by the use of detector controls. In the analysis of proteins and peptides, the data collection rate will depend on the chromatographic technique used for analysis. In most cases when liquid chromatography is used, the narrowest peak will be approx 15 s wide. Size-exclusion chromatography usually yields very broad peaks (on the order of minutes), whereas capillary electrophoresis or capillary electrochromatography produce very narrow peaks.

2.2.5. Detection Wavelength, Bandwidth, and Reference Wavelength

For detection in the UV or visible light, the choice of detection wavelength, reference wavelength, and the bandwidth will influence sensitivity, baseline noise, and baseline slope. Ideally, the sample wavelength and the bandwidth should be chosen based on the UV-Vis spectrum of the analyzed compound. If the compound of interest has a well-defined absorbance peak, the sample wavelength should be set at the maximum absorbance of the compound, and the bandwidth should be set at the absorbance peak width (in nanometers) at the half-peak height. This kind of setting will produce rela-

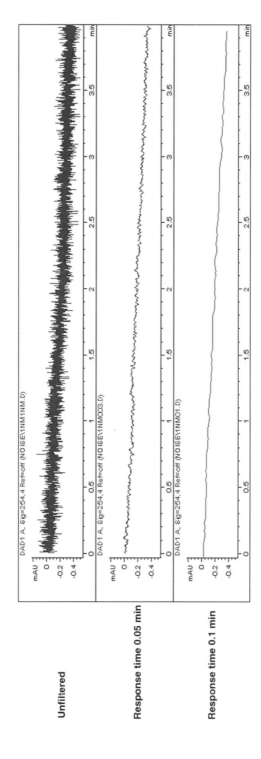

Fig. 1. Influence of response time selected on diode array detector on baseline noise. (From Agilent, *1100 Series Diode Array and Multiple Wavelength Detector Reference Manual*, part G1315-90004, with permission.)

tively low baseline noise, good response, and good linearity over the range of different concentrations. The reference wavelength does not have to be selected. In fact, not all DAD detectors have the option of setting the reference wavelength. Absorbance at the reference wavelength is subtracted from the absorbance at the sample wavelength to minimize the influence of refractive index and/or temperature changes. Selecting the reference wavelength may be helpful when gradient elution is used in the chromatographic method. The influence of changes in temperature of the flow cell on baseline slope is usually very low. For this reason, selecting the reference wavelength for isocratic separations will have little or no effect on the outcome. Reference wavelength and its bandwidth should be selected well outside the absorbance peak of interest and the bandwidth should be relatively broad. Selecting the sample and reference wavelengths may be more complicated if chromatographic process yields multiple components with different absorbance maxima. The most critical factor in such a case is selecting the reference wavelength outside of the absorbance peaks of all compounds of interest or not selecting it at all. It should be stressed that incorrectly selected reference wavelength may produce negative peaks and/or greatly reduce the detector response for some (or all) analyzed compounds.

Use of a DAD allows for collection of full spectral information of the compounds of interest during the initial stage of chromatographic method development. Such information is sufficient to determine the best settings for the sample and the reference wavelength. In the case of proteins and peptides, the sample wavelength of 215 nm provides the best sensitivity of detection and is most frequently used. It is important to remember that absorbance of solvents such as water, acetonitrile, and additives such as trifluoroacetic acid (which is frequently used in separations of proteins/peptides) increase below 215 nm. For these reasons selecting a broad bandwidth for the sample wavelength of 215 nm may result in the increased baseline noise and decreased signal-to-noise ratio. Detection of proteins/peptides at 280 nm is based on a well-defined absorption maximum and broader bandwidth may be advantageous, although the presence of DNA with absorption maximum at 260 nm may limit the bandwidth to avoid interference. In most cases, proteins and peptides are detected at both 215 and 280 nm with a bandwidth of approx 4 nm at 215 nm and bandwidth of approx 16 nm at 280 nm. Selecting the reference wavelength at 360 nm with the bandwidth of 100 nm helps to reduce the baseline slope related to water/trifluoro-acetic acid (TFA)–acetonitrile/TFA gradients commonly used in the reversed-phase chromatography of proteins and peptides. Variable-wavelength detectors do not allow the bandwidth adjustments and the bandwidth is usually set to several (4–10) nanometers.

2.2.6. Slit Width

This setting is strictly related to the collection of UV spectra with the DAD detector. A very narrow slit width results in a spectrum with more details and provides sufficient resolution for identification of fine spectra, such as the one of benzene. It should be noted that narrower slit width limits the amount of light that is received by photodiodes and, for this reason, increases the baseline noise. In general, proteins and peptides do not have fine spectra requiring the use of a narrow slit width. In such cases, depending on the instrument, the slit should be selected at 4 nm, or if the option is not available, it should not be a reason for concern.

2.2.7. Detection of Proteins With Variable-Wavelength Detectors

As described earlier, the critical parameters influencing the peak resolution are flow cell volume, data rate, and selection of sample wavelength and bandwidth. Flow cell volume should be selected based on the flow rate of mobile phase in the chromatographic method used for the separation. Data rate should be selected on the recording device (or on the detector) according to the chromatographic profile. The sample wavelength has to be selected based on the UV-Vis spectrum of the compound(s) of interest and desired outcome (such as sensitivity, specificity, etc.), and the sample bandwidth, if available, should be (in most cases) set to approx 16 nm. The reference wavelength and its bandwidth settings are not available in variable-wavelength detectors. The setting(s) for slit width is also not available, because this is related only to the collection of UV-Visible spectrum. This is also true for the multiple-wavelength detectors (MWD). It should be noted that the older types of detector may have a limited choice of settings, and in the case of detectors based on filters, the only parameters available for control are the flow cell volume, externally set data rate, and the sample wavelength defined by the selected filter. The sample bandwidth will be given with the filter without any option to change it.

Ultraviolet detection is relatively simple, sensitive, and straightforward. The spectra obtained with DAD may provide additional information about analyzed compounds and permit determination of the peak purity. Modern detectors provide multiple features to enhance their performance and are relatively easy to maintain. In protein chromatography, the UV detection is used for the majority of applications. The main limitation is related to compounds that do not contain any chromophores (such as polyethylene glycols or carbohydrates) and cannot be detected by UV-Vis detectors. The use of solvents transparent in the UV-Vis range used for the detection of analyte(s) is another important factor, although with the existing selection of solvents, it rarely limits the applicability of UV detection.

3. Fluorescence

3.1. Fluorescence Detection

The ability of chemical compounds to emit photons after excitation with light of higher energy is called fluorescence (when the emission of photons occurs almost instantaneously) or phosphorescence (when the emission of photons lasts much longer than the excitation) *(6)*. Both fluorescence and phosphorescence may be used as a very selective and sensitive method of detection of chemical compounds. The option of chemical "tagging" of many nonfluorescent compounds with highly fluorescent chemical tags greatly enhances the applicability of fluorescence detection in chemical and biochemical analysis using chromatographic techniques. A large variety of highly fluorescent tags, which react selectively with specific chemical functional groups, creates a number of options for an analytical chemist *(7,8)*.

In fluorescence detection, the signal is obtained by the measurement of light emitted by analyzed compound(s). This is different from the UV-Vis detectors, for which the signal is obtained by measurement of the decrease in light intensity as the light of specific wavelength is absorbed by the sample. Emitted photons can be detected with much higher sensitivity than the difference in light intensity, and for this reason, fluorescence detectors usually offer sensitivity of detection superior to the UV-Vis-absorbance-based detectors. Fluorescence detection has a limited use for the detection of unmodified proteins because it is related to the content of aromatic amino acids (tryptophan, tyrosine, and phenylalanine). Fluorescence of phenylalanine is not detected in the presence of tyrosine, and fluorescence of tyrosine may be almost completely quenched by ionization and/or close vicinity of the tryptophan, amino, and/or carboxyl group. For this reason, tryptophan is the only amino acid that may be used for reliable fluorescence detection of proteins. Tryptophan content in proteins is usually low, which results in low sensitivity of protein detection and possibly no detection if a protein does not contain tryptophan at all. The use of fluorescence detection may be greatly enhanced if a protein is modified by "tagging" with highly fluorescent tags prior to chromatographic separation. Sometimes, such tagging may change the chromatographic properties of a protein or peptide and limit its usefulness, and in this case, it is also possible to add the "tag" post-column prior to the detector.

Analysis of carbohydrates bound to proteins as well as analyses of process residuals in proteins manufactured by recombinant technology seem to be the areas where fluorescence detection is very useful. Carbohydrate content in proteins may be relatively low, and process residuals have to be detected (usually) with very high sensitivity, giving the fluorescence detection a clear advantage over UV detection. It has to be noted that in the case of carbohydrates as well as process residuals, initial chemical modification of the analyzed compounds with fluorescent chemical tags is required because neither carbohydrates nor most of the process residuals common for biopharmaceutical processes are fluorescent. This may be a limiting factor in the case of chemically nonreactive compounds such as quaternary ammonium salts or nonreducing carbohydrates, which may be difficult to modify with fluorescent tags.

For the efficient detection of analytes using fluorescence detector, its settings should be carefully optimized *(9,10)*. The selection of a flow cell is an important factor and was described previously for the UV-Vis detectors. Other parameters specific to fluorescence detectors are described in **Subheadings 3.2.–3.9.** It should be noted that the type of available fluorescence detector might be a limiting factor in the optimization of sensitivity. Filter-based detectors provide very little flexibility because the choice of filters is limited. Other detectors let the operator select exact excitation wavelength, but the emission side still depends on the limited selection of filters. More advanced detectors provide the wavelength selection for both the excitation and emission light, and most of the modern detectors provide additional scanning capability for the determination of excitation and emission spectra.

3.2. Fluorescence Quenching

Fluorescence of an analyte may be decreased (quenched) by a large variety of chemical compounds *(11)*. In some cases, strong quenching of an analyte may eliminate fluorescence as a choice of detection. Quenching of fluorescence of phenylalanine and/or tyrosine described earlier is a good example of this problem in relation to proteins. It should be noted that fluorescence quenching could be utilized for detection of nonfluorescent compounds. In such cases, the mobile phase is prepared with a fluorescent additive that is quenched by the analyte. The decrease in fluorescence results in a "negative" peak used for identification and quantitation of the analyte *(12)*.

3.3. Excitation and Emission Wavelengths

For the best sensitivity and signal-to-noise ratio, the detector should be set for the optimal wavelengths for both the excitation and emission. Most of the time, the optimal values may be found in literature, especially when related to specific fluorescent tags, but in specific cases, the exact excitation and emission profile of a tagged compound may be substantially different from the reported values. The use of a scanning fluorescence detector permits an accurate determination of excitation and emission maxima and is highly recommended. If such a detector is not available, the excitation maximum may be determined by the UV scan of the analyzed sample. In almost all cases, UV absorbance maximum matches the excitation profile very closely and can be used to establish the excitation wavelength. It should be noted that fluorescence detectors use a much broader bandwidth than UV detectors and this may result in noticeable differences between excitation maximum found by UV and spectrofluorimeter scans and between scans obtained on different instruments. In addition, it should be noted that most of the fluorescence detectors are optimized for excitation wavelengths in the range of 200-350-nm and will perform noticeably worse if the excitation wavelength is above this range. Unfortunately, the UV scan cannot be of help in the determination of emission spectrum. This spectrum has to be determined by a scanning fluorescence detector or by time-consuming multiple measurements of emission intensity of the same sample at different wavelengths. To assure the best sensitivity of detection, excitation and emission settings on the detector should correspond to the respective maxima found during the spectral analysis of the analyte.

3.4. Bandwidth

Fluorescence detectors use a much broader bandwidth compared to the UV-Vis detectors. Some of the fluorescence detectors permit selection of the bandwidth, whereas in others, the bandwidth is set and cannot be changed. In detectors that use filters (for excitation and emission or emission only), the bandwidth will depend on the quality and characteristics of the filters. Selecting the broader bandwidth may decrease the baseline noise and thus improve sensitivity of detection. The main limitation of broader bandwidths is the necessity to separate the excitation and emission wavelengths. In fluorescence detectors, the emitted light is measured through the pathway perpendicular to the excitation beam, but the excitation light is much stronger in intensity than the emission light. Dispersion of excitation light in the flow cell could create a serious problem for the proper detection unless the excitation light is filtered off from the emission light. This means that the wavelengths of excitation and emission light should not overlap. In the case of samples where the excitation and emission maxima are relatively close, selection of a very broad bandwidth may significantly decrease the sensitivity of detection. An additional tool to minimize the stray light is the cutoff filter. Some of the fluorescence detectors use such a filter placed between the excitation and emission gratings.

3.5. Amplification (Gain)

Fluorescence detectors measure light emitted by the sample, and the signal created by photons reaching the detecting element (photomultiplier) may be amplified electronically. Many detectors provide the selection of several levels of signal amplification

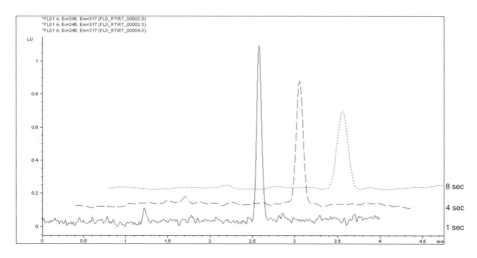

Fig. 2. Influence of response time selected on fluorescence detector on baseline noise, peak width, and sensitivity of detection. (From Agilent, *1100 Series Fluorescence Detector Reference Manual*, part G1321-90002, with permission.)

and the setting may improve the signal-to-noise ratio. In almost all cases, amplification of the signal will improve the signal-to-noise ratio, although for each sample, there will be an amplification value (limit) beyond which the signal-to-noise ratio cannot be further improved. If maximum sensitivity of detection is desired, this limit will have to be found empirically by measurements of the same sample at different amplification levels and calculations of the signal-to-noise ratio.

3.6. Response Time

Similar to the UV-Vis detectors, the response time (data rate) will influence the sensitivity, baseline noise, and peak resolution. In all cases, the lower response time (higher data frequency) will improve the sensitivity and resolution but at the same time will increase the baseline noise (*see* **Fig. 2**). In the case of narrow peaks (peak width of up to approx 15 s), a long response time will decrease the sensitivity and affect peak resolution and may lead to the false conclusion that the performance of selected HPLC column and mobile phase are unsatisfactory. It should be noted that even when the response time is too high, the retention time and peak areas will still be reproducible. Compounds that elute as narrow peaks may require selection of a relatively short response time.

3.7. Raman Band of Water

Water disperses UV light, producing excitation and emission spectra with maximum at 350 and 397 nm, respectively. This phenomenon is frequently used for the calibration of fluorescence detectors. It should be noted that many solvent systems used as the mobile phase in HPLC separations contain water and therefore will interfere with any analysis requiring excitation and emission wavelength, which are close to excitation/emission observed for the Raman band of water.

3.8. Xenon Flash Lamp Frequency

Some of the fluorescence detectors provide several modes of operation for the xenon lamp. The high frequency of flashes of the xenon lamp results in a lower baseline noise and may improve the sensitivity. At the same time, the high frequency of flashes decreases the lamp's life. In many cases, the highest frequency of flashes is not necessary.

3.9. General Recommendations

Using a fluorescence detector with optimal sensitivity requires the detector's optimization for the specific sample(s) and chromatographic conditions. A good starting point for optimization may be selecting the amplification at the mid-point of the detector range, a response time of 4 s, and a bandwidth of 20 nm.

4. Light Scattering

4.1. Light-Scattering Detectors

Light-scattering detectors have been around since the 1940s, when the first one was built by one of the pioneers in the field, Dr. Bruno Zimm. It is one of the less commonly used detectors during protein purification and its main application is the determination of absolute molecular weight of proteins and oligonucleotides. The other two techniques for determining the absolute molecular weight of biopolymers in solution are osmometry and analytical ultracentrifugation. Light-scattering detection is the easiest of the three modes of detection for absolute molecular-weight determination. The detector can be employed in the stand-alone mode using a manual syringe or pump for sample introduction, or it can be used on-line with column chromatography. The detector measures light scattered by the biopolymer, and for this reason, careful filtration of protein solution and elution buffers is necessary to remove particulates. To facilitate removal of particulate these detectors are commonly connected with a liquid chromatography (LC) system for on-line molecular-weight determination during protein purification. In the stand-alone mode, the protein solutions and buffers should be filtered through a 0.2-μm filter. The advantage of using the light-scattering mode of detection for molecular-weight determination is that it provides absolute molecular weight without recourse to instrument or column calibration, as is common in size-exclusion/gel permeation chromatography. If the determination of molecular weight of an analyte is desired, its concentration has to be known. This can be achieved by the use of a UV-Vis detector placed in series with a light-scattering detector.

The light scattering covers a broad range of detectors predominantly using a laser as the light source. The class includes the historical low-angle laser light scattering (LALLS), multiangle laser light scattering (MALLS), and dynamic light-scattering (DLS) detectors. The low-angle light-scattering detector detects light scattered by the biopolymer at a low angle of 3°–7° to the incident beam. However, with the advent of computers for data processing, the low-angle detector has been completely replaced by multiangle light-scattering detectors. The latter detector utilizes scattered light at as few as three angles of 45°, 90°, and 135° (miniDAWN®) (*see* **Fig. 3**) or at as many as 18 angles ranging from 15° to 165° (DAWN®) for the molecular-weight determination. For the theory of molecular-size and molecular-weight determination by light scattering, the reader is directed to the

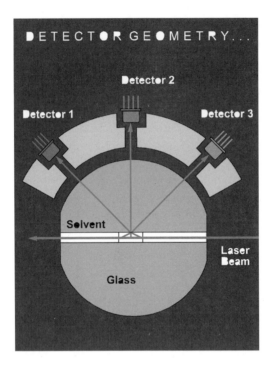

Fig. 3. Schematic representation of multiangle laser light-scattering detection. (Courtesy of Wyatt Technologies.)

reviews by Kratochvil *(13)* and Jackson et al. *(14)*. The MALLS detector was pioneered by Dr. Philip Wyatt of Wyatt Technologies (Santa Barbara, CA) and is available in two models: miniDAWN and DAWN. All of these detectors require careful installation and maintenance. The response from all light-scattering detectors depends on the size of the protein. Proteins with larger hydrodynamic radii (higher molecular weights) yield higher detector response compared with low-molecular-weight proteins. As a result, high-molecular-weight proteins are detected with higher sensitivity. One of the most frequent applications of MALLS detectors is the determination of protein aggregate during purification, formulation, or on storage *(15)*.

4.2. Evaporative Light-Scattering Detector

The evaporative light-scattering detector (ELSD) was first introduced by Dr. Michel Dreux in the 1980s. Since its introduction, this class of detector has seen significant improvement in technology such that they have become routine detectors in most analytical labs involved in the analysis of carbohydrates, sugars, amino acids, lipids, polymers, and UV transparent solutes. The ELSD can be considered a true concentration detector because the response is directly proportional to the amount of solute, and in this respect, it is different from the UV detector, for which the response is dependent on the molar absorptivity of solute. The principle of the operation of the ELSD is relatively straightforward *(16,17)*. There are three steps in the detection of column effluent in the ELSD. In the first step (*see* **Fig. 4**), the HPLC column effluent is mixed with an inert gas and passed through a narrow orifice to generate a homogenous plume of

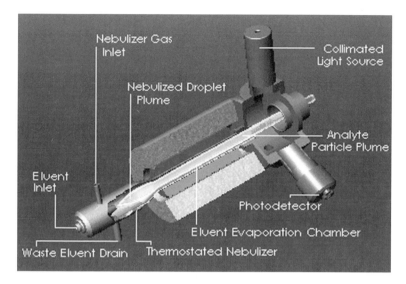

Fig. 4. Schematic representation of the evaporative light-scattering detector. (Courtesy of Polymer Laboratories.)

mobile phase mixed with the solute of interest. This step is referred to as *nebulization*. In the second step, called *evaporation*, the nebulized mobile phase is passed through a heated drift tube to evaporate the mobile phase. In the third step, referred to as *detection*, the nonvolatile particles enters a flow cell that consists of a light source and a photomultiplier tube for the detection of light scattered by the nonvolatile solute. The response of scattered light is directly proportional to the concentration of solute. The response is completely independent of the presence or absence of chromophores or fluorophores.

Briefly, there are four main processes by which an incident light beam is "scattered" when passing through a medium containing a suspended particulate. These are Rayleigh scattering, Mie scattering, refraction, and reflection *(18)*. The size of the particulate and, hence, the ELSD response could be varied by altering the nebulizing gas velocity, the eluent flow rate, the temperature of the evaporator, and the solute concentration. Based on published results, the ELSD instrument has a maximum sensitivity when the ratio of particulate radius to wavelength (r/λ) is approx 4. At this optimum ratio, the reflection and refraction are the predominant mechanisms of light scattering, resulting in increased sensitivity. For a detailed discussion of the theory and principle of the ELSD, two review articles are recommended *(16,17)*.

4.3. Practical Considerations in the Use of the ELSD

The nebulization of the mobile phase is best achieved by inert gases like nitrogen or helium. The inert gas is introduced at a flow rate of approx 2–5 L/min at a pressure of 60–100 psi. The column effluent is passed through the heated nebulizer and is fed perpendicular to the incoming gas stream. The gas shears the liquid droplets from the narrow orifice as they begin to form, atomizing the solution into a uniform dispersion of droplets or mist that then pass as a continuous stream into the evaporator. The size of the droplets is dependent on the gas flow rate. The lower the gas flow rate, the larger the resulting droplets will be. Larger droplets scatter more light and increase the sensitivity of the analysis but they are more difficult to evaporate in the evaporation tube (*see* below). The objective of optimizing the flow rate would be to produce the highest signal-to-noise ratio. A lower column inner diameter of a HPLC column would result in lower gas flow rates and increased sensitivity. The atomized spray is propelled into the heated evaporation chamber assisted by the carrier gas. In the evaporator, the solvent is evaporated, leaving a dry particle plume. A diffuser located in the evaporator assists in the drying of particles, by acting as an efficient heat exchanger. The most critical part of the ELSD is the appropriate evaporation of nebulized droplets. If the temperature in the evaporation chamber is too high, the thermally labile solutes may decompose and will not be detected. If the temperature in the evaporator is too low, the droplets may not evaporate completely when the plume enters the detector flow cell, resulting in poor light scattering and high noise, leading to poor detection. The proper temperature in the evaporation tube depends on eluent volatility. Highly volatile mobile phases or eluents delivered at a low flow rate require lower temperature. The optimum temperature is determined by observing the signal-to-noise ratios at different temperatures. A light source from a tungsten–halide lamp or laser is passed at right angles to the direction of particle plume. When pure solvent is evaporated, only the vapor passes through the light path and the amount of light scattered is small, giving a constant response. However, when a nonvolatile solute is present in the light path, the light is scattered much stronger. This scattered light enters the photomultiplier tube of the detector and generates a signal response that is directly proportional to the solute concentration and solute particle-size distribution.

One of the important requirements of the ELSD is the need for volatile mobile phases devoid of nonvolatile salts and modifiers. The nonvolatile salts will not evaporate and may collect inside the evaporator tube, resulting in poor sensitivity. However, the mobile phase may include different percentages of water. The other common mobile-phase solvents could be acetonitrile, tetrahydrofuran (THF), chloroform, acetone, dimethyl-formamide (DMF), ammonia and its volatile salts, ethylenediamine, triethylamine, butylamine, pyridine, formic and acetic acids and their volatile salts, and dilute HCl and TFA as ion-pairing agents. Because HPLC solvents with corrosive acids and bases are being evaporated, the ELSD must be appropriately vented. Also, the ELSD could be used as part of a series of detector for analysis. As the eluent is being evaporated in the course of analysis, the ELSD must be the last in a series if used in conjunction with other detectors. Care should be taken to ensure not to exceed the recommended backpressure in detector cells in other detectors.

4.4. Applications of the ELSD

The ELSD has been successfully used for the determination of fatty acids, carbohydrates, polyethyene glycol (PEG)-derivatized proteins and peptides, lipids, polymers, detergents, and surfactants *(19,20)*. The ELSD could be obtained from several different vendors. Sedere Inc. (Alfortville Cedex, France) makes three models of ELSD. These are Sedex 55, Sedex 65, and Sedex 75. Sedex 65 and Sedex 75 models are both microprocessor-controlled ELSD instruments, whereas Sedex 75 model features a new nebulizer and redesigned evaporation tube. Polymer Laboratories (Shropshire, UK) makes two models of ELSD. These are PL-ELS 1000 and PL-ELS 1000μ, and Alltech Associates Inc. (Deerfield, IL) makes ELSD 2000 model of the evaporative light-scattering detector.

5. Refractive Index

5.1. Refractive Index Detectors

The refractive index (RI) is one of the simplest detectors utilizing light refraction. It is based on the principle that when the light beam passes from one medium to another of different density, its velocity and direction changes. This is called refraction. If the second medium is optically denser than the first, the ray will become more nearly perpendicular to the dividing surface. The angle between the ray in the first medium and the perpendicular to the dividing surface is called the angle of incidence, i, whereas the corresponding angle in the second medium is called angle of refraction, r. The ratio sin i/sin r is the Snell's law of refraction and is called the index of refraction, n. Commonly, n is taken as greater than 1 when the ray is passing from optically rarer medium (usually air) to the denser. Temperature, the wavelength of incident light, and the density of the medium affect the refractive index *(21)*.

5.2. Practical Considerations in the Use of Refractive Index Detectors

The RI detector commonly used as an analytical instrument in protein analysis is a differential refractometer that measures the deflection of a light beam because of differences in refractive index between the liquids in the sample and reference compartments of a flow cell. Initially, both sample and reference cells are flushed with the mobile phase. The reference cell is then closed and solvent flows through the sample cell. When the analyte elutes from the column into the sample cell, the refractive index of the cell contents changes. The change in refractive index deflects the light beam as it passes through the flow cell, resulting in an unequal amount of light falling on sample and reference diodes. The change in current from the diodes is amplified and produces a detector response. The signal is expressed in nano-Refractive Index Unit (nRIU) and it corresponds to the difference between the refractive index of the analyte in the sample compartment and the mobile phase in the reference compartment.

The RI detector measures one of the basic properties of analyte in solution. Similar to the ELSD discussed in **Subheading 4.**, the RI detector could be considered as a true concentration detector, for which the response is independent of any chromophoric or fluorescent group and depends solely on refractive index of the analyte. The RI detected is a simple detector to operate and calibrate. It can be used to determine the contents of complex physiological fluids utilizing refraction properties of different components. In protein purification, the RI detector can determine elution of unknown proteins and impurities in a single determination without the need for measurements of absorption maxima of different analytes. Also, the analyte does not need to be large or small in mass or optically chromogenic. One of the limitations of the RI detector is its poor sensitivity compared to UV or fluorescence detectors. Because the RI detector compares the refractive index of analytes in the mobile phase and compared the mobile phase in reference compartment, this detector works best under isocratic elution conditions, which is its other limitation. However, unlike for ELSD, the mobile phase does not need to be volatile, and unlike UV-Vis detectors, the mobile phase does not need to be UV transparent.

6. Electrochemical Detection

6.1. Electrochemical Detectors

Electrochemical detection may be used for the analysis of many different groups of chemical compounds such as carbohydrates, aliphatic alcohols, aminoalcohols, sulfur-containing compounds, and others. In some cases, these compounds do not absorb UV light and may be difficult to "tag" with fluorescent and/or UV absorbing tags. Electrochemical detection may be used for analysis of many different compounds, but is rarely used if UV and/or fluorescence detection are possible. One notable exception is carbohydrates, which are frequently analyzed by electrochemical detection despite the fact that fluorescent tagging permits for very sensitive determination.

Both UV-Vis and fluorescence detection are undeniably the most common modes of detection in protein chemistry. These modes of detection permit the analysis of most compounds related to protein characterization and the determination of most compounds that are potential process residuals in protein manufacturing. In some specific cases, the compounds of interest may have very low (or no) UV absorbance and may be very difficult to chemically "tag" with the UV absorbing or fluorescent tag. Examples of such compounds are oxidized dithiothreitol (DTT), isopropyl-β-galactosidase (IPTG), quaternary ammonium salts, urea, guanidine, and polysorbates (Tweens), all of which may be process-related impurities in protein manufacturing. In such cases, electrochemical detection may provide an easy and very sensitive means of detection. Moreover, in the case of electrochemical detection, most samples can be analyzed "as is" without initial sample preparation and/or modification simplifying analytical procedures. Carbohydrates are another example of compounds frequently detected electrochemically, although in this case, fluorescent tags and chemistry for tagging in most cases is readily available.

Electrochemical detection utilizes reduction/oxidation of compounds enforced by an externally applied voltage. The detector measures the current (charge) changes that result from the electron flow resulting from the oxidation or reduction that occurs at the electrode. The full theory of electrochemical detection is well beyond the scope of this chapter and may be found elsewhere *(22,23)*. In practice, the most popular modes of electrochemical detection are pulsed amperometric detection and integrated pulsed amperometric detection (IPAD). In these modes, voltage is delivered to the electrodes in pulses (cycles) and it changes according to the preprogrammed pattern (waveform). The changes in the current flow are integrated over the selected part of the waveform and sent to the recording device. Each pulse (cycle) may last from a fraction of a second to several seconds.

Integrated pulsed amperometric detection sometimes utilizes complex waveforms that permit sensitive detection and "self-cleaning" of the electrode. Most electrochemical detectors are limited to three-step waveforms permitting for time-controlled three changes in voltage and selection of the integration part of the cycle. The voltage changes correspond to the detection, oxidation, and reduction potential selected for the analyte. Some electrochemical detectors such as ED50 manufactured by Dionex permit for programming of complex waveforms apparently increasing the sensitivity and reliability of detection. It should be noted that the development and optimization of a waveform is time-consuming, requires specialized instrumentation, and frequently needs additional optimization by "trial and error."

Most modern electrochemical detectors offer very low-volume flow cells with exchangeable electrodes. Most common electrodes are gold, platinum, silver, and glassy carbon. Some manufacturers offer porous carbon electrodes designed for the detection of selected groups of compounds such as catecholamines.

The linear dynamic range and the accuracy of electrochemical detectors are comparable to UV-Vis detectors and the detection limit is in the low-nanogram range. An additional advantage of electrochemical detectors is a possibility of selective detection of analytes that may be achieved by manipulating potential of the electrode(s).

6.2. General Considerations for Electrochemical Detection

High-performance liquid chromatography system used for the separation should provide solvent(s) flow free of metal ions. Some HPLC systems used with electrochemical detectors may require initial cleaning/passivation. The electrode should be chosen based on the published reports. For many applications related to protein analysis, a gold electrode is the most universal and versatile. Proper waveform should be selected based on existing information reported for the specific or structurally similar compounds. The electrode and flow cell should be clean. This is a very critical issue and both the electrode(s) and the cell should be frequently cleaned and polished (mechanically) using very fine sandpaper (400 or greater) and polishing pads. Reference pH electrode, if used, should be checked frequently for possible mechanical damage. The electrode should be partially filled with electrolyte. After each cleaning, the flow cell should be rinsed with alcohol and water to remove any solid particles, which could affect the cell's performance. Any spacers and gaskets used to assemble the cell should be inspected and have to be in excellent condition. A freshly cleaned cell may require several blank HPLC runs until it fully equilibrates. After completion of the analysis, the cell should be disconnected from power to avoid damage by extensive oxidation. Properly maintained electrochemical cell should provide very reproducible and reliable results even with HPLC runs lasting several days. In many cases, electrochemical detectors will be able to detect analytes at a very low concentration (parts per billion [ppb] level), which may be useful for analysis of process residuals.

7. Conductometric Detection

Conductometric detection is usually a separate mode of detection available with most electrochemical detectors (24,25). This mode of detection requires special instrumentation and supplies and is applicable to ionizable compounds. The most common application is detection of low levels of anions or cations. An example may be the detection of chloride or bromide ions. Conductometric detection has limited use in the analysis of protein and protein-related compounds. An example of conductometric detection in protein-related analyses might be the detection of guanidine. Guanidine is frequently used for protein isolation and its viral deactivation and can be detected at a low level as process impurity.

8. Mass Spectrometry

Mass spectrometry (MS) as applied to protein analysis is a rapidly evolving technology. Initially, mass spectrometers were designed as very specialized (and expensive) analytical tools for structural analysis of proteins. Technological progress led to a large variety of mass spectrometers suitable for many tasks related to protein analysis. At the same time, the cost of relatively simple instruments kept going down and the physical dimensions of the mass spectrometer have decreased. This created the possibility of using mass spectrometers as universal detectors in chromatographic separations. Total ion current measured by mass spectrometers yields a profile similar to that obtained with UV or other detectors and molecular-mass information is invaluable for protein chemist. For this reason, mass spectrometers have a good chance of becoming universal detectors in protein analysis and chromatography in general. Among their several applications in protein analysis is the determination of molecular mass of intact proteins, determination of posttranslation modifications like glycosylation, truncation, deamidation, and so forth, and the determination of primary structure of recombinant proteins *(26,27)*.

The full theory of operation of mass-spectrometry-based detectors is well beyond the scope of this chapter. Multiple vendors (Perkin Elmer Sciex, Micromass, Thermo Finnigan, etc.) offer a large variety of detectors based on different principles of operation (triple–quadrupole, ion trap, time-of-flight, etc.). Therefore, it would be very difficult to provide any universal set of suggestions regarding implementation and optimization of mass-spectrometry-based detectors in chromatographic separations. Properly installed and optimized MS detectors provide chromatographers with a great deal of information, and within several years, they will probably become standard detectors for many HPLC applications. It should be noted, though, that most mass spectrometry detectors have very low tolerance to salts, detergents, and certain solvents and chemicals (e.g., organic amines or TFA), limiting the choice of chromatographic systems that may be used with such detectors.

9. Specialized Detectors

A separate group of detectors originated from gas chromatography and detect specific elements. Examples of such detectors are HPLC nitrogen and sulfur detectors offered by ANTEK (www.antekhou.com/product/chrom/hplc). These detectors may be very useful in specific cases where UV, fluorescence, and other detectors cannot be readily used, especially if the percent content of nitrogen (or sulfur) in the compound of interest is relatively high. However, these detectors require oxygen and inert gas (argon or helium) supplies and cannot be used with solvent systems containing nitrogen (and/or sulfur) and/or high salt concentration. In addition, these detectors cannot accept flow rates exceeding 0.3 mL/min, which further limits chromatographic options. It is highly recommended that the HPLC system used with nitrogen/sulfur detectors be dedicated for this specific purpose and not be used with any nitrogen- and/or sulfur-containing solvents (acetonitrile, DMSO, trimethylamine, etc.). Despite all of the limitations, these detectors offer a selective and sensitive means of detection of specific compounds difficult to determine by any other technique.

Another example of specialized detectors is the Radiomatic series of detectors for measuring radioactivity in isotopically labeled proteins and peptides. These detectors are made by Perkin-Elmer and are available in three models (150TR, 610TR, and 625 TR). The detectors can work in stand-alone or in-line with HPLC system and are capable of monitoring single or dual-labeled radioisotopic samples. The detectors are equipped with preset energy windows for ^3H, ^{14}C, ^{35}S, ^{32}P, ^{125}I, and several other radioactive isotopes. This class of detectors are especially useful for identifying and quantifying sites of radioactive isotope incorporation in a protein. For example, when a protein is labeled with the ^{125}I isotope using IODO-GEN chemistry, the site most commonly labeled is tyrosine. The specific tyrosine residues that contain radioactive iodine and the extent of its incorporation could be determined using this detector. Another common application of this detector is in drug metabolism and pharmacokinetic analysis, where isotopically labeled metabolites are detected in physiological fluids.

References

1. Moore, S., and Stein, W. H. (1963) Chromatographic determination of amino acids by the use of automated recording equipment, in *Methods in Enzymology*, vol. 6 (Colowick, S.P. and Kaplan, N., eds.), Academic, San Diego, CA, pp. 819–831.
2. Snyder, L. R., Kirkland, J. J., and Glajch, J. L. (1997) *Practical HPLC Methods Development*, 2nd ed., Wiley, New York.
3. Kok, W. T. (1998) *Principles of Detection*, in Chromatographic Science Series, vol. 78, Marcel Dekker, Inc., New York, NY, pp. 43–168.
4. Grandhee, S. K.and Monnier,V. M. (1991) Mechanism of formation of the Maillard protein cross-link pentosidine. *J. Biol. Chem.* **266**, 11,649–11,653.
5. Pace, C. N., Vajdos, F., Fee, L., Grimsley, G., and Gray, V. (1995) How to measure and predict the molar absorption coefficient of a protein. *Protein Sci.* **4**, 2411–2423.
6. Lakowicz, J. R. (1999) Instrumentation for fluorescence spectroscopy, in *Principles of Fluorescence Spectroscopy*, 2nd ed., Kluwer Academic/Plenum, New York, pp.25–60.
7. De Antonis, K. M., Brown, P. R., and Cohen, S. A. (1994) High performance liquid chromatographic analysis of synthetic peptides using derivatization with 6-aminoquinolyl-*N*-hydroxysuccinimidyl carbamate, *Anal. Biochem.* **223**, 191–197.
8. Goto, J. (1990) Fluorescence derivatization, in *Detection-Oriented Derivatization Techniques in Liquid Chromatography* (Cazes, J., ed.) Marcel Dekker, New York, pp. 323–358, and references therein.
9. Vickrey, T. M. (1983) *Liquid Chromatography Detectors*, Marcel Dekker, New York.
10. Lingeman, H., Underberg, W. J. M., Takadate, A., and Hulshoff, A. (1985) Fluorescence detection in high performance liquid chromatography. *J. Liquid Chromatogr.* **8**, 789–874.
11. Eftink, M. R. (1991) Fluorescence quenching: theory and applications. *Topics Fluoresc. Spectrosc.* **2**, 53–126.
12. Yang, M., and Tomellini, S. A. (2000) An HPLC detection scheme for underivatized amino acids based on tryptophan fluorescence recovery. *Analy. Chim. Acta* **1–2**, 45–53.
13. Kratochvil, P. (1987) *Classical Light Scattering from Polymer Solutions*, Elsevier, Amsterdam.
14. Jackson, C., Nilsson, L. M., and Wyatt, P. J. (1989) Characterization of biopolymers using multi-angle light scattering detector with size exclusion chromatography. *J. Appl. Polym. Symp.* **43**, 99–114.
15. Gombotz, W. R., Pankey, S. C., Phan, D., et al. (1994) The stabilization of human IgM monoclonal antibody with poly(vinylpyrrolidone). *Pharm. Res.* **11**, 624–632.

16. Stolyhwo, A., Colin, H., and Guiochon, G. (1983) Use of light-scattering as a detector principle in liquid chromatography. *J. Chromatogr.* **265**, 1–18.

17. Guiochon, G., Moysan, A., and Holley, C. (1988) Influence of various parameters on the response factors of the evaporative light scattering detector for a number of non-volatile compounds. *J. Liquid Chromatogr.* **11**, 2547–2570.

18. Polymer Laboratories (2001) *PL-ELS 1000 and PL-ELS 1000μ Operator's Manual*, Polymer Laboratories, Amherst, MA.

19. Kermasha, S., Kubov, S., Safari, M., and Reid, A. (1993) Determination of positional distribution of fatty acids in butter fat triglycerides, *J. Chromatogr. Sci.* **70**,169–173.

20. Tommey, A. B., Dalrymple, D. M., Jasperse, J. L., Manning, M. M., and Schulz, M. V. (1997) Analysis of quaternary ammonium compounds by high performance liquid chromatography with evaporative light scattering detection. *J. Liquid Chromatogr. Related Technol.* **20**, 1037–1047.

21. Willard, H. H., Merritt, L. L., Dean, J. A., and Settle, F. A. (1981) Refractrometry and interferometry; polarimetry, circular dichroism, and optical Rotatory dispersion, in *Instrumental Methods of Analysis*, 6th ed., Wadsworth, Belmont, CA, pp. 403–407.

22. Lisman, J. A., Underberg W. J. M., and Lingeman, H. (1990) Electrochemical derivatization, in *Detection-Oriented Derivatization Techniques in Liquid Chromatography* (Cazes, J. ed.), Marcel Dekker, New York, pp. 283–322.

23. LaCourse, W. (1997) Pulsed Electrochemical Detection in High Performance Liquid Chromatography, Wiley, New York.

24. Smith, R. E. (1998) *HPLC of ions: Ion-Exchange Chromatography*, in Chromatographic Science Series, vol. 78, Marcel Dekker, New York, NY, pp. 365–411.

25. Dionex Corp. (1995) ED40 Electrochemical Detector Operator's Manual Dionex Corp document# 034855, Sunnyvale, CA Rev. 03.

26. Reinhold, V. N. (1998) Evaluation of glycosylation, in *Development in Biological Standardization* (Brown, F, Lubiniecki, A., and Murano, G., eds.), Karger, Basel, Vol 96, pp. 49–53.

27. Harris, R. J., Molony, M. S., Kwong, M. J., and Ling, V. T. (1996), Identifying unexpected protein modifications, in *Mass Spectrometry in the Biological Sciences* (Burlingame, A. L. and Carr, S. A. eds.), Humana, Totowa, NJ., pp. 333–350.

40

Peptide Proteomics

Dean E. McNulty and J. Randall Slemmon

1. Introduction

Proteomics involves the global analysis of protein expression in an organism, attained through linking the quantitation and characterization of the proteins in a cell and their temporal and spatial relationships within biological pathways, networks, and complex systems. Two-dimensional polyacrylamide gel electrophoresis (2-D PAGE) has been the predominant method for resolving complex protein mixtures; its interface with mass spectrometry has enabled its successful application in answering a wide range of important biological questions (1,2).

Several technical advancements have ensured the continued popularity of 2-D PAGE in proteomics applications (3–5). Significant improvements in resolution have been gained through the use of prefractionation techniques, narrow isoelectric focusing (IEF) zoom gels providing first-dimension separation within a single pH unit, and large gel formats up to 40 cm in size. Likewise, sensitivity enhancements have been achieved by employing immobilized pH gradient strips, allowing greater load capacities, superior protein solubilization cocktails for difficult protein classes, fluorescent staining and visualization methods with greater dynamic range, and improved image analysis and archiving software.

Despite these improvements, several shortcomings pertaining to 2-D PAGE methodology are repeatedly cited. Of primary concern is the visual underrepresentation of the vast complexity of the proteome on the gel. Contributing to this are poor dynamic range, which severely restricts the ability to detect low-copy-number and low-abundance proteins, selective losses of entire important protein classes such as hydrophobic integral membrane species, and the inability to separate and/or detect proteins at the pI and molecular weight extremities of the gel (polypeptides <10 kDa in molecular weight, highly basic species, etc.) (6). The robustness of the technique is significantly compromised by the inherent variability in the quality and consistency of 2-D PAGE separations between individuals and labs. Also, the technique remains relatively low throughput because of bottlenecks preventing seamless process automation, of which image analysis and archiving may be the greatest hindrance.

From: *Methods in Molecular Biology, vol. 244: Protein Purification Protocols: Second Edition*
Edited by: P. Cutler © Humana Press Inc., Totowa, NJ

These challenges have prompted the development by several independent laboratories of multidimensional chromatographic approaches to complement and perhaps eventually supplant 2-D PAGE as the preferred visualization of the proteome *(7–12)*. These techniques exploit the superb resolving power of peptide separations by liquid chromatography, as well as its inherent quantitative and automated nature, especially when directly interfaced to modern mass spectrometric analyzers. Typically, these experiments are performed as "shotgun proteomic"-type studies that identify proteins in a complex mixture via complete proteolytic digestion followed by the separation and analysis of large numbers of relatively small-sized peptide fragments. Peptide digests may be directly fractionated by reversed-phase liquid chromatography–mass spectrometry–mass spectrometry (LC-MS-MS), or alternately prefractionation schemes involving ion exchange chromatography *(10,13)* or affinity chromatography *(7,11)* have been used to reduce the complexity of the mixtures. Of particular note is the isotope-coded affinity tag (ICAT) approach, which as an enrichment strategy and quantitative tool for differential expression proteomics has generated considerable interest *(7)*. This system employs an ICAT reagent containing a thiol-reactive functional group, an isotopically heavy or light linker, and a biotin group used as an affinity capture tag. Initial reports also suggest that multidimensional chromatography of large, intact cellular proteins facilitated by the use of reversed-phase chromatography columns containing 1.5 μm nonporous silica C_{18}-coated beads (NP-C18) in the second dimension may be feasible *(14,15)*.

This chapter deals with the extension of multidimensional chromatographic methods to the global profiling of endogenous small proteins and peptides of <0.5 kDa to 8 kDa molecular weight from whole tissues—the classes of molecules particularly ill-suited to 2-D PAGE analysis because of their anomalous electrophoretic mobility and staining properties. Considering the recent trends toward chromatographic-based studies of the proteome, it is somewhat surprising, in light of their biological relevance and suitability to analysis, that the study of "peptidomes" has not yet attracted much attention. This class of molecules is represented by biological messengers such as peptide hormones, neuropeptides, cytokines, and enzyme inhibitors that permit communication between remote cells. Included are major histocompatibility (MHC) class I and II ligands that are presented on the cell surface for antigenic determinants in the T-cell-mediated immune system.

"Peptidomics" can be used to monitor selective protein degradation and turnover by the ubiquitin–proteasome, endosome–lysozome, or calpain–caspase systems. This may allow insight into changes in the expression of important regulatory proteins that are too short-lived to be viewed as intact molecules by 2-D PAGE *(16)*. The limited research presented has shown that peptide analysis can be used to identify biological markers associated with accelerated proteolysis in Alzheimer's disease and neurological insult *(17,18)*. In addition to work with whole tissues, studies using peptide banks prepared from hemofiltrate of blood plasma and other bodily fluids have proven rich in biologically interesting regulatory molecules *(19,20)*. It has been demonstrated that positive identification of *de novo* sequence data generated by MS-MS using expressed sequencing tag (EST) and genomic search databases is feasible, even without *a priori* knowledge of posttranslational modification or proteolytic processing of the endogenous peptide or protein precursor *(21)*. The following chapter introduces the quantitative aspects of endogenous peptide isolation and comparative profiling from whole-tissue samples.

2. Materials

2.1. Apparatus

High-performance liquid chromatography (HPLC) equipment should include an autosampler, elution gradient programmer with data processing software, binary solvent delivery system, ultraviolet (UV) absorbance detector, and automated fraction collector. The present study was performed using two HPLC systems comprised of Bio-Rad model AS-100 autosamplers, Beckman 126 Solvent Module, and Beckman 166 Detectors controlled by System Gold software, and Pharmacia Frac-100 fraction collectors. External contact closures to enable communication between the various components were established using a Beckman Remote Interface Module.

2.2. Tissue Extraction and Polypeptide Capture

1. Tekmar Tissuemizer with S25KR probe or equivalent.
2. 50 mM H_3PO_4, pH 2.1.
3. Sterivex-HV 0.45-μm filter unit (Millipore).
4. Sep-Pak Plus C_{18} bulk environmental resin (55–105 μm, 120 Å) (Waters).
5. C_4 Wide-pore butyl bulk resin (40 μm, 275 Å) (Baker).
6. HPLC-grade acetonitrile (Baker).
7. HPLC-grade water (Baker).
8. Trifluoroacetic acid (TFA), sequanal grade (Pierce).
9. Econopak disposable column (Bio-Rad).
10. Vacuum extraction manifold (Waters).
11. Bath-type sonicator (Bransonic).
12. Nylon Acrodisc 13 0.45-μm filter unit (Gelman).

2.3. Preparative Size-Exclusion Chromatography

1. Superdex 30 prep grade media (Amersham Biosciences).
2. XK 16/70 (1.6 × 70-cm) column (Amersham Biosciences).
3. 50 mM H_3PO_4, 150 mM NaCl, pH 2.5.
4. 1 g Sep-Pak Plus C_{18} environmental cartridge (55–105 μm, 120 Å) (Waters).
5. HPLC-grade acetonitrile (Baker).
6. HPLC-grade water (Baker).
7. Trifluoroacetic acid (TFA), sequanal grade (Pierce).
8. Vacuum extraction manifold (Waters).
9. Nylon Acrodisc 13 0.45-μm filter unit (Gelman).

2.4. Preparative Strong Cation-Exchange Chromatography

1. Vydac 400VHP 5 × 25-mm (0.5-mL), 5-μm, 900-Å sulfonic acid cation exchanger.
2. Buffer A: 10 mM Sodium phosphate, 25% (v/v) CH_3CN, pH 2.7.
3. Buffer B: 10 mM Sodium phosphate, 25% (v/v) CH_3CN, 1 M NaCl, pH 2.7.
4. 1 g Sep-Pak Plus C_{18} environmental cartridge (55–105 μm, 120 Å) (Waters).
5. HPLC-grade acetonitrile (Baker).
6. HPLC-grade water (Baker).
7. Trifluoroacetic acid (TFA), sequanal grade (Pierce).
8. Vacuum extraction manifold (Waters).

2.5. Profiling of Peptides by Analytical Reversed-Phase HPLC

1. Vydac C18 4.6 × 250 mm, 5μm, 300 Å.
2. Buffer A: 0.06% (v/v) TFA.

3. Buffer B: 0.054% 9 (v/v) TFA, 80% (v/v) CH_3CN in H_2O.
4. 1g Sep-Pak Plus C18 environmental cartridge (55–105 μm, 120 Å) (Waters).
5. HPLC grade acetonitrile (Baker).
6. HPLC grade water (Baker).
7. Trifluoroacetic acid (TFA), sequanal grade (Pierce).

3. Methods

3.1. Tissue Extraction and Polypeptide Capture

The successful profiling of endogenous peptides from biological samples first requires peptide capture in a quantitative and reproducible manner from the source material. For whole-tissue samples, this typically involves homogenization and extraction of polypeptides in an acidic aqueous solution, followed by capture and elution from a reversed-phase, solid-phase extraction support in batch-mode fashion. This approach can also be readily extended to the analysis of biological fluids, cell culture lysates, and conditioned media through acidification followed by clarification and direct peptide capture. The procedure reported has been optimized for the analysis of peptides derived from whole rat liver tissue samples.

1. One gram Wet weight of frozen rat liver tissue is homogenized as a 2% (w/v) extract in ice-cold 50 mM H_3PO_4, pH 2.1, using a Tekmar Tissuemizer equipped with a miniprobe at 20,000 rpm for 3 min. The extract is centrifuged at 5500g for 90 min at 4°C (*see* **Notes 1** and **2**).
2. The clarified supernatant is filtered through a Millipore Sterivex-HV 0.45-μm unit. One gram of Waters Sep-Pak Plus C_{18} bulk environmental resin (55–105 μm, 120 Å) is wetted with 5 mL ethanol and washed with an addition of 45 mL of 0.1% (v/v) TFA. The slurry is centrifuged briefly at 5500g to pellet the resin. The wash buffer is carefully decanted off and the clarified, filtered tissue extract is added. The mixture is batch incubated overnight with end-over-end rocking at 4°C. The slurry is packed a Bio-Rad Econopak disposable column and the unbound material is reserved. An upper frit is added, and using a vacuum manifold apparatus, the column is washed with 25 mL of 0.1% (v/v) TFA, and the bound peptides are eluted in 5 mL of 60% (v/v) CH_3CN, 0.1% (v/v) TFA. A slow flow is maintained such that individual drops are visible in the effluent. The eluent is concentrated by Speedvac to dryness.
3. One gram of Baker C_4 wide-pore butyl bulk resin (40 μm, 275 Å) is wetted with 5 mL ethanol and washed with an addition of 45 mL of 0.1% (v/v) TFA. The slurry is centrifuged briefly at 5500g to pellet the resin. The wash buffer is carefully decanted off, and the unbound extract from **step 2** is added. The mixture is batch incubated for 6 h with end-over-end rocking at 4°C. The slurry is packed into a Bio-Rad Econopak disposable column and the unbound material from the dual resin capture is discarded. An upper frit is added, and using a vacuum manifold apparatus, the column is washed with 25 mL of 0.1% (v/v) TFA and the bound peptides are eluted in 5 mL of 60% (v/v) CH_3CN, 0.1% (v/v) TFA. A slow flow is maintained such that individual drops are visible in the effluent. The eluent is combined with the dried peptide extract from **step 2** and the pooled peptides are concentrated by Speedvac to dryness (*see* **Note 3**).
4. The peptide extract is resuspended in 5mL of 0.1% TFA and vortexed and sonicated aggressively to dissolve completely. Then, 1.2 mL of the sample is microcentrifuged at 15,000g for 2 min, and the supernatant is filtered through a Gelman Nylon Acrodisc 13 0.45-μm filter in preparation for size-exclusion chromatography.

3.2. Preparative Size-Exclusion Chromatography

Polypeptides captured from rat liver homogenate extracts are further enriched for low molecular weight components by size exclusion chromatography using Superdex 30

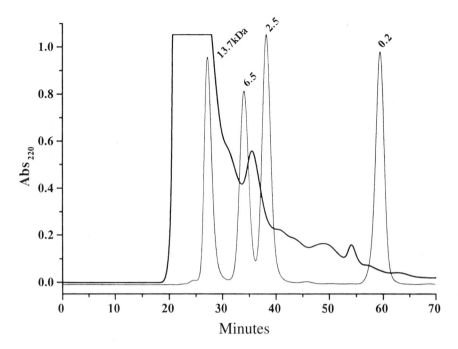

Fig. 1. Superdex 30 size-exclusion chromatography of polypeptides. Thick line: UV absorbance profile from preparative fractionation of captured polypeptides from 200 mg wet-weight equivalent of crude rat liver extract. The area of the chromatogram from 30 to 70 min represents the pooled region of interest containing peptide components of mass distribution <0.5 to 8 kDa. Thin line: Molecular-mass standards represented by RNAse (13.7 kDa), aprotinin (6.5 kDa), insulin chain A oxidized (2.5 kDa), and Phe-Ala (0.2 kDa).

prep grade media. Judicious selection of a single pool from the column results in a well defined polypeptide library containing components ranging from <0.5kDa to 8kDa in molecular mass.

1. One hundred ten milliliters of Superdex 30 prep-grade chromatography media are packed into a 70 × 1.6-cm column housing using standard procedures. The column is equilibrated in buffer containing 50 mM H$_3$PO$_4$, 150 mM NaCl, pH 2.5. Then, 1.0 mL of sample containing captured polypeptides from **step 4** of **Subheading 3.1.** (the equivalent of 200 mg wet-weight tissue homogenate) is injected via autosampler onto the HPLC. The column is developed isocratically for 70 min (1.1 column volumes) as detailed below. The void volume from 20 to 30 min containing high-molecular-mass polypeptides (>8 kDa) is diverted via waste valve on the fraction collector, and a single pool from 30 to 70 min (0.5–1.1 column volumes) representing the included polypeptide volume is retained (*see* **Fig. 1** and **Notes 4** and **5**).

   Size-Exclusion HPLC:
   Column: Superdex 30 Prep 70 × 1.6-cm column (110 mL).
   Buffer: 50 mM Sodium phosphate, 150 mM NaCl, pH 2.5.
   Flow: 1.8 mL/min (55 cm/h).
   Detection: A$_{220\ nm}$.

2. The fraction containing the low-molecular-weight-sized pool is applied to a prewetted (ethanol) and equilibrated (0.1% [v/v] TFA) 1-g Waters Sep-Pak Plus C$_{18}$ environmental cartridge (55–105 μm, 120 Å) using a vacuum manifold apparatus. A slow flow is maintained such that individual drops are visible in the effluent. The cartridge is washed with 10

mL of 0.1% (v/v) TFA and eluted in 3.5 mL of 60% (v/v) CH_3CN, 0.1% (v/v) TFA. The eluent is concentrated by a Speedvac to dryness.

3. The low-molecular-weight pool is resuspended in 1.150 mL of 10 mM sodium phosphate, 25% (v/v) CH_3CN, pH 2.7, and filtered through a Gelman Nylon Acrodisc 13 0.45-μm filter unit in preparation for strong cation-exchange chromatography.

3.3. Preparative Strong Cation-Exchange Chromatography

The pooled material from the preceding size-exclusion chromatography step results in a well-defined polypeptide library containing components of molecular mass <0.5 to 8 kDa. The sized polypeptide pool is further fractionated into pools of differing charge by high-performance cation-exchange chromatography. The gradient is optimized to provide similar amounts of polypeptide mass in each pool. Five individual pools (cuts) of increasing basicity are produced for each sample. Analogous to 2D gels, the charge pools can be thought to represent narrow ampholyte ranges within the isoelectric focusing first dimension.

1. One milliliter of the resuspended, filtered low-molecular-weight pool from **step 3** of **Subheading 3.2.** (the equivalent of 175 mg wet-weight tissue homogenate) is injected for cation-exchange HPLC via autosampler. The column is developed using the conditions detailed below, and a fraction collector is programmed to automatically collect five pools (7.5 mL cuts each) of increasing basicity over the duration of the 75-min gradient: (1) 0–15 min (unbound—22 mM NaCl), (2) 15–30 min (22–55 mM NaCl), (3) 30–45 min (55–88 mM NaCl), (4) 45–60 min (88–188 mM NaCl), and (5) 60–75 min (188–1000 mM NaCl) (*see* **Fig. 2** and **Note 6**).

    Strong Cation-Exchange HPLC:
    Column: Vydac 400VHP 5 $\times$ 25-mm (0.5-mL), 5-μm, 900-Å sulfonic acid cation exchanger.
    Buffer A: 10 mM Sodium phosphate, 25% (v/v) CH_3CN, pH 2.7.
    Buffer B: 10 mM Sodium phosphate, 25% (v/v) CH_3CN, 1 M NaCl, pH 2.7.
    Flow: 0.5 mL/min (1 column volume per minute).
    Gradient:
    0% B—5 min.
    0–10% B—45 min.
    10–25%B—15 min.
    25–100% B—5 min.
    Detection: $A_{220\ nm}$

2. The pools eluted from the cation exchanger are partially concentrated by a Speedvac to remove CH_3CN and are diluted to 10 mL with 0.1% (v/v) TFA. The pools are desalted by applying to a prewetted (ethanol) and equilibrated (0.1% [v/v] TFA) 1-g Waters Sep-Pak Plus C_{18} environmental cartridge (55–105 μm, 120 Å) in parallel using a vacuum manifold apparatus. A slow flow is maintained such that individual drops are visible in the effluent. The cartridge is washed with 10 mL of 0.1% (v/v) TFA and eluted in 3.5 mL of 60% (v/v) CH_3CN, 0.1% (v/v) TFA. The eluents are concentrated by a Speedvac to dryness.

3. The five pools (cuts) from the strong cation-exchange chromatography step are resuspended in 1.150 mL of 0.1% (v/v) TFA in preparation for peptide profiling by analytical reversed-phase HPLC.

3.4. Profiling of Peptides by Analytical Reversed-Phase HPLC

The five charge-separated cation-exchange pools are profiled by analytical C_{18} reversed-phase HPLC. Peptides are eluted with an increasing TFA/CH_3CN-containing gra-

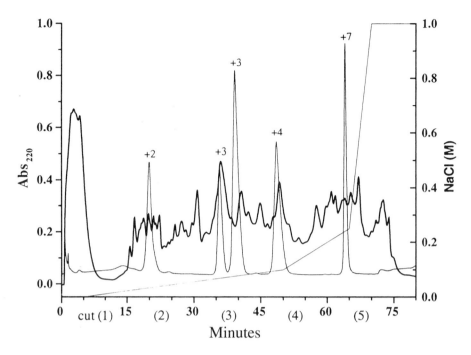

Fig. 2. Strong cation-exchange chromatography of polypeptides. Thick line: UV absorbance profile from preparative fractionation of captured and sized polypeptides from 175 mg wet-weight equivalent of crude rat liver extract. The five peptide pools (cuts) of increasing basicity collected in 15-min increments are annotated at the bottom of the chromatogram. Thin line: Peptide standards of known positive charge at pH 2.7 represented by Arg(8) vasopressin (+2), bradykinin (+3), angiotensin III (+3), angiotensin I (+4), and ACTH fragment (1–17) (+7).

dient that has been optimized to provide the greatest peak resolution for each pool (cut). The elution time of individual components is based on increasing hydrophobicity. Analogous to gel-based proteomics, the chromatogram trace is the "readout" similar to a 2D gel image. Individual peaks and their associated elution times correspond to a "spot," whereas the height/area absorbance values represent spot intensity. The chromatography data analysis software is able to integrate approx 750–1000 discrete UV-absorbing peaks over the set of five runs per tissue sample. For small study sets, chromatograms of corresponding cuts from differentially treated samples may simply be inspected visually to detect changes. To aid interpretation of larger sample sets, statistical analysis packages such as those described in **ref. *18*** can be used. Elution time and absorbance height/area identifiers can be converted to DIF file format and placed in spreadsheet form as single observations. Statistical analysis software packages such as SAS can be used to sort and group like observations from analogous chromatograms representing different treatments, disease states, and so forth. The primary criteria for placement rely on retention time and nearest-neighbor peaks.

1. One milliliter of each of the five pools (cuts) from the strong cation-exchange chromatography **step 3** of **Subheading 3.3.** (150 mg wet-weight equivalent) are injected via autosampler onto the HPLC. The gradients detailed below have been optimized for each cut to provide maximum resolution of chromatographic peaks (*see* **Fig. 3** and **Notes 7–10**).

Analytical reversed-phase HPLC:
   Column: Vydac C$_{18}$ 4.6 × 250 mm, 5 μm, 300 Å.
   Buffer A: 0.06% (v/v) TFA.
   Buffer B: 0.054% (v/v) TFA, 80% (v/v) CH$_3$CN in H$_2$O;
   Flow: 0.5 mL/min.
   Detection: A$_{214 nm}$.
   Gradient: Cuts 1 and 2:
      5% B—10 min
      5–15% B—10 min
      15–50% B—90 min
      50–90% B—10 min;
   Cut 3:
      5% B—10 min
      5–20% B—10 min
      20–50% B—90 min
      50–90% B—10 min;
   Cut 4:
      5% B—10 min
      5–25% B—10 min
      25–50% B—90 min
      50–90% B—10 min;
   Cut 5:
      5% B—10 min
      5–30% B—10 min
      30–50% B—90 min
      50–90% B—10 min.

4. Notes

1. Fifty millimolars of H$_3$PO$_4$, pH 2.1, was found to be an excellent solvent for the extraction of acid-soluble, low-molecular-weight polypeptides from liver tissue samples. Peptide library representation and recovery exceeded those obtained using classical acid–organic extraction techniques (e.g., acidified ethanol [ethanol, 0.8 *N* HCl, 3:1 v/v] or 70% v/v acetone, 1 *M* acetic acid, 20 m*M* HCl]) that may be better suited to particular peptide subclasses. It is recommended that upon initiating a peptide profiling study, various extraction buffers be screened for optimal yields.

2. Minimization of endogenous protease activity is a consideration when preparing soluble protein lysates from tissue samples. Stability studies indicated that negligible proteolytic activity was observed when liver tissue polypeptides were homogenized and extracted in ice-cold 50 m*M* H$_3$PO$_4$, pH 2.1. Boiling of the tissue extract, a technique often employed to deactivate proteases, resulted in substantial losses of peptides based on amino acid analysis and the complexity of the HPLC chromatogram traces. This may be the result of coprecipitation with irreversibly denatured insoluble protein aggregates.

3. The use of solid-phase extraction media in cartridge format resulted in inefficient polypeptide capture. Linking multiple cartridges demonstrated that binding capacity was not exceeded, suggesting that complete binding equilibrium and internal pore access to the resin was not realized in a single passage of the clarified crude extract. The overnight batch incubation process outlined resulted in approximately fourfold improvement in peptide recovery. The dual resin capture approach was developed based on studies showing that the Waters Sep-Pak Plus C$_{18}$ bulk environmental resin (55–105 μm, 120 Å) favored the capture of small, relatively hydrophilic peptides of <2 kDa mass, whereas the Baker C$_4$ wide-

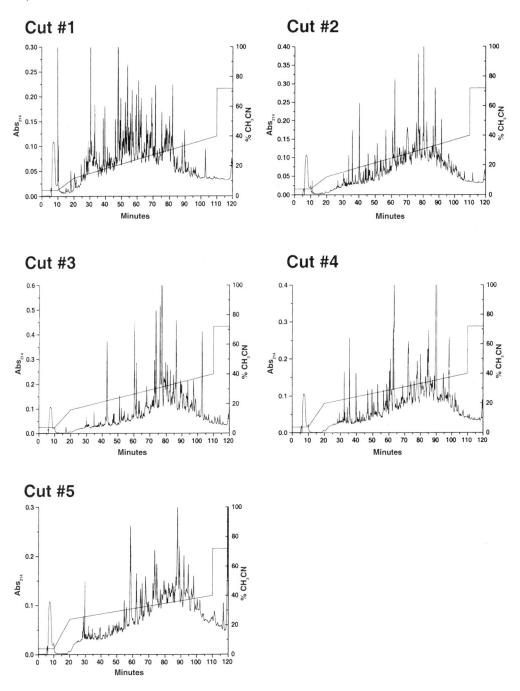

Fig. 3. Profiling of endogenous liver polypeptides by analytical C_{18} reversed-phase chromatography. UV absorbance profiles of five cation exchange cuts of increasing basicity from 150 mg wet-weight equivalent of processed rat liver extract. Typically, 750–1000 discrete UV absorbing peaks are detected and integrated across the five chromatograms.

pore butyl bulk resin (40 μm, 275 Å) excelled in the capture of larger polypeptides of mass approx 6–20 kDa. Both resins performed well in the intermediate mass range. Following production of the peptide pool by preparative sizing, no losses were observed when desalting and concentration steps were performed using solid-phase extraction cartridges. Amino acid analysis indicated that phosphoric acid extraction of rat liver tissue yielded an average of 160 mg/g wet-weight total soluble protein. Of this, typical yields of low-molecular-weight polypeptides following batch resin capture were approx 35 mg/g wet weight prior to fractionation, and approx 8 mg/g wet weight of defined peptide library was obtained following preparative size-exclusion chromatography.

4. Superdex 30 prep grade is a preparative gel filtration media containing a composite matrix of dextran and agarose. The useful fractionation range is <10,000 Daltons. It is recommended that the buffer used for gel filtration chromatography contain an ionic strength equivalent to 0.15 M NaCl or greater to avoid ionic interactions with the gel matrix. Prior to profiling studies, the column should be carefully calibrated using known molecular-mass standards in order to select the appropriate included volumes for peptide library refinement.

5. If desired, several pools of differing molecular mass may be collected from the Superdex 30 gel filtration step. Processing in parallel would result in a three-dimensional chromatography profile, with maps differing by size, charge, and hydrophobicity properties. The entire profiling process may be linked and largely automated through use of a chromatography system equipped with multiple column switching valves and buffer ports such as those found on an Integral 100Q Multidimensional Biospecific HPLC system (Perseptive Biosystems) or equivalent.

6. Vydac VHP ion-exchange resins are polymeric matrices composed of spherical polystyrene-divinylbenzene (PS-DVB) containing a proprietary hydrophilic surface derivatized with sulfonic acid. Alternatively, silica-based cation-exchange supports such as PolySULFOETHYL Aspartamide may be suitable. The inclusion of 25% (v/v) CH_3CN was found to reduce interactions of noncharged residues with the matrix, dramatically improving resolution and providing a separation mode based solely on net charge. The use of multiply charged peptide standards to evaluate column performance is advisable. Gradient conditions were optimized for resolution and the separation of multiple-charge pools containing similar peptide mass.

7. Analytical C_{18} reversed-phase HPLC gradients have been optimized for resolution based on the number of assigned integrated peaks for each cut. Because the strong cation-exchange HPLC separation is based on increasing absolute charge, it follows that the more highly basic pools (cuts) generally contain higher-molecular-weight polypeptides. These display greater hydrophobicity on the C_{18} matrix, so gradients of successive cuts are started at a progressively higher concentration of organic modifier (buffer B).

8. Buffers should be thoroughly degassed and maintained under helium to maximize retention time reproducibility. Less than 0.15 min variability in peak time assignment should be readily obtainable even for extended gradient runs. Height/area reproducibility of integrated peaks was found to be <10% for single homogenates processed in parallel (measure of system/analyst reproducibility), <15% for processing of multiple tissue sections from a single subject (measure of intratissue heterogeneity), and <25% for profiles from multiple tissue samples (measure of biological variability between normal subjects (*see* **Fig. 4.**)

9. The eluent from the analytical C_{18} reversed-phase profiling step may be fractionated for off-line analyses such as MALDI-MS (matrix-assisted laser desorption ionization mass spectrometry), Edman sequencing, amino acid analysis, immunoassay, and so forth. Alternately, direct interface to mass spectrometers using an electrospray ionization source (ESI) provides an ideal means to simultaneously obtain molecular-mass information. Systems employing quadrupole, time-of-flight (TOF), and ion trap mass analyzers are most common.

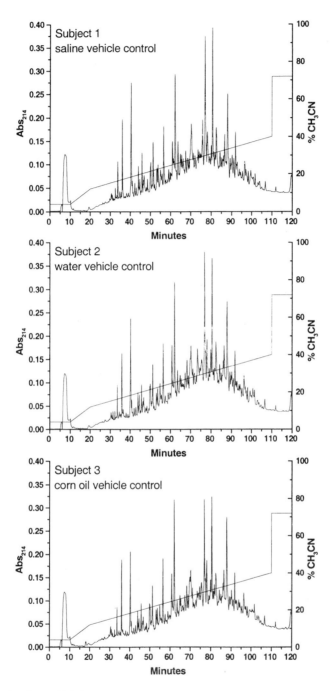

Fig. 4. Determination of biological variability between normal animal subjects. Comparison of rat liver RP-HPLC profiles from strong cation exchange cut 2 for saline-, water-, and corn oil-treated control vehicle animals. Peak height/area percent variability was typically less than 25% for all integrated peaks. Retention time reproducibility for individual peaks was consistently within 0.15 min (*see* **Note 8**).

LIVERPOOL
JOHN MOORES UNIVERSITY
AVRIL ROBARTS LRC
TEL. 0151 231 4022

Additionally, mass selection followed by CID (collision-induced dissociation) of peptides can be used to obtain primary structural information, which can be searched against data-bases, enabling positive identification.

10. Edman sequencing data of selected fractions from the C_{18} reversed-phase HPLC profile in-dicate that the major peaks can contain greater than 100 pmol of the most abundant peptide components. This suggests that when interfaced with a mass spectrometer, profiling with a dynamic range of four to five orders of magnitude may be obtainable, permitting detection and quantitation of rare, low-abundance polypeptide species.

References

1. Aebersold, R. and Goodlett, D. R. (2001) Mass spectrometry in proteomics. *Chem. Rev.* **101**, 269–295.
2. Mann, M., Hendrickson, R. C., and Pandey, A. (2001) Analysis of proteins and proteomes by mass spectrometry. *Annu. Rev. Biochem.* **70**, 437–473.
3. Ong, S. E. and Pandey, A. (2001) An evaluation of the use of two-dimensional gel elec-trophoresis in proteomics. *Biomol. Eng.* **18**, 195–205.
4. Herbert, B. R., Harry, J. L., Packer, N. H., Gooley, A. A., Pedersen, S. K., and Williams, K. L. (2001) What place for polyacrylamide in proteomics? *Trends Biotechnol.* **19(10)**, S3–S9.
5. Issaq, H. (2001) The role of separation science in proteomics. *Electrophoresis* **22**, 3629–3638.
6. Corthals, G. L., Wasinger, V. C., Hochstrasser, D. F., and Sanchez, J. C. (2000) The dynamic range of protein expression: A challenge for proteomic research. *Electrophoresis.* **21**, 1104–1115.
7. Gygi, S. P., Rist, B., Gerber, S. A., Turecek, F., Gelb, M. H., and Aebersold, R. (1999) Quan-titative analysis of complex protein mixtures using isotope-coded affinity tags. *Nature Biotechnol.* **17**, 994–999.
8. Link, A., Eng, J., Schieltz, D. M., et al. (1999) Direct analysis of protein complexes using mass spectrometry. *Nature Biotechnol.* **17**, 676–682.
9. Washburn, M. P., Wolters, D., and Yates III, J. R. et al. (2001) Large-scale analysis of the yeast proteome by multidimesional protein identification technology. *Nature Biotechnol.* **3**, 242–247.
10. Davis, M. T., Beierle, J., Bures, E. T., et al. (2001) Automated LC-LC-MS-MS platform using binary ion-exchange and gradient reversed-phase chromatography for improved pro-teomic analysis. *J. Chromatogr. B.* **752**, 281–291.
11. Riggs, L., Sioma, C., and Regnier, F. E. (2001) Automated signature peptide approach for proteomics. *J. Chromatogr. A.* **924**, 359–368.
12. Moseley, M. A. (2001) Current trends in differential expression proteomics: isotopically coded tags. *Trends Biotechnol.* **19(10)**, S10–S16.
13. Wolters, D. A., Washburn, M. P., and Yates, J. R., III. (2001) An automated multidimensional protein identification technology for shotgun proteomics. *Anal. Chem.* **73**, 5683–5690.
14. Wall, D. B., Kachman, M. T., Gong, S., et al. (2000) Isoelectric focusing nonporous RP HPLC: a two-dimensional liquid-phase separation method for mapping of cellular proteins with identification using MALDI-TOF mass spectrometry. *Anal. Chem.* **72**, 1099–1111.
15. Unger, K. K., Racaityte, K., Wagner, K., et al. (2000) Is multidimensional high performance liquid chromatography (HPLC) an alternative in protein analysis to 2D gel electrophoresis? *J. High Resolulot. Chromatogr.* **23(3)**, 259–265.
16. Nakai, K. (2001) Review: Prediction of in vivo fates of proteins in the era of genomics and proteomics. *J. Struct. Biol.* **134**, 103–116.

17. Slemmon, J. R., Wengenack, T. M., and Flood, D. G. (1997) Profiling of endogenous peptides as a tool for studying development and neurological disease. *Biopolymers* **43**, 157–170.
18. Slemmon, J. R. and Flood, D. G. (1992) Profiling of endogenous brain peptides and small proteins: methodology, computer-assisted analysis, and application to aging and lesion models. *Neurobiol. Aging* **13**, 649–660.
19. Schulz-Knappe, P. and Schrader, M. (1996) Systematic isolation of circulating human peptides: the concept of peptide trapping. *Eur. J. Med. Res.* **1**, 223–236.
20. Schrader, M. and Schulz-Knappe, P. (2001) Peptidomics technologies for human body fluids. *Trends Biotechnol.* **19(10)**, S55–S60.
21. Verhaert, P., Uttenweiler-Joseph, S., de Vries, M., Loboda, A., Ens, W., and Standing, K. G. (2001) Matrix-assisted laser desorption/ionization quadrapole time-of-flight mass spectrometry: an elegant tool for peptidomics. *Proteomics* **1**, 118–131.

41

Multidimensional Liquid Chromatography of Proteins

Rod Watson and Tim Nadler

1. Introduction

The term "multidimensional liquid chromatography" (MDLC) simply means the linking together of more than one column step. It is common to associate linking of column steps with combining two or more steps in a purification procedure into one simple protocol, but this belies both the range of possibilities that this technique provides and also the technical difficulties associated with method linking. In general, the use of MDLC can be split into three main categories:

- Analytical: the automatic analyses of fractions collected from a column run.
- Preparative: the use of two or more linked columns as part of a purification procedure.
- Procedural: the combining of column technologies to facilitate experimental procedures.

Each of these will be considered separately, as they all have their own individual requirements; however there are basic principals and needs that apply to all three.

The first procedural need that has to be addressed is that of buffer compatibility; it is essential when directing the flow from one column to another that the eluant used on the first column be suitable for use as a loading buffer with the second column. It is this apparently simple but vital need that often proves the stumbling point in designing MDLC procedures. A variety of columns such as size exclusion and ion exchange can be used to manipulate buffer compositions between columns (1) and there are ways of manipulating the composition of buffer streams (2) these are far from simple and require complex instrumentation and, as such, are often impractical.

The second general requirement is that of buffer manipulation. In order to be able to properly equilibrate, elute, and clean multiple, often-diverse columns, a range of buffer conditions will be required. The capacity of the chromatography instrument to utilize multiple buffers and blend them appropriately is paramount, if buffer limitations are not going to limit the experimental design. Again, there are often ways of increasing the number of buffer options available on an instrument, usually involving additional cost and increasing complexity, but the software has to have the flexibility to be able to make use of the extra buffers available.

The third general requirement is that of column switching. Any two complementary columns will have very different equilibration, elution, and cleaning requirements and

From: *Methods in Molecular Biology, vol. 244: Protein Purification Protocols: Second Edition*
Edited by: P. Cutler © Humana Press Inc., Totowa, NJ

it may be that the buffers used for one column may be destructive if used on the other. For this reason alone, it is essential that the columns used can be independently switched in and out of the flow path at will. Many chromatography instruments now allow for the addition of extra switching valves, but, again, an extra cost is involved and often the control of these valves is sometimes less than ideal. The best solution is to use instruments already configured with multiple switching valves.

Keeping these basic requirements in mind, we will now discuss the additional needs of the three main categories of use described earlier.

1.1. Analytical Use of MDLC

Essentially, this is utilizing a highly defined column procedure to identify the protein of interest in the collected fractions from an initial column run. An analytical column method is used, whereby a small aliquot of each collected fraction is loaded separately onto the analytical column and then eluted. The presence or absence of a specific eluted peak indicates whether or not the fraction contains the target protein, eliminating the need for time-consuming manual analysis of the fractions. Although simple in principal, the technique can only be used with highly sophisticated instrumentation that has the ability to collect fractions and then automatically reinject a portion of each fraction onto a second column. This robotic sample handling and method-linking capability is only available from a very limited number of manufacturers.

1.2. Preparative Use of MDLC

Regardless of whether the intention is to target the purification of a specific molecule or to identify every component in a complex sample, the same principles apply and have to be given due consideration.

Many purification procedures designed to purify specific proteins contain a series of column steps to produce the target protein in sufficient quantity and purity, but these are frequently done as a series of independent procedures. The ability to combine two or more steps into one procedure has obvious benefits in terms of time and labor and also, in the case of labile targets, an increase in yield. Although simple, in principle, the application of multiple column steps to a single procedure requires careful planning if it is to be successful. For example, the directing of a particular peak or bunch of peaks to a second column while eliminating other closely eluting species requires an exact knowledge of the elution pattern from column one for this to be achieved successfully. Careful selection of the order of use of chromatography modes is important. The use of two sequential ion exchange steps is unlikely to succeed, because the salt used in eluting the first column will probably prevent the target binding to the second column. However the use of a salt-tolerant mode, such as affinity following an ion-exchange step, is likely to be much more successful.

1.3. Procedural Use of MDLC

The advancement of column technologies, allowing a wider and wider range of ligands to be coupled to increasingly robust media, has enabled the development of chromatographic techniques that replace or augment traditional bench-based techniques. A classic example of this can be found in the area of peptide mapping; traditionally, pro-

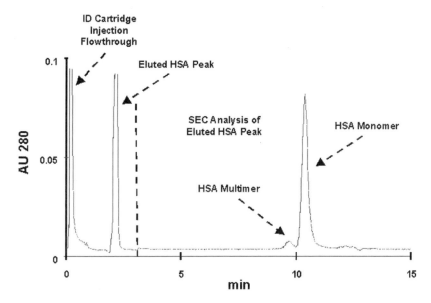

Fig. 1. Analysis of HSA monomers and dimers by affinity chromatography and SEC. Column: ultrasphere OG, 7.5 mm diameter/300-mm long; ID cartridge: anti-HSA, 2.1-mm diameter/30 mm long; flow rate: 1 mL/min; buffer/eluent: phosphate-buffered saline (PBS)/12 mM HCl; Sample: 20 μL (100 μg) HSA standard. Use of both affinity and SEC steps in an MDLC manner allows the separation of HSA monomers and dimers, which is not seen using the techniques independently.

teins were digested overnight by dilute enzyme solutions prior to separation by reversed-phase chromatography. The immobilization of proteolytic enzymes onto chromatography media has resulted in a technique whereby proteins can now be digested on-column in a matter of minutes, the digest is directed onto a reversed-phase column, and peptide separation is achieved in 2 h as opposed to 2 d (*3*). The use of immobilized antibodies has led to the development of chromatographic immunoassays that, combined with other analytical columns, give information on isotypes or variants (*4*) (e.g., analysis of human serum albumin [HSA] dimers from a complex serum by affinity capture and size-exclusion chromatography [SEC]). SEC alone and affinity capture alone is incapable of separating HSA dimers and monomers, but if you affinity capture all of the HSA and elute onto the SEC column, you can classify HSA into monomers and multimers with ease (*see* **Fig. 1**) (Nadler, personal communication).

A rapidly expanding area of biology is that of proteomics and it is in this area that MDLC will have a crucial role to play. Currently, the principal technique used to separate proteins for mass spectrometry is that of two-dimensional (2-D) gel electrophoresis. Although a powerful and widely used technique, it has major drawbacks when looking at certain protein classes (*5*). Being based on gel electrophoresis, it is unable to separate so-called "gel-incompatible" proteins (i.e., those that are very small, large, acidic, basic, hydrophobic, etc.), which clearly leaves a large gap in the proteome jigsaw. The ability of chromatography to handle all protein classes means that it can be used to handle the gel-incompatible proteins, and the ability to link two or more column steps automatically allows for the rapid simplification of often very complex mixtures. In the high-throughput world of current proteomics, this level of automation is important if timescales are to be met. The careful pairing of orthogonal separation modes allows the targeting and separation of specific protein types; for example, using a cation-exchange step at very high pH will select for extremely basic proteins and coupling of this directly to an analytical technique such as reversed phase often provides sufficient separation for the direct injection into a mass spectrometer *(6)*.

2. Materials

2.1. Buffers

There is no difference in principle between buffers used for single-column chromatography and those used for MDLC. All have to be made up fresh, from the highest-quality reagents and filtered through a 0.2-μm filter before use. Those buffers that are going to be used for reversed phase also need helium sparging before use. The buffers required for each separation have to be determined empirically.

2.2. Columns

Although virtually any column can, in theory, be used for MDLC, there are some important practical considerations to be taken into account when choosing columns.

Two columns in series will generate a higher back-pressure than either on its own, and when linking two columns in series, only one pressure limit can be set, which, in order to protect the columns, is normally the lower of the two column values. This can impact the flow rates that can be used forcing the operator to reduce flow rates and subsequently increasing run times. Attempts to disregard the pressure setting and run at "the normal rate" leaves the lower pressure column unprotected should a blockage occur. The best results tend to be obtained when columns of roughly equal pressure limits are used, try to avoid combining a column with a 100-psi pressure limit with a column of 5000-psi limit. Many high-pressure medias are now available so balancing column types should not present a problem.

Another important consideration is that of buffer compatibility; not all media are suitable for use with all buffers. In order to prolong column performance, care must be taken to ensure that the media used is stable in all of the buffers with which it will come into contact. For example, if you are going to elute from an anion column at pH 8.5 directly onto a reversed-phase column, then the high pH would preclude the use of silica in the second column.

Column capacities are also important and care should be taken to ensure that the relative capacities of the columns used are compatible. Column size alone is not a good guide, as media from different manufacturers have vastly different capacities, nor is the declared capacity of the media, as capacities determined by bulk capture can be up to 10 times higher than those seen under flow conditions. An MDLC experiment will generally use decreasing column sizes as you proceed through the procedure. The first column must have enough capacity for everything. The second column only needs to be large enough to accommodate the elution from the first column. It is important to decrease the size of the column so that the sample remains concentrated for detection and is subjected to the minimum surface area to reduce nonspecific binding.

2.3. Instrumentation

It is the instrument that determines whether or not MDLC is even possible. The best instruments to use are those that already have multiple column-switching valves, multiple buffer lines, complex method writing capability, and, in some cases, robotic sample handling, as standard. The great advantage of these systems is that they are already set up to do MDLC-type experiments and all the operator has to do is write the method. That said, MDLC experiments can very successfully be carried by combining elements of different systems to provide the required hardware; however, the operation of these combinations tends to be much less straightforward.

The use of MDLC in the analytical mode to identify fractions that contain the target protein from an initial preparative run can only be done on fully integrated chromatography and robotic sample handling systems. The requirement that the analytical method be linked to the preparative method in order to track fraction collection means that the linking together of different instruments will not work. Instruments designed specifically for this type of analysis are available, but from a very limited number of manufacturers.

There are basic requirements that have to be met for any MDLC instrument, whether it be bought specifically or constructed in house. First is the issue of buffers and buffer blending; there needs to be sufficient buffer lines to attach all of the required solvents. Clearly, the more column steps involved, the more buffers that will be needed. In certain cases, the use of buffer concentrates that can be blended in varying proportions can help with buffer provision, but not all instruments have the blending capability to utilize this approach.

Second, there has to be sufficient column attachment and switching ability. For each column used, there needs to be the ability to switch it into and out of the flow path at will. Systems designed to perform MDLC tend to have multiple 6- or 10-port switching valves to which 1 or more columns can be attached to each valve. Many modern single-valve systems have the option of attaching additional valves, which can be controlled by the software, thereby increasing the column-carrying capability of the systems. A thorough understanding of the valve operation is essential to ensure proper plumbing of additional valves. The switching of valves to move columns in and out of the flow path is equally crucial. Ideally, switching will be based on time and also detector threshold values such as ultraviolet (UV), conductivity, or pH, but clearly this depends on the instrument having appropriate detectors.

3. Methods

3.1. Plumbing of Valves

Proper plumbing of switching valves is the key to successful multidimensional liquid chromatography. Systems designed for MDLC usually come with established plumbing configurations that simplify their use. For those systems that have been put together by the user, this type of help is not available, so it is essential that the valves be plumbed properly; a good understanding of how valves work is useful at this point. It is not within the scope of this chapter to describe the workings of the numerous valves available on the market, so it will be limited to the most common valve types currently used. The majority of valves used for this type of work are 6-port switching valves, with some systems using 8- or 10-port valves, although for most valves, the principal of operation is the

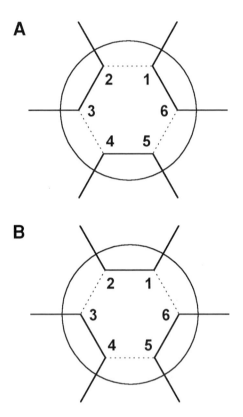

Fig. 2. Operation of a six-port valve. (**A**) Rotor seal in the clockwise position. Ports 2 and 3 connect, ports 4 and 5 connect, and ports 6 and 1 connect. (**B**) Rotor seat in the counterclockwise position. Ports 1 and 2 connect, ports 3 and 4 connect, and ports 5 and 6 connect.

same. **Subheading 3.2.** is a brief description of the operating mode of a common six-port valve and plumbing suggestions that allow the use of two or more columns in multidimensional mode.

3.2. Mode of Operation of a Six-Port Valve

This type of valve has six ports arranged in a circle. The ports run through to the back of the valve where they meet the rotor seal. The rotor seal has three semicircular channels on the front face, each of which can connect adjacent ports together, and switches one position clockwise or counterclockwise, thereby connecting different ports (*see* **Fig. 2**). Eight- and 10-port valves operate similarly with additional channels in the rotor seals in order to connect adjacent ports.

3.3. Use of Six-Port Valves for Connecting Columns

This is probably the most simple and effective plumbing configuration for use with MDLC (*see* **Fig. 3**). By using one valve for each column, it is possible to select either column independently, or both in line or neither in line, giving maximum flexibility. Each column is placed in-line by setting the appropriate valve to the clockwise position and taken out of line by selecting the counterclockwise position. The possibility exists

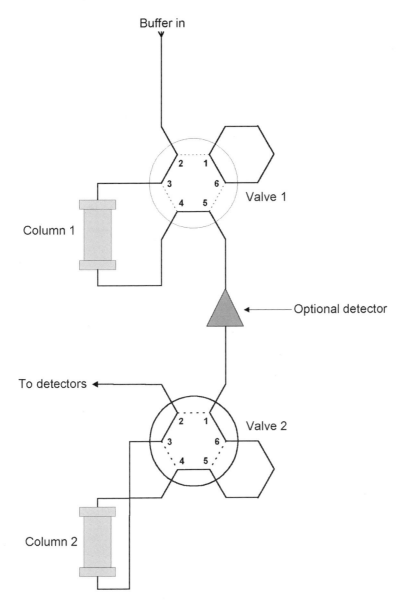

Fig. 3. Use of six-port valves for attaching columns.

for the attachment of additional detectors between the valves to give additional switching options (*see* **Subheading 3.4.**).

Using this configuration, it is possible to attach two columns on the second valve giving a choice of secondary columns (*see* **Fig. 4**). The disadvantage of this is that there is no column bypass available, so all buffer flows have to go through one of the columns on valve 2.

If these valves are used as additional valves, added onto an existing system, then activation is usually via transistor-transistor logic (TTL) or contact closure signaling, depending on the individual valve types used.

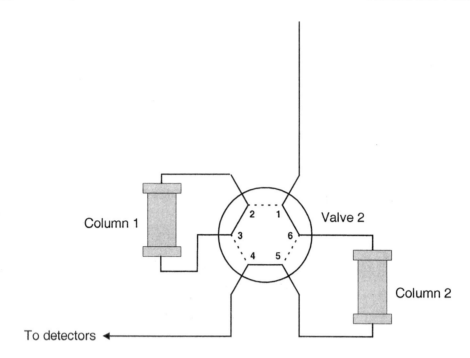

Fig. 4. Alternative two-column option for valve 2. Columns are selected by switching of the valve to the appropriate position.

3.4. Use of 10-Port Valves for Connecting Columns

The basic principal of operation of a 10-port valve is the same as that of a 6-port valve. The advantage that these valves have is that the higher number of ports allows much more complexity in the plumbing configurations, thereby providing options not available with the smaller valves. If the switching of columns in or out of line is going to be based on a (UV) signal, then the detector has to plumbed such that it can measure the outflow of first one column and then the second column independently. The use of 10-port valves gives the level of flexibility required for this plumbing to be achieved (*see* **Fig. 5**). Although the plumbing appears complex, by following the plumbing diagram and simple switching of the valves, the outflow from each column can be monitored by the UV detector.

The use of multiple valves with large numbers of ports can give remarkable complexity in terms of attaching and controlling columns. For example, three 10-port valves can provide linking and switching capability for up to 5 columns (for a good example of this, *see* **ref. *1***). It is not possible to go into detail on all possible plumbing configurations here, but help is usually available from instrument or valve manufacturers.

3.5. Attaching Additional Valves

Although this part of the methods has concentrated on the use of two switching valves with one or two columns per valve, there is clearly the opportunity to use additional valves for additional columns. The limit is dictated by the ability of the instrumentation to control added valves, but should the capability exist to attach extra valves, there is no

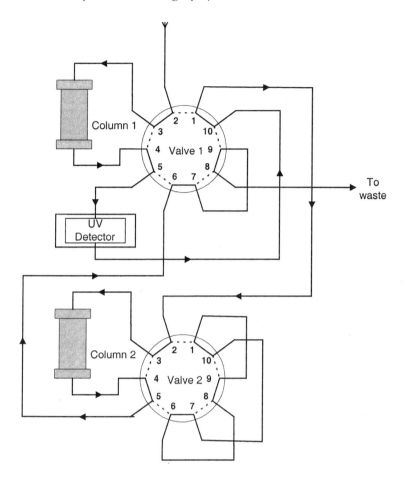

Fig. 5. Use of 10-port valves enabling UV readings from both columns.

reason not to do so. The valves can be plumbed in exactly the same way as already described. It is often methodological difficulties that ultimately limit the number of columns that can be linked together, and these are discussed later. Note that each additional valve doubles to potential number of plumbing states within an MDLC system, such that the total number of valve states is 2^n, where n is the number of valves (*see* **Note 5**). However, when using valves with only six or eight ports, the true number of distinct options may be fewer.

3.6. Switching Criteria

Although the ability to switch columns in and out of the flow path is the basis of MDLC, of equal importance is timing of the switching to ensure the correct part of the effluent from one column is directed onto the next column. A failure of this timing will lead to a failure of the whole separation procedure.

The point at which to switch a column in or out of line has to be decided empirically for each protocol based on a thorough knowledge of each separation step. This can only be gained by running the individual steps and analyzing the resultant separation. Once the switching point has been determined, the trigger to activate the switch can be one of a number of parameters—the most common being time, or a physical characteristic of the buffer stream such as UV (or other spectroscopic) absorbance, conductivity, or pH.

3.7. Time-Based Valve Switching

Time is the simplest type of trigger, as it does not require input from any sort of detector. A timed switch command can simply be positioned anywhere in a method to accommodate a predetermined action. Usually, the timing is aimed at directing a specific part of the elution onto another column, based on a known elution pattern. Clearly, reproducibility is critical for this to be successful, so elution based on step rather than linear gradients tend to be more effective (*see* **Note 1**).

The disadvantage of using time as a trigger is that it is invariable; once set in a method, it will activate regardless of what happens during the running of the method. Consequently, even small variations in timing can cause the target protein to be missed if the time period used is very tight.

3.8. Detector-Based Valve Switching

The use of detector readings to trigger valve switches offers a degree of flexibility not available with time-based switching but has a prerequisite that the equipment being used has the capability to respond to detector readings. Because these triggers respond to the physical nature of the buffer stream and are not time restricted, they are useful when protocols are frequently changed in a way that would invalidate timed switching. The most commonly used detectors for switching valves are UV absorbance and conductivity, which respond to protein and salt concentrations, respectively. The disadvantage of detector-based switching is that you must have a good idea of what signal is expected. For example, if the expected signal was one absorbance unit (AU) and the threshold was set to 0.1 AU but the actual signal was only 0.095 AU, then the threshold event would be missed, even though the signal could be easily seen visually. Likewise, if the threshold is set too low, then a baseline drift or disturbance may be mistaken for a true signal and the event triggered in error.

The detectors have to be plumbed into an appropriate point in the flow path, which may be problematic on some systems, as detectors are usually positioned at the end of a chromatography system in order to provide readings prior to fraction collection. If the readings from these detectors are required earlier in the procedure, a degree of customized plumbing may be required. This is particularly true with regard to UV detection, as the elution from the final column will need to be monitored separately from the previous columns, a solution such as that shown in **Fig. 5** will have to be adopted should the system used not have this ability built-in. In the example in **Fig. 5**, the detector observes the effluent of column 1 whenever column 1 is in-line. As a peak elutes from column 1, it may be diverted to column 2 without affecting the chromatogram from the first column. At this point, when column 1 is in-line, the effluent of column 2 is not monitored. However, when column 2 is placed in-line by itself, the detector effectively monitors outflow from column 2. In this way, a peak captured from column 1 onto column 2 will pass through the detector a second time.

A typical example of where a detector-based switch may be used is when a desalting column is used prior to a capture column. As the protein elutes from the bottom of the desalt column, an UV absorbance reading can be used to switch the next column into

Table 1
Compatibility of Separation Modes

Initial separation mode	Typical elution conditions	Compatible secondary modes	Possible secondary modes	Incompatible secondary modes
Anion exchange (AEX)	High ionic strength	SEC, RP[a,b], affinity	HIC[c]	AEX, CEX
Cation exchange (CEX)	High ionic strength	SEC, RP[b], affinity	HIC[c]	AEX, CEX
Hydrophobic interaction (HIC)	Low ionic strength	SEC, RP[b]	AEX[d], CEX[d], affinity[e]	HIC
Size exclusion (SEC)	Moderate to high ionic strength	SEC, RP[a], Affinity	AEX[d], CEX[d], HIC[c]	
Affinity: protein A/G, immobilized antibodies	High ionic strength, low pH	SEC, RP[b]	HIC[c]	AEX, CEX, affinity
Affinity: specialized ligand	Variable	SEC, RP,	AEX[d], CEX[d], HIC[c]	Affinity
Reversed phase[g] (RP)	Polar organics, low pH	SEC	CEX[f]	AEX, HIC, affinity

[a]When using buffers at a pH above 7.0, avoid silica-based RP columns, as silica is soluble in high-pH buffers.

[b]Very high levels of salt and other buffer components in the sample can interfere with binding to RP columns, so some manipulation of the column load may be necessary (*see* **Subheading 3.4.**).

[c]HIC usually requires protein to be loaded in the presence of salt at near precipitating concentrations. At the lower salt levels of a column eluate, it is likely that only very hydrophobic proteins will bind effectively.

[d]If the proteins can be eluted at sufficiently low salt concentrations, it is possible that an ion-exchange step will capture eluted proteins (*see* **Note 2**).

[e]Care must be taken to ensure the level of salt remaining in the eluted sample will not precipitate the ligand on the affinity column.

[f]CEX following RP is usually only successful if the protein can be eluted in relatively low levels of methanol or isopropanol.

[g]Reversed phase is rarely used as an initial step in a separation protocol, being almost exclusively reserved for final-stage analytical separations.

line to capture the proteins. As the protein peak comes to an end, the same detector can be used to switch the desalt column out of line, allowing the later capture column to be run independently.

3.9. Combining Separation Modes

It is important when designing the separation protocol that linked column steps are compatible; that is, the effluent from one column is in a buffer that is suitable for the next column step. There are ways of changing the buffer composition between columns, and they are discussed in **Subheading 3.10.**, but it is very much easier if the effluent from one column can simply be directed straight onto the next column.

There are certain modes of chromatography that link together very well, others that may work, and some that definitely will not (**Table 1**). Unless you are using separation modes with known compatibility, the success or failure of sequential column steps has to be determined empirically. It is sometimes worth compromising absolute efficiency at each step, by altering buffer conditions, in order to build a successful MDLC approach that will save time and manpower (*see* **Note 3**).

3.10. Changing Buffer Composition

Often the requirement of an MDLC procedure is simply to automate an existing protocol by combining column steps, which, at first sight, appear incompatible. The ability to manipulate the buffer composition of an eluted sample before loading onto another column becomes essential. There are a number of ways of manipulating buffer compositions—some of which are very simple and others much more difficult. In this simplest case, if the volume of sample transferred from one column to the second is small relative to the column volume, the incompatible solvent may be diluted sufficiently for direct transfer without further manipulation. However, this is rarely the case because one of the objectives of MDLC is enrichment and concentration of a dilute sample. A key factor in many buffer manipulation techniques is the use of auxiliary columns for protein capture and buffer exchange, which is dependent on having sufficient column linking and switching capability. Without this capability, experimental design becomes more complex.

3.11. Desalting Columns

The most obvious way of changing a buffer is to use a desalting column between the two preparative columns. This works very well if there is sufficient capability on the system to accommodate three independently switched columns: the first column has to be eluted onto the desalt column, which, in turn, has to be independently eluted, and the second column is switched into line as the protein peak emerges. If there is insufficient switching capability, then large desalt columns cannot be used.

If the total volume of sample transferred from one column to another is small (e.g., a specific peak), then the use of a small desalting guard column prior to the second column may be all that is needed. The guard column desalts the sample, allowing it to bind to the main column, and although the original buffer components are eluted from the guard column, the dilution effects often prevent them from dislodging the bound proteins. Because the guard column is small, it does not adversely affect the subsequent elution of the main column.

3.12. Ion-Exchange Columns for Buffer Exchange

When eluting from an affinity column, such as immobilized antibodies, the elution is nearly always done using an acid pH step at typically pH 2–3. At this pH, all proteins will be positively charged. If the affinity step can be run in little or no salt (*see* **Note 3**), an in-line cation column can then be used to capture the effluent, where the pH can be neutralized and the sample eluted with a salt step. Clearly, the following column step has to be tolerant of the eluting salt concentration.

Ion-exchange columns can also be used to remove organic solvent. For example, proteins or peptides eluting from a reversed-phase column often contain an acidic ion-pairing agent (e.g., TFA) that facilitates binding to a cation column. The organic solvent may then be washed away with a low pH buffer (e.g., 0.1% TFA or 0.2% formic acid) and, again, the ion-exchange column may be eluted with salt or increased pH.

The same approach can be used with elution done at high pH and captured on an anion-exchange column. However, there are few situations where high pH is used for protein elutions, so this remains a rare solution.

3.13. Reversed-Phase Capture Columns

A particular problem arises with the use of microbore reversed-phase columns, which are frequently used prior to injection into electrospray mass spectrometers. These columns are often preceded by, for example, cation-exchange or tryptic digest columns. The high level of salt and other buffer components in the effluent of these columns is unsuitable for direct injection onto a microbore column prior to mass spectrometry.

The differential hydrophobicity of different media types can be used for a solution. The effluent can be captured on a polymeric reversed-phase trapping column where it is desalted with 0.1% TFA prior to elution onto a microbore C_{18} column. Polymeric reversed-phase media tends to be slightly less hydrophobic than conventional C_{18} media, so the peptides will elute from the polymeric trap at a given point in the reversed-phase gradient and recapture onto the C_{18} column prior to elution at a later point in the gradient (*see* **Note 4**).

3.14. Manipulating Buffer Streams

When all else fails, there is always the possibility of manipulating the buffer stream itself. Essentially, this involves the blending in of an additional buffer stream to the flow from one column to another, usually via a mixing tee (*see* **Fig. 6**). Simple in theory, it involves the utilization and control of an external pumping system to provide the additional buffer.

The important considerations revolve around relative mixing proportions and flow rates. For example, if the requirement is to reduce the salt concentration of the effluent from the first column from 500 to 100 mM, you cannot simple add an additional diluting buffer at four times the existing volumetric flow rate, as this will almost certainly exceed the maximum operating flow rate of the second column. The maximum operating flow rate (in mL/min) of the second column should be determined, the relative mixing proportions calculated, and then the relative flow rates of both the system and external pumps determined. Normally, control of external pumps is limited to ON and OFF, with little or no ability to control flow rate. Hence, the flow delivered by the external pump will normally have to be preset manually before starting the procedure. At the point at which the external buffer flow begins, the flow rate delivered by the system pump will have to be adjusted to ensure that the overall flow rate and mixing is correct. This may well have the affect of slowing the elution of the first column, but this cannot be helped.

This sort of approach can be used to alter virtually any parameter of a buffer stream, but it requires the user to purchase an appropriate additional pump that can be controlled by TTL-type signaling, which involves a significant cost. Also, the establishment of correct flow parameters and buffer compositions requires a significant input in time and effort and is often much harder than the theory suggests.

Another approach is to use a gradient pumping system that has multiple solvent-selection for each pump. One or more columns may be placed in the flow path prior to the mixing tee. For example. with this technique, two antibody affinity columns may be

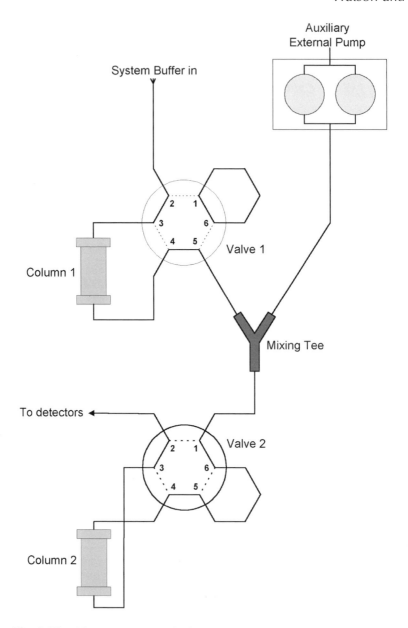

Fig. 6. Plumbing arrangement for in-line dilution using an external pump.

placed in tandem: one before and one after the mixing tee. The first affinity column may
be placed in the path of only one of the solvent pumps. First, the pump is used to deliver
the loading buffer, which is used to inject the sample and wash the affinity column. Then,
the solvent on that pump is switched to the elution buffer. The first affinity column may
be placed back in-line with the second affinity column. To neutralize the effluent of the
first column, the second pump is teed into the flow stream. The blend is set to 50% pump
1 (elution buffer: e.g., 12 mM HCl) and 50% pump 2 (neutralization buffer: e.g., 200 mM
Tris-HCl, pH 8). As the sample elutes, it is neutralized and captured on the second affinity column. To elute the second affinity column, the blend may be switched back to 100%

pump 1. It is important to note that gradient elutions may only be performed on columns downstream of the mixing tee, and only step elutions may be performed on the upstream columns. It is important that the different plumbing lengths between the pumps and the mixing tee do not have detrimental effects on gradient production and note that the delay volume of the gradient is only the delay between the mixing tee and the column inlet.

Because many systems are limited to three or four solvents per pump, another method of delivering multiple step elution compositions is to inject them from an autosampler (e.g., a series of salt steps may be applied to an ion-exchange column by injection of 1–2-mL aliquots of ever higher salt solutions from sample vials stored in the autosampler).

3.15. Combining Existing HPLC Systems

Labs often have existing high-performance liquid chromatography (HPLC) systems that on their own are not suitable for MDLC, often because of the limited number of buffers that they can handle (*see* **Subheading 3.9.**). However, combining two such HPLC systems together can produce functional MDLC systems that can handle two or three columns perfectly well. The way the combined systems are controlled is important; normally, one system is the controller and the other acts as a slave system. Modern HPLC systems are often computer controlled with the ability to run more than one system from the same controller, which provides an ideal solution. The advantage of this type of combination is that it provides a duplicate set of detectors and thus can provide extensive information during the running of the columns and provide the ability to use an UV signal as a column-switching trigger, for example.

The systems are set up so that one pump provides the buffers for the first column and the other pump the buffers for second column (*see* **Fig. 7**). Switching valves move the columns in and out of line as normal. There are examples in the literature of innovative system design using combined HPLC systems *(7,8)*.

3.16. Peptide Mapping

Peptide mapping gets a section to itself because it is a classic example of where advances in column technology coupled with an MDLC approach has simplified and dramatically improved a traditional technique.

The technique revolves around the use of immobilized enzyme cartridges, where proteolytic enzymes, such as trypsin, are covalently linked to high-flow media at high concentrations (up to 20 mg/mL) and packed into cartridges, typically with a bed volume of approx 100 μL. Protein samples can be digested by passing them through the cartridges at a flow rate that gives sufficient contact time; this is usually in the order of 10–20 min for immobilized trypsin.

A trypsin cartridge is connected in line with a C_{18} peptide mapping column. The trypsin cartridge has to be equilibrated with an appropriate digest buffer, such as 50 mM Tris, pH 8.0, 10 mM Ca^{2+}, and the C_{18} column with 0.1% TFA. Reduced and alkylated protein sample is applied to the trypsin cartridge at a flow rate that will give the appropriate contact time for digestion; for example, if the optimum contact time is 10 min, then a flow rate of 10 μL/min should be used for a 100-μL cartridge. The effluent from the trypsin cartridge is directed onto the C_{18} column to capture the digested peptides, and the trypsin cartridge is switched out of the flow path. The C_{18} column can then

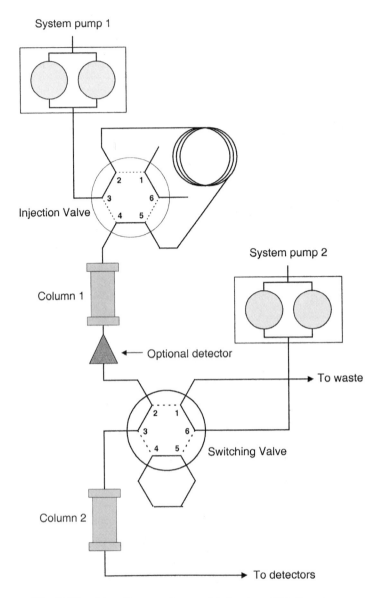

Fig. 7. Plumbing diagram for combining two HPLC systems.

be eluted with an eluant gradient as required, typically 0–80% acetonitrile over 20–30 column volumes. The effluent from the C_{18} column can then be spotted onto a MALDI plate prior to MALDI mass spectrometry, or with the appropriate use of flow splitting, it can be directed straight into an electrospray mass spectrometer. The whole procedure takes approx 2 h, as opposed to around 2 d by conventional methods.

If the chromatography instrument being used has the capability of being linked to an autosampler, then the reduction and alkylation of the protein sample can also be automated, resulting in a completely hands-free procedure. Apart from the saving in time and effort, the use of immobilized enzymes also has other benefits: There is no enzyme contamination of the sample, the digest pattern for any particular protein is extremely reproducible, and the cartridge can be used repeatedly with no loss of function *(3)*.

3.17. Biphasic Column Technology

The use of columns containing two different types of medium, giving multidimensional-type separations but from a single column, has recently been described *(9,10)*. In these experiments, the authors combined a strong cation resin, followed by a C_{18} reversed-phase resin in a single capillary column. Loaded sample was captured by the cation resin and eluted by a series of salt steps onto the C_{18} resin. After each step elution from the cation resin, the salt was washed from the column, and the proteins captured by the C_{18} resin were eluted with an acetonitrile gradient, followed by re-equilibration and the next salt step. The authors termed this approach "multidimensional protein identification technology" (MudPIT).

This approach has obvious benefits in that it reduces the need for column switching when eluting from one column to the other. There are however important considerations for anyone wanting to use this approach. First is the issue of buffer compatibility. Packing two types of media in one column means that both media are exposed to all of the buffers used for equilibrating, eluting, and cleaning; clearly, the medias must be chemically resistant to all these. In addition, the eluants used on the first medium must not prevent adsorbtion to the second medium; similarly, the buffers used to equilibrate and elute the second medium must not have any effect on the proteins bound to the first medium and, especially, must not have any eluting effect.

The media used in any biphasic column must have compatible pressure ratings. Biphasic columns are not available commercially, so they will have to be packed by the user and care must be taken to ensure that the medias used can be packed together effectively—if there is a difference in packing pressures, the first medium packed must be able to tolerate the pressure needed to pack the second medium or it will simply collapse. Under operation, the column will generate different back-pressures depending on which buffers are being used; both media must be able to withstand the highest back-pressures the column will generate.

The use of a single column means that there is no way of monitoring the elution from the first medium and there is no way of ignoring part of the elution. Everything that elutes from the first medium will be captured on the second medium and so this technology appears best employed when complete profiling of a sample is required rather than targeting of a specific component.

A minor note: If the column has to be depacked the medias will not be recoverable, as they will almost certainly mix. Although not a problem for capillary columns, should larger columns be employed this could incur a significant cost.

3.18. Reinjection of Collected Fractions

This technique is only possible with specialized equipment that has this specific capability, but the means to automatically reinject all, or part of, collected fractions onto another column has some potentially enormous benefits.

3.19. Partial Fraction Reinjection

The principal use of injecting partial fractions is for analysis. During a purification procedure, a column run can be analyzed by injecting part of each fraction onto an analytical column to determine the presence or absence of the target protein in each fraction. Ideally, the analytical column will be an immobilized antibody column, specific for the target protein. Following the injection of a portion of a collected fraction, the elution of the column with a low pH buffer will produce a peak if the target protein has bound, giving a clear-cut result as to which fractions contain the target protein. The downside of using affinity columns is that they give little indication of quantity or the number of contaminants. Also, some antibodies are not stable and will not work for more than a few runs; thus, selecting an appropriate antibody may be difficult.

In the absence of appropriate antibodies, reversed phase makes a good option. Here, a peak appearing with a specific retention time (determined previously for the target protein) determines the presence or absence of target protein. Advantages of this technique are that it has a higher dynamic range and thus, gives information on amounts of protein and also an indication of the level of contaminants. The downside is that other components with very similar retention times can be confusing, as can changes in retention time with column age.

Other techniques such as size exclusion can also be used, where, again, retention time is the guiding parameter, but this relatively slow technique results in long analysis times.

Partial fraction reinjection can also be used as a quality control measure. For example, the efficiency of purification of synthetic oligonucleotides can be checked by analyzing a portion of the collected peak by reversed phase.

3.20. Whole Fraction Reinjection

Whole fraction reinjection can also be a useful tool. Profiling the components of a complex mixture by collecting fractions from the first column run, rather than by directing the flow straight to another column, gives a complete chromatogram of the first-dimension separation. The fractions are then reinjected onto the second column for further separation, but the operator has the ability to view the chromatogram from the first column and eliminate any fractions that are not needed from the run queue.

Advances in proteomics technology is now producing protein-labeling kits that contain simple multiple-column steps. The requirement of these kits is each sample is passed through one column, and eluted as a single fraction, which is then applied to a second column and again eluted as a single fraction. This kind of procedure lends itself ideally to fraction collection and reinjection and the use of these instruments can dramatically speed up a very labor-intensive operation.

4. Notes

1. One of the principal problems of using time for triggering valve switches is that of changes in the retention time of the target component. Retention time and peak width both change with column age and degree of fouling, resulting in the target component drifting out of the time window set for its transfer to the second column. This obviously seriously affects both purity and recovery of the target. One solution to this problem of drifting retention times is the use of step gradients rather than linear gradients to elute the first column. For example, if a column is eluted with a linear 0- to 500-mM NaCl gradient, this could be replaced by a series of say 20 small step gradients, each of 25 mM. As the elution of a component will always occur at the same salt concentration, the valve can be switched at the beginning of the step in which the target is eluted. This also has the added advantage that a wider time period can be used for directing the effluent onto the second column. In theory, this may lead to having a slightly less efficient separation on the first column, but with good method development, it should be possible to achieve a highly comparable elution. As a general rule, any elution onto another column will be more reliable if elution is done stepwise rather than linearly.

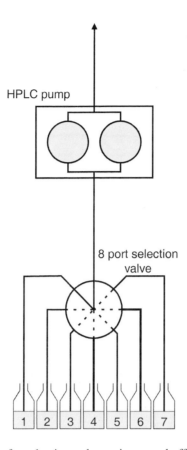

Fig. 8. Use of a selection valve to increase buffer capacity.

2. If a system is limited by the number of buffer lines attached but has the capacity to control additional valves, this can be used to add extra buffer capacity by using a multiport selection valve. Selection valves have a variable number of ports, any of which can be connected to a single common port. The use of these valves on an existing buffer line can increase the number of buffers attached to that line from one to as many as needed (the maximum currently available is 24). The buffer required is selected by switching the valve to the appropriate port (*see* **Fig. 8**).

3. It is sometimes necessary to manipulate buffer conditions to effectively link certain modes of chromatography effectively. For example, if the need is to link an affinity step with an ion-exchange step, then, clearly, the affinity step cannot be run with high salt or the effluent will not adsorb to the ion-exchange column. Frequently, however, high salt is required to minimize nonspecific binding to the affinity column. The same reduction in nonspecific binding can sometimes be achieved by increasing the concentration of the buffer itself, using 200 m*M* Tris-HCl rather than 20 m*M* and/or using a low concentration of a nonionic detergent such as 0.05% NP-40 or Tween-20 and lowering the salt concentration to a usable level. Whether or not this sort of manipulation will work depends on the individual case, but it is sometimes well worth trying.

4. When using gradient elution, it is sometimes useful to employ a less retentive trap column for initial capture of a sample. In these cases, the idea is to use a shorter and larger bore column than the final analytical column (usually a capillary). This lowers the pressure drop and enables faster loading times. Once captured, the trap column may be washed prior to

placing it in-line with the final analytical column. The gradient is run through both columns in series. At any given point in the gradient, an analyte that is eluted from the first column will still be retained at that gradient composition on the second column (e.g., the trap column might be a C_4 and the analytical column a C_{18}). The analyte is then recaptured and focused in the narrower analytical column. As the gradient continues, the composition finally reaches a point where the analyte is eluted from the analytical column, but, this time, in a much smaller volume and thus sharper peak.

Because the trap is meant to capture analytes quickly and is not required to have high resolution, it is also advisable to use a larger-diameter particle (e.g., 10–50 µm versus the 3- to 5-µm particles in the analytical column) or to use a perfusive polymeric media that may be operated at high flow rates. In the case of peptides, the polymeric medias behave similarly to a C_8 so that they are appropriate for capture columns in front of C_{18} analytical columns.

5. Notes on MDLC plumbing configurations:
   a. Note that the number of valve states is 2^n, where n is the number of valves (i.e., you double your flexibility with each additional valve).
   b. With multiposition valves, you must pass though other positions on your way to the selected position. This may overpressure a column or introduce the wrong solvent to a column.
   c. Stopping the pump does not necessarily stop the flow immediately, nor does flow start the instant that the pump is started.
   d. Consider whether two columns cannot be placed in tandem and operated as a single column.
   e. Columns that are to be mutually exclusive may be placed on the same valve.
   f. Columns totally independent of each other (i.e., one only; two only; both together) must be on different valves.

If linking two columns is unavoidable, where the first column has a much lower pressure limit than the second, then the flow rate will have to be reduced to protect the lower-pressure column. This may well make the procedure unusably slow. If there is spare switching capability on the instrument, one solution must place a sample loop between the two columns. The effluent from column one is collected in the sample loop, with the second column off-line. Once elution is complete, the first column is switched out of line, the loop is switched into line with the second column, and the collected effluent is injected onto the second column. As the two columns are never in line together, they can both be run at their optimum flow rates.

It is not the intention of this chapter to give exhaustive plumbing diagrams for all valve types. Rather, it is intended as an introduction to the way plumbing can be designed and to indicate that with a good knowledge of valve operation and a bit of imagination, most plumbing problems can be resolved. As said earlier, system and valve manufacturers are usually willing to help with designing plumbing solutions; so, if stuck, just ask.

References

1. Hsieh, Y. L. F., Wang, H. Q., Elicone, C., Mark, J., Martin, S. A., and Regnier, F. (1996) Automated analytical system for the examination of protein primary structure. *Anal. Chem.* **68(3),** 455–462.
2. Kassel, D. B., Consler, T. G., Shalaby, M., Sekhri, P., Gordon, N., and Nadler, T. (1995) Direct coupling of an automated 2-dimensional microcolumn affinity chromatography-capillary HPLC system with mass spectrometry for biomolecule analysis. *Tech. Protein Chem.* **VI,** 39–46.

3. Nadler, T., Blackburn, C., Mark, J., Gordon, N., Regnier, F., and Vella G. (1996) Automated proteolytic mapping of proteins. *J. Chromatogr. A* **743,** 91–98.

4. de Frutos, M. and Regnier F. E. (1993) Tandem chromatographic–immunological analyses. *Anal. Chem.* **65(1),** 17–25.

5. Gygi, S. P., Corthals, G. L., Zhang, Y., Rochon, Y., and Aebersold, R. (2000) Evaluation of two-dimensional gel electrophoresis-based proteome analysis technology. *Proc. Natl. Acad. Sci. USA* **97(17),** 9390–9395.

6. Barth, H. G., Barber, W. E., Lochmuller, C. H., Majors, R. E., and Regnier, F. E. (1988) Column liquid chromatography. *Anal. Chem.* **60(12),** 387R–435R.

7. Opitech, G. J., Lewis, K. C., and Jorgenson, J. W. (1997) Comprehensive on-line LC/LC/MS of proteins. *Anal. Chem.* **69(8),** 1518–1524.

8. Opitech, G. J. and Jorgenson, J. W. (1997) Two-dimensional SEC/PRLC coupled to mass spectrometry for the analysis of peptides. *Anal. Chem.* **69(13),** 2283–2291.

9. Washburn, M. P., Wolters, D., and Yates, J. R. (2001) Large-scale analysis of the yeast proteome by multidimensional protein identification technology. *Nature Biotechnol.* **19,** 242–247.

10. Wotters, D. A., Washburn, M. P., and Yates J. R. (2001) An automated multidimensional protein identification technology for shotgun proteomics. *Anal. Chem.* **73(23),** 5683–5690.

42

Mass Spectrometry

David J. Bell

1. Introduction

Toward the end of the 1980s, two advances in the mass spectrometry (MS) of peptides and proteins were announced independently and, with ongoing developments, are still at the center of current methodologies *(1)*. Both developments were the refinement and application of ionization techniques, namely electrospray and matrix-assisted laser desorption and ionization (MALDI), and both offered increased sensitivity and increased the upper mass limit of peptides or proteins amenable to mass spectrometry. Such was the impact of these techniques that the originators were awarded Nobel Prizes in 2002.

1.1. Electrospray

Although the process of ionization is not fully understood, electrospray is simple in practice in that the method only requires a solution of the analyte to be sprayed across an electric field. The resulting analyte ions are then sampled into a mass analyzer and the sample usually observed as protonated molecules in the positive ion detection mode. Although initially used to measure intact proteins, electrospray was also found to detect lower-molecular-mass peptides with newfound sensitivity. A fortuitous advantage of the technique is that as the molecular mass of the peptide increases, so do the dimensions of the molecule and number of basic sites within the molecule; hence, higher mass samples can collect more protons per molecule. Thus, a peptide of molecular mass 500 Daltons may be only singly protonated, whereas a protein of mass 20 kDa may average 20 protons per molecule and the sample ions would appear at mass-to-charge 500 and 1000, respectively, in an electrospray mass spectrum.

1.2. Matrix-Assisted Laser Desorption and Ionization

By cocrystallizing a peptide or protein with an ultraviolet (UV) absorbing matrix, such as α-cyano-4-hydroxycinnamic acid, and hitting this mixture with a laser pulse of appropriate wavelength, analyte ions are formed that can be sampled into a suitable mass analyzer. The technique results in predominantly singly charged ions of proteins, either protonated or deprotonated up to molecular masses in excess of 100 kDa, and although chemical noise from matrix adduct ions can be a limitation, lower-mass peptides are also detected with high sensitivity.

From: *Methods in Molecular Biology, vol. 244: Protein Purification Protocols: Second Edition*
Edited by: P. Cutler © Humana Press Inc., Totowa, NJ

2. Instrumentation

2.1. Original Instrumentation

Although electrospray and MALDI result in ionized peptides and proteins, the mass analysis associated with each technique was originally quite different. Electrospray required atmospheric pressure inlet and was suited to quadrupole mass analysis because this continuous ionization process allowed scans to be performed. Additionally, the low voltages in a quadrupole were more compatible with the higher pressures than electromagnetic mass analyzers. Electrospray ionization on a triple quadrupole instrument provided the option of fragmenting the protonated peptide ions to generate sequence data.

As a pulsed ionization technique, MALDI was combined with time-of-flight (TOF) mass analysis. With subsequent modification, namely reflectron energy compensation and delayed extraction, high-resolution mass spectra of peptides and proteins were readily generated.

2.2. Developments in Instrumentation

From the original situation of electrospray ionization on a triple quadrupole and MALDI with TOF analysis, with the two techniques provided complementary data, there has been an array of combinations of ionization techniques and mass analyzers aimed at improving one or more of sensitivity, resolution, selectivity, speed of analysis and cost-effectiveness. The techniques applied to electrospray and/or MALDI included ion traps (capable of MS/MS and high sensitivity), single quadrupoles (cost-effective) and Fourier transform ion cylcotron resonance (sensitivity, resolution, and mass accuracy). In addition to these existing technologies, emerging methods of mass analysis were hybrid instruments.

Quadrupole–TOF (Q-TOF) instrumentation was a major development that facilitated the accumulation of molecular mass and MS/MS spectra, with both high resolution and increased sensitivity in each mode, by pulsing packets of ions into a TOF analyzer orthogonal to the initial quadrupole separation, rather than using another quadrupole with the resulting loss of sensitivity of scanning technique. A product of Q-TOF technology was the interfacing of electrospray or MALDI with orthogonal TOF analysis, which allowed the pulsing of ions from the respective sources into a TOF to yield high sensitivity, high mass accuracy, and a short cycle time per spectrum acquisition. Ongoing instrumental developments include quadrupole–ion trap technologies and MALDI-TOF/TOF.

2.3. Scan Functions

As both electrospray and MALDI ionization impart little energy into the analyte, intact molecules are observed using either technique; thus, the first information generated in peptide and protein analysis is usually molecular mass information, acquired by detecting all ions formed in a mass-to-charge range of interest. In positive ion mode, the ions detected are predominantly protonated, unless other cations, such as ammonium, sodium, or potassium, are present in the sample, which may result in these adducting with the peptide. In the negative ion detect mode, deprotonated species are detected.

To complement the molecular-mass information, ionized peptides can be fragmented by collision with inert gas molecules to yield information revealing the amino acid sequence of the molecule. This is possible because, although the ionized peptides can cleave in three positions and the charge remain on either side of the cleavage, one fragmentation pathway usually dominates, resulting in a series of ions separated by the mass of each amino acid residue. As each amino acid has a different molecular mass, the sequence can be read off the spectrum, except the two pairs of amino acids leucine and isoleucine plus lysine and glutamine cannot be readily differentiated because their nominal molecular masses are the same in each case.

The detection of the fragment ion spectrum, or product ion MS/MS, is the most commonly used scan function, although precursor ion scanning is used to detect peptides producing specific fragment ions, such as phosphate or ions from selected amino acids. Furthermore, neutral loss acquisitions can monitor the ejection of specified molecules from analytes. Some mass analyzers, such as triple quadrupole instruments, are more suited to these types of acquisitions, although they may lack sensitivity; thus, such acquisitions are emulated on more sensitive analyzers such as Q-TOFs and ion traps, although emerging Q-trap technology is bringing both functionality and sensitivity.

3. Methods

Despite the array of techniques available, two complementary techniques can be utilized to provide much of the data molecular mass and sequence data for the determination or confirmation of the structure of proteins and peptides. Practical steps in each of these approaches, MALDI-TOF and LC/Q-TOF MS/MS, are detailed in the following subsections.

3.1. MALDI

3.1.1. Sample Preparation

There are many specific methods of sample preparation, but the general method is that analyte solutions are deposited with a solution of UV absorbing matrix solution on the sample target plate and the components are allowed to cocrystallize. Ideally, the sample is dissolved in aqueous/organic solvents (10–100 fmol/μL) without buffers and the matrix is dissolved in organic solvent with acid added, typically 2–10 μg/μL α-cyano-4-hydroxycinnamic acid in ethanol:acetonitrile (1:1, v/v) plus 1% of 0.1% aqueous trifluoroacetic acid. Sample and matrix solutions are added in equivolumes (e.g., 0.5 μL of each) and allowed to dry naturally.

3.1.2. Sample Introduction

Samples are deposited on target plates that are made of steel marked typically with between 48 and 384 sample positions and individual positions 2–5 mm in diameter. The plate may be treated with hydrophobic material surrounding the spot position or an absorbent material in the target center, to concentrate the sample and also allow washing with water to remove salts and buffers. Once prepared and dried, the sample target plate is introduced into the mass spectrometer through a vacuum lock.

3.1.3. Optimization of Instrument Settings

Although modern instruments are stable, some optimization of the instrument parameters prior to analysis is recommended. A standard sample may be deposited on a target position and the signal from this spot is observed while instrument settings are adjusted to achieve acceptable sensitivity and separation of ions by mass-to-charge (i.e., resolution). Key variables are laser position, laser power in addition to extraction, focusing, and detection voltages.

3.1.4. Calibration

Highest mass accuracy is achieved by converting the TOF into mass-to-charge by calibration of the data with a recently acquired reference acquisition using identical instrument settings. Ideally a multipoint calibration is used, with reference ions acquired across the mass-to-charge region of interest. Optimal mass accuracy may then be achieved by having an ion of known composition in the sample spectra (e.g., a standard is added or trypsin autodigestion peptides), so these can be used to lock the calibration and thus correct for any mass drift.

3.1.5. Data Acquisition

Data are acquired by accumulation of spectra from laser shots. Highest quality data are generated by using a laser power just above the appearance threshold and finding positions where the matrix:sample ratio is optimal. This is achieved by experimentation.

3.1.6. Interpretation of Data

The MALDI spectra are inherently simple to interpret, as only protonated molecules are detected; hence, the molecular mass of each component is one mass unit less than the ions detected (*see* **Fig. 1**). In practice, some additional cation adduction may occur; thus, the sodiated molecule may appear 22 Daltons higher than a protonated molecule.

3.1.7. Automation

As a reflection of the relative simplicity and robustness of the technique, all stages of the MALDI process can be automated. Sample spot preparation, including matrix solution addition, can be automated by liquid-handling systems and some mass spectrometers now allow automated sample plate changing. Automated data acquisition can be programmed by selecting either a predetermined or random laser firing positions coupled with automated variation of parameters. These parameters include increasing laser power until a set number of spectra have met acceptance criteria, usually the sample signal becoming greater than a specified level.

3.2. Electrospray

3.2.1. Sample Preparation

In addition to determining molecular mass, electrospray is used to sequence either endogenous peptides or those generated from proteins too large for direct sequencing by MS. For maximum efficiency of ionization, nanospray is the method of choice, but high-performance liquid chromatography (HPLC) separation may be required for complex mixtures or sample preconcentration using a trapping cartridge. The technique to be used will determine sample preparation required (i.e., nanospray requires salt- and buffer-free sample solutions, whereas HPLC will clean up samples on-line. Using LC/MS with a trapping system, larger volumes of more dilute solutions may be analyzed (e.g., up to 100 µL can be trapped and then eluted through a 75-µm LC column compared with a maximum of 1–2 µL if the sample is to be introduced directly onto the column. Although the limit of detection will depend on the instrumentation and exact method used, 10 fmol of

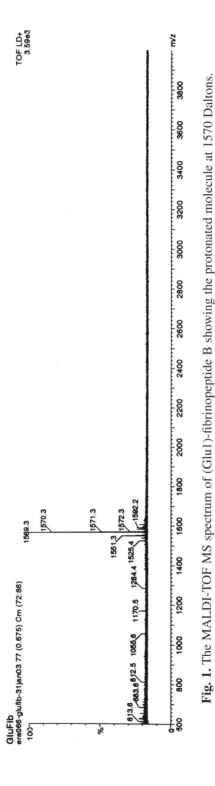

Fig. 1. The MALDI-TOF MS spectrum of (Glu1)-fibrinopeptide B showing the protonated molecule at 1570 Daltons.

Table 1
Capillary LC/MS Gradient

Time	% Solvent A
0	95
6	95
36	60
37	1
44	1
45	95
66	95

Note: Solvent A = 98% water, 2% acetonitrile plus 0.1% formic acid; solvent B = 5% water, 95% acetonitrile plus 0.1% formic acid.

peptide should be readily detected and generate an MS/MS spectrum using ion trap or Q-TOF technology. As the most tolerant technique with respect to sample presentation, capillary LC/MS/MS with sample trapping is detailed here.

3.2.2. Sample Introduction

With an LC system incorporating a sample trap (C_{18} 300-μm packing inner diameter × 5 mm), relatively dilute samples can be introduced into the system via a chilled autosampler allowing the analysis to be automated. HPLC separations are achieved using a reversed-phase packing (C_4–C_{18}) in a column of 75 μm inner diameter and 5–15 cm in length. A gradient of acetonitrile is used to elute peptides, with 0.1% formic acid used as an ion-pairing agent and a flow rate of 250 nL/min used (*see* **Table 1**). Fused-silica nanospray needles capable of operating at such flow rates are connected to the column outlet and form the ion source.

3.2.3. Optimization of Instrument Settings

Although considerable time may be invested in adjusting instrument parameters for optimal sensitivity and resolution, Q-TOF or ion trap instruments generally require little day-to-day tuning. It is advisable to analyze a standard sample solution prior to batches of samples, to ensure that instrument performance has been maintained. This may be a peptide at an appropriate concentration in water.

3.2.4. Calibration

Spectrum calibration is achieved by use of a previously acquired reference spectrum, which may be the MS/MS spectrum of the standard peptide used to check instrument performance. Again, for greatest mass accuracy, an internal lock mass is preferred and this may now be achieved using emerging ion sources that allow the introduction of a reference sample into the instrument during sample analysis.

3.2.5. Data Acquisition

Analyses may be set up manually, but for greatest throughput and the accumulation of data at the highest rate, automated MS/MS is used. This entails creating an MS method that acquires an MS spectrum followed by the software selecting precursor ions of interest (e.g., the eight most intense ions), and then automatically accumulating MS/MS data. This can be done for either a set time or until the MS/MS spectra reach a

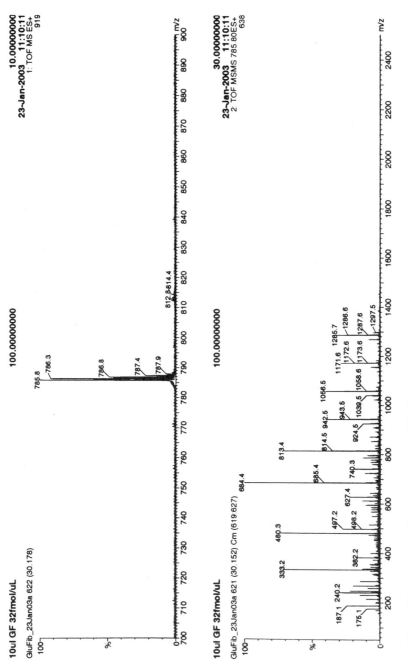

Fig. 2. Electrospray-MS (upper) and -MS/MS spectra (lower) of (Glu1)–fibrinopeptide B showing molecular mass and sequence information, respectively.

preselected intensity. The instrument will return to the MS mode following these experiments and then repeat the cycle. The maximum total cycle time should be less than the width of the chromatographic peaks to ensure that all components are detected. Precursor ions used for MS/MS are usually excluded from further acquisition of MS/MS for a period defined in the MS experiment (e.g., 3 min) so that MS/MS spectra from other co-eluting components are generated.

3.2.6. Interpretation of Data

The data set accumulated contains MS spectra and MS/MS data as determined by the experiment and sample composition. These are automatically separated by the data system and then may be interpreted manually to generate molecular-mass information from the MS data and sequence data from the MS/MS acquisitions (*see* **Fig. 2**). Alternatively, software may be used to sequence individual components or, in the case of digested proteins, used to search the combined data against databases to identify the proteins originally present.

3.2.7. Automation

The above-detailed LC/MS/MS process is routinely automated by the combination of autosampler introduction of samples, interactive data acquisition methods, and automated processing.

Reference

1. Cook, K. D. (2003). ASMS members John Fenn and Koichi Tanaka share Nobel: the world learns our "secret." *J. Am. Soc. Mass Spectrom.* **13**, 1359.

43

Purification of Therapeutic Proteins

Julian Bonnerjea

1. Introduction

Purified proteins have been used as human medicines for many decades. Before recombinant DNA technology was developed, hormones (e.g., insulin and human growth hormone) and other proteins were extracted and purified from blood and other tissues. Because of the low levels of the protein of interest in the tissues, the purification was technically very difficult and often it was more of an "enrichment" than a purification to homogeneity. Since the 1970s, the ability to genetically engineer *Escherichia coli* and other organisms to produce large quantities of therapeutically interesting molecules has led to the development of dozens of therapeutic proteins *(1)*. Some of these proteins (e.g., erythropoetin) are hugely successful drugs with annual worldwide sales of several billion dollars *(2)*.

Other than the very high purity required, there is no fundamental technical difference between the purification of proteins for therapeutic uses and the purification of proteins for other applications. However, all manufacturers of therapeutic products, irrespective of whether they produce proteins or chemical entities, must comply with the principles and guidelines of cGMP (current Good Manufacturing Practices). The cGMP requirements, as described in the UK "Orange Guide" *(3)*, explain the requirements the manufacturer must fulfill in terms of facilities, equipment, personnel, quality systems, documentation/records, laboratory testing, and so forth. Therefore, there are certain aspects of protein purification that are of great importance for therapeutic applications but may be of minor or no importance for other applications. Topics such as virus inactivation, endotoxin removal, process validation, use of materials derived from animal components, and so forth are key considerations in the development of a purification process for a therapeutic protein *(4)*.

2. Purity Requirements

Most proteins are broken down in the acidic conditions in the human stomach and, therefore, therapeutic proteins cannot be administered orally. They need to be injected into the bloodstream or into muscle, and the purity requirements for a protein intended to be injected into a human being are very high. There are a large number of different

From: *Methods in Molecular Biology, vol. 244: Protein Purification Protocols: Second Edition*
Edited by: P. Cutler © Humana Press Inc., Totowa, NJ

types of impurity that need to be removed (e.g., DNA, host cell proteins, fermentation media components, antifoams, leached affinity chromatography ligands, etc.) *(5)*. Many different analytical techniques are used to measure purity because no single technique can measure the many different types of molecule that may be present as impurities. A complication often encountered is that many therapeutic proteins are microheterogeneous and it may not be obvious which molecular species is product and which is impurity. For example, many highly purified "homogeneous" proteins appear as multiple bands on an isoelectric focusing gel or as multiple peaks on a reversed-phase chromatogram. Therefore, it is important to determine whether a molecular species detected using a sensitive analytical technique is an impurity that needs to be removed or a product variant that needs to be characterized but may not need removal.

The overall purity of a protein is usually measured by techniques that can detect a range of different components (e.g., high-performance liquid chromatography [HPLC] techniques using reversed-phase or size-exclusion columns, or sodium dodecyl sulfate-polyacrylamide gel electrophoresis [SDS-PAGE] and other electrophoretic techniques). In general, typical specifications for these tests are frequently "greater than 98% or 99% product." In addition to the assays designed to detect a range of different molecules, it is also essential to have specific assays designed to measure the concentration of certain key impurities. For example, enzyme-linked immunosorbent assay (ELISA)-based assays are used to measure the levels of any proteins that may be added to the cell culture medium to promote cell growth, any ligands that may leach off affinity resins, and host cell proteins (HCPs). Although specifications vary for different therapeutic products, the maximum permitted protein impurity levels are frequently in the parts-per-million range (i.e., nanograms of impurity per milligram of product). A typical specification for a bulk therapeutic protein, such as a monoclonal antibody, is shown in **Table 1**.

There are also further impurity specifications that are tested on a few lots of product, but then are not repeated for every lot thereafter. For example, it is necessary to know what leachables are released from filters and membranes that are used during the manufacturing process, but these are not included in the specification for each individual lot of therapeutic protein.

3. Method

3.1. Process Steps

Virtually all therapeutic proteins are purified using a sequence of chromatography operations (*see* **Fig. 1**). Typically, three of these steps are sufficient to meet the purity requirement laid down in the "product specification" document. Normally, the first step serves to concentrate the product and remove the bulk of the main impurity. This step needs to have a high capacity and should also rapidly remove potentially deleterious impurities such as protease enzymes released from the host cells. Rigid or semirigid ion-exchange resins are often chosen for the first step of a purification process for a therapeutic protein. These resins have high capacities and can be operated at high flow rates. Generally, affinity chromatography using protein ligands is not well suited for the first step of a purification process for therapeutic proteins, with one notable exception. Monoclonal antibodies are often purified by a sequence of chromatography steps starting with protein A affinity chromatography (e.g., **ref. 6**). Unlike many other protein affin-

Table 1
Typical Specification for a Bulk Therapeutic Protein

General characteristics	Comments
Appearance	For example, "clear colorless solution" as determined spectroscopically or by visual examination.
Protein concentration	Generally determined by absorbance at 280 nm using a predetermined extinction coefficient.
pH	Specify acceptable range (e.g., pH 6.5 ± 0.2).
Identity	
(e.g., isoelectric focusing or capillary electrophoresis)	Can be replaced by sequence determination or some other method that confirms the identity of the product.
Purity	
SDS-PAGE	General-purpose method used to measure overall purity.
GP-HPLC	Often used to measure the percentage of aggregated product.
RP-HPLC	Useful for the detection of both impurities and product variants derived from proteolysis, crosslinking, etc.
Impurities	
DNA	Specification are often in the "picogram of DNA per milligram of protein" range and therefore very sensitive assays are required (e.g., hybridization or Q-PCR).
Host cell proteins	As it is necessary to measure the levels of a very wide range of HCP proteins, antibodies are raised against a mixture of HCP proteins and used in an ELISA or similar format.
Fermentation medium proteins	Addition of BSA and other proteins to fermentation culture media used to be common before chemically defined, protein-free media became widespread. Where proteins are added, their concentration in the purified product must be determined.
Integrity	
(e.g., peptide mapping)	Provides confirmation of the primary sequence of the protein. Other tests may be included to measure secondary and tertiary structure.
Activity	
(e.g., cell-based assays)	An assay is required to demonstrate that the protein has biological activity.
Safety	
Endotoxin	Commercial colorimetric assays are available based on Lymulus amoebocyte lysate (LAL) tests.
Sterility	As described in European and other Pharmacopoeia.
Virus tests	Where mammalian cell lines are used, various virus tests are required at different stages of the manufacturing process.

ity ligands, the protein A molecule is extraordinarily stable and the resins are available in large volumes at a reasonable cost. Consequently, protein A chromatography resins are used for the purification of many therapeutic antibodies. The very high-purification factors obtained with protein A chromatography outweigh the disadvantages it has in terms of its instability on prolonged exposure to high pH (and the resulting difficult cleaning procedures required) and its high price relative to nonaffinity resins.

The development of the first chromatography step is often the most difficult and time-consuming aspect of developing a purification process, as a highly efficient and robust

	Step	Function
1	Protein A affinity chromatography	Purification to >95% purity and concentration of the product
2	'Hold' step at low pH	Inactivation of retroviruses and other enveloped viruses
3	pH adjustment (and dilution if necessary), or buffer exchange	Required to ensure product is applied to the subsequent ion exchange step at the correct pH and conductivity
4	Ion Exchange chromatography	Removal of low levels of impurities including leached Protein A residues.
5	Nanofiltration	Physical removal of virus particles
6	Formulation	To ensure a consistent final product and to add any excipients etc required for stability

Fig. 1. Typical purification process for a therapeutic monoclonal antibody

first step is critical to a successful manufacturing process. Automated chromatography workstations (e.g., the Akta Explorer© from Amersham Biosciences) are often used to aid the development of a chromatographic step as they enable a large number of conditions to be tested in a short period of time. For any chromatographic step, and especially for the first step in a purification sequence, it is important to test the sensitivity of the step to key variables by undertaking a Process Limits Evaluation (PLE) study. Parameters such as the concentration of the product in the feedstream, the amount of product loaded per liter of resin, the pH and conductivity of buffers, the flow rate, and so forth need to be investigated *(7)*. Variations inevitably arise in all biological processes, and it is essential that the purification process performs as expected despite such deviations. Having achieved a volume reduction and a reasonable degree of purification by the first step, the second, third, and any subsequent chromatography steps are used to remove relatively small amounts of the remaining impurities. For the purification of proteins from mammalian cell culture or other animal source, virus inactivation and removal steps must also be included in the purification sequence.

The requirements for the chromatography steps or "unit operations" to purify a therapeutic protein are similar to those for any high-value, large-scale purification operation [e.g., high capacity, high throughput, reusable, ability to withstand harsh cleaning conditions, etc. *(8)*]. There are a few specific requirements that are important for therapeutic but not for nontherapeutic applications. For example, human or animal components should not be used in the manufacture of the chromatographic resin or, indeed, of any component of the manufacturing process, as these complicate the regulatory approval of the therapeutic protein because of the potential for transmission of infectious agents. Also, the resin manufacturer should have a package of data detailing the manufacture and testing of the resin, its physical and chemical stability, the identity and toxicology of any leachable material, the test methods used, and so forth. This information should be assembled in the form of a "Regulatory Support file" or a "Drug Master File."

3.2. Virus Issues

Proteins purified from mammalian cell culture or from animal tissue could potentially contain viruses. These viruses may be derived from latent viruses within the genome of the mammalian cells used to produce the protein or they may be accidentally introduced into the manufacturing process from contaminated raw materials or from operators. Because of the serious risk to patient safety posed by viruses, a manufacturing process for a therapeutic protein purified from a mammalian source must contain at least two "robust" virus inactivation or removal steps. Steps that are considered robust include filtration using nanofilters, chemical treatment of the protein with solvents and/or detergents, and incubation of the protein under acidic conditions (e.g., between approx pH 3.0 and 4.0) that inactivates certain classes of enveloped viruses. Validation studies must be performed to demonstrate that these treatments do indeed inactivate or remove a range of different viruses and that the laboratory validation studies are representative of the manufacturing process *(9,10)*. A calculation of the risk that one dose of product contains a virus must be performed. This calculation is based on the number of viruslike particles observed in the unpurified product entering the purification process and on the cumulative virus removal and inactivation potential of the purification process (expressed in terms of logs of virus reduction). Detailed guidelines are available on the testing and viral safety of products derived from human or animal cell lines *(11)*.

3.3. Regulatory Compliance

Any medicine intended for human use must have a Product Licence from a regulatory authority. To obtain such a licence for a therapeutic protein, the applicant must submit a very detailed dossier about the product that may run to many tens of volumes. One section of this dossier, called the CMC or Chemistry, Manufacturing and Controls section, must describe the product itself, the cell line producing the product, the complete manufacturing process, the tests performed during the manufacturing process as well as on the final purified product, and so forth. Other sections describe the facility that the product is manufactured in, the preclinical and clinical data, and so on, to give an overall picture of the safety and efficacy of the product. Once all of the data are assembled and submitted, the regulatory authority will review the data and will normally inspect the facility where the product is made. Regulatory authorities look for documented evidence that the manufacturing process can be operated reproducibly within specified limits and that the product will have consistent efficacy, safety, and purity. This requires the manufacturing process to be fully characterized. For example, limits need to be set on the number of times a chromatography resin can be reused, and there must be documented evidence to show that resin nearing the end of its specified lifetime still functions as intended. Similarly, limits need to be set on the storage time for intermediate, part-purified product. This too must be documented to show that storage up to the maximum specified period at the appropriate temperature and in the appropriate vessel is acceptable and is not detrimental to the product. In addition to such "process limits studies," a validation study must be performed to demonstrate that at least three lots of product can be manufactured consistently. These lots should be produced in the same manufacturing facility, using the same procedures, and the same scale and type of equipment to be used for in-market supply of the product. Once all of these process validation studies are completed and the manufacturing process is validated, changes to the process are difficult to make and are likely to require a repeat of certain validation studies or a demonstration of product equivalence before and after the change.

Another aspect of the manufacturing process for a therapeutic protein that always receives the attention of regulatory authorities is the prevention of contamination *(12)*.

Microbiological contamination and particulate contamination of injectable products are major concerns in the manufacture of therapeutic products and a great deal of effort is expended to ensure that neither type of contamination enters the production process from the environment, from raw materials, or from personnel. Biopharmaceuticals are particularly prone to contamination, as bacteria can grow in the aqueous solutions used in the purification operations. Any bacterial contamination is a concern not only because the contaminating organism could be pathogenic but also because bacteria can release enzymes that could degrade the product or endotoxins and other toxins that can cause undesired biological effects when injected into humans. Therefore, although purification operations for therapeutic products are not generally conducted in a completely sterile manner, nonetheless a high level of cleanliness and a very low bacterial count or "bioburden" is required throughout the purification process. It is not adequate to solely perform a final sterile filtration to remove micro-organisms at the end of the purification process, as the micro-organisms may have released enzymes or toxins while they were in contact with the product.

Thus, although a terminal sterile filtration of the bulk purified product is generally performed at the end of the purification process, the emphasis is on preventing the introduction of micro-organisms into the manufacturing process, rather than removing them once they have entered. Often, the specification for bioburden gets tighter as a product progresses through the purification process with a stricter requirement at the end of a process than at the beginning. Chemical sanitization of equipment, chromatography resins, membranes, and so forth (e.g., with sodium hydroxide solutions) is very widely used to ensure bioburden levels are generally no more than a few tens of bacteria (or cfu, colony-forming units) per milliliter. Sodium hydroxide is a popular cleaning agent because it is inexpensive, relatively effective as a depyrogenation and disinfectant agent, and easy to remove and dispose of *(13)*. The strength of the sodium hydroxide solution, the contact time, and the temperature are all important variables in determining the effectiveness of the sanitisation operation. Solutions in the concentration range 0.1–1.0 M with contact times of at least 1 h are frequently specified. The equipment needs to be specifically designed to ensure that it is easy to clean with no crevices, dead legs, rough surface finishes, and so forth. For example, threaded connectors for piping are not permitted because bacteria and other deposits can accumulate in the threads; therefore, sanitary "tri-clamp"-

type connectors are used instead. Provided that careful attention is paid to such details, the process used to purify therapeutic proteins will produce product that consistently meets the required specification and also meets the approval of the regulatory authorities.

4. Notes

Points to note when developing a purification strategy for manufacture of a therapeutic protein:

1. **Resins, membranes and filters**: To ensure consistency of supply at the manufacturing scale, choose chromatographic resins, membranes, filters, and so forth from established suppliers that are committed to making that product for many years and that have Drug Master Files or Regulatory Support Files for their products. Once a purification process is validated, the withdrawal of a resin or membrane, or even just a change to its manufacturing process, can require a great deal of additional work to demonstrate process and product equivalence.

2. **Batch-to-batch reproducibility**: Before selecting a chromatography resin, obtain samples of three batches of resin to test the batch-to-batch reproducibility. Select the resins to cover the acceptable range for the key parameters quoted in the specification. For example, if the specification quotes a ligand density of 5–15 mg/mL of resin, then select lots of resin that span the range quoted.

3. **Source of manufacturing components**: Avoid the use of animal-derived raw materials in the manufacturing process for a therapeutic product. For example, affinity ligands should be synthetic or from a recombinant source, not purified from an animal component or purified using animal components. The protein A used in the purification of monoclonal antibodies should itself be purified using conventional techniques, not using IgG molecules from human or animal blood.

4. **Resin reuse**: When developing a purification process, keep a record of the number of times you use and repack a resin or membrane and the storage time and condition. Although a formal reuse study will be required later to demonstrate reusability, data from early process development studies can be useful to set process conditions and limits.

5. **Assay development**: The design and validation of sensitive assays for product purity/integrity, as well as for specific impurities such as DNA, host cell proteins, and so forth, are critical to the success of process development. This should be in place before the start of purification development.

6. **Composition of raw materials**: Consider carefully the chemical composition of buffers and the grade of chemicals. Do not automatically select the highest grade of chemical available for buffer components. Chose a "Pharmacopeal" grade, which is often less expensive than a "research" grade and is frequently used for pharmaceutical purposes. Similarly, select simple buffers such as acetates and phosphates in preference to the more specialized and inevitably more expensive "Goods" buffers. These may be very expensive and difficult to source once the process is goes into routine production.

7. **pH Stability**: Choose resins, membranes, and filters that are stable at high pH. Ideally, they should withstand 1 M NaOH. Although this is not a strict requirement, it will simplify cleaning procedures greatly.

8. **Additives**: Avoid the use of additives such as protease inhibitors, DNase enzymes, and so forth. If these are used, it will be necessary to develop and validate specific and sensitive assays to measure the additives at very low levels.

9. **Product collection**: Processes that require the collection of fractions across an elution peak with subsequent analysis of the fractions prior to pooling of the suitable fractions and then

proceeding to the subsequent step are difficult and time-consuming in a cGMP-regulated environment. It is preferable to collect one elution fraction only with the collection based on the A_{280} signal. Although this may result in a lower overall purity, the process will be much simpler and more robust.

10. **Operating temperature**: The purification process should ideally be performed at ambient temperature. Avoid cold-room operation unless it proves essential because of product instability.

References

1. Birch, J. R. (1997) Review of Biotechnology-derived products in use and in development. *Eur. J. Parenteral Sci.* **2(3),** 3–10.
2. Ramakrishnan, A. and Sadana, A. (2000) Economics of bioseparation processes, in *Handbook of Bioseparations* (Ahuja, S., ed.), Academic, New York, pp. 667–685.
3. Medicines Control Agency (2002) Rules and Guidance for Pharmaceutical Manufacturers and Distributors 2002, The Stationery Office, London.
4. Sofer, G. and Zabriskie, D. W. (eds.) (2000) *Biopharmaceutical Process Validation*, Marcel Dekker, New York.
5. O'Keefe, D. O. (2000) Analysis of protein impurities in pharmaceuticals derived from recombinant DNA, in *Handbook of Bioseparations* (Ahuja, S., ed.), Academic, New York, pp. 23–70.
6. O'Leary, R. M., Feuerhelm, D., Peers, D., Xu, Y., and Blank, G. S. (2001) Determining the useful lifetime of chromatography resins: prospective small-scale studies. *BioPharm* **14(9),** 10–18.
7. Kelly, B. D., Jennings, P., Wright, R., and Briasco, C. (1997) Demonstrating process robustness for chromatographic purification of a recombinant protein. *BioPharm* **10(10),** 36–47.
8. Wisniewski, R., Boschetti, E., and Jungbauer, A., (1996) Process considerations for large-sale chromatography of biomolecules, in *Biotechnology and Biopharmaceutical Manufacturing, Processing and Preservation* (Avis, K. E. and Wu, V. L., eds.), HIS Healthcare Group, Englewood, CO.
9. Roberts, P. (1996) Efficient removal of viruses by a novel polyvinylidene fluoride membrane filter. *J. Virol. Methods* **65,** 27–31.
10. Darling, A. J. (2000) Validation of the purification process for viral clearance, in *Biopharmaceutical Process Validation* (Sofer, G. and Zabriskie, D. W., eds.), Marcel Dekker, New York, pp.157–196.
11. The European Agency for the Evaluation of Medicinal Products CPMP Biotechnology Working Party (1997) Notes for Guidance on Quality of Biotechnological Products: Viral Safety Evaluation of Biotechnology Products Derived from Cell lines of Human or Animal Origin, CPMP/ICH/295/95
12. Sherwood, D. (2000) Cleaning: multiuse facility issues, in *Biopharmaceutical Process Validation* (Sofer, G. and Zabriskie, D. W., eds.), Marcel Dekker, New York, pp. 235–249.
13. Amersham Biosciences Application note, Process Chromatography: Use of Sodium Hydroxide for cleaning and Sanitizing Chromatography media and systems (Code no. 18-1124-57).

44

Purification Process Scale-Up

Karl Prince and Martin Smith

1. Introduction

The term "scale-up" is used to describe a transition of size, volume, or output for any given process, or sequence of operational steps. It is generally used when transferring a process from development to pilot or manufacturing scale but can be equally applied to any change that is intended to increase the quantity of final product produced in a given time frame. The intention of this chapter is to provide an overview of the technology and methods used to increase the output of a protein purification process from a developed laboratory process to a scale many times larger.

The requirement of process scale-up is to allow more product to be made in any given time period; this can be referred to as increased output or process throughput. In general, the simplest way to increase the output from any purification system is to provide a faster turnover of that system. This can be achieved by automation of the laboratory technique or process. The process of automation in most laboratories can be simply achieved using computer-controlled equipment. This is especially true for chromatography operations, where sophisticated systems for gradient formation, column selection, sample processing, and elution collection are common place. Systems such as the AKTA™ (Amersham Biosciences) (1) allow for the preparation of complex buffer systems, column identification and selection, operation, and preprogrammed control. It is a simple matter to program such systems to complete the same step multiple times to provide greater output in a given time period. This approach does not work well for more manual systems, where operator instruction, intervention, or manipulation is required or in the case of processes with poor reproducibility. For this type of operation (e.g., dialysis) the selection of an analog system is required; typically, the method of increasing product output by increasing equipment capacity and changing technology is referred to as scale-up. Scaling-up a purification process by increasing throughput is limited by the time period allowed for processing and in-process stability of the product, but it can provide an easily achieved route to increased output. If the quantities required are so great that the time involved exceeds the available processing time, then the equipment itself must be increased in capacity or number.

Scale-up by increasing equipment number is also an easy method for increasing

From: *Methods in Molecular Biology, vol. 244: Protein Purification Protocols: Second Edition*
Edited by: P. Cutler © Humana Press Inc., Totowa, NJ

process throughput in a short time, with minimal process development. To increase equipment number does not require additional process development or change in process steps and the equipment should be identical to the original system in operation. This system for process scale-up also reduces the amount of training required for operators. However, the system is limited in that there is very little improvement in process scale economies. In fact, all inefficiencies inherent in the original process will be multiplied by the increased items of equipment. This type of scale-up can be applied to processes for which there is a short-term demand, or a single larger quantity is required, or where equipment is flexible enough to provide many different processing options. Increasing equipment number will also increase demand for space, which, in some instances, can be problematic and is often more costly than increasing equipment capacity. Because of this increase in space costs and the cost of the individual instruments, this method of process scale-up must be carefully considered prior to implementation, the increased throughput must be weighed against the equipment and associated costs and any future demands. Increasing equipment number and automation can be combined to yield throughput increases many times the original capacity of a laboratory-scale process. However, to achieve increases in process capacity beyond the ability of throughput increases, the scale-up technique is to increase system capacity with larger equipment and systems. Increasing system size and capacity allows for larger single batches of product and, hence, provides greater product output. For purification operations, the limit of this method for scale-up is often the cost associated with raw materials supply; however, this is the most well known and common form of process scale-up. The concepts introduced in the rest of this chapter are intended to provide an overview of the considerations for scale-up of processes by the method of increased equipment capacity.

The first step in scale-up for any protein purification process is to determine the target of the scale-up *(2)*. The various techniques that are available at laboratory scale have several analog or direct equivalents at larger scales and the selection of each is dependent on the goal and scale of the operation (*see* **Note 1**). During the scale-up of a process, the developer must ask several questions about the process and the available techniques or equipment for larger-scale manufacture, the limitations and requirements of any regulatory bodies, and the conservation of product quality and efficacy *(3–5)*. Some process techniques, such as column chromatography, are well defined and relatively simple for scale-up, whereas others, such as solvent extraction, may be much less common at manufacturing scales. The basic sequence for a purification system is shown in **Fig. 1** *(6)*. The techniques used throughout the purification process are similar in nature and have nearly identical scale-up requirements; the following text provides some techniques and pointers to the possible problems involved in scale-up of purification processes through increasing equipment capacity.

2. Materials

The materials used in the scale-up of a process should be determined during the initial development of that process. However, there should be some accounting for the availability and financial costs of materials at the larger scale. Typically, the use of raw materials at the large scale should provide the same operating principals as for smaller scales but allow for a lower cost of goods wherever possible. Thus, in most situations the development of aqueous processes with robust purification steps will provide simpler and

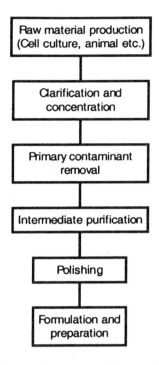

Fig. 1. Basic operational sequence for a purification process.

less expensive scale-up than complex processes using polar or organic solvents. The best scale-up programs begin early in the development of a process and provide support for both the development program and the eventual commercial operation of the process *(7)*.

In most situations, the materials of the scale-up scientist are pen, paper, and the application of previous experience. The scale-up of a manufacturing process should include not only the science of the process but also the physical and operational aspects of the process *(8)*. To achieve this, a typical scale-up project for a manufacturing process requires the involvement of development scientists, engineers, controls specialists, quality and compliance groups, materials supply and logistics specialists and, ideally, a strong project management presence. Even projects with seemingly smaller scopes, such as an increase in scale for a laboratory experiment, include many participants, including scientists, lab technicians, and materials supply. In this way, a successful process scale-up is rarely if ever an individual achievement; rather, it is almost exclusively the culmination of the efforts of a coordinated and effective team *(9)*.

3. Methods

3.1. Clarification and Concentration

All biological manufacturing systems (e.g., cell cultures or animal and plant sources) provide a complex mixture of proteins, waste materials, growth media, extracellular and intracellular products in which the target molecule is generally a small proportion *(10)*. To enable the purification of the target molecule, the first step is often a removal of the particulate matter in the process stream. The process by which this removal is achieved is called clarification.

Much laboratory-scale clarification is achieved by centrifugation of small volumes of the process stream. Centrifugation is based on the difference in density between particulate matter and the liquid in which the particles are contained. Centrifuges can be used to remove particulate matter down to approx 0.5 μm in size prior to filtration for preparation prior to chromatography. There are both continuous and discontinuous centrifuges available for large-scale operations that can be selected to suit the scale-up factors derived for your process (*see* **Note 2**). One of the most important considerations for centrifuge scale-up is that of suitability for CIP and SIP (cleaning and steaming in-place); the requirement for CIP or SIP compatibility can increase the cost of a large centrifuge system and increase the space requirements for the equipment and ancillary services.

After clarification, many processes use a concentration step to improve the quality of the feedstream for subsequent process. This concentration step can be performed by chemical or physical means. A popular technique in the laboratory is solvent extraction; this is often known as liquid–liquid extraction and can be performed in the aqueous or solvent phase. The process is based on the solubility and partitioning of the target molecule in the different liquids (*see* **Note 3**). This process is common in many industries but is less useful in protein purification because of the risk of protein denaturation and because the equipment required for solvent recovery and recycling is often highly specialized. A concentration method closely related to liquid–liquid extraction is called precipitation; here, the addition or removal of a chemical is used to precipitate the proteins present in the solution. Similar chemicals to those used in liquid–liquid extraction are used to reduce the solubility of the protein until the proteins salt-out as a solid (*see* **Note 4**). Chemicals such as ammonium sulfate can be used as precipitants for many proteins. A simple method for protein precipitation that is used at larger scales is polyethylene glycol (PEG) precipitation. PEG is simple to prepare, presents no chemical or waste hazards, and can be used to provide high recoveries and purifications. In addition to this, PEG is nonionic and therefore does not interfere with many downstream operations.

An issue with all precipitation methods for proteins is the possibility of causing protein denaturation. This must be examined on a case-by-case method, as even denaturation of the protein may not be problematic for some purposes. If renaturation is required, this can be achieved in some cases by simple solublization of the precipitate or the use of a specific method such as salt exchanging *(11)*. This should therefore be reflected in the process development and characterization of the product to ensure that the final protein is suitable for use.

To preserve the integrity and quality of target molecules, a common laboratory practice is to add a selection of protease inhibitors to crude protein solutions. In industrial settings, the use of such inhibitors is less common, as they can be prohibitively expensive in large quantities, and the removal of such inhibitors may require additional validation or processing steps for certain products. It is more common in large-scale processing to provide additional protection from protease activity via rapid or low-temperature processing. However, both of these options provide for additional design complexity at large scale and can be avoided by providing a step with high efficiency early in the purification process.

The concentration of product is achieved by the selective removal of solvent, typically water from the protein solution; methods for protein precipitation as described earlier can provide for solvent removal. However, the use of membrane technology to remove solvents is preferable in many applications. The most common method for concentration at large scale is cross-flow filtration (*see* **Note 5**). This method uses a membrane with controlled porosity to allow passage of small molecules (water, buffers, salts, etc.) while retaining larger molecules and particulates *(12,13)*. The selection of the membrane will control both the functionality of the step and its efficiency. Membranes should be selected to have minimal interaction with the components of the solution being concentrated, thus reducing membrane fouling and increasing process efficiency. In general, the selected pore size should be approx 10 times smaller than the molecule of interest to provide good process recoveries *(14)*. This is required because although protein size and mass are linked, the tertiary structure of proteins can allow passage through membrane pores where none is expected. The ionic conditions for the step also need to be closely controlled to ensure that the process is efficient and that the sieving characteristics of the protein–membrane system are controlled. As cross-flow filtration is often used to provide a buffer exchange step, it is important to ensure that the start and end conditions are compatible with all species present in the solution. This will reduce the possibility of the formation of precipitates or protein polymers that could otherwise affect the efficiency of the step. There are many methods for the scaling of cross-flow filtration systems; however, for most practical applications, the simplest method is to maintain fluid flows and path lengths across all of the scales intended for operation (*see* **Note 6**) *(15)*. Membrane cleaning and preparation is also important to ensure that maximum membrane life is obtained in large-scale operations (*see* **Note 7**). In addition to the selection of membrane type and processing conditions, the design of the cross-flow filtration system is of vital importance. The positioning of feed and retentate piping, correct tank mixing, and temperature control are all aspects of the physical design that will affect system performance (*see* **Note 8**).

3.2. Primary Contaminant Removal

The goal of primary contaminant removal is to provide a significant increase in the purity and concentration of the target molecule with respect to the starting material. In many systems, the target molecule represents only a minute portion of the total material present and so the efficiency and importance of this step is a primary consideration for the development scientist. Most purification processes for the isolation of proteins will use column chromatography for the primary contaminant removal step and these steps are developed with laboratory-scale equipment in mind. Several differences are evident during the scale-up of processes to pilot and manufacturing scales *(16)*. These differences can reduce chromatographic performance during scale-up unless provision is made during the developmental program. Typical small-scale column chromatography uses techniques that are close to the theoretical ideal for chromatography with well-defined and predictable behavior. In most cases, the process conditions for large-scale chromatography are not close to ideal and so do not adhere to the theoretical ideals of column behavior *(17)*. Typical column operations at large scale are aimed to provide maximum use of the matrix (often one of the most expensive raw materials in purification). For this reason, it is important to describe the capacity of the column for the target molecule; this is generally done via the generation of a breakthrough curve at small scale (*see* **Note 9**). This curve is then used to determine the maximum product load for the large-scale column.

Column scale-up is generally limited by the sizes of columns commercially available. Each of the many manufacturers of large chromatography columns (Amersham Biosciences, Millipore Corporation, Bio-Rad Laboratories, etc.) have standard large-scale columns that increase in diameter from 5 to 200 cm. Selection of the column for the process should be based on the expected output from the system and the mode of scale-up used (*see* **Notes 10** and **11**). In all scale-up modes for column chromatography it is important to identify which parameters must be increased to meet the target. It is often a simple issue to increase column volume and hence capacity to meet a perceived need of scale-up (more product throughput in a given time), but this may not always be the best case. In situations that have constraints on space, or raw materials, or where the costs of the materials in process are so high that physical limits are placed on volume, the most effective scale-up technique could be to increase liquid velocities. There must always be a balance in the mind of the process scientist between cost, time, scale, and process simplicity to allow the most efficient system to be developed.

Chromatographic scale-up is not without problems whichever method is chosen. Artifacts that at small scale have little or no impact on the system can be amplified many times to cause problems at larger scales. The most common of these factors to affect scale-up are column wall effects, distribution problems, matrix packing and unpacking *(18)*, outgassing, and temperature changes. All of these can cause significant deviations from the expected operation of a chromatographic step if not addressed (*see* **Notes 12–15**).

3.3. Intermediate Purification

Intermediate purification for proteins is usually achieved by a combination of column chromatography *(19,20)* and ultrafiltration and buffer-exchange steps. These are scaled with the same parameters as described in the preceding subsections. However, as the product is purified and normally reduced in volume, it becomes more vital to address the issue of product recovery. All purification processes involve a certain loss in product at each step, an ultrafiltration step might be expected to operate at 95–99% recovery, a column chromatography step at ≤80%, a filtration step at 90%, and so on. The total of these losses is actually the product of them all *(21)* and so each one should be addressed.

When addressing issues of recovery, four questions should be asked and answered before making process changes.

- Is the product truly lost or are assay variations affecting measurements? This is quite common where two different assay techniques are used to assess the product recovery. The most common of these inaccuracies arises from cases where different assay techniques are used because of changes in the product medium or changes in the product itself.
- Is the lost product of equivalent quality to the desired product? The very nature of chromatographic separation will allow differentiation between different isoforms or structures of the same product. Before these "losses" are returned to the product pool, an assessment of their characteristics compared to the desired product should be made.
- Where in the process is the loss occurring? Because of the sequential nature of purification processes, a small improvement at the beginning of a process can often be lost by the end. Conversely, a small change of a few percent in the final stage of a process will yield the same increase in total product output. Therefore, in the real world with limited resources,

it is more efficient to focus on the latest largest loss in a process to provide yield improvement.

• What process validation is required prior to implementing the change? As most processes are validated to some degree, or have expected outcomes, it is often required to address the impact of process changes with scaled-down experiments. These should be carefully planned and documented to ensure that the desired results and changes can be supported.

If all of these questions can be answered satisfactorily, then the process changes should be investigated and made during scale-up to yield an improved process (*see* **Note 16**). A single area of analysis that can provide improvements in the process yield is that of process control *(22)*. It is often possible to study the past history of a process regarding the control of parameters for each step, such that critical parameters affecting yield can be identified. The data for this kind of analysis are often available as part of the normal processing outputs and careful statistical analysis for trends and cause/effect relationships can be performed. Many software packages (e.g., JMP; SAS Institute, Cary, NC) can be used to investigate this kind of process improvement.

3.4. Polishing

Polishing is the portion of the process used to provide final product of the correct quality and safety prior to formulation. For many protein purification processes, this step is performed by column chromatography; however, the target for the separation is often an isoform, aggregate, or polymer of the final product. One of the most common column chromatography techniques used for polishing is size-exclusion chromatography (SEC). The use of SEC can provide a highly purified single species of protein but does have some scale-up problems (*see* **Note 17**). Addressing these problems along with a correctly selected control strategy can provide for an efficient SEC step in the polishing sequence. Another process step that is often used in polishing, and throughout the purification process, is normal flow filtration. At small scale, this system is most well known to scientists as the vacuum-assisted filtration device (e.g., Stericup™; Millipore Corp., Bedford, MA). These devices provide for the rapid filtration of liquids at the lab scale; they are suitable for systems up to 1000 mL in volume, but impractical above this. Filtration of larger volumes requires a change in equipment to filter cartridges and pressure vessels or pumps (*see* **Note 18**). These systems should be sized to provide filtration of the fluid at the desired flow rate with minimal area. To establish the area required, filtration tests called V_{max} and P_{max} can be performed, V_{max} is most useful for membrane (surface) filters *(23)* and P_{max} is most useful for depth or charged filters. To perform these tests requires specialist equipment and analysis; however, most filter manufacturers will provide this service free of charge with the potential of filter supply. Accurate filter sizing is important for processes that are likely to be operated for extended periods; here, there is eventually a drive to reduce costs to a minimum so effective filter use is important in reducing filter costs.

3.5. Conclusion

The techniques required for scale-up of a purification process are varied and, in some cases, conflicting. The scientist must balance all of the conflicting choices and options to provide the most cost-effective, efficient method of production in as short as possible timescales. These contrary forces and demands mean that the science of scale-up, although being able to be modeled and predicted, will always fall foul of the fiscal demands placed on the system. The true scale-up scientist will be continuously striving to improve and renew the processes that are developed and implemented, constantly applying both experience and theory to gain that extra percent recovery or reduce operating time or costs. This chapter was intended to provide an insight into the various aspects that are considered during a scale-up process and, hopefully, to provide a simplified view of the complexity that this area of process science holds.

4. Notes

1. The equivalence of process steps at different scales is dependent on the final objective of the step and not on the physical, chemical, or other method used. Hence, a system at the laboratory scale that uses a benchtop centrifuge (physical particle-density-based separation) to provide separation of solids from solution could be replaced by a filtration (physical apparent particle size separation) step at larger scales. **Table 1** provides a list of common laboratory techniques and some common analog for these at larger scales.

2. Centrifuge scale-up is based on the rate of separation of particulates from the liquid stream. Calculations can be performed to determine the acceleration of the separation process based on the selected centrifuge geometry, fluid, and particle properties. A series of experiments called spin tests, where a sample of the process fluid is spun at the expect process conditions and sequentially measured for cake formation, is used to scale-up for each manufacturer's centrifuge model. It is recommended that these experiments be performed in conjunction with the selected manufacturer, as each has developed scale-up curves for their own equipment.

3. The difference in protein solubility can be used to partition the target molecule into a suitable solvent stream. This process is based on the addition of salts or other components that combine with the areas of hydrophobicity contained in every protein. Common additives used are ammonium sulfate and PEG. Depending on the concentration of the additive used, the protein can be specifically targeted to one fluid or another, or entirely precipitated out of solution.

4. Simple salt additions (20–80% of ammonium sulfate, potassium phosphate, or sodium sulfate) are often used in laboratory conditions to create a protein precipitate. This method is susceptible to interference from temperature affects on the saturation of the salt and on the variable mixing, and hence localized concentrations, in large tanks. In addition, residual salt in subsequent steps can cause interference with later purification steps. Large-scale application of salt precipitation requires that a highly characterized feedstream be used to minimize interference and variability in processing and that careful attention be applied to waste streams to accommodate local waste requirements.

5. Cross-flow filtration is a technology used to separate molecules based on their size. In large-scale applications, the cross-flow filtration step can be used to replace buffer-exchanging steps such as dialysis and gel filtration de salting, size-exclusion chromatography for molecules with wide size separation, and product concentration steps such as evaporation or stirred cell concentrators. Cross-flow filtration technology can also be applied to upstream operations such as cell or particulate removal from product streams. The portion of the process stream excluded from the membrane is generally termed the "retentate," and in most applications, this is recycled to a retentate/feed vessel that, in turn, feeds the cross-flow filtration operation. The portion of the process stream that is allowed to pass through the separating membrane is termed the "permeate"; this permeate stream may or may not be product-free, depending on the parameters and technology used for the step. A typical cross-flow filtration design scheme is shown in **Fig. 2**.

Table 1
Laboratory Techniques and Larger-Scale Analog

Laboratory-scale technique	Larger-scale analog	Application (with example suppliers)
Ultracentrifugation (particle removal)	Low-speed centrifugation	Large-particulate removal (e.g., cells), available at scales from bench top to 100 L/min continuous systems (Carr Industries).
	Depth filtration	Large-particulate removal (e.g., cells and organelles), various scales from 50 cm² to many square meters (Millipore).
	Membrane filtration	Small-particulate removal (e.g., down to 0.1 μm size), various scales from 5 cm² to many square meters (Millipore).
Protein concentration	Cross-flow filtration	Cross-flow filtration systems can be used to concentrate proteins by selecting a membrane that excludes the protein of interest. Systems can be scaled from 50 cm² to many square meters (Sartorius AG).
	Liquid–liquid extraction	Large-scale systems are available for protein concentration by this method; this is particularly common in the albumin industries.
	Stirred cell systems	The application of these systems is limited to laboratory scales; systems can are available to handle up to 500 mL (Amicon).
	Precipitation (PEG or salt)	Although less commonly used in industry than in the laboratory, this is applied in some cases.
Dialysis (buffer exchange)	Size-exclusion chromatography	Chromatography resins such as G-25 can be used to provide a buffer-exchange step when packed in large columns (Amersham Biosciences).
	Tangential flow filtration	TFF systems can be used with appropriate membrane exclusions to provide a buffer exchange step. Systems can be scaled from 50 cm² to many square meters (Sartorius AG).
Column chromatography	Column chromatography	Simple linear scale-up techniques can be used to increase column size. Columns are available at diameters from 5 to 2000 mm and greater (Amersham Biosciences).
	Batch adsorption	Simple adsorption steps can be performed with chromatographic resins in large stirred vessels. Here, the resin is added directly to the system and mixed thoroughly. After several washes, the target can be eluted from the resin with a suitable buffer.
	Filter chromatography	Several systems exist for rapid chromatographic separations based on filter technology (Pall).

Membrane selection for cross-flow filtration is determined by the properties of the target molecule and the aim of the step. Microfiltration is a form of cross-flow filtration technology used to remove particulates, cells, and cell debris from process streams. The approximate pore size range for microfiltration is 0.02–10 μm; systems such as this are often used in the earlier stages of processing to clarify product streams prior to purification

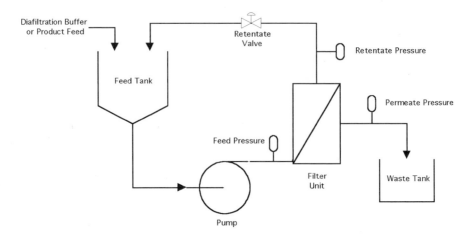

Fig. 2. Basic cross-flow filtration system setup.

operations. In these microfiltration systems, the product is processed through the membrane and appears in the permeate stream; the retentate stream contains the particulate contaminants. It is generally accepted practice to apply a washing buffer to the retentate stream, often three to four times the retentate volume, as the volume decreases to recover product retained in this stream. Microfiltration can also be used to wash or concentrate particulate matter such as cells, organelles, or micelles; this can be of importance when the target product is contained in these systems. Ultrafiltration is a form of cross-flow filtration technology used to separate molecules in the 10^3- to 10^6-kDa molecular-weight range. In most ultrafiltration operations, the target molecule is rejected by the membrane and hence retained in the retentate. In this mode of operation, the permeate contains those contaminants of lower molecular weight than the product; these could be organic molecules or buffer components. Ultrafiltration is often used in both laboratory-scale and large-scale processes to provide a buffer exchange step or to provide a consistent control for product concentration. Reverse osmosis is a form of cross-flow filtration technology used to separate molecules below 10^3 kDa; the most common application for reverse osmosis is in the purification of water.

6. The scale-up approach for cross-flow filtration operations that is most commonly followed is to maintain constant path length through the channels of the system. This approach ensures a constant pressure drop across the membranes over the various sizes. Several manufacturers provide laboratory-scale equipment (e.g., Millipore Pellicon series) that can be scaled in this manner. The development of flow rates, pressure, and diafiltration constraints can be performed on systems as small as 50 cm², and then transferred across equipment of 0.1 m², 0.5 m², and 2.5 m² using such equipment. Such systems can be purchased with varying levels of automation to aid process development by monitoring and/or controlling each of the parameters. Pump scaling is also an important factor during scale-up of cross-flow filtration processes; many suppliers recommend very high flow rates for cleaning (up to 1000 L/m²h of membrane). These flow rates do not present a large problem at small scales, but with large systems, the pumping demands (hence, pump size, cost, and heat input) can be considerable.

7. Membrane cleaning is important for large-scale operations because of the costs associated with membrane replacement, both for the purchasing of the membranes and the operational

time required to install, test, and prepare large-system membranes. Typical cleaning regimes require the use of solutions with salt concentrations of 0.5 *M* or higher, possibly with the addition of similar concentrations of caustic chemicals for compatible membranes. These solutions are often applied at elevated temperatures (over 35°C), with high flow rates (typically twice the operating cross-flow rate) to increase sheer forces at the membrane surface, and in some applications, even applied in the reverse-flow mode. These extreme conditions exert a selecting force over the membrane material that can be used for the processing. Many membrane materials have been developed that allow processing with low-protein-binding properties and high chemical and physical resistance to the rigors of cleaning; such membranes should be considered when selecting the correct processing equipment. Examples of such membrane materials are polyethersulfone, regenerated cellulose, and ceramics; each membrane type has advantages and disadvantages unique to their construction and composition. These should be thoroughly reviewed prior to selection of the final membrane type. Each of the major suppliers of cross-flow filtration membranes will readily provide both technical and practical help during this selection and scale-up phase via their technical specialist groups.

8. When building mid-scale ultrafiltration systems, it is important to remember that the systems should be well mixed. At the small to medium scale (0.005–0.1 m² of membrane area, systems are often of bespoke design resulting from the "Heath–Robinson" amalgamation of lots of different pieces of equipment, the primary purposes of which are often forgotten. Design considerations for such systems include the following. Within the retentate vessel, the retentate feed and return lines should be physically separated by some distance (e.g., to opposite sides of the retentate vessel). If the vessel has a bottom outlet, the retentate return should arrive close to the feed outlet but not too close. In both cases, failure to separate these two lines adequately can lead to poor mixing within the retentate vessel, which, in turn, has major consequences for the performance of the ultrafiltration step. If the feed and return lines are too close together, then "short-cutting" can occur, with the retentate return actually forming a major part of the fluid in the feed outlet line, rather than mixing with the bulk of the retentate pool.

During concentration, shortcutting can result in localized hot spots of very high product concentration within the vessel, which may lead to product aggregation. Additionally, during diafiltration, when fresh buffer is introduced into the retentate tank to replenish the volume lost to the permeate stream, short-cutting leads to an overprediction of the amount of diafiltration buffer needed to reach a certain pH and/or conductivity. As a design guide, it is recommended that ultrafiltration retentate vessels incorporate some form of mixing device, be it a bottom-driven magnetic close-coupled drive, a top-driven agitator, or simply a suitably sized magnetic stirrer bar combined with a stirrer plate.

Fortunately, it is very easy to determine whether or not an ultrafiltration system is well mixed and the test uses common laboratory chemicals and equipment. Based on typical process conditions for cross-flow velocity, permeate flux and retentate volume, simply diafilter a set volume of 1 *M* sodium chloride against 5–10 diafiltration volumes (DV) of deionized water, measuring the conductivity of the retentate stream every 0.5 DVs from 0 to 5–10 DVs in total. If the system is well mixed, then a semilog plot of retentate conductivity versus diafiltration volume should yield a straight line according

$$\frac{C}{C_0} = e^{-\text{DV}} \qquad (1)$$

where C is retentate conductivity, C_0 is the initial retentate conductivity, and DV is the number of diafiltration volumes.

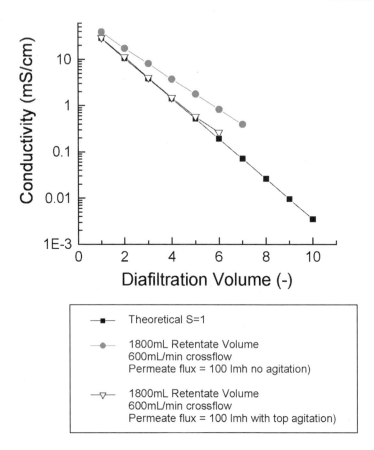

Fig. 3. Diagnosis of poor mixing in ultrafiltration systems using water diafiltration against 1 *M* sodium chloride in the ultrafiltration retentate vessel, mimicking process conditions of cross-flow, flux, and volume.

Where a system experiences short-cutting, the resultant graph will deviate from the theoretical well-mixed system performance, as illustrated in **Fig. 3**. In this particular example, when a top-driven agitator was added to the tank, short-cutting was eliminated as illustrated in **Fig. 3**.

9. In most cases, large-scale operation of chromatography columns is such that the binding of product to the column matrix is close to the total capacity of the matrix; this differs from laboratory conditions, where only a small percentage of the column capacity is used. This technique requires that the conditions for product breakthrough from the column be determined during the scale-up of any process. The typical method for determining this breakthrough is to continuously load product to a column and to monitor the column effluent for product concentration. A plot of product concentration per unit volume of column bed (e.g., g/L) against column effluent product concentration expressed as a fraction of load concentration (C/C_0) is used to determine the acceptable product load for a column with known product loss (*see* **Fig. 4**).

   As the available binding capacity of the column is consumed by the product, the flowthrough contains more and more of the incident product. It is normal to select a maximum binding capacity for a column that shows less than 5% breakthrough ($C/C_0 \leq 0.05$); this choice is balanced against cost and supply of materials and process economics to provide maximum column usage and, hence, output.

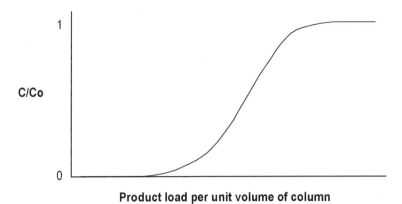

Fig. 4. Example of a typical product breakthrough curve.

Table 2
Comparison of Factors Involved in Chromatographic Scale-Up

Scale-up technique	Constant parameters	Varied parameters
Constant bed height	Packed-bed height of column Superficial flow rate of fluids	Column diameter
Constant aspect ratio	Ratio of bed height to column diameter	Column diameter and bed height Operational time Superficial flow rate of fluids
Constant residence time	Superficial flow rate of fluids	Column diameter and bed height
Constant sample-to-column volume ratio	Sample-to-column volume ratio Superficial flow rate of fluids	Column diameter and bed height

10. The scale-up principles generally applied to chromatography are to ensure the maintenance of a single constant parameter and to increase other parameters to provide a change in scale. The most common parameter that is held constant is that of bed height. Keeping this parameter constant and increasing the column diameter provides the simplest way to increase column scale; this is the technique with which most small-scale chromatographers are familiar. This method combined with a constant superficial flow rate at all scales will provide the most similarity for chromatographic conditions and results at all scales. This mode of scale-up works well and should be used for most applications.

Other possibilities for scale-up of chromatography columns include constant aspect ratios (bed height to diameter), constant residence time for process fluids (often used when column length is changed during scale-up), and constant sample volume-to-column volume ratio (*see* **Table 2**).

11. Scale-up by constant bed height is the most common and simplest scale-up technique used for column chromatography. The technique is dependent on the increasing volume of a cylinder as the diameter increases. Using this system will allow the scientist to maintain and easily predict the behavior of all other parameters, such as volumes for buffers, retention times, void volumes, mixing times, and elution volumes. As stated in **Note 11**, it is critical to keep both the bed height and the superficial flow rate applied to the column constant.

The simplest way to achieve constant bed height scale-up is to back-calculate the desired column diameter from the capacity of the column matrix and the available time for processing. With a known column capacity (*see* **Note 10**) and a known desired processing time for a fixed volume of material, it is possible to calculate the desired final column volume and, hence, diameter at a fixed bed height. The following is an example of the calculations required to complete this step.

- Target process scale-up: 10000 L of 5 g/L in 10 h.
- Existing process scale: Column capacity is 100 g/L with a 20-cm bed height and 200-cm/h superficial flow rate.
- Flow rate at scale-up desired $= V_t/T$, where V_t is the total volume and T is the time. In this case, 10,000/10 = 1000 L/h (1 m³/h).
- Surface area required to reach this $= F_t/F_s$, where F_t is the target flow rate (m³/h) and F_s is the superficial flow rate (m/h).
- In this case, 1/2 = 0.5 m².
- Column diameter to provide this surface area $= 2 \cdot \sqrt{A/\pi}$, where A is the area (m²). In this case, $2 \cdot \sqrt{0.5/\pi}$ = 0.80 m (80 cm).
- Large column capacity $= C(AH_t \times 1000$, where H_t is the bed height (m) and C is the column capacity (g/L). In this case, $100 \times (0.5 \times 0.2 \times 1000)$ = 10.0 kg. Therefore the column will need to be cycled five times to complete the processing of all 50 kg in 10 h.

12. It has long been a recognized feature of column chromatography that so-called "wall effects" can cause serious failures during scale-up. The wall effect acts to provide extra structural support to packed chromatography columns. It was long thought that these effects were limited to the small scale. However, recent work has shown that certain bead sizes and structures can display wall effects in columns up to 40 cm in diameter. The practical outcome of wall effects can be increased back-pressures during operation, possible bed failures resulting from channeling or cracking, long equilibration or retention times, and poor packing and testing efficiencies. Although it is not possible to eliminate all wall effects, the best method to alleviate the problems associated with this effect is to maintain operating conditions for flow and pressure within the compressibility limits for the matrix being used *(23)*. It is also advisable to provide changes in flow in a gradual manner, to limit physical changes in the bed that may be exacerbated by wall effects.

13. Column chromatography relies on a complete and even distribution of applied fluids at the top and bottom of the packed bed. Even small defects in the design or completion of the flow distribution system can have large effects on perceived column performance. In particular, the liquid dispersion and bed support portion of the column should be carefully designed, installed, and maintained to provide maximal efficiency. Each of the column manufacturers has the ability and technical prowess to aid in the design of a system to meet the exact needs of a customer; however the basic requirements should be as follows:

- Minimal dead volume. This will reduce mixing and dilution hence promoting good flow dynamics.
- Efficient flow dispersal. This will reduce column artifacts that will affect perceived column performance.
- Inert material. It is essential that the material of construction does not interact with the process fluids.

14. Temperature is often a neglected process parameter in the scale-up of chromatography, and it represents a true variable in making life difficult at the large scale. Temperature can dramatically affect the viscosity of buffer and process solutions to the point that processes become inoperable because of excessive pressure drops at even moderately larger scales than laboratory columns. This effect is worsened by buffers exhibiting exothermic and endothermic reactions upon preparation such as 20% (v/v) ethanol and 6 *M* guanidine–HCl, respec-

Table 3
Common Tubing Diameters and Flowrates

Standard system description	Column inlet tubing diameter (mm)	Pump inlet tubing diameter (mm)	Volumetric flow rate (L/h)
3 mm	3	6	2–30
6 mm	6	10	7.5–120
10 mm	10	15	30–400

tively. It is always best to operate at the same temperature within a given chromatographic process and remember to specify the temperature during scale-up and remember to never position your chromatography system or columns next to the window on hot summer days.

Active control of temperature during process development can actually result in more complications than neglecting the parameter. Columns and even entire chromatography system can be mounted inside deli-style cold cabinets with pane-glass doors for viewing purposes with temperatures controllable to ±1°C down to just above freezing. Although it is relatively easy to operate in this manner and to chill 5- to 10-L quantities of buffer in the deli-cabinet, it is a far harder proposition at intermediate and larger scale.

First, larger volumes of buffer simply require more cooling duty to chill them to lower temperatures. Above 10 L, this is difficult without either walk-in cold rooms or specially designed jacketed vessels and chiller units. Chromatography systems and especially rotary lobe pump heads generate heat that must be removed using heat exchanges and columns must be specially design to incorporate jackets that increases their cost. Put simply, temperature-controlled chromatography must have a sound reason to justify its incorporation during scale-up, and its avoidance is worth every additional small-scale experiment.

15. Related to the control of temperature is the phenomena of mobile-phase degassing. Every researcher familiar with HPLC or low-pressure chromatography should be equally familiar with the practice of helium sparging of mobile phases to remove dissolved gases or at the larger-scale, up to 10 L, vacuum degassing with filtration. Both degassing techniques help prevent dissolved gasses coming out of solution, or gassing, as they experience small changes in pressure when moving from narrow-bore tubing to the wider-bore column body. If degassing is not performed, bubbles can be observed to accumulate in the column body, usually at the column walls and may, if unchecked over time, result in channeling and poor column flow dynamics. When buffer and process volumes exceed 10 L, degassing and vacuum filtration is unpractical. The effects of gassing may be minimized by allowing buffers to reach the same temperature as the column prior to loading. Tubing to the pump head on the chromatography system should be as wide as possible while minimizing holdup volumes.

    As a guide, Amersham Biosciences standard systems posses the following tubing diameters relative to the flow rates stated in **Table 3**.

16. Although it is impossible to identify all of the possible areas for product loss and a specific remedy for each that is applicable to all processes, **Table 4** provides some common areas for investigation.

17. Size-exclusion chromatography is based on the separation of molecules on their apparent diameters. It can be used to provide extremely sensitive separations but is highly dependent on the quality of the equipment and packed column bed being used. For large-scale operations, packing columns of length over 50 cm can be problematic; this is especially true of columns where the diameter of the bed exceeds the column height. SEC columns are very susceptible to column wall effects (*see* **Note 13**) and can be notoriously difficult to pack at large scale. When considering the use of SEC columns, the following criteria should be assessed:

Table 4
Common Areas for Investigation of Product Loss in Scale-Up Processes

Step type	Areas to investigate
Clarification and concentration	Potential binding of product to filter media
	Losses resulting from shear in pumping/separation systems
	Incomplete product recovery from raw material
	Solubility effects in two-phase systems
Chromatography	Incomplete column equilibration, elution, or washing
	Overloading of product
	Buffer composition for product elution
	Resin fouling
	Product aggregation or precipitation
	Flow distribution and dilution in the column system
	Flow rates
	Collection criteria for product eluates
	Step vs gradient elutions
Ultrafiltration/Diafiltration	Potential binding of product to filter media
	Losses resulting from shear in pumping/separation systems
	Incomplete mixing in retentate vessels
	Product aggregation or precipitation
	Correct flushing of membranes post use to recover product
	Gel point investigations
Normal flow filtration	Correct filter sizing based on product/process loads
	Appropriate preuse treatment

- Column-packing technique: SEC columns should always be packed while held in a vertical position; using a spirit level and plumb line to orient columns prior to packing is advised.
- Some SEC matrices swell and shrink with changes in the ionic conditions of buffers; if used, such matrices should be packed in solutions of similar ionic conditions to the buffers used in processing.
- Packing should be performed at higher flow rates and pressures than expected during operation. This will ensure that the packed bed reaches a higher level of compression than during operation. This reduces the possibility of damaging packed beds during operation.
- Selection of multiple smaller columns for SEC may improve performance over single large units. Providing both the required length and cross-sectional area for SEC with multiple smaller columns can avoid some of the problems seen with very large systems.
- Extracolumn dilution effects from fluid distribution and chromatography control systems can have a large effect on peak resolution and shape. This is especially true of post-column mixing problems, which have the same effect as peak broadening on the column. System piping and volumes should be minimized at all times.

18. A simple arrangement for filter operations at large scale is shown in **Fig. 5**. The system is comprised of a pressure can supplying the liquid to be filtered, a filter capsule (as available from Pall, Millipore or Sartorius), a receiving vessel, and connecting silicone hosing. The motive force in this system is supplied by air pressure at the supply vessel. An alternative to this arrangement is to use a pump to provide the motive force for the liquid. The filtration occurs as the liquid passes from the supply to receiving vessels. Care must be taken to vent the filter at the start of filtration to remove trapped air; this ensures that the maximum filter area is used during the operation.

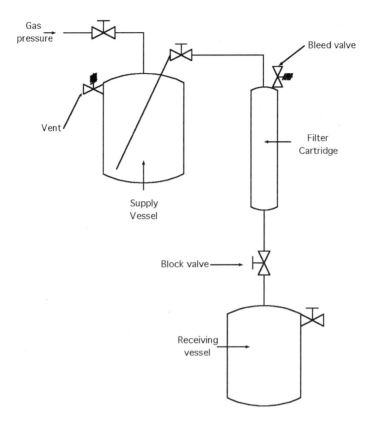

Fig. 5. Simplified filtration system diagram.

References

1. Orr, T. (1999) Bioprocessing technology options. *Gen. Eng. News*, April 1, 16.
2. Olson, W. P. (1995) Separations in pharmaceutical manufacturing, in *Separations Technology, Pharmaceutical and Biotechnology Applications* (Olson, W. P., ed.), Interpharm Buffalo Grove, IL.
3. Ladisch, M. R. (2001) Principles of bioseparations for biopharmaceuticals and recombinant protein products, in *Bioseparations Engineering Principles, Practice, and Economics* (Ladisch, M. R., ed.), Wiley, New York, pp. 514–518.
4. Doblhoff-Dier, O. and Bliem, R. (1999) Quality control and assurance from the development to the production of biopharmaceuticals. *TIBTECH* **17,** 266–270.
5. Hesse, F. and Wagner, R. (2000) Developments and improvements in the manufacturing of human therapeutics with mammalian cell cultures. *TIBTECH* **18,** 173–180.
6. Fulton, S. P. (1994) Large-scale processing of macromolecules. *Curr. Opin. Biotech.* **5,** 201–205.
7. Asenjo, J. and Leser, E. (1996) Process integration in biotechnology, in *Downstream Processing of Natural Product: A Practical Handbook* (Verrall, M., ed.), Wiley, New York, pp. 123–138.

8. Wisniewski, R., Boschetti, E., and Jungbauer, A. (1996) Process design considerations for large-scale chromatography of biomolecules, in *Biotechnology and Biopharmaceutical Manufacturing, Processing, and Preservation* (Avis, K. E. and Wu, V. L., eds.), Manufacturing and Technology Series, Vol. 2, IHS Health Group, Englewood, CO.

9. Pepper, C., Patel, M., and Hartounian, H. (1999) cGMP pharmaceutical scale-up, Part 1: design, *BioPharm.* **Sept,** 26–34.

10. Berthold, W. and Kempen, R. (1994) Interaction of cell culture with downstream purification: a case study. *Cytotechnology* **15,** 229–242.

11. De Bernandez Clark, E. (2001) Protein refolding for industrial processes. *Curr. Opin Biotechnol.* **12,** 202–207.

12. Shiloach, J., Martin, N., and Moes, H. (1988) Tangential flow filtration, in *Downstream Processes: Equipment and Techniques. Advances Biotechnology Processes* vol. 8, Alan R. Liss Inc. New York, 97–125.

13. van Reis, R. and Zydney, A. (2001) Membrane Sseparations in biotechnology. *Curr. Opin. Biotechnol.* **12,** 208–211.

14. Cherkasov, A. N. and Polotsky, A. E. (1996) The resolving power of ultrafiltration. *J. Membr. Sci.* **110,** 79–82.

15. van Reis, R., Goodrich, E. M., Yson, C. L., et al. (1997) Linear scale ultrafiltration. *Biotechnol. Bioeng.* 55, 732–746.

16. Dream, R. F. (2000) Process design considerations for large-scale chromatography of biologics. *Pharm. Eng.* **July/August** 60–82.

17. Lode, F. G., Rossenfeld, A., Yuan, Q. S., et al. (1998) Refining the scale-up of chromatographic separations. *J. Chromatogr. A.* **796,** 3–14.

18. Hofmann, M. (1998) A novel technology for packing and unpacking pilot and production scale columns. *J. Chromatogr. A* **796,** 75–80.

19. Sofer, G. (1995) Preparative chromatographic separations in pharmaceutical, diagnostic and biotechnology industries: current and future trends. *J. Chromatogr. A* **707,** 23–28.

20. Ladisch, M. R. (2001) Principles of bioseparations for biopharmaceuticals and recombinant protein products, in *Bioseparations Engineering Principles, Practice, and Economics* (Ladisch, M. R., ed.), Wiley, New York, pp. 539–540.

21. Ladisch, M. R. (2001) Principles of bioseparations for biopharmaceuticals and recombinant protein products, in *Bioseparations Engineering Principles, Practice, and Economics* (Ladisch, M. R., ed.), Wiley, New York, p. 527.

22. Knaack, C. A. and Hawrylechko, A. M. J. (1998) A systematic approach to the validation of monoclonal antibody manufacturing processes. *Pharma. Sci. Technol. Today* **1(7),** 300–308.

23. Badmington, F., Wilkins, R., Payne, M., et al. (1995) V_{max} testing for practical microfiltration train scale-up in biopharmaceutical processing. *PharmTech,* **Sept,** 64–76.

24. Stickel, J. J. and Fotopoulos, A. (2001) Pressure–flow relationships for packed beds of compressible chromatography media at laboratory and production scale. *Biotechnol. Prog.* **17,** 744–751.

Index